ADVANCES IN SEMICONDUCTOR LASERS AND APPLICATIONS TO OPTOELECTRONICS

SELECTED TOPICS IN ELECTRONICS AND SYSTEMS

Editor-in-Chief: **M. S. Shur**

Published

Vol. 1: Current Trends in Integrated Optoelectronics
ed. *T. P. Lee*

Vol. 2: Current Trends in Heterojunction Bipolar Transistors
ed. *M. F. Chang*

Vol. 3: Current Trends in Vertical Cavity Surface Emitting Lasers
ed. *T. P. Lee*

Vol. 4: Compound Semiconductor Electronics: The Age of Maturity
ed. *M. Shur*

Vol. 5: High Performance Design Automation for Multichip Modules and Packages
ed. *J. Cho and co-ed. P. D. Franzon*

Vol. 6: Low Power VLSI Design and Technology
eds. *G. Yeap and F. Najm*

Vol. 7: Current Trends in Optical Amplifiers and Their Applications
ed. *T. P. Lee*

Vol. 8: Current Research and Developments in Optical Fiber Communications in China
eds. *Q.-M. Wang and T. P. Lee*

Vol. 9: Signal Compression: Coding of Speech, Audio, Text, Image and Video
ed. *N. Jayant*

Vol. 10: Emerging Optoelectronic Technologies and Applications
ed. *Y.-H. Lo*

Vol. 11: High Speed Semiconductor Lasers
ed. *S. A. Gurevich*

Vol. 12: Current Research on Optical Materials, Devices and Systems in Taiwan
eds. *S. Chi and T. P. Lee*

Vol. 13: High Speed Circuits for Lightwave Communications
ed. *K.-C. Wang*

Vol. 14: Quantum-Based Electronics and Devices
eds. *M. Dutta and M. A. Stroscio*

Vol. 15: Silicon and Beyond
eds. *M. S. Shur and T. A. Fjeldly*

Forthcoming

Vol. 17: Frontiers in Electronics: From Materials to Systems
eds. *Y.-S. Park, M. S. Shur, S. Luryi, J. M. Xu and A. Zaslavsky*

Vol. 18: Sensitive Skin
eds. *M. S. Shur and S. Wagner*

Selected Topics in Electronics and Systems – Vol. 16

ADVANCES IN SEMICONDUCTOR LASERS AND APPLICATIONS TO OPTOELECTRONICS

Editors

Mitra Dutta

Michael A. Stroscio

U.S. Army Research Office, U.S. Army Research Laboratory

World Scientific

Singapore • New Jersey • London • Hong Kong

Published by

World Scientific Publishing Co. Pte. Ltd.

P O Box 128, Farrer Road, Singapore 912805

USA office: Suite 1B, 1060 Main Street, River Edge, NJ 07661

UK office: 57 Shelton Street, Covent Garden, London WC2H 9HE

British Library Cataloguing-in-Publication Data
A catalogue record for this book is available from the British Library.

ADVANCES IN SEMICONDUCTOR LASERS AND APPLICATIONS TO OPTOELECTRONICS

ISBN 981-02-4257-3

Printed in Singapore.

JANICE LYNCH

U.S. Army Materiel Command

FOREWORD

The continuing down scaling of dimensional features in optoelectronic and microelectronic devices coupled with the dramatic advances in semiconductor growth and processing technologies are augmenting current heterojunction device technology and opening many new avenues for extending the capabilities of semiconductor lasers. To be sure, semiconductor lasers and optoelectronics are undergoing dramatic changes as a result of the rapidly evolving field of heteroepitaxy. Indeed, the recent stunning advances in nanofabrication technology facilitate band-engineering and atomic-level structural tailoring not possibly previously except through molecular and atomic systems found in nature. As the technologies for growing new semiconductor heterostructures mature, more and more heterojunction device options become realizable. Driven by these dramatic technological advances, there have been ever increasing numbers of novel heterostructure-based semiconductor lasers conceived over the last fifteen years.

New subdisciplines of semiconductor physics and optoelectronics are emerging as a result of these dramatic advances in semiconductor physics and device technologies. These rapid developments — many of them ongoing developments — make it essential to understand both the physical phenomena and the technological advances underlying novel semiconductor lasers. This volume includes highlights of the theories underlying the essential phenomena occurring in novel semiconductor lasers as well as the principals of operation of selected heterostructure lasers. To understand scattering processes in heterostructure lasers and related optoelectronic devices its essential to consider the role of dimensional confinement on both charge carriers as well as on acoustical and optical phonons in quantum structures. Indeed, it is important to consider the confinement of both phonons and carriers in the design and modeling of novel semiconductor lasers such as the tunnel injection laser, quntum well intersubband lasers, and quantum-dot lasers. The full exploitation of dimensional confinement leads to the exciting new capability of scattering-time engineering in novel semiconductor lasers. As an example, band engineering of multiple quantum well structures facilitates the fabrication of novel tunnel injection lasers with intrinsic modulation bandwidths near one-hundred billion cycles per second. Moreover, highly advanced epitaxial growth of complex superlattice structures makes it possible to fabricate unipolar semiconductor lasers achieving the long-awaited goal of lasing at far infrared wavelengths. Likewise, self-assembled quantum-dot lasers exhibit enhanced differential gain. Moreover, new III-V nitride-based semiconductor lasers operate in the short-wavelength portion of the optical spectrum and potent many applications in optoelectronics. As a result of continuing advances in techniques for growing quantum heterostructures, these recent developments are likely to be followed in coming years with many more advances in semiconductor lasers and optoelectronics. As our understanding of these devices and abilities to fabricate them grow, so does our need for more sophisticated theories and simulation methods bridging the gap from quantum to classical transport. This volume surveys these and other recent exciting developments underlying advanced semiconductor lasers and their applications to optoelectronics.

The authors have many professional colleagues, friends and family members who must be acknowledged for without their contributions and sacrifices this work would not have been undertaken or completed. MD and MAS are both grateful to the editor, Prof. M. Shur, for making this effort a highly professional and meaningful undertaking. MD acknowledges the discussions, interactions and work of many colleagues and friends who have had an impact on the work in this book. These include Drs. Doran Smith, K. K. Choi, and Paul Shen of the Army Research Laboratory, Professor Athos Petrou of the State University of New York at Buffalo, Professors K. W. Kim, M. A. Littlejohn, and R. J. Nemanich and Dr. Leah Bergman of the North Carolina State University. MD would also like to thank Dhiren Dutta without whose encouragement she would never have embarked on a career in science and to Michael and Gautam Stroscio who everyday add meaning to everything. MAS acknowledges the essential roles that several professional colleagues and friends played in the events leading to his contributions to this book; these people include: Professor Gerald J. Iafrate of Notre Dame University, Professors M. A. Littlejohn and K. W. Kim of the North Carolina State University, Professor Vladimir Mitin of the Wayne State University, Professors G. Belenky and S. Luryi of the State University of New York at Stony Brook, Professors George I. Haddad, Pallab K. Bhattacharya, and Jasprit Singh and Dr. J.-P. Sun of the University of Michigan, Professors Karl Hess and J.-P. Leburton University of Illinois, Professor L. F. Register now of the University of Texas at Austin, and Professors H. Craig Casey and Steven Teitsworth of Duke University. MAS also thanks family members who have been attentive during the period when this book was being edited; these include: Anthony and Norma Stroscio, Mitra Dutta, as well as Gautam. Marshall, and Elizabeth Stroscio.

Dr. Michael A. Stroscio earned a Ph.D. in Physics from Yale University and held research positions at the Los Alamos Scientific Laboratory and the Johns Hopkins University Applied Physics Laboratory, before moving into the management of federal research and development at a variety of U.S. government agencies. Dr. Stroscio has served as a policy analyst for the White House Office of Science and Technology Policy, and as Vice Chairman of the White House Panel on Scientific Communication. He has taught and lectured on Physics and Electrical Engineering at several universities including Duke University, the North Carolina State University and the University of California at Los Angeles. Dr. Stroscio is currently the Senior Scientist in the Office of the Director at the U.S. Army Research Office as well as an Adjunct Professor at both Duke University and the North Carolina State University. He has authored about 500 publications, presentations and patents covering an exceptionally wide variety of topics in the physical sciences and electronics. His professional publications have appeared in over 20 different journals, has written one book and edited another. He is a Fellow of the Yale Science and Engineering Association, Fellow of the Army Research Laboratory, as well as a Fellow of the Institute of Electrical and Electronics Engineers (IEEE). He received the 1998 IEEE Harry Diamond Memorial Award. He is a member of the Phi Beta Kappa honorary society and a long-time member of the American Physical Society.

Dr. Mitra Dutta received a B.Sc. and an M.Sc. in physics (first class honors) from the University of Delhi. She then spent three years on the faculty at the College of Arts, Science and Technology in Kingston, Jamaica, as well as lecturing part-time at the Physics Department of the University of the West Indies. She received her PhD in physics from the University of Cincinnati, Ohio, following which she was a research associate at Purdue University and at City College, New York, as well as a visiting scientist at Brookhaven National Laboratory. She then worked at the Electronics Technology and Devices Laboratory (ETDL), Fort Monmouth, which was incorporated into the Army Research Laboratory, first as team leader for the optoelectronics team, then as branch chief and finally as director of the physics division. After moving the laboratory to Adelphi, Maryland, she joined the Electronics Division of the Army Research Office where she is currently the associate director. She has over one hundred and sixty refereed publications, one hundred and seventy conference presentations, ten book chapters, edited one book and has had twenty-four US patents issued. She is a Fellow of the Optical Society of America, a Fellow of the Institute of Electrical and Electronic Engineers and a member of the American Physical Society. She has received the Army R&D Achievement Awards in 1990, 1992 and 1994, the ETDL Harold Jacobs Award in 1991, one of the ten outstanding AMC Personnel of the Year Award in 1992, led the team that won the prestigious Paul A. Siple Memorial Award (First Prize) at the 19th Army Science Conference held in Orlando Florida in June 1994, and was elected one of the twenty Fellows of Army Research Laboratory. She is an Adjunct Professor of the Electrical and Computer Engineering and Physics Departments of North Carolina State University, and has had adjunct appointments at the Electrical Engineering Departments of Rutgers University and the University of Maryland.

Dr. Michael A. Stroscio earned a Ph.D. in Physics from Yale University and held research positions at the Los Alamos Scientific Laboratory and the Johns Hopkins University Applied Physics Laboratory, before moving into the management of federal research and development at a variety of U.S. government agencies. Dr. Stroscio has served as a policy analyst for the White House Office of Science and Technology Policy and as Vice Chairman of the White House Panel on Scientific Communication. He has taught and lectured on Physics and Electrical Engineering at several universities including Duke University, the North Carolina State University and the University of California at Los Angeles. Dr. Stroscio is currently the Senior Scientist in the Office of the Director at the U.S. Army Research Office as well as an Adjunct Professor at both Duke University and the North Carolina State University. He has authored about 300 publications, presentations and patents covering an exceptionally wide variety of topics in the physical sciences and electronics. His professional publications have appeared in over 20 different journals. He has written one book and edited another. He is a Fellow of the Yale Science and Engineering Association, Fellow of the Army Research Laboratory, as well as a Fellow of the Institute of Electrical and Electronics Engineers (IEEE). He received the 1998 IEEE Harry Diamond Memorial Award. He is a member of the Phi Beta Kappa honorary society and a long-time member of the American Physical Society.

Dr. Mitra Dutta received a B.Sc. and an M.Sc. in Physics (first class honors) from the University of Delhi. She then spent three years on the faculty at the College of Arts, Science and Technology in Kingston, Jamaica, as well as lecturing part-time at the Physics Department of the University of the West Indies. She received her PhD in physics from the University of Cincinnati, Ohio, following which she was a research associate at Purdue University and at City College, New York, as well as a visiting scientist at Brookhaven National Laboratory. She then worked at the Electronics Technology and Devices Laboratory (ETDL), Fort Monmouth, which was incorporated into the Army Research Laboratory, first as team leader for the optoelectronics team, then as branch chief and finally as director of the physics division. After moving the laboratory to Adelphi, Maryland, she joined the Electronics Division of the Army Research Office where she is currently the associate director. She has over one hundred and sixty refereed publications, one hundred and twenty conference presentations, ten book chapters, edited one book and has had twenty-four US patents issued. She is a Fellow of the Optical Society of America, a Fellow of the Institute of Electrical and Electronic Engineers and a member of the American Physical Society. She has received the Army R&D Achievement Awards in 1990, 1992 and 1994, the ETDL Harold Jacobs Award in 1992, one of the ten outstanding AMC Personnel of the Year Award in 1992, led the team that won the prestigious Paul A. Siple Memorial Award (First Place) at the 19th Army Science Conference held in Orlando Florida in June 1994, and was selected one of the twenty Fellows of Army Research Laboratory. She is an Adjunct Professor at the Electrical and Computer Engineering and Physics Departments of North Carolina State University, and has had adjunct appointments at the Electrical Engineering Departments of Rutgers University and the University of Maryland.

PREFACE TO "ADVANCES IN SEMICONDUCTOR LASERS AND APPLICATIONS TO OPTOELECTRONICS"

Charles H. Townes

This volume provides valuable summaries of many aspects of semiconductor lasers. The variety of semiconductor lasers involved is remarkable, as are the fascinating ideas and possibilities suggested for the future. This field is bursting into flower — already semiconductors provide the majority of lasers in production, and the possibilities which are reviewed or suggested for the future offer much more. We can expect the variety of such lasers and their applications to continue to grow beyond what can presently be envisioned. But the excellent summaries here provide an excellent view of what we now know and bases for further developments.

Semiconductors allow detailed and remarkable structural designs on nanometer scales — essentially equivalent to the design atoms or molecules with desired energy levels and relaxation processes. The variety of phenomena involved — structures and their production, design of electron and photon states, along with the more macroscopic design of reflections and energy inputs — produce a rich and challenging field for both physicists and engineers. They can yield some lasers of high efficiency, and ones which are compact and relatively inexpensive. They can cover a wide range of frequencies, and some can be nicely tunable. One can hope for and expect extension to wavelengths beyond those of the mid-infrared, into a region where we presently have very few lasers and essentially no tunable ones. And further technical possibilities continue to come into view, for example carbon tubes and new ways of making quantum dots such as the recently announced use of DNA for forming them. How far will biological-type processes go in doing our delicate manufacturing? Multiple small semiconductor lasers for multichannel computer board interconnections already seem practical. There is also the intriguing but rather speculative possibility of using semiconductors for photon or laser computing, perhaps with quantum mechanics providing complex logical processing.

It is interesting that the mathematician John von Neumann was the first to work seriously on semiconductor laser phenomena. As early as 1953, even before the first maser was operating, he envisioned a semiconductor excited by intense neutron fluxes from a nuclear reactor, and then producing a strong cascade of photons as stimulated emission took place. He worked this out mathematically in some detail and wrote Edward Teller about the possibility. Von Neumann did not envisage a resonator nor recognize the possibility of coherence and monochromaticity, but he did ask me what I thought about this when he heard of our first maser, which utilized these properties. My reply was yes, excitation of semiconductors in principle could work, but at that time I thought the energy levels were so widely spread that generation of net gain and oscillators would be difficult.

With the many ideas and contributions which have happened since the 1950's, in particular the advances in semiconductor technology, how things have changed! The subtleties of design and control of the many parameters manifested in semiconducting structures provide a very rich and exciting field for science and technology. Authors of this volume have given us excellent reviews of what we presently know and discussions of challenging new areas which will help us move still further forward.

CONTENTS

Recent Developments in Theory of Semiconductor Lasers

International Journal of High Speed Electronics and Systems, Vol. 9, No. 4 (1998) 847–866

TUNNEL INJECTION LASERS

PALLAB BHATTACHARYA
Department of Electrical Engineering and Computer Science,
University of Michigan, 1301 Beal Avenue,
Ann Arbor, Michigan 48109-2122, USA

Carrier heating in conventional quantum well lasers can lead to several deleterious effects and are related to the transport and thermalization characteristics of injected carriers. The properties of a quantum well laser in which the electrons are directly transported to the lasing subband by tunneling is described here. The resulting device — a tunnel injection laser — is shown to have negligible gain compression, superior high-temperature performance, lower Auger recombination and wavelength chirp, and better high frequency modulation characteristics when compared to conventional lasers. All these improvements are attributed to the reduction of hot-carrier population in the active region of the laser. Results are presented here for lasers made with GaAs — and InP — based heterostructure systems.

1. Introduction

Multiquantum well (MQW) lasers are versatile coherent light sources which are being used in a host of demanding applications. Improvements in modulation bandwidth, optical output power, maximum operating temperature, and CW and dynamic linewidths require a large intracavity photon density, easily in excess of 10^{17} cm^{-3}. At the high injection levels required to achieve this photon density, several deleterious effects are observed. They are chiefly related to the production of hot carriers in the active region due to quantum carrier heating, radiation heating, slow carrier relaxation rates into the quantum well and carrier leakage over heterobarriers. In particular, as optical phonon scattering rates are reduced in lower-dimensional quantum confined structures, electron relaxation can take up to 40 ps, particularly if the intersubband energy is not in resonance with phonon energies. The carrier distribution in the gain region therefore changes quickly from a quasi-Fermi to a hot carrier distribution, as shown in Fig. 1. Because of their higher effective mass and higher scattering rate due to multiple hole states, the thermalization times of holes are ~ 1 ps. A substantive part of the injected carrier density therefore ends up as wasted charge and gives rise to a higher diffusion capacitance, a low-frequency roll-off in the modulation bandwidth, saturation of the differential gain, increase in Auger recombination and high-frequency chirp and a larger temperature dependence of the threshold current.

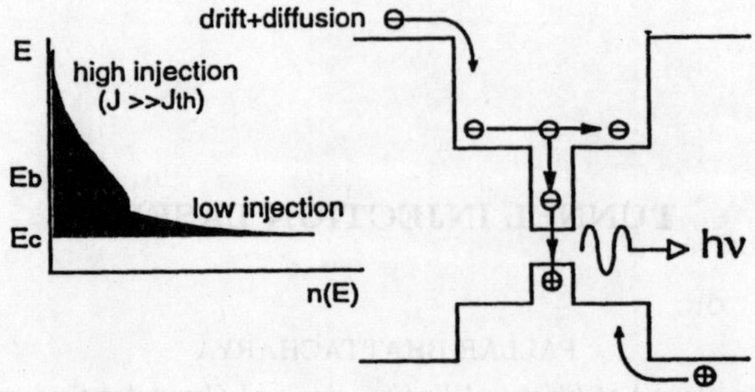

Fig. 1. Band diagram of a single-well tunneling injection laser. Also shown is the hot carrier distribution in a SCH laser.

Under high levels of injection, the stimulated emission time is rather small, and ~ 10 ps. If carriers with lower initial energies, or "cold" carriers can be supplied at the same rate, then quantum carrier heating is minimized and the carrier description will remain quasi-Fermi. As we have demonstrated earlier, cold electrons can be supplied by tunneling and we call the device a tunneling injection laser.

2. Hot Carriers in Quantum Well Lasers

There are two aspects to the problem of hot carriers in quantum well lasers. As mentioned earlier, carrier scattering rates are lowered and this may enhance the thermalization time, particularly of electrons, to the lasing ground state. However, even as injected carriers lose energy in the active region, there is an increase in the quantum carrier temperature, forcing carriers out of the lasing states and to higher energies in the well. This heating saturates the differential gain. The gain depends on the occupation probability of the lasing states, which is proportional to

$$\exp\left(\frac{qE_{fn}}{kT_n}\right) \tag{1}$$

where E_{fn} is the electron quasi-Fermi level, and T_n is the electron temperature. Grupen and Hess[1] have shown that after an injection bias cycle, both E_{fn} and T_n increases at the same rate. In addition to carrier thermalization and hot carriers, diffusion effects also have to be considered.

We have studied the carrier injection process both theoretically and experimentally. For efficient laser performance, the carriers should be able to enter the quantum well and reach thermal equilibrium with the quasi-Fermi level. Carriers that cross the active region are lost, as far as lasing is concerned, and are a significant source of poor temperature performance of lasers. If carriers entering the quantum well region do not thermalize, the carrier distribution will become hot, resulting in two negative effects. The hot carrier distribution will provide lower gain at the lasing energy which arises from the ground states of the quantum well. Also,

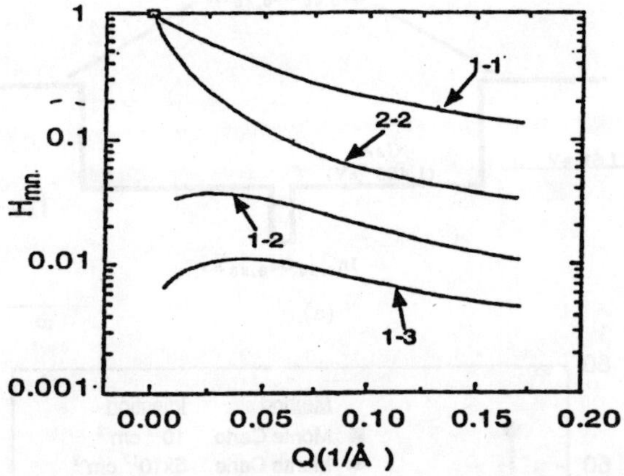

Fig. 2. Calculated multi-subband coupling coefficient H_{mn} as function of phonon wave vector $Q(= k_1 - k_2)$.

Auger recombination will be enhanced. We are primarily concerned with electron relaxation, since hole relaxation is very fast due to the high scattering rates for holes. In direct bandgap bulk semiconductors the electron thermalization time is ~ 1 ps due to the high phonon scattering rate.

We have used a Monte Carlo approach[2] to study the thermalization of an electron distribution injected into the three-dimensional inner cladding layer of the separate confinement heterostructure (SCH) shown in Fig. 1. Both inter- and intra-subband scattering with polar optical and acoustic phonons are considered. The multi-subband coupling coefficients, $H_{mn}(Q)$, in a 50 Å GaAs/$Al_{0.3}Ga_{0.7}As$ quantum well are depicted in Fig. 2. The source for slow thermalization in narrow quantum wells can be identified. The intra-subband scattering rate remains high, but the inter-subband rate decrease sharply. The limiting step in the carrier thermalization is the transition from subband 2 to subband 1. This rate is $\sim 10^{11}$ s^{-1} for a 100 Å well and $\sim 5 \times 10^{10}$ s^{-1} for a 50 Å GaAs/$Al_{0.30}Ga_{0.70}As$ well. The key factor here is the energy separation of these subbands. The relaxation time to enter the second subband is quite small $\sim 2{-}3$ ps. Note that H_{mn} decreases exponentially with increasing 1/Q-large electron momentum changes have a lower probability of occurring. Note also the much higher intra-subband scattering rate. The calculated values of H_{mn} imply that transitions between discrete electron levels in a quantum confined system are more difficult than in bulk semiconductor due to the energy and momentum conservation requirements. This behavior has been shown to create a "bottleneck" to the relaxation of carriers from the bulk-like confinement layers to the confined quantum well states in (graded and ungraded) SCH and structures.[3] Similar thermalization times have been measured by us[2] and others[4,5] using time resolved photoluminescence (TRPL) and pump-probe techniques. Our measurements were performed with quantum well heterostructures shown in Fig. 3(a) and the thermalization times, derived from analysis of TRPL data from various well

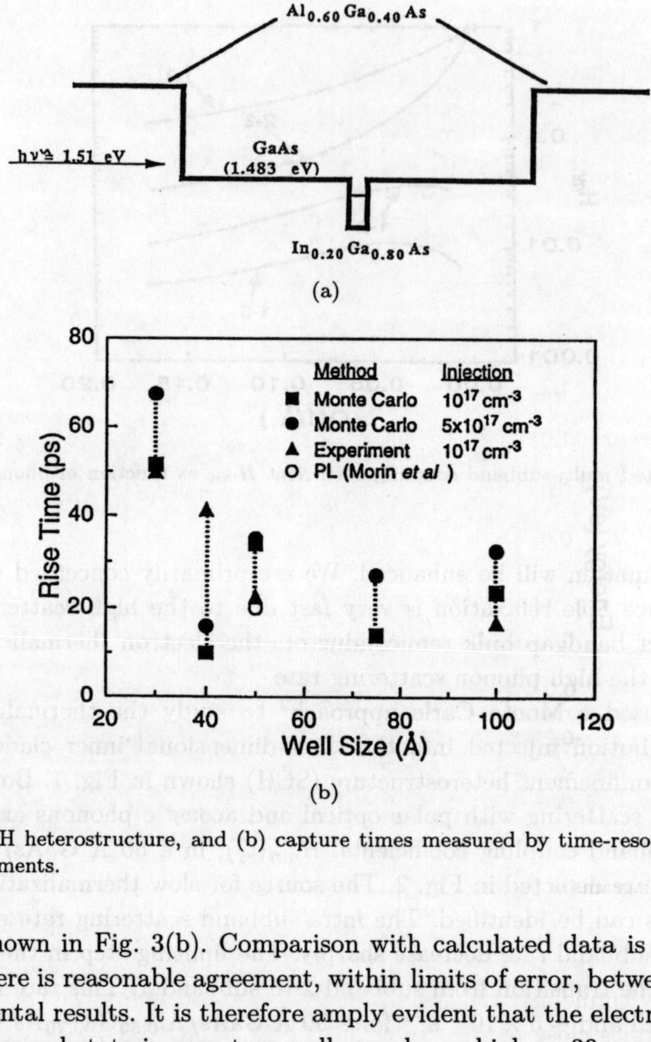

Fig. 3. (a) SCH heterostructure, and (b) capture times measured by time-resolved photoluminescence experiments.

widths are shown in Fig. 3(b). Comparison with calculated data is also made in Fig. 3(b). There is reasonable agreement, within limits of error, between calculated and experimental results. It is therefore amply evident that the electron relaxation time to the ground state in quantum wells can be as high as 30 ps, depending on size.

Since the total time taken by an injected carrier to reach the lasing subband can become longer than the stimulated emission time, which is $\sim$ 10 ps, for large injection currents the carrier distribution will no longer remain quasi-Fermi and instead becomes a hot carrier distribution. In fact, in a laser operated beyond threshold, the carrier temperature can be $\sim 40°$ higher than the lattice temperature.

3. Design of a Tunneling Injection Laser

The tunneling injection scheme for electrons in a quantum well laser, experimentally demonstrated by us[6] for the first time, is illustrated in Fig. 4(a). If electrons are introduced into the lasing subband of the active quantum well at or near the

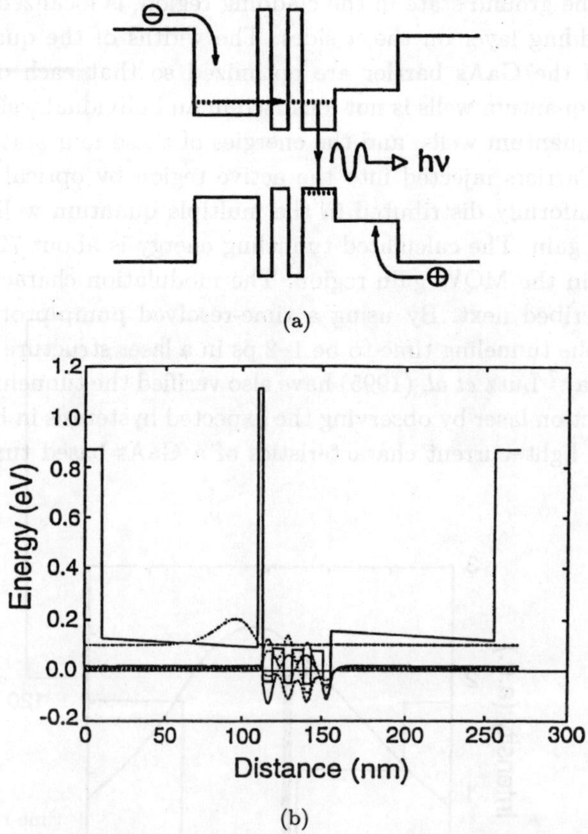

Fig. 4. Band diagrams of (a) single quantum well, and (b) multiquantum well tunneling injection lasers.

Fermi level by tunneling, then hot carrier effects are minimized. Electrons are injected at energies close to the lasing energy at a rate faster than or equal to the stimulated emission rate, and the electron distribution remains quasi-Fermi even at large drive currents. The injection of holes is done in a manner similar to that in conventional lasers. This is because, as noted above, there is no problem in hole thermalization process in quantum wells. To achieve higher values of gain and differential gain, it is common practice to incorporate a multiquantum well (MQW) gain region. Use of a MQW also increases the confinement factor. Therefore, to achieve the highest levels of performance, a MQW tunneling injection scheme needs to be adopted.

Figure 4(b) shows the energy band diagram of the MQW tunneling injection laser structure under an applied forward bias, which was obtained by solving the time-independent Schroedinger equation. The dashed lines are the wave functions of the first five states. It can be seen that the wave functions of the first four states are localized in the four quantum wells and the wave function of the fifth state, which is the ground state in the cladding region, is localized in the injection layer (inner cladding layer on the n side). The widths of the quantum wells and the thickness of the GaAs barrier are optimized so that each of the four wave functions in the quantum wells is not localized in an individual well but distributed in the multiple quantum wells, and the energies of these four states are very close to each other. Carriers injected into the active region by optical phonon-assisted tunneling are uniformly distributed in the multiple quantum wells, resulting in a high differential gain. The calculated tunneling energy is about 72 meV above the lasing subband in the MQW gain region. The modulation characteristics of these devices are described next. By using a time-resolved pump-probe technique, we have measured the tunneling time to be 1–2 ps in a laser structure identical to that shown in Fig. 4(a).[7] Lutz *et al.* (1995) have also verified the tunneling mechanism in a tunneling injection laser by observing the expected hysteresis in both the current-voltage and the light–current characteristics of a GaAs-based tunneling injection (TI) laser.

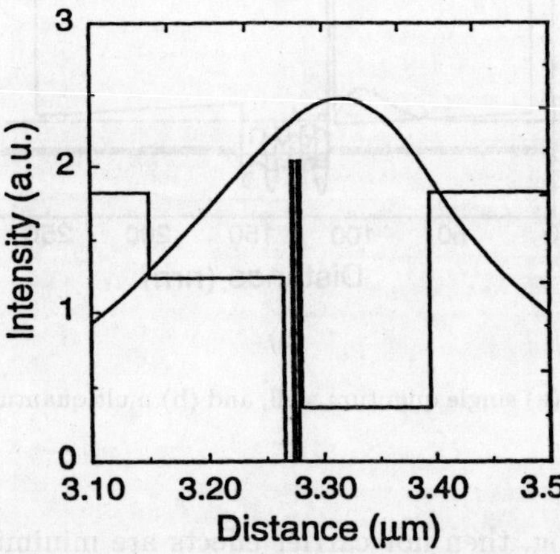

Fig. 5. Conduction band diagram and optical mode profile of InP-based 1.55 μm tunneling injection laser.

It is also important to note that the optical mode profile in the waveguide is not distorted due to the tunnel design. This is evidenced in Fig. 5, which shows the calculated mode profile for a 1.55 μm InP-based tunnel injection laser. Only a very slight asymmetry, which does not increase the waveguide losses, is observed.

4. GaAs-Based Tunneling Injection Lasers

A typical heterostructure of a single-quantum well (SQW) tunneling injection laser is shown in Fig. 6(a). Such heterostructures can be grown by MBE or MOCVD. Broad-area or single mode lasers are fabricated by conventional processing techniques. GaAs-based TI lasers generally exhibit an emission peak at $\lambda \cong 1$ μm.

0.1 μm	CONTACT GaAs	$p^+(5 \times 10^{18}\ cm^{-3})$
1.0 μm	OUTER CLAD $Al_{0.60}Ga_{0.40}As$	$p(5 \times 10^{17}\ cm^{-3})$
0.1 μm	INNER CLAD $Al_{0.30}Ga_{0.70}$	i
80 Å	ACTIVE WELL $In_{0.10}Ga_{0.90}As$	i
20 Å	Resonant Tunneling Barrier AlAs	i
40 Å	Tunneling Well $In_{0.10}Ga_{0.90}As$	i
20 Å	Tunneling Barrier AlAs	i
0.1 μm	INNER CLAD GaAs	i
1.0 μm	OUTER CLAD $Al_{0.60}Ga_{0.40}As$	$n^+(5 \times 10^{17}\ cm^{-3})$
0.7 μm	CONTACT GaAs	$n^+(5 \times 10^{18}\ cm^{-1})$
	S. I. GaAs (100) Substrate	

(a)

0.1 μm	CONTACT $In_{0.53}Ga_{0.47}As$	$p^+(2 \times 10^{19}\ cm^{-3})$
1.85 μm	OUTER CLAD InP	$p\ (5 \times 10^{17}\ cm^{-3})$
0.115 μm	INNER CLAD InGaAsP ($\lambda = 1.2\ \mu m$)	i
70 Å	ACTIVE WELL $In_{0.10}Ga_{0.4}As$	i
25 Å	Resonant Tunneling Barrier $In_{0.52}Al_{0.48}As$	i
30 Å	Tunneling Well $In_{0.6}Ga_{0.4}As$	i
25 Å	Tunneling Barrier $In_{0.52}Al_{0.48}As$	i
0.115 μm	INNER CLAD $In_{0.53}Ga_{0.47}As$	i
1.5 μm	OUTER CLAD InP	$n\ (5 \times 10^{17}\ cm^{-3})$
0.25 μm	CONTACT $In_{0.53}Ga_{0.47}As$	$n^+(5 \times 10^{18}\ cm^{-3})$
	S. I. InP (100) Substrate	

(b)

Fig. 6. Typical heterostructures for (a) GaAs-based and (b) InP-based tunneling injection lasers.

It is generally believed that spectral hole burning and carrier heating are the two main causes of gain suppression in lasers. We have studied gain-compression effects in a GaAs-based SQW tunneling injection laser by observing the spontaneous emission spectra during lasing as a function of bias current. A broad-area laser, 125 μm wide and 400 μm long, is specially made with a window in the top *p*

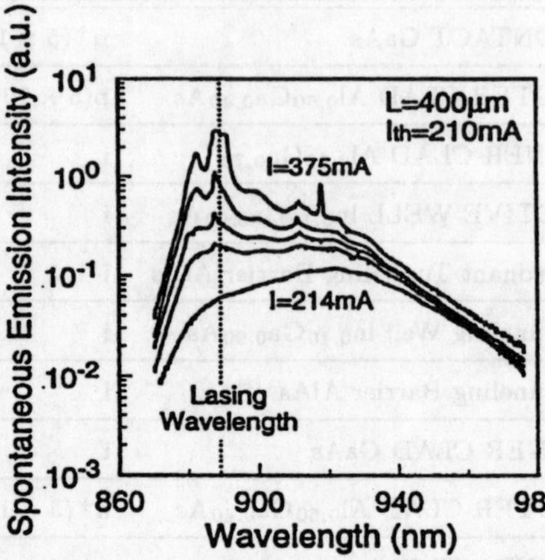

Fig. 7. Spontaneous emission spectra of the tunneling injection lasers as a function of bias above threshold.

ohmic contact through which the spontaneous emission exists. The laser is driven at different injection levels and the spontaneous emission is collected and analyzed by a scanning spectrometer. The emerging light is detected with a Ge detector. The measured data are shown in Fig. 7. The intensity of the spontaneous emission increases continuously at the lasing wavelength with increase of injection current, indicating absence of gain compression at the lasing wavelength. If gain compression effects are severe, then little or no increase in the spontaneous emission intensity would be expected at the lasing wavelength, as recently reported by Girardin *et al.*[8] for a 1.5 μm laser. The relative contributions of spectral hole burning and carrier heating in a semiconductor laser are strongly dependent on the laser type and material in the active region.[9–12] Theoretical studies indicate that strained quantum-well lasers exhibit enhanced carrier heating effects due to the extremely low threshold carrier density.[13] Carrier heating effects are manifested in the spontaneous emission spectra as a gain suppression at wavelengths *below* the gain maximum and closer to the peak of the spontaneous emission spectrum.[8,14] Although the device used in this study is a $In_{0.2}Ga_{0.8}As$-GaAs strained quantum-well laser, it has only one lasing well and I_{th} is quite high. Therefore, n_{th} is expected to be high. It is difficult, therefore, to distinguish spectral hole burning and carrier heating effects and it is probably accurate to state that both effects contribute to a reduction of gain compression in the TI laser in an indistinguishable manner.

The reduction of gain compression achieved in the tunneling injection lasers promises higher bandwidth and lower chirp. We have measured[15] the high-frequency dynamic linewidth (wavelength chirp) of a GaAs-based 0.91 μm TI laser and have compared them with SCH and DFB lasers emitting at 1.07 and 1.3 μm, respectively. The cavity lengths and threshold currents of the SCH, TI, and DFB lasers are 300, 350, and 400 μm, and 11.8, 11.5, and 11.3 mA, respectively. Chirp was measured at a fixed value of bias current (20 mA) and peak-to-peak modulation

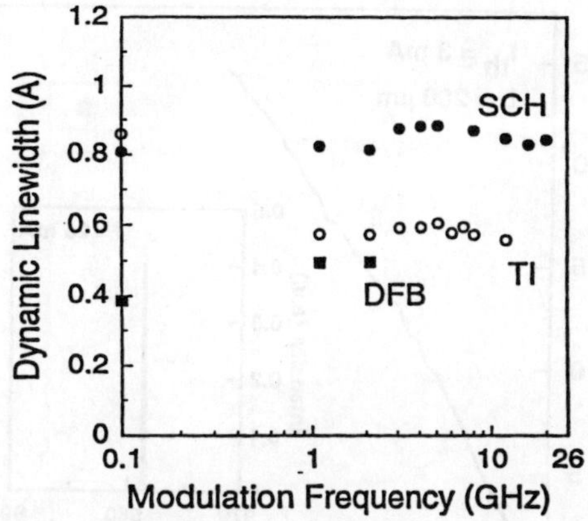

Fig. 8. Dynamic linewidth versus modulation frequency at fixed bias current (20 mA) and peak-to-peak modulation current (14 mA).

current (14 mA). The chirp was estimated to be the full width at half maximum of the dominant lasing longitudinal mode measured by a low-speed photodiode form a 0.75-m scanning spectrometer. The measured dynamic linewidth of the three lasers are shown in Fig. 8. The chirp in the TI laser is lower than that in the SCH laser and almost comparable to that of the DFB laser, whose dynamic linewidths are characteristically low due to the mode pulling mechanism. We attribute this significant reduction of the chirp in the TI laser to a reduction in hot carrier population and the resulting reduction of the gain compression factor ε.[15]

We will now describe the characteristics of GaAs-based $\sim 1\ \mu$m MQW tunneling injection lasers. The conduction band diagram of the waveguide region of such a device was shown in Fig. 4(b). Measurements have been done as a function of temperature in a variable temperature cryostat to determine the d.c. and dynamic characteristics. The optical modulation response was typically measured with a network analyzer, high speed photoreceiver and amplifier. The measured response was corrected for the frequency response of the cables, amplifier and photoreceiver. The facets of our device are uncoated.

The light–current characteristics of a MQW TI laser under pulsed bias conditions are shown in Fig. 9. The threshold current at room temperature is 3 mA and the slope efficiency is 0.33 mW/mA. The transparency current density, determined from a plot of $J_{\rm th}$ versus inverse cavity length, is only 167 A/cm^2. The emission spectrum of the device is shown in the insert of Fig. 9. The peak of the laser emission is at 0.98 μm, which confirms lasing from the MQW region.

Figure 10(a) shows the modulation frequency response of the undoped MQW tunneling injection laser at room temperature under pulsed and CW bias conditions. The maximum modulation bandwidth was achieved at a drive current of 80 mA under pulsed bias conditions. The extrapolated -3 dB modulation bandwidth is ~ 45 GHz. From a plot of the damping factor γ vs. f_r^2 shown in Fig. 10(b) the

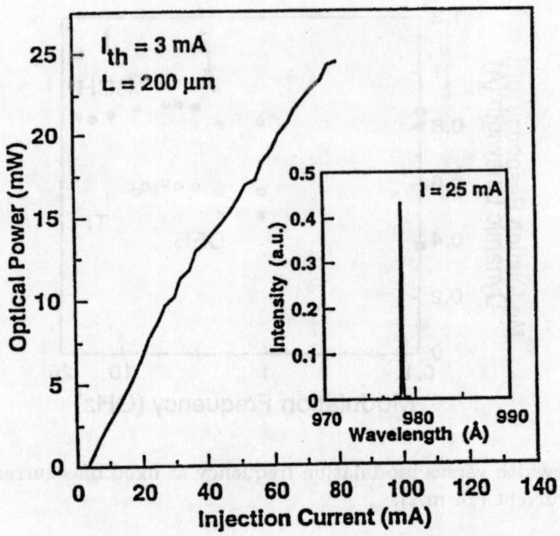

Fig. 9. Light–current characteristics of 4-QW 0.98 μm tunneling injection laser. The inset shows the output single mode characteristics.

K factor is 0.105 ns, representing a maximum intrinsic bandwidth, $f_{-3\text{ dB (max)}}$ of 84 GHz. For CW operation the modulation characteristics are $f_{-3\text{ dB}} = 43$ GHz, $K = 0.116$ ns, and $f_{-3\text{ dB (max)}} = 76$ GHz. These high-speed modulation characteristics are better than those normally achieved with conventional SCH quantum well lasers.[16]

It is believed that the frequency response of these lasers is limited by heating effects and with better mounting and heat-sinking higher bandwidths can be measured. It is also clear that the value of $f_{-3\text{ dB (max)}}$ may approach 100 GHz in these devices which exceeds the predicted gain-compression limited bandwidths of 60–70 GHz in high performance GaAs-based quantum well lasers. Large bandwidths can only be achieved at the expense of a very high photon density, which would definitely induce facet damage. It is therefore apparent that to achieve high modulation bandwidths, the laser has to be operated with a lower photon density, the gain compression factor ε has to be reduced and dg/dn has to be enhanced. The tunneling mechanism and "cold" operation reduce spectral hole burning and gain compression effects and increase the differential gain. Similarly, the differential gain is extracted form the relationship between resonance frequency and net injection current, $I - I_{\text{th}}$, and the internal conversion efficiency and optical confinement.[17] Values for internal conversion efficiency and optical confinement factor calculated at room temperature were used in the calculation at lower temperatures. Values for differential gain calculated at various temperatures are shown in Fig. 11. It is seen that the differential gain increases from $\sim 7 \times 10^{-15}$ cm^2 (which is in itself a very high value) to 3×10^{-14} cm^2 at 120 K. The gain compression factor in these devices is also very small, $\sim 10^{-23}$ m^3 for pulsed or cw operation.

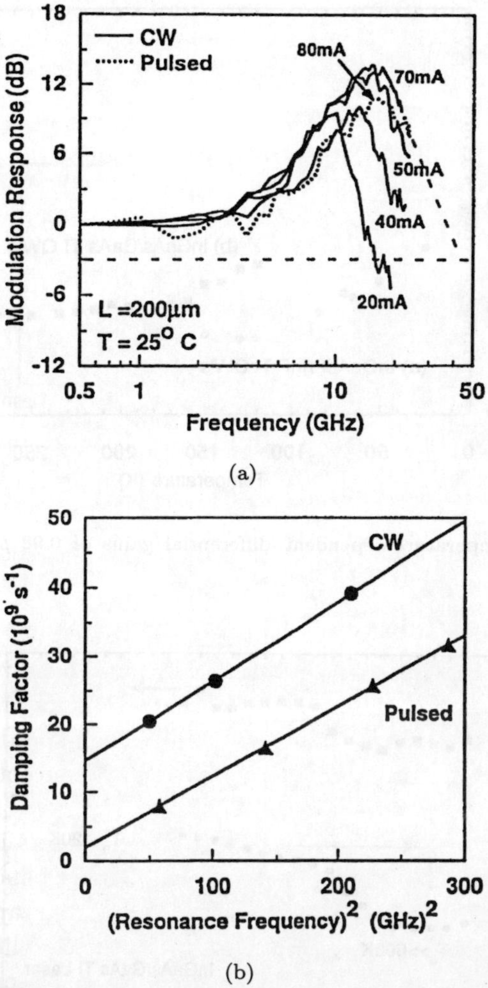

Fig. 10. (a) Modulation frequency response of the undoped MQW tunneling injection laser under CW and pulsed conditions, and (b) plots of damping factor versus resonance frequency squared.

The temperature dependent threshold current and slope-efficiency of single mode GaAs-based 0.98 μm TI lasers are shown in Fig. 12. It is interesting to note that there is a nearly flat threshold current region for temperatures less than about 60 K and even a slight increase in threshold current at the very lowest measurement temperature. The characteristics temperature T_o, defined as

$$I = I_o \exp(T/T_o) \tag{2}$$

is $\sim$ 120 K in the temperature range 100–300 K and is (not shown) 170 K in the temperature range 298–343 K for several devices.[18] The latter values are considerably higher than SCH devices, in which T_o is usually $\sim$ 125 K.

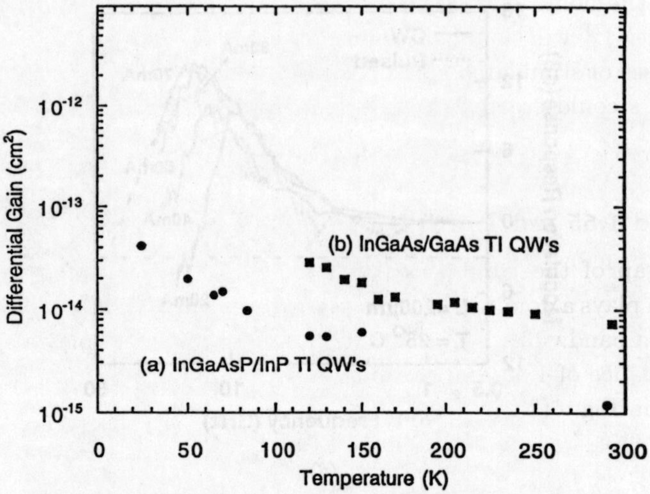

Fig. 11. Measured temperature-dependent differential gains of 0.98 μm and 1.55 μm tunnel injection lasers.

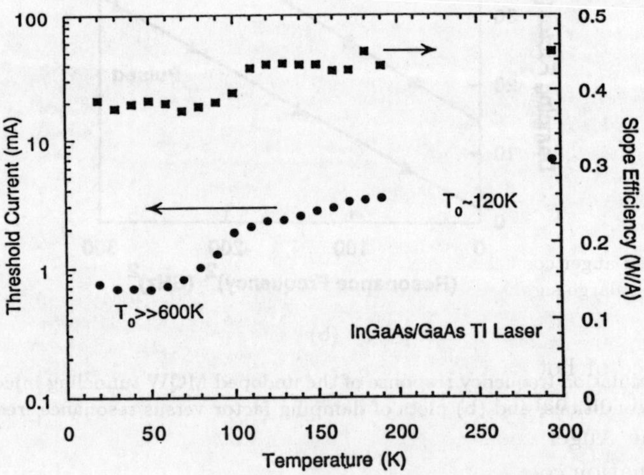

Fig. 12. Measured variation of threshold current and slope-efficiency of GaAs-based TIL with temperature.

The temperature independent threshold current at low temperatures is qualitatively similar to what has been observed in quantum dot lasers[19] and which has been attributed to the temperature independent density-of-states, and atom-like behavior. However, the threshold current is also affected by carrier transport and lifetime-limited linewidth broadening. At low temperatures, the decreased ability of carriers to equilibrate within the wells will tend to increase the threshold current[20] counteracting the reduced thermal distribution of the carriers. Additionally, at low temperatures the homogeneously broadened linewidth due to carrier-carrier scattering ($\sim$ 3 meV)[21] will dominate the thermal width and the carrier distribution, for the purposes of stimulated emission, will be independent of temperature. Both of these factors could create a nearly temperature-independent threshold current region.

5. InP-Based 1.55 μm Tunneling Injection Lasers

As the bandgap of the gain region of a semiconductor laser is reduced, Auger recombination plays a dominant role in increasing the threshold current and limiting the modulation bandwidth. Auger recombination is known to play a significant role in the performance of 1.55 μm lasers. The typical heterostructure for a InP-based TIL is shown in Fig. 6(b).

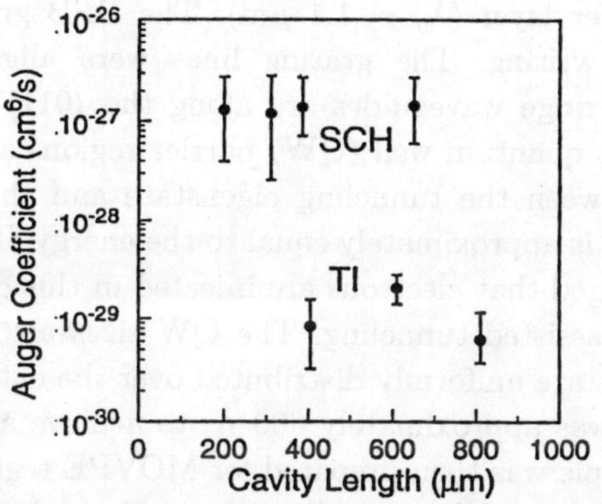

Fig. 13. Measured auger coefficients in 1.55 μm SQW-TI lasers and MQW-SCH lasers. The data were obtained from large-signal modulation experiments.

The presence of hot carriers enhances the rate of the conduction-heavy hole-heavy hole-split off band (CHHS) Auger process, given by $R_{\text{Auger}} = C_a\ \text{np}^2$, where C_a is the Auger coefficient. We have made extensive measurements of the Auger recombination coefficients in the TI lasers and compared them with 1.55 μm MQW-SCH lasers.[22] The measured Auger coefficients, in the two types of lasers of different lengths, are shown in Fig. 13. The coefficients were derived from detailed analysis of large-signal modulation data. Compared to an average value of $C_a = 1 \times 10^{-27}$ cm^6 s^{-1} for the SCH devices (averaged over the cavity lengths), an average value of $C_a = 8 \times 10^{-31}$ cm^6 s^{-1} was obtained for the TI lasers. This represents a reduction of over three orders of magnitude in the value of C_a. Furthermore, considering the fact that Auger recombination in MQW lasers is generally smaller than in SQW devices, due to a lower hot carrier density in the former, this result is all the more significant. We have also measured the small-signal modulation response of these lasers and 3-dB bandwidths $\sim$ 21 GHz have been measured at

300 K. The temperature-dependent differential gain of these lasers are also shown in Fig. 11, for comparison.

Lasers with a distributed feedback (DFB) of 1.55 μm have emerged as the most promising sources for coherent communication systems. Not only have linewidths as small as 70 kHz[23] been measured in 1.55 μm DFB lasers, but the ability to tune the output precisely by varying and grating period makes the devices very attractive for wavelength-division-multiplexed (WDM), fiber-optical communication systems. We have investigated the properties of 1.55 μm tunneling injection DFB lasers grown by MOCVD in two steps.[24] The laser heterostructure is shown in Fig. 14.

The active (gain) region consists of five 1% strain-compensated 80 Å InGaAsP quantum wells with −1% 70 Å InGaAsP barriers exhibiting a room temperature photoluminescence (PL) peak at 1.53–1.54 μm. The compressively strained wells reduce intervalence band absorption, as a loss mechanism, to some extent and provide a wider optical gain spectrum.[25] The sample was first grown up to the 0.2 μm InGaAsP buffer layer ($\lambda_g = 1.1$ μm). The DFB grating was formed by direct electron-beam writing. The grating lines were aligned along the $\langle 011 \rangle$ direction so that the ridge waveguides are along the $\langle 011 \rangle$ direction. The inner cladding layer and the quantum well (QW) barrier regions are so chosen that the energy separation between the tunneling eigenstate and the multiquantum well (MQW) ground states is approximately equal to the energy of an optical phonon in InP. Thus, it is envisaged that electrons are injected in this device by longitudinal optical (LO) phonon assisted tunneling. The QW sizes are so designed that the electron wavefunctions are uniformly distributed over the entire active region. The depth of the grating was approximately 800 Å, to achieve a coupling constant of ~ 150 cm^{-1}. The sample was then prepared for MOVPE regrowth and the 1.5 μm *p* type InP layer, the graded layer and the $In_{0.53}Ga_{0.47}As$ *p* contact layer were successively grown. A standard single-mode ridge waveguide laser was then fabricated by a combination of photolithography techniques, wet and dry etching and contact metallization. The ridge was etched to a depth of ~ 1.5 μm from the top.

The measured temperature-dependent light versus current (L–I) characteristics for a 300 μm long device is shown in Fig. 15(a). The slope efficiency is only slightly lower than Fabry–Perot lasers and the output power is larger by a factor of 3.5. The temperature-dependent threshold current density is plotted in Fig. 15(b). The temperature T_o is estimated to be 120 K around room temperature and below and decreases to 70 K at higher temperatures. Again, part of this degradation is due to lack of appropriate heat sinking. Nonetheless, the measured values of T_o are significantly higher than those reported earlier for 1.55 μm lasers[26,27] and is believed to predominantly result from the tunneling injection mechanism. It is worthwhile to note that the value of T_o measured in 1.55 μm Fabry–Perot tunneling injection lasers is 70 K for the temperature range 298–328 K, without any heat sinking. The value of T_o is in reasonable agreement with the value obtained for the present DFB

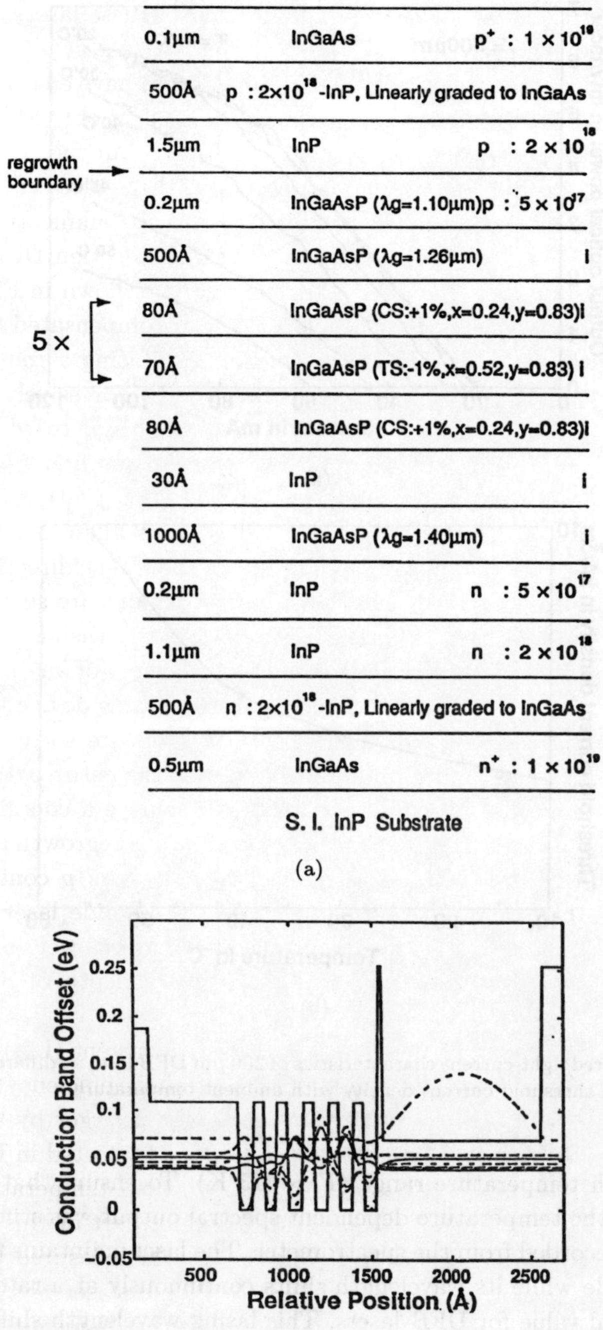

Fig. 14. (a) InP-based 1.55 μm strain-compensated MQW tunnel injection laser grown by MOVPE; (b) conduction band profile of active region with calculated wavefunctions of the first five quasi-bound states.

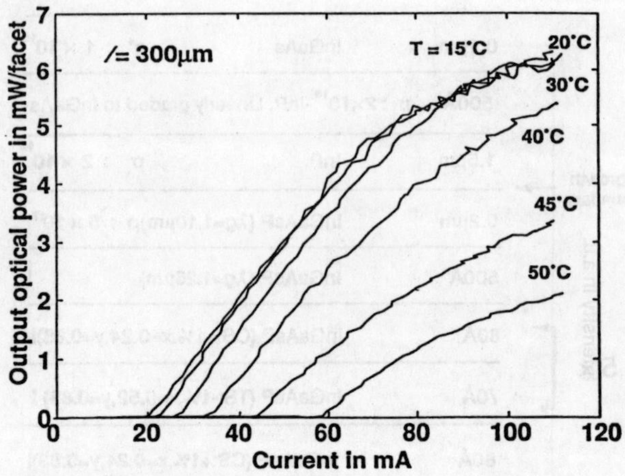

(a)

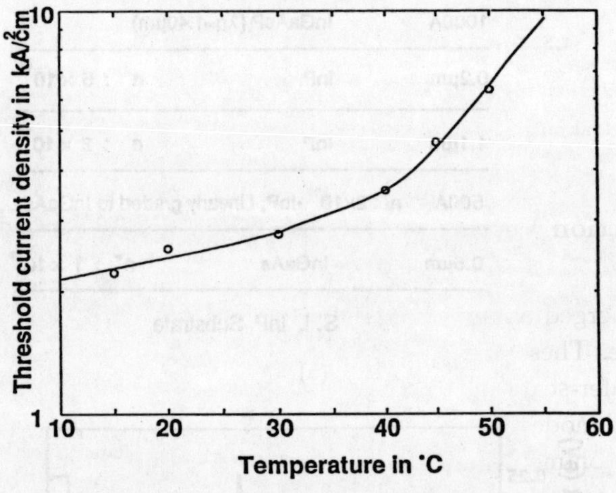

(b)

Fig. 15. (a) Measured light-current characteristics of 300 μm DFB laser at different temperatures, and (b) variation of threshold current density with ambient temperature.

lasers in the high temperature range ($T > 313$ K). To ensure that lasing is from the DFB mode, the temperature dependent spectral output was studied. Figure 16 shows the data recorded from the spectrometer. The laser maintains the same single longitudinal mode while its wavelength shifts continuously at a rate of ~ 1 Å/°C, which is a typical value for DFB lasers. This lasing wavelength shift is due to the variation of mode effective index with temperature. In other words, the material effective bandgap shrinks as the temperature increases leading to a red-shift in lasing modes.

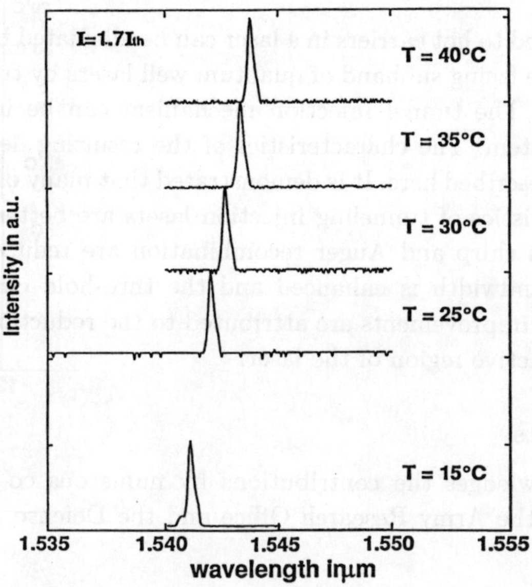

Fig. 16. Spectral output of 300 μm laser above threshold at different temperatures.

6. Tunnel Injection Vertical Cavity Surface Emitting Lasers (VCSELs)

VCSELs have emerged as important light sources for several inherent advantages that they provide. These are: a low diffraction output beam suitable for efficient fiber coupling, wafer-scale and two-dimensional array fabrication and easy probing, single longitudinal mode operation, low threshold current (low power consumption), and small modal volume. There has been a drive to reduce the threshold current of VCSELs in order to reduce heat dissipation in array applications and to enhance modulation bandwidth. The lowest threshold currents have been observed in oxide-confined VCSELs by Huffaker and Deppe.[28] It was observed by these authors and others[29] that as the size of the modal volume is reduced, the threshold current steadily decreases, and then begins to increase for cavity diameters ≤ 3 μm. The latter is accompanied by a reduction in the slope efficiency. These phenomena have been attributed to hot carrier effects and loss of the injected electrons over the heterobarriers.

In order to alleviate this problem, Huffaker *et al.*[30] have incorporated the tunnel injection scheme in the design of the active region, so that the carriers injected remain cold. A lasing threshold of 36 μA with a slope efficiency of 23% was measured for a 2.5 μm device, and a 52% slope efficiency was achieved for a 4 μm device with a 92 μA threshold current. These tunnel injection VCSELs demonstrate the highest slope efficiencies yet achieved for low threshold current.

7. Conclusion

The problems related to hot carriers in a laser can be alleviated by the cold injection of electrons into the lasing subband of quantum well lasers by tunneling or phonon-assisted tunneling. The tunnel injection mechanism can be incorporated in any heterostructure system. The characteristics of the resulting device, the tunneling injection laser, is described here. It is demonstrated that many of the important performance characteristics of tunneling injection lasers are better than conventional SCH devices. Thus chirp and Auger recombination are reduced, T_o is increased, the modulation bandwidth is enhanced and the threshold current is reduced in VCSELs. All these improvements are attributed to the reduction of the hot-carrier population in the active region of the laser.

Acknowledgments

The author acknowledges the contributions by numerous co-workers. The work was supported by the Army Research Office and the Defense Advanced Research Projects Agency.

References

1. M. Grupen and K. Hess, "Severe gain suppression due to dynamic carrier heating in quantum well lasers", *Appl. Phys. Lett.* **70** (1997) 808–810.
2. L. Davis, Y. L. Lam, Y. C. Chen, J. Singh, and P. K. Bhattacharya, "Carrier capture and relaxation in narrow quantum wells", *IEEE J. Quantum Electron.* **30** (1994) 2560–2564.
3. R. Kersting, R. Schwedler, K. Wolter, K. Leo, and H. Kurz, "Dynamics of carrier transport and carrier capture in In_{1-x} Ga_xAs/InP heterostructures", *Phys. Rev.* **B46** (1992) 1639–1648.
4. B. Deveaud, J. Shah, T. C. Damen, and W. T. Tsang, "Capture of electrons and holes in quantum wells", *Appl. Phys. Lett.* **52** (1988) 1886–1888.
5. S. Morin, B. Deveaud, F. Clerot, K. Fugiwara, and K. Mitsunaga, "Capture of photoexcited carriers in a single quantum well with different confinement structures", *IEEE J. Quantum Electron.* **27** (1991) 1669–1675.
6. H. C. Sun, L. Davis, S. Sethi, J. Singh, and P. Bhattacharya, "Properties of a tunneling injection quantum-well laser: Recipe for a 'cold' device with a large modulation bandwidth", *IEEE Photon. Technol. Lett.* **5** (1993) 870–872.
7. C. Y. Sung, T. B. Norris, X-K. Zhang, Y. L. Lam, I. Vurgaftman, J. Singh, and P. Bhattacharya, "Studies of carrier relaxation in low dimensional structures", *Solid State Electron.* **40** (1996) 751–754.
8. F. Girardin, G. Duan, P. Gallion, A. Talneau, and A. Qugazzaden, "Experimental investigation of the relative importance of carrier heating and spectral-hole-burning on nonlinear gain in bulk and strained multi-quantum-well 1.55 μm lasers", *Appl. Phys. Lett.* **67** (1995) 771–773.
9. R. Frankenberger and R. Schimpe, "Origin of nonlinear gain saturation in index-guided InGaAsP laser diodes", *Appl. Phys. Lett.* **60** (1992) 2720–2722.
10. K. L. Hall, G. Lenz, E. P. Ippen, U. Koren, and G. Raybon, "Carrier heating and spectral hole burning in strained-layer quantum-well laser amplifiers at 1.5 μm", *Appl. Phys. Lett.* **61** (1992) 2512–2514.

11. C. K. Sun, H. K. Choi, C. A. Wang, and J. G. Fujimoto, "Femtosecond gain dynamics in InGaAs/AlGaAs strained-layer single-quantum-well diode lasers", *Appl. Phys. Lett.* **63** (1993) 96–98.
12. J. Zhou, N. Park, J. W. Dawson, K. J. Vahala, M. A. Newkirk, U. Koren, and B. I. Miller, "Highly nondegenerate four-wave mixing and gain nonlinearity in a strained multiple-quantum-well optical amplifier", *Appl. Phys. Lett.* **62** (1993) 2301–2303.
13. M. Willatzen, T. Takahasi, and Y. Arakawa, "Nonlinear gain effects due to carrier heating and spectral holeburning in strained-quantum-well lasers", *IEEE Photon. Technol. Lett.* **4** (1992) 682–685.
14. M. Willatzen, A. Uskov, J. Mork, H. Olesen, B. Tromborg, and A. P. Jauho, "Nonlinear gain suppression in semiconductor lasers due to carrier heating", *IEEE Photon. Technol. Lett.* **3** (1991) 606–609.
15. H. Yoon, H. C. Sun, and P. Bhattacharya, "Dynamic linewidth of tunneling injection laser", *Electron. Lett.* **30** (1994) 1675–1677.
16. J. D. Ralston, S. Weisser, E. C. Larkins, A. Schoenfelder, J. Daleiden, K. Dzotscher, M. Maier, J. Fleissner, W. Benz, and J. Rosenzweig, presented at *Photonics West '96*, San Jose, California, 27 January–2 February, 1996.
17. R. Nagarajan, M. Ishikawa, T. Fukushima, R. S. Geels, and J. E. Bowers, "High speed quantum-well lasers and carrier transport effects", *IEEE J. Quantum Electron.* **28** (1992) 1990–2008.
18. P. Bhattacharya, J. Singh, H. Yoon, X. Zhang, A. Gutierrez-Aitken, and Y. Lam, "Tunneling injection lasers: A new class of lasers with reduced hot carrier effects", *IEEE J. Quantum Electron.* **32** (1996) 1620–1629.
19. P. Bhattacharya, K. Kamath, J. Singh, D. Klotzkin, J. Phillips, H-T. Jiang, N. Chervela, T. B. Norris, T. Sosnowski, J. Laskar, and M. Ramana Murty, "In(Ga)As/GaAs self-organized quantum dot lasers: DC and small-signal modulation properties", *IEEE Trans. Electron. Devices* **46** (1999) 871–883.
20. R. Jambunathan and J. Singh, "Effects of non-uniform charge injection on gain, threshold current, and linewidth enhancement factor for a 1.55 μm InP-based multiple quantum well laser", *J. Appl. Phys.* **80** (1996) 6875–6879.
21. P. Zory, *Quantum Well Lasers*, Academic Press, Boston, MA, 1993.
22. H. Yoon, A. L. Gutierrez-Aitken, R. Jambunathan, J. Singh, and P. Bhattacharya, "A 'cold' InP-based tunneling injection laser with greatly reduced Auger recombination and temperature dependence", *IEEE Photon. Technol. Lett.* **7** (1995) 974–976.
23. H. Bissessur, C. Starck, J.-Y. Emery, F. Pommereau, C. Duchemin, J.-G. Provost, J.-L. Beylat, and F. Fernier, "Very narrow-linewidth (70 kHz) 1.55 μm strained MQW DFB lasers", *Electron. Lett.* **28** (1992) 998–999.
24. P. Bhattacharya, Y. Yuan, T. Brock, C. Caneau, and R. Bhat, "Temperature dependence of the threshold current in 1.55 μm strain-compensated multiquantum-well distributed-feedback lasers", *IEEE Photon. Technol. Lett.* **10** (1998) 778–780.
25. J. S. Osinski, Y. Zou, P. Grodzinski, A. Mathur, and P. D. Dapkus, "Low-threshold-current-density 1.5 μm lasers using compressively strained InGaAsP quantum wells", *IEEE Photon. Technol. Lett.* **4** (1992) 10–13.
26. C. E. Zah, P. S. D. Lin, F. Favire, B. Pathak, R. Bhat, C. Caneau, A. S. Gozdz, N. C. Andreadakis, M. A. Koza, T. P. Lee, T. C. Wu, and K. Y. Lau, "1.5 μm compressive-strained multiquantum-well 20-wavelength distributed-feedback laser arrays", *Electron. Lett.* **28** (1992) 824–826.
27. Y. Zou, J. S. Osinski, P. Grodzinski, P. D. Dapkus, W. C. Rideout, W. F. Sharfin, J. Schelafer, and F. D. Crawford, "Experimental study of Auger recombination, gain, and temperature sensitivity of 1.5 μm compressively strained semiconductor lasers",

IEEE J. Quantum Electron. **29** (1993) 1565–1575; G. L. Belenky, R. F. Kazarinov, J. Lopata, S. Luryi, T. Tanbun-Elk, and P. A. Garbinski, "Direct measurement of the carrier leakage out of the active region in InGaAsP/InP laser heterostructures", *IEEE Trans. Electron Devices* **42** (1995) 215–218.

28. D. L. Huffaker, D. G. Deppe, K. Kumar, and T. J. Rogers, "Native-oxide defined ring contact for low threshold vertical-cavity lasers", *Appl. Phys. Lett.* **65** (1994) 97–99.
29. K. D. Choquette, W. W. Chow, G. R Hadley, H. Q. Hou, and K. M. Geib, "Scalability of small-aperture selectively oxidized vertical cavity lasers", *Appl. Phys. Lett.* **70** (1997) 823–825.
30. D. L. Huffaker and D. G. Deppe, "Improved performance of oxide-confined vertical-cavity surface-emitting lasers using a tunnel injection active region", *Appl. Phys. Lett.* **71** (1997) 1449–1451.

International Journal of High Speed Electronics and Systems, Vol. 9, No. 4 (1998) 867–899

QUANTUM WELL INTERSUBBAND LASERS

J. P. SUN and G. I. HADDAD
Center for High Frequency Microelectronics,
Department of Electrical Engineering and Computer Science,
The University of Michigan, Ann Arbor, Michigan 48109-2122, USA

M. DUTTA and M. A. STROSCIO
U.S. Army Research Office,
P.O. Box 12211, Research Triangle Park, NC 27707-2211, USA

This chapter/paper provides an overview of quantum well intersubband lasers including quantum cascade and step quantum well lasers. A simplified model of the quantum cascade laser is presented which provides a good estimate of major device parameters and illustrates the principles of operation and physical processes. Device schemes of other intersubband lasers such as step quantum wells are also presented and analyzed.

1. Introduction

The idea for a unipolar semiconductor laser was first proposed in 1971 by Kazarinov and Suris[1] following the seminal work of Esaki and Tsu on superlattices. However, it took more than 20 years for its realization in the form of the quantum cascade laser (QCL).[2] The successful demonstration and the subsequent refinement of the QCL are one of the milestones in semiconductor laser research and development since the invention of the injection laser because of its significance both in device realization and applications.

The QCL represents a great achievement in combining quantum physics, optics, and crystal growth technology. It was a long-standing proposal to utilize electron transitions between subbands in quantum well structures for laser action. However, due to the complexity of intersubband physics involving efficient electron injection and control of intersubband transition rates to establish and maintain carrier population inversion, an optimal intersubband engineering is required for the device design and performance improvement. Moreover, it would not have been possible for the implementation of the device structures if the state-of-the-art technology of molecular beam epitaxy (MBE) were not available.

For device application, the QCLs are prime candidates of solid state sources operating at room temperature with tens to hundreds of milliwatts of optical power in the mid-infrared wavelengths. Presently, mid-infrared (MIR), far-infrared (FIR), and terahertz frequencies (THz) remain among the most underdeveloped part of the electromagnetic spectrum. For a growing number of applications at these frequencies, such as laser-based portable environmental monitoring systems, chemical gas spectroscopy, and space optical communications, a solid state source with good, room temperature performance is essential. Semiconductor quantum well lasers based on interband transitions can generate coherent radiation into the infrared, but the longest possible wavelengths are limited by the available bandgaps of useful semiconductor materials. The existing technology based on lead-salt lasers and the HgCdTe alloys can operate up to 8 μm, but is not satisfactory because of their low optical power and poor temperature performance associated with defects or Auger relaxation effects. In the QCLs and the other intersubband lasers, the lasing wavelengths are determined by the quantum confinement. With proper device design, the wavelengths can be tailored from the MIR to the THz range.

The QCL realization at the FIR and THz frequencies can be problematic, however, due to the strong free electron absorption at the long wavelengths. In addition, QCLs demand the state-of-the-art control of MBE growth and sophisticated design for the efficient electron injector as well as the coupled quantum wells and superlattices. Therefore, there is a continued interest to explore alternative laser schemes especially optically pumped intersubband lasers at the longer wavelengths. Some device schemes of those lasers will be outlined and discussed.

We start our discussions with a brief description of the quantum cascade laser (QCL), its device structure and operation principles in Sec. 2. The main features of QCLs and their recent performance are enumerated. For illustration and understanding purposes, we present a simplified model calculation of the major physical parameters in the active region of the QCL device in Sec. 3, including electron quasi-bound states, intersubband relaxation rates (times) due to polar-optical-phonon scattering, electron population inversion, tunneling injection and escape times, dipole transition matrix elements, lineshape function, and optical gain spectra. The device threshold current density and optical output power are also estimated. Other device approaches of quantum well intersubband lasers, particularly optically-pumped intersubband lasers, are discussed in Sec. 4.

2. Device Structure and Operation

Following the description of Faist *et al.*[2] in their first demonstration of QCL, the conduction band diagram of the device active region is schematically shown in Fig. 1(a); Fig. 1(b) shows the major device regions of the QCL including the contact, the active and graded regions which constitute the device core. Outside the core there are the cladding layers and the substrate. The active region of the QCL structure will be used to perform our model calculation.

The QCL structure was grown with the $Al_{.48}In_{.52}As$ (barrier) and $Ga_{.47}In_{.53}As$ (well) heterosystem which is lattice matched to an InP substrate using MBE. The active lasing region is made up of 3 undoped quantum wells designed such that the quasi-bound states in the wells form a four-level laser system,[2] since there exists

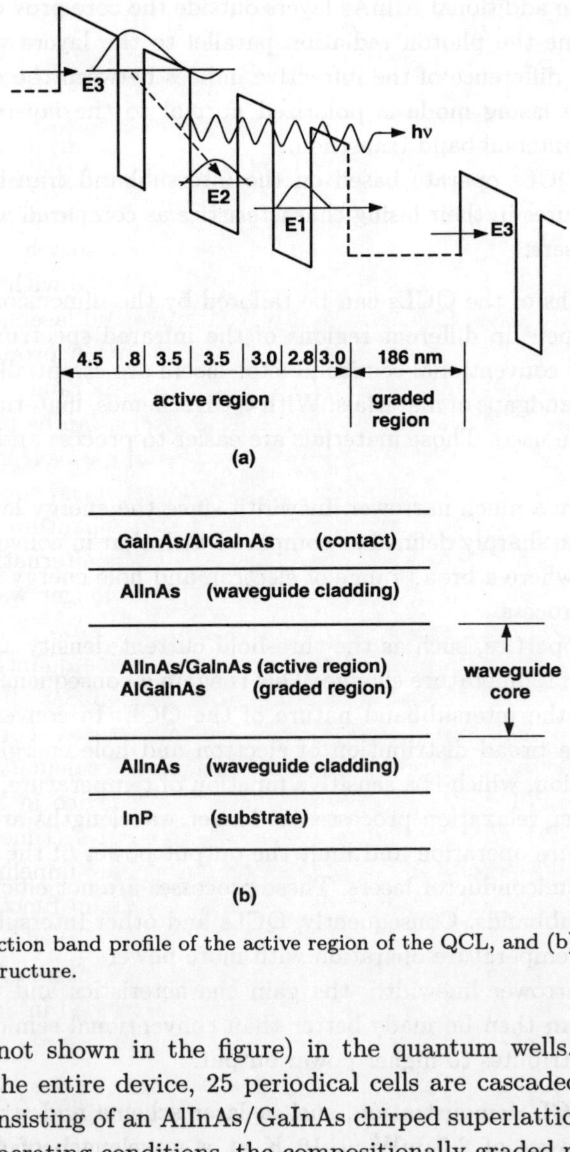

Fig. 1. (a) Conduction band profile of the active region of the QCL, and (b) the main regions of the QCL layered structure.

a fourth level (not shown in the figure) in the quantum wells. To enhance the lasing through the entire device, 25 periodical cells are cascaded by the digitally graded layers consisting of an AlInAs/GaInAs chirped superlattice with a constant period. Under operating conditions, the compositionally graded regions reach near flat-band condition so the electron injection is facilitated from one cell to the next. In the laser operation, electrically pumped (injected) electrons from the contact tunnel through the first potential barrier, filling the energy level E_3. In the active region, electron population inversion is achieved between the two excited states at the energies of E_3 and E_2, by controlling the intersubband transition or relaxation rates. The lasing transition occurs with the wavelength corresponding to the energy difference $E_{32} = E_3 - E_2$. Under the applied bias voltage, electrons will stream down the potential staircase and emit photons sequentially at each cell, which will further stimulate radiative transitions in the first and other cells to enhance optical power output. The additional AlInAs layers outside the core provide the waveguide cladding to confine the photon radiation parallel to the layers within the active region due to the difference of the refractive indices between the cladding and the core regions. The lasing mode is polarized normal to the layers because of the selection rule for intersubband transition.

Because the QCLs operate based on the intersubband transitions, they have the following features in their lasing characteristics as compared with conventional semiconductor lasers:

1. The wavelengths of the QCLs can be tailored by the dimensions of the layered structure to open up different regions of the infrared spectrum. The emission wavelengths of conventional semiconductor lasers are essentially determined by the available bandgaps of materials. With QCL schemes, materials of much wider bandgap can be used. Those materials are easier to process and more mature in technology.
2. The QCLs have a much narrower linewidth since the energy levels or subbands in the QCLs are sharply defined as compared with that in conventional semiconductor lasers, where a broad range of electron and hole energy levels is involved in the lasing process.
3. The device properties, such as the threshold current density and linewidth are less sensitive to temperature change. This is again a consequence of the unipolar character and the intersubband nature of the QCL. In conventional semiconductor lasers, a broad distribution of electron and hole energies is involved in the laser emission, which is a sensitive function of temperature. In addition, the enhanced Auger relaxation processes at longer wavelengths are detrimental to high temperature operation and limit the output power of the long wavelength bipolar type semiconductor lasers. These processes are not efficient between the near-parallel subbands. Consequently, QCLs and other intersubband lasers can achieve room temperature operation with more power.
4. Due to the narrower linewidth, the gain characteristics and the overall laser performance can then be made better than conventional semiconductor lasers, which also contributes to higher power output.

In the first QCL demonstration, such a laser scheme and structure delivered a peak optical power of 8.5 mW at 10 K at a wavelength of 4.26 μm, with a threshold current density of 15 k A/cm^2, and the laser action was evident up to a temperature of 90 K. The subsequent QCL development achieved by the group at Bell Laboratories has been impressive in the subband engineering, configuration of QCL structures, device performance, as well as the practical development of QCL applications. Within the past five years, the wavelengths of the QCLs have been extended to cover a wide range, from 3.4 μm to 17 μm.[3] The QCL with the active region consisting of chirped superlattices has generated pulsed power of 12 mW at low temperatures and operated up to 150 K at 17 μm.[4] QCL devices operating in the 5–11 μm range have achieved pulsed peak power up to 0.5 W at 8 μm at room temperature, and cw power of 200 mW at 77 K at 8 and 5.2 μm wavelengths, respectively. Other noticeable examples of development include the high power interminiband superlattice mid-infrared lasers,[5] which attained 800 mW peak power at 80 K; the dual wavelength intersubband emitters,[6] which can emit mW optical power in two distinct portions of the mid-infrared (8 μm and 10.5 μm); the tunable distributed feedback (DFB) QCLs,[7] which achieve tuning operation in the atmosphere windows of 5.3 μm and 8.5 μm, with single-mode cw tuning range of 20 nm and 70 nm, and pulsed tuning range of 140 nm between 8.38 and 8.61 μm; and finally the quantum cascade microdisk lasers[8] with chaotic resonators for increased power. QCLs have also been demonstrated by a research group of North Western University using gas source MBE growth.[9] In the device applications, the QCL-based sensors to detect minute amounts of trace gases or pollutants operating at room temperature has been demonstrated and its practical development of application is underway.

3. A Simplified Model of the QCL

In this section, we present a model calculation of QCL physical parameters to illustrate the principle of device operation and gain some insight of the major physical processes that lead to the stimulated photon emission. The physical and device parameters of interest here include the electron quasi-bound states in the QCL structure, the optical-phonon limited intersubband relaxation rates among the QCL subbands, the electron population inversion between the lasing subbands, the transition matrix element, the lineshape function, all of which lead to an evaluation of the optical gain spectra of the QCL. The processes considered in the calculations are restricted within the QCL active region. In addition, for the device terminal properties, the threshold current density and output optical power are estimated. These calculations presented here are mainly for illustration purposes and are considered to be semi-quantitative for some device characteristics. Present efforts to understand the physics of the QCLs have been focused on the design of the active region and are on-going. Complete theoretical models of QCLs and other intersubband lasers are still in their preliminary stages. Substantial work is needed for accurate modeling and design optimization of these lasers.

3.1. *Electron quasi-bound states*

The electron quasi-bound states in the QCL active region can be calculated approximately by solving the single band, time-independent, effective mass Schrödinger equation in the growth direction x,

$$\frac{-\hbar^2}{2}\frac{\partial}{\partial x}\left[\frac{1}{m^*(x)}\frac{\partial\psi(x)}{\partial x}\right] - [E - V(x)]\psi(x) = 0\,, \tag{1}$$

where the electron potential energy term $V(x)$ is specified from the conduction band profile once a device structure and its material composition are determined at a specific applied voltage across the active region. Equation (1) can be solved with appropriate boundary conditions to obtain the electron eigenenergies E and the associated wavefunctions ψ. For a rough approximation, a closed-system solution can be employed, in which the wavefunctions are set to zero at the boundaries. The closed-system approach often yields a reasonably good estimate of the electron resonant states in the quantum well for electron energies not very close to the top of the barriers. On the other hand, it is conceptually incorrect to set the electron wave functions to vanish at the device boundaries during the laser operation since electric current is flowing in the entire device structure. Using the closed-system method tends to give higher values of eigenenergies since it takes the quantum well structure as an ideally isolated quantum system. Moreover, it does not give correct wavefunctions for a device under operational conditions. In the open-system approach, the interactions of the electrons in the contact regions with the device environment are properly treated using the open-system boundary conditions such as the quantum transmitting boundary conditions. A spectrum of electron waves with their known eigenenergies are incident upon the quantum well regions, and the solution of Eq. (1) subject to the open-system boundary conditions then gives the associated wavefunctions at any spatial point in the quantum well region. Once the electron wavefunctions are obtained, one can calculate virtually any quantity of interest in the quantum well structure, such as the electron transmission coefficients and tunneling current density. For example, the resonant quasi-bound energies are those at which the electron transmission coefficients have their peak values. The reader is referred to Ref. 10 for a comprehensive review of open-system boundary conditions, and Refs. 11 and 12 for the quantum transmitting boundary method and an example of its implementation in a quantum device calculation.

Another aspect in the determination of quasi-bound states is the effect of impurities that exist due to intentional and unintentional doping. These fixed charges in combination with mobile electron charge will lead to band bending, changes of effective barrier shape and effective voltage drop in the barrier/well regions. This is the well-known self-consistent problem. To determine the resonant electron states in the quantum wells, Poisson's equation and the Schödinger equation are solved simultaneously to give more accurate values of the energies and their associated wavefunctions. In addition, it is also desirable to incorporate effects of non-parabolicity of electron dispersion in the calculation.

In Fig. 2, the calculated first three quasi-bound state energies and the electron envelope wavefunctions are shown to have the energy separations $E_{32} = 289.20$ meV and $E_{21} = 30.95$ meV for a field strength of approximately 10^5 V/cm. The energy separation between the lasing subbands corresponds to the emission wavelength of

$$\lambda = \frac{1.2398}{\hbar\omega[eV]} = 4.28\ \mu\text{m}\,, \tag{2}$$

which is consistent with the measured emission wavelength of 4.26 μm.[2]

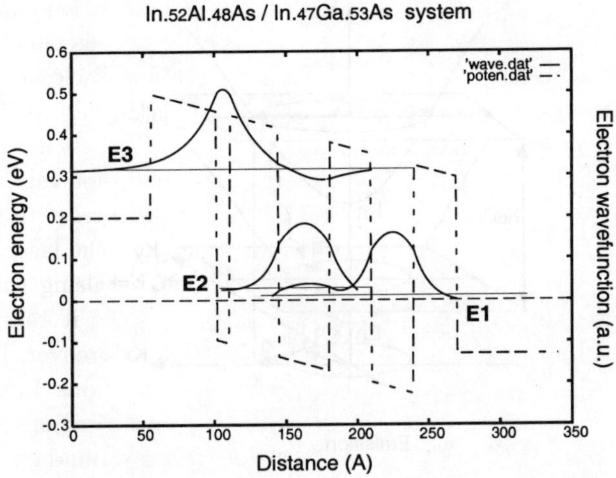

Fig. 2. Calculated quasi-bound states and the associated electron wavefunctions in the active region under an applied field strength of 10^5 V/cm. Self-consistency is not implemented.

3.2. *Intersubband relaxation rates*

The control of relaxation rates among the energy subbands in the quantum wells is critical in the design of the QCLs and other intersubband lasers for the establishment and maintenance of carrier population inversion between the lasing subbands required for successful laser operation. The electron transitions we discuss in this section are the spontaneous transitions as they are not induced by photon emission. The photon-induced stimulated electron transitions will be treated later when we discuss the transition matrix element in the dipole approximation. The electron transition processes between the subbands are explained in some detail here. An approximate calculation of the relaxation rates due to polar-optical-phonon scattering in the QCL structure is given.

Taking the three-level of the QCL active region as an example, i.e., the upper and lower lasing subbands at E_3 and E_2, and the lowest subband at E_1, the following design is typically sought:

1. Efficient carrier injection or capture into the upper lasing subband.
2. A long relaxation time or a low transition rate from the upper lasing subband to the lower subbands.
3. A short relaxation time or a high transition rate from the lower lasing subband to the lowest subband.
4. Fast, efficient removal of the carriers from the lowest subband to the next cell.

The electron transition processes between two electronic states can be visualized in Fig. 3. In the evaluation of the transition rates, only the dominant scattering mechanism due to polar-optical phonons is considered, neglecting effects of other phonon scattering mechanisms and electron-electron interactions. For phonon

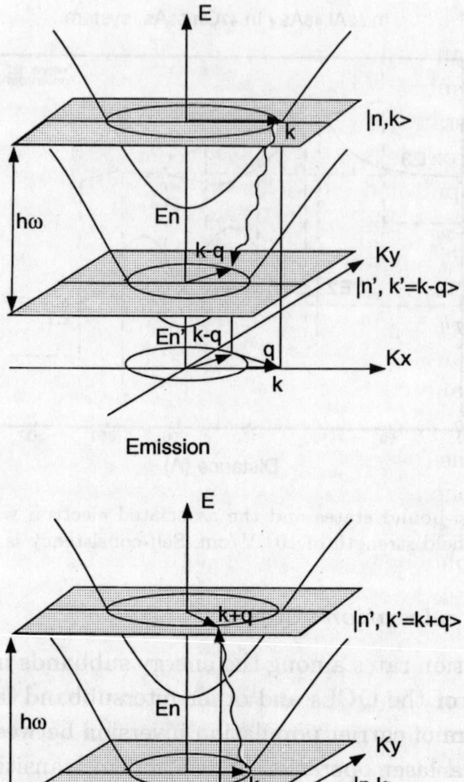

Fig. 3. Electron transition between an initial state $|n, \mathbf{k}\rangle$ and a final state $|n', \mathbf{k}'\rangle$, with a phonon emission or absorption process.

emission, an electron in the initial state $|n, \mathbf{k}\rangle$ in the upper subband with in-plane wavevector $\mathbf{k}$ can emit a phonon and jump to one of the accessible final states $|n', \mathbf{k}'\rangle$ in the lower subband (shown as the equal-energy circle in the lower subband). For phonon absorption, an electron in the lower subband absorbs a phonon and jumps to one of the states on the circle in the upper subband. The phonon-induced transition rate of a single transition event between these two electronic states is given by Fermi's golden rule based on first-order perturbation theory,

$$W_{n,n'}^{\left\{ \begin{matrix} e \\ a \end{matrix} \right\}}(\mathbf{k}, \mathbf{k}') = \frac{2\pi}{\hbar} |\langle n'\mathbf{k}', N_q \pm 1 | H_{e-ph} | n\mathbf{k}, N_q \rangle|^2 \times \delta(E_{n'} + E_{\mathbf{k}'} \pm \hbar\omega - E_n - E_{\mathbf{k}}), \qquad (3)$$

where the superscripts of the transition rate denote the related phonon process: e for phonon emission and a for absorption, E_n and $E_{n'}$ are the band-edge energies of the nth and n'th subbands, $\mathbf{k}$ and $\mathbf{k}'$ are the in-plane electron wavevectors of the initial and final states, respectively, $E_{\mathbf{k}} = \hbar^2 k^2/(2m^*)$ is the in-plane energy in the parabolic approximation, ω is the frequency of a given phonon mode, and N_q is the phonon population based on the Bose–Einstein statistics. The interaction Hamiltonian H_{e-ph} represents the electron-phonon scattering potential and is given by

$$H_{e-ph} = -e \sum_{\mathbf{q}} \frac{1}{\sqrt{A}} e^{i\mathbf{q}\cdot\mathbf{r}_{||}} \phi_s(\mathbf{q}, x)(a^\dagger_{-\mathbf{q}} + a_{\mathbf{q}}) , \tag{4}$$

where A is the in-plane cross section area of the structure, and $\mathbf{r}_{||}$ the in-plane position vector, $\phi_s(q, x)$ is the electrostatic potential function of the phonon mode s, with in-plane wavevector $\mathbf{q}$; $a^\dagger_{\mathbf{q}}$ and $a_{\mathbf{q}}$ are the creation and annihilation operators of phonons. The emission and absorption rates are generally of the same order and have similar in-plane energy dependence. For given states $|n, \mathbf{k}\rangle$, $|n', \mathbf{k}'\rangle$ and a given phonon mode, the ratio of the emission and absorption rates $W^e_{n,n'}(\mathbf{k}, \mathbf{k}')/W^a_{n,n'}(\mathbf{k}, \mathbf{k}') = (N_q+1)/N_q = e^{\hbar\omega/k_B T}$ is a constant at a specified temperature. The difference between these two processes lies in whether the emission or absorption process is forbidden or requires an onset in-plane energy, required by energy conservation. When the subband separation is smaller than the phonon energy, an electron emitting a phonon has to possess a certain amount of in-plane energy $E_{\mathbf{k}}$ for the transition to occur. For phonon absorption, the transition can always happen when $E_{n'} - E_n \leq \hbar\omega$. If the subband separation is larger than the phonon energy, then the emission can always happen while the absorption requires the electron possessing a certain amount of initial in-plane energy. Since we are concerned with electron energy relaxation, the focus hereafter will only be on the emission process and we drop the superscripts of W, but indicate the difference in the absorption case when appropriate.

The electron wave function can be written as

$$|n, \mathbf{k}\rangle = \frac{1}{\sqrt{A}} e^{i\mathbf{k}\cdot\mathbf{r}_{||}} \psi_n(x) . \tag{5}$$

In writing ψ_n independent of k we have neglected the effect of kinetic energy confinement. $\psi_n(x)$ in the quantum well structures are obtained numerically by solving the Schrödinger equation as outlined previously.

From Eqs. (3) and (5), the emission rate is given by

$$W_{n,n'}(\mathbf{k}, \mathbf{q}) = \frac{2\pi}{\hbar}(N_q + 1)|F(q)|^2 \times \delta(E_{n'} + E_{\mathbf{k}-\mathbf{q}} + \hbar\omega - E_n - E_{\mathbf{k}}) , \tag{6}$$

with $F(q) = \int dx\ h(q, x)\psi^*_n(x)\psi_{n'}(x)$ and the two-dimensional in-plane wavevector conservation relation $\mathbf{k} = \mathbf{k}' + \mathbf{q}$. $h(q, x) = -e\phi(q, x)$, as explained following Eq. (4). Effects of different phonons or different modes of a specific phonon can be evaluated once the phonon potential is known, for example, from solutions of the electrostatic equations in a macroscopic continuum model or a microscopic analysis.

Summing over all allowable final states (or summing over $\mathbf{q}$), we calculate the emission rate of an electron in an initial state $|n, \mathbf{k}\rangle$ to a lower subband $|n', \mathbf{k}'\rangle$,

$$W_{n,n'}(\mathbf{k}) = \sum_{\mathbf{q}} W_{n,n'}(\mathbf{k}, \mathbf{q}) . \tag{7}$$

As discussed above, due to energy conservation, some transitions are forbidden. For emission process, when $E_{\mathbf{k}} < \hbar\omega' \equiv \hbar\omega - (E_n - E_{n'})$ and $\omega' > 0$, i.e., the two subband edges are closer than a phonon energy and the electron does not possess enough initial kinetic energy, there are no final states available and the process is forbidden. For $\omega' > 0$ and $E_{\mathbf{k}} > \hbar\omega'$,

$$W_{n,n'}(\mathbf{k}) = \frac{A}{4\pi^2}\frac{2\pi}{\hbar}(N_q + 1)\int_{-\theta_{\max}}^{\theta_{\max}} \times d\theta \frac{2m}{\hbar^2}\frac{|F(q_+)|^2 q_+ + |F(q_-)|^2 q_-}{|q_+ - q_-|} , \tag{8}$$

where

$$q_\pm = k \cos\theta \pm \sqrt{k^2 \cos^2\theta - \frac{2m}{\hbar}\omega'}, \theta_{\max} = \cos^{-1}\left(\sqrt{\frac{2m\omega'}{\hbar k^2}}\right).$$

For $\omega' < 0$,

$$W_{n,n'}(\mathbf{k}) = \frac{A}{4\pi^2}\frac{2\pi}{\hbar}(N_q + 1)\int_0^{2\pi} d\theta \frac{2m}{\hbar^2}\frac{|F(q_+)|^2 q_+}{q_+} . \tag{9}$$

The above argument also holds for the absorption process, as long as changes are made for $\omega \to -\omega$, $\omega' \to -\omega'$, $N_q + 1 \to N_q$ and $\mathbf{q} \to -\mathbf{q}$. That is, for the absorption process,

$$W_{n,n'}(\mathbf{k}, \mathbf{q}) = \frac{2\pi}{\hbar} N_q |F(q)|^2 \delta(E_{n'} + E_{\mathbf{k+q}} - \hbar\omega - E_n - E_{\mathbf{k}}) , \tag{10}$$

when $E_{\mathbf{k}} < -\hbar\omega'$ and $\hbar\omega' \equiv \hbar\omega - (E_{n'} - E_n) < 0$, i.e., the two subband edges are farther away than a phonon energy and the electron does not have enough initial in-plane energy, there are no final states available and the absorption process is forbidden. For $\omega' < 0$ and $E_{\mathbf{k}} > -\hbar\omega'$,

$$W_{n,n'}(\mathbf{k}) = \frac{A}{4\pi^2}\frac{2\pi}{\hbar} N_q \int_{-\theta_{\max}}^{\theta_{\max}} \times d\theta \frac{2m}{\hbar^2}\frac{|F(q_+)|^2 q_+ + |F(q_-)|^2 q_-}{|q_+ - q_-|} , \tag{11}$$

where

$$q_\pm = k \cos\theta \pm \sqrt{k^2 \cos^2\theta + \frac{2m}{\hbar}\omega'}, \theta_{\max} = \cos^{-1}\left(\sqrt{-\frac{2m\omega'}{\hbar k^2}}\right).$$

For $\omega' > 0$,

$$W_{n,n'}(\mathbf{k}) = \frac{A}{4\pi^2}\frac{2\pi}{\hbar}N_q\int_0^{2\pi}d\theta\frac{2m}{\hbar^2}\frac{|F(q_+)|^2 q_+}{q_+}\,. \tag{12}$$

The calculated relaxation times among the three-subbands using the formulations presented above are shown in Fig. 4 as a function of the electron in-plane wave number. The relaxation rates depend on the spatial overlap of the relevant electron wave functions, the interaction Hamiltonian into which the specific phonon mode and other material parameters are entered. These results are calculated with the bulk phonon mode of polar-optical phonon and the bulk Fröhlich interaction Hamiltonian, at the field strength of 10^5 V/cm through an applied bias voltage included in the solution for the electron wave functions.

It can be seen from Fig. 4 that the optical-phonon-assisted transition time from subband 2 to subband 1 is very fast. $\tau_{21} \sim 0.5$ ps at a near-zero in-plane momentum; the small amount in-plane momentum is necessary for an electron at E_2 to make a transition to E_1 with an optical-phonon emission. The relaxation is fast because of the close proximity of the electron wavefunctions due to the narrow energy separation and the strong coupling of the adjacent GaInAs quantum well. On the other hand, the relaxation from subbands 3 and 2 to subband 1 takes much longer time. Our estimate gives $\tau_{32} = 4.86$ ps and $\tau_{31} = 7.54$ ps at zero in-plane momentum, respectively. As indicated in Ref. 2, for the QCL structure in operation, the transition between subbands 2 and 1 occurs with a near-zero momentum transfer, while the transition process from subband 3 to subband 2 is between the electron states of reduced spatial overlap and a large momentum transfer, associated with a large subband separation. This situation is depicted in Fig. 5(a), with the parabolic

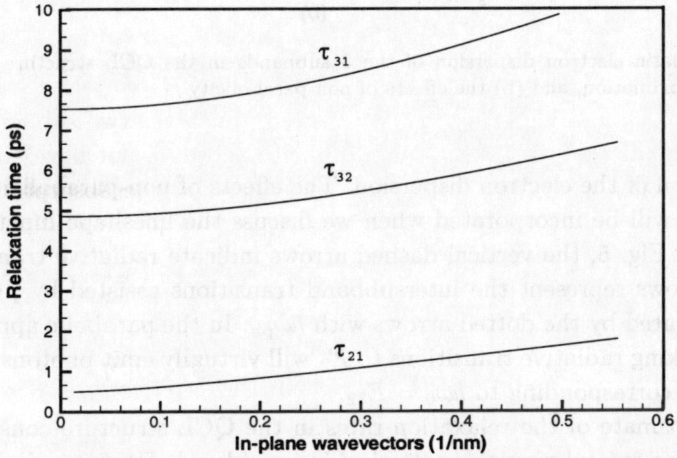

Fig. 4. Electron relaxation times as a function of electron in-plane wave number among the energy subbands in the QCL structure.

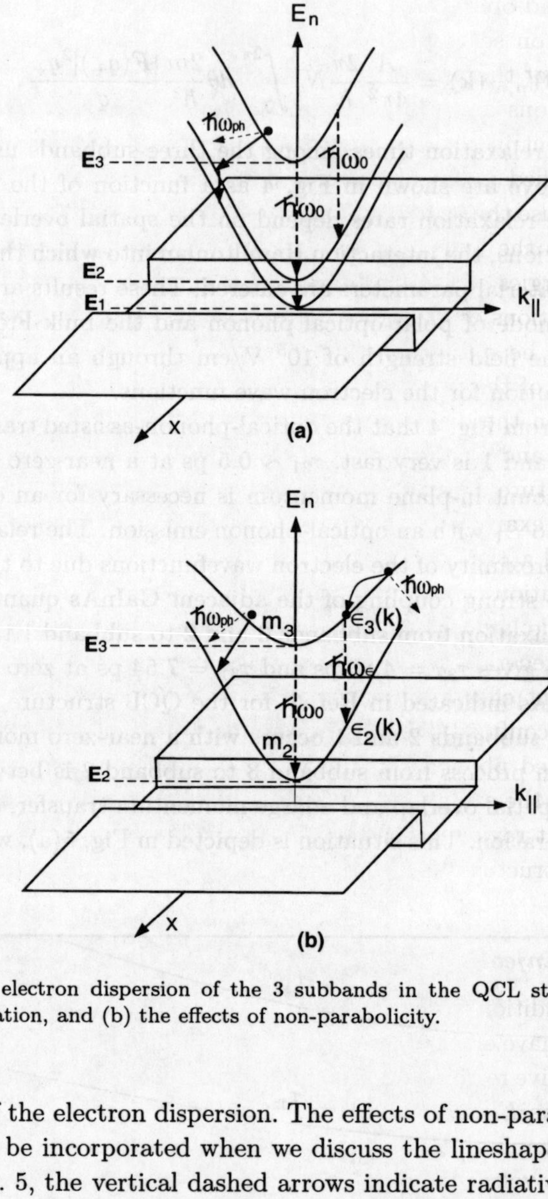

Fig. 5. Schematic electron dispersion of the 3 subbands in the QCL structure, with (a) the parabolic approximation, and (b) the effects of non-parabolicity.

approximation of the electron dispersion. The effects of non-parabolicity, as shown in Fig. 5(b), will be incorporated when we discuss the lineshape function and optical gain. In Fig. 5, the vertical dashed arrows indicate radiative transitions, and the solid arrows represent the intersubband transitions assisted by phonon emission, as indicated by the dotted arrows with $\hbar\omega_{\rm ph}$. In the parabolic approximation, electrons making radiative transitions to E_2 will virtually emit photons of the same frequency ω, corresponding to $\hbar\omega_0 = E_{32}$.

For an estimate of the relaxation rates in the QCL structure considered here, only the polar-optical-phonon scattering is considered. However, for a complete evaluation of the relaxation rates, other scattering mechanisms related to the the non-radiative relaxation may have to be investigated, depending on the specific device structure and operating conditions. These include acoustic-phonon scattering, electron-electron scattering, and impurity scattering. Another simplification in the calculation of the relaxation rates is that we have used the bulk mode for polar-optical phonons and the bulk Fröhlich interaction Hamiltonian. Similar to electron confinement in the low-dimensional structures, both acoustic and optical phonons are modified in quantum well structures, especially in narrow quantum wells. This is because the lattice vibrations are now subject to boundary conditions very different from the bulk case, resulting in various confinement phonon modes. Moreover, the energies of the modified optical phonons are comparable to the intersubband separations in the quantum well structures. Therefore, these confined phonon effects can be significant on the relaxation rates and in turn on the other design parameters of the intersubband lasers. The confined phonon modes can be calculated based on the dielectric continuum model and the microscopic analysis by Lucas *et al.*,[13] and Licari *et al.*[14] These confinement phonon modes are both material and structure dependent and need to be calculated for individual device structures. For an example of the confined phonon effects on the relaxation rates, we have performed a preliminary investigation on step quantum well structures,[15] in which the relaxation rates due to the interface phonon scattering are one or two orders of magnitude larger than those calculated with the bulk modes. The above formulations are generally valid for calculation of the relaxation rates when the phonon confinement effects are considered. We will not describe more details on the calculation of confined phonons in the quantum well structures. However, the influence of confined phonons on the spectral lineshape and optical gain spectra will be included in the evaluation of the lineshape functions and optical gain of the QCLs. For a recent review on the tailoring of acoustic and optical phonon modes in quantum well structures for engineering relaxation rates, see Ref. 16.

3.3. *Tunneling injection and escape*

For the QCL, in addition to the control of the relaxation times among the subbands, it is important to have efficient electron injection from the contact through the first barrier into the active region, and from one cell to the next. A fast escape of electrons tunneling out of the E_1 subband is thus required.

For efficient pumping, the current injection efficiency of the QCL must be maximized at the bias voltage that leads to the flat-band condition for the compositionally graded region. At the flat-band voltage the ground subband states in the injector should be approximately aligned with the bottom of the upper lasing subband in the active region to facilitate resonant tunneling. The is similar to the design of resonant tunneling devices where a high current density is desired. An accurate calculation of the injection current is complicated as the injection electrons will see an extended potential profile in the QCL regions. Since the contact doping, barrier shape, and bias voltage are critical parameters pertaining to the tunneling current density, Poisson's equation should be solved together with the Schödinger equation for self-consistent potential and carrier distributions in the structure so that the calculated current density includes the space charge effects. With such a calculation, one can identify and tailor the physical parameters to optimize the injection performance in the QCL devices. We will not discuss more details of the current density calculation. For a recent review of modeling of resonant tunneling current-voltage characteristics, see Ref. 17.

The electron tunneling times of interest can be calculated either from a direct solution of the time-dependent Schrödinger equation, or estimated by solving the time-independent Schrödinger equation. In the former approach,[18] the following equation is solved in the time domain for the electron state ψ_i in the particular device structure.

$$j\hbar\frac{\partial\psi_i}{\partial t}=\frac{\hbar^2}{2m^*(x)}\frac{\partial^2\psi_i}{\partial x^2}+V(x)\psi_i, \tag{13}$$

which gives the electron concentration $n(x,t)$ and the electron current density $J_n(x,t)$ at any point in space and time in the device structure,

$$n(x,t)=\sum_i p_i\psi_i^*(x,t)\psi_i(x,t)\,, \tag{14}$$

and

$$J_n(x,t)=\frac{-q\hbar}{m^*}\sum_i p_i\ \mathrm{Im}\left[\psi_i^*(x,t)\frac{\partial\psi_i(x,t)}{\partial x}\right], \tag{15}$$

where the probability p_i or a weighting function associated with the electron state ψ_i can be obtained by integrating the Fermi-Dirac distribution over the in-plane momenta:

$$p_i=\frac{m^*K_BT\Delta k}{2\pi^2\hbar^2}\log\left[1+\exp\frac{(E_F-\hbar^2k_i^2/(2m^*))}{K_BT}\right], \tag{16}$$

where Δk is the discretization in k space, E_F the Fermi level in the contact, K_B the Boltzmann constant, and T the absolute temperature. Note that the solution of Eq. (13) requires an initial condition, obtainable from the static solution of the time-independent Schrödinger equation subject to the open-system boundary conditions, as discussed previously. The electron tunneling time through a particular potential barrier or a quantum barrier/well structure can then be evaluated from values of the corresponding points in space and time, as can be extracted from Eqs. (14) and (15).

Alternatively, in the second approach, the tunneling time of interest can be estimated by solving the time-independent Schrödinger equation Eq. (1) to calculate the current density through the barrier or a barrier/well region, and the tunneling or traverse time is then estimated by dividing the current density by the electron charge. The expression for the current density is similar to Eq. (15) but without the time variable.

Applying both approaches described above, the tunneling time through the first trapezoidal barrier at the field of 10^5 V/cm is calculated to be 0.27 ps in the time-dependent calculation and 0.31 ps in the time-independent calculation, which are consistent to fall within a small fraction of a picosecond. The tunneling escape time out of the E_1 state is calculated to be 0.40 and 0.55 ps, respectively. These estimates are also consistent with Ref. 2. However, the accuracy of these times is difficult to determine since they depend to a large extent on how these times are defined and on the overall potential profile used in the solution procedure. Nonetheless, the calculation of these times does give a reasonable estimate and is helpful for an understanding of the related design considerations. The effect of the electron escape time on the population inversion will be clearly seen in the next section.

A theoretical analysis of resonant tunneling in quantum cascade lasers has been carried out by Sirtori *et al.*[19] using a tight-binding approach with experimental evidence to show that resonant tunneling does control the electron injection into the active region. Critical physical parameters such as the doping concentration and injection barrier thickness are identified and optimized for QCL injector design for maximum output power.

3.4. *Population inversion*

The carrier population inversion between the lasing subbands in the QCL is determined by the electron transitions among the subbands, the pumping level, and the interaction of the electrons with the photon field when lasing is sustained. This is a complicated many-body interactive problem and difficult to model accurately. However, we can estimate the population inversion using the semi-classical rate equations with simplifications to gain some insight into this critical device parameter.

Considering the three-level system being electrically pumped, the single mode rate equations for a three-subband laser can be written below by counting electrons into and out of each subband,

$$\frac{dn_3}{dt} = P - \frac{n_3}{\tau_{31}} - \frac{n_3}{\tau_{32}} - vgS\,, \tag{17}$$

$$\frac{dn_2}{dt} = P + \frac{n_3}{\tau_{32}} - \frac{n_1}{\tau_{21}} + vgS\,, \tag{18}$$

$$\frac{dn_1}{dt} = \frac{n_3}{\tau_{31}} + \frac{n_2}{\tau_{21}} - \frac{n_1}{\tau_{\text{out}}}\,, \tag{19}$$

$$\frac{dS}{dt} = vgS\,, \tag{20}$$

where the n's are electron densities at the subbands, P the rate of pumping, and the τ's are the relaxation times among the subbands as described above, with τ_{out} being the time of carrier removal from the bottom subband, v is the speed of light in the lasing medium, g the optical gain, and S the optical field or the photon density.

For an estimate, the electron densities at the QCL subbands are calculated below the threshold, the effect of the photon field is then neglected. For electrical pumping as in the case of the QCLs, $P = J/ed$, where J is the injection current density, e the magnitude of electron charge, and d is the thickness of the active region. The rate equations are thus greatly simplified for an estimate of the electron densities at the three subbands under the threshold. The equations can be solved analytically to give

$$\begin{aligned} n_3 &= P\tau_3(1 - e^{-t/\tau_3}), \\ n_2 &= P\frac{\tau_3\tau_{21}}{\tau_{32}}\left(1 + \frac{\tau_{21}}{\tau_3 - \tau_{21}}\, e^{-t/\tau_{21}} - \frac{\tau_3}{\tau_3 - \tau_{21}}\, e^{-t/\tau_3}\right), \\ n_1 &= n_1(0)\, e^{-t/\tau_{\text{out}}} + P\,\frac{\tau_3\tau_{\text{out}}}{\tau_{32}}\Bigg[(1 - e^{-t/\tau_{\text{out}}}) \\ &\quad + \frac{\tau_{21}^2}{(\tau_3 - \tau_{21})(\tau_{21} - \tau_{\text{out}})}(e^{-t/\tau_{21}} - e^{-t/\tau_{\text{out}}}) \\ &\quad - \frac{\tau_3^2}{(\tau_3 - \tau_{21})(\tau_3 - \tau_{\text{out}})}(e^{-t/\tau_3} - e^{-t/\tau_{\text{out}}})\Bigg], \end{aligned} \tag{21}$$

where $1/\tau_3 = 1/\tau_{31} + 1/\tau_{32}$, τ_3 can be considered to be the lifetime of the electrons in the third subband.

Figure 6 shows an example of the calculated electron population at the QCL subbands as a function of time. The results are obtained using the intersubband relaxation times at the zero in-plane momenta as calculated in Sec. 3.2, and an injection current density of $J = 5$ k A/cm^2. An initial value of the electron concentration of 5.0×10^{16} cm^{-3} is also assumed to account for the ground state charge

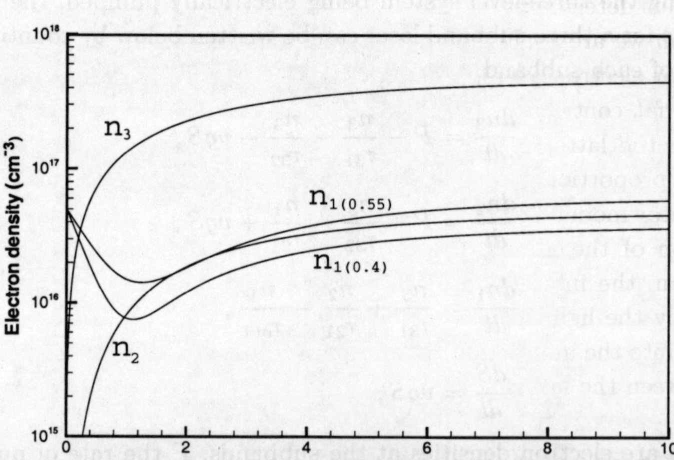

Fig. 6. Estimated electron population at the subbands under the threshold. The two n_1 curves are obtained with different electron escape times (0.4 ps and 0.55 ps) as indicated on the curves.

in the quantum well. It can be seen from the results that a high ratio of population inversion is reached in the steady state, and that the effect of the electron escape time on the population inversion is clearly seen from the two curves of the electron density at the bottom subband, obtained using 0.40 and 0.55 ps for τ_{out}, respectively. The electron density at the bottom subband can be either higher or lower than that of the lower lasing subband, depending on the the escape time. Note that the population inversion is over-estimated in this approximation since the effect of the photon field is not included, which would have reduced the electron density at the upper lasing subband, and increased the electron density at the lower lasing subband due to radiative transitions. The injection current density used in the above calculation is between the experimental threshold current density and the theoretical estimate in Ref. 2, to yield a sheet electron density of approximately 8.9×10^{10} cm^{-2} at threshold. The estimate of the electron population inversion discussed above shows the importance of control of the intersubband relaxation rates among the subbands in the QCL active region and the electron escape time from that region since population inversion is a result of the interplay of these characteristic times.

3.5. *Transition matrix element*

To understand laser characteristics, it is essential to study the optical gain spectra of the device. Before we discuss the gain formulation, it is helpful to recall that the laser operation is the result of interaction between the electromagnetic field and the lasing medium, or the electron-photon interaction in the semiconductor gain medium. To represent the strength of electron transition between the lasing states induced by stimulated radiation, the transition matrix element is employed. Its square modulus determines the transition probability between the lasing subbands. The transition we are concerned with here is different from those discussed in the last section for electron energy relaxation assisted by polar-optical phonons, and is the process that contributes to the laser emission. The former is the spontaneous process while the latter is the stimulated process. Therefore, we expect that the laser gain is proportional to the squared transition matrix element. If we use a simple oscillator model to describe the laser medium, for the cases of ionic crystals, the oscillation of the atoms corresponds to a set of dipoles. Taking this dipole approximation, the interaction of the photon field and the gain medium is then represented by the light-induced electric dipole moment of an electron, $\boldsymbol{\mu} = -e\mathbf{r}$, which enters into the interaction Hamiltonian responsible for the electron transition to occur between the lasing states:

$$H_{int} = e\mathbf{r} \cdot \mathbf{E} = -\boldsymbol{\mu} \cdot \mathbf{E}\,, \tag{22}$$

where $\mathbf{r}$ is the space coordinate and $\mathbf{E}$ the electric field strength. For the transition between the state $|2\rangle$ and state $|3\rangle$ in the QCL structure, choosing the polarization direction of the electric field to lie in the plane of $\mathbf{r}_{23}$ and electron wave vector $\mathbf{k}$, the transition probability is given by

$$|x_{23}|^2 = |\langle 2|x_{23}|3\rangle|^2$$

$$= \left[\int((\mathrm{Re}(\psi_2)\mathrm{Re}(\psi_3) - \mathrm{Im}(\psi_2)\mathrm{Im}(\psi_3))x\,dx\right]^2$$

$$+ \left[\int(\mathrm{Im}(\psi_2)\mathrm{Re}(\psi_3) + \mathrm{Im}(\psi_3)\mathrm{Re}(\psi_2))x\,dx\right]^2. \quad (23)$$

With the wave functions at the subbands obtained by numerically solving the effective mass equation given previously, Eq. (23) yields the transition matrix element $|x_{23}| \simeq 1.10$ nm for the QCL structure. The transition strength is determined by the dimensions of the device structure and the electron state functions. The spatial overlap and parity of the wavefunctions of the subbands play an important role. In general, for symmetric structures, as the state $|3\rangle$ is even and state $|2\rangle$ is odd, the partial cancellation of the state functions may give small values of the transition matrix element. The matrix element for asymmetric structures can be made larger. Equation (23) can be used to examine the transition strength for different device structures. Design of the laser structures should be engineered to enhance the transition matrix element, but compromises may have to be made with other design considerations. In the QCL structure we consider here, the radiative transition is called "diagonal" from the higher initial state to a lower final state in the adjacent quantum well, if we loosely consider the quantum well where the electron wavefunction is localized is the location of the electron state. Later designs of the QCLs have used "vertical" transition in that the transition occurs virtually in the same quantum well with stronger spatial overlap between the wavefunctions, resulting in larger values of the transition matrix element as well as the oscillator strength. The oscillator strength is a semi-classical analogy for atomic systems: if we take the pair of the lasing subbands E_j and E_i as an oscillator, with the associated resonant frequency $(E_i - E_j)/\hbar$, the oscillator strength is used to represent the interaction of light with the oscillator, with the strength proportional to the transition probability,

$$f_{ij} = \frac{2m^*}{\hbar^2}(E_j - E_i)|x_{ij}|^2. \quad (24)$$

An excellent example of exploring the oscillator strength to add flexibility in QCL design is shown in Fig. 7. This QCL features two quantum wells in the active region, where the radiation transition occurs between state 1 in the first well and state 1′ in the adjacent well. The thick potential barrier between the two wells helps keep the electron density essentially constant in state 1 while the thin barrier on the right of the second well helps maintain the electron population inversion between the two lasing states. At a sufficiently high bias voltage (> 50 kV/cm), the bandedges in the quantum wells are slanted, resulting in an enhancement of the oscillator strength through two mechanisms: (1) the energy separation between states 1 and 1′ is shifted to a higher value due to the Stark effect, and (2) the

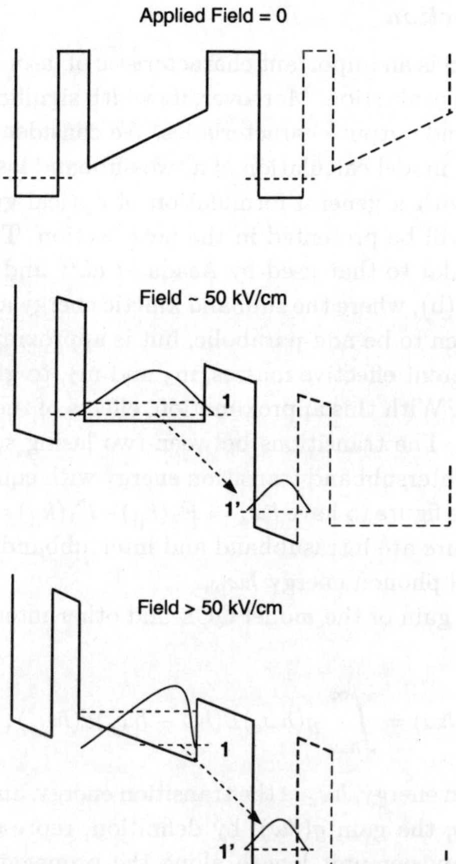

Fig. 7. Schematic energy band diagram of the active region of a QCL at various bias conditions (adapted from Ref. 20). The radiation transition occurs between the states 1 and 1′, and the oscillator strength is tuned by the electric field.

transition matrix element is increased due to the enhanced spatial overlap of the electron wavefunctions between the states. An additional feature of the laser device is that the gain spectrum and hence the lasing wavelength is tunable by applied bias voltage. For example, with two cells having significantly different length, a QCL can lase at two wavelengths simultaneously, making this kind of QCL very appealing for applications such as chemical gas spectroscopy.[20]

Sirtori *et al.* have performed a detailed study of the oscillator strength and the associated sum rule in multiple quantum well structures, based on Kane's model. In particular, the effects of non-parabolicity of different effective masses in the quantum well materials were studied to show an overall reduction of the oscillator strength as compared to the parabolic case. The interested reader is referred to Ref. 21 for details.

3.6. *Lineshape function*

The lineshape function is an important characteristic of laser devices, which directly enters the optical gain evaluation. Moreover, its width signifies the temperature sensitivity of laser gain and output characteristics. We consider the lineshape function of the QCL through a model calculation of a two-subband laser, and its significance is introduced along with a general formulation of optical gain. The evaluation of optical gain spectra will be presented in the next section. The approach employed here is in general similar to that used by Asada *et al.*[22] and Gelmont *et al.*[23]

Referring to Fig. 5(b), where the subband kinetic energy as a function of electron in-plane energy is taken to be non-parabolic, but is approximated using a parabolic formulation with different effective masses, m_3 and m_2, to give the dispersion relations in the subbands. With this approximation, effects of the subband nonparabolicity can be included. The transitions between two lasing subbands are expressed with $\hbar\omega_\varepsilon$, an allowed intersubband transition energy with equal in-plane wavevector $(k_{||})$. It is shown in the figure to have $\hbar\omega_\varepsilon = E_2(k_{||})-E_1(k_{||}) = \hbar\omega_0+\varepsilon_2(k_{||})-\varepsilon_1(k_{||})$. Also shown in the figure are intrasubband and intersubband transitions with emission of a polar-optical phonon energy $\hbar\omega_{\rm ph}$.

The intersubband gain of the model QCL and other intersubband lasers can be expressed as

$$G(\hbar\omega) = \int_{\hbar\omega_0}^{\infty} g(\hbar\omega_\varepsilon)L(\hbar\omega - \hbar\omega_\varepsilon)d(\hbar\omega_\varepsilon)\,, \tag{25}$$

where $\hbar\omega$ is the photon energy, $\hbar\omega_\varepsilon$ is the transition energy, and $\hbar\omega_0$ is the separation of the subband edges; the gain $g(\hbar\omega)$, by definition, represents the fractional net increase in photon flux per unit length along the propagating direction x of the electromagnetic wave, determined by the transition matrix element and the carrier distributions f_3 and f_2 in the subbands; in addition, $L(\hbar\omega - \hbar\omega_\varepsilon)$ is the lineshape function when the spectral broadening is considered. The lineshape function is originated from the transitions which take place from a range of initial states in the upper lasing subband due to various broadening mechanisms, such as phonon scattering and other collisions. For the lineshape function, Eq. (25) means that the gain g at the transition energy $\hbar\omega_\varepsilon$, which is in general not equal to the photon energy $\hbar\omega$, contributes to the gain at $\hbar\omega$ with the weight $L(\hbar\omega - \hbar\omega_\varepsilon)$. Since the peak gain occurs when the photon energy is equal to the transition energy, the gain spectra will be smoothed and reduced around the gain peak by the lineshape function. In the case of conventional semiconductor lasers, the laser transitions are between broad distributions of carriers with an exponential dependence on temperature, leading to their temperature-sensitive characteristics.

Based on the density matrix theory with a phenomenological intrasubband relaxation time, $\tau_{\rm intra}$, the lineshape function is given by

$$L(\hbar\omega) = \frac{1}{\pi}\,\frac{\gamma(\varepsilon)}{(\hbar\omega - \hbar\omega_\varepsilon)^2 + (\gamma(\varepsilon))^2}\,, \tag{26}$$

where $\gamma(\varepsilon) = \hbar/\tau_{\text{intra}}(\varepsilon)$ is the normalized intrasubband relaxation rate. If τ_{intra} is assumed to be a constant independent of energy, Eq. (26) gives the well-known Lorentzian lineshape. The factor π came out as a result of normalization such that the area under the lineshape curve from $-\infty$ to $+\infty$ equals unity. In the present calculation, the integration limits of $(\hbar\omega - \hbar\omega_0)$ are chosen to cover $-5\ K_BT$ to $+5\ K_BT$, which should introduce a negligible error. When the intrasubband scattering rate is taken to be dependent on energy, or when the other broadening mechanisms are included, the lineshape function becomes non-Lorentzian in shape.

In the present calculation of the QCL lineshape, effects of confined phonons are considered in terms of their contributions to the intrasubband relaxation as well as to the intersubband relaxation, since both the intra- and inter-subband scattering reduce the lifetime of a lasing electron state in the upper subband, which will in turn modify the spectral lineshape. Assuming that the intra- and inter-subband electron-phonon scattering events are independent of each other, the intrasubband relaxation rate in Eq. (26) is replaced by a joint scattering rate

$$\gamma(\varepsilon) = \frac{1}{\hbar}\,\frac{\tau_{\text{intra}}(\varepsilon) + \tau_{\text{inter}}(\varepsilon)}{\tau_{\text{intra}}(\varepsilon)\,\tau_{\text{inter}}(\varepsilon)}\,, \tag{27}$$

in which the intrasubband scattering rate as a function of the electron in-plane subband energy is close to a step function in the range of $1.0 - 5.0 \times 10^{12}$ sec^{-1}.[23] A rigorous calculation of intrasubband scattering rates requires knowledge of non-parabolic electron states in the subband, which can be calculated using a tight-binding or the $\mathbf{k} \cdot \mathbf{p}$ approach. For a simplified treatment, the intrasubband relaxation rate is given by

$$\gamma_{\text{intra}}(\varepsilon) = \gamma_0 \times \begin{cases} N_{\text{ph}} & \text{phonon absorption}\,, \\ (N_{\text{ph}}+1)\theta(\varepsilon - \hbar\omega_{\text{ph}}) & \text{phonon emission}\,, \end{cases} \tag{28}$$

with the phonon Planck function

$$N_{\text{ph}} = \frac{1}{\exp}\left(\frac{\hbar\omega}{K_BT} - 1\right), \tag{29}$$

and

$$\gamma_0 = \frac{\pi}{2}\frac{e^2}{\hbar}\left(\frac{1}{\kappa_0} - \frac{1}{\kappa_\infty}\right)\sqrt{2m_e\omega_{\text{ph}}/\hbar}\,, \tag{30}$$

where κ_0 and κ_∞ are the static and optical dielectric constant, respectively, and the phonon frequency can be obtained from calculations based on a confined phonon model.[15]

Figure 8 shows the calculated lineshape function plotted as $L(\hbar\omega - \hbar\omega_\varepsilon)$ vs. the normalized photon energy, $(\hbar\omega - \hbar\omega_\varepsilon)/KT$ for the two-subband QCL model. A Lorentzian lineshape with a constant $\tau_{\text{intra}} = 0.1$ ps is also plotted for comparison. It can be seen that the peak value of the lineshape function of the QCL (L2) is shifted to a lower photon energy as compared to that of the Lorentzian lineshape

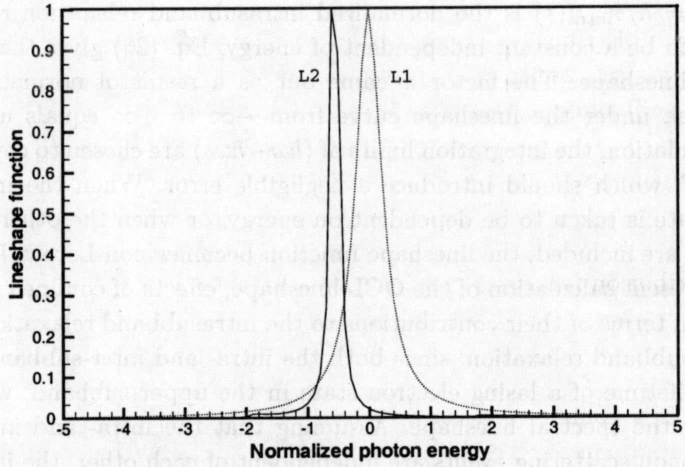

Fig. 8. Calculated lineshape function (L2) at $T = 300$ K with an energy-dependent intrasubband relaxation rate, as compared with the Lorentzian lineshape function (L1). The lineshapes are plotted as $L(\hbar\omega - \hbar\omega_\varepsilon)$ vs. $(\hbar\omega - \hbar\omega_\varepsilon)/KT$.

(L1), which occurs at the subband edge separation, $\hbar\omega = \hbar\omega_0$. This is because, as seen from Eq. (26), the denominator reaches its minimum when the photon energy coincides with the subband separation in the Lorentzian model. Moreover, the width of the non-Lorentzian lineshape is narrower as compared to the Lorentzian with the constant intrasubband relaxation rate typically used in calculations of quantum well lasers. In general, due to the averaged constant scattering rate used for an approximation to the energy-dependent scattering rates, the commonly used Lorentzian lineshape will either overestimate or underestimate the effect of scattering and will subsequently either underestimate or overestimate the gain magnitude, depending on the value of the constant scattering time used. The narrow lineshape function is one of the features of the QCL, which implies that the QCLs are expected to have less performance variation with temperature changes than the conventional semiconductor lasers.

3.7. *Distribution functions*

Before we can carry out an estimate of optical gain of QCLs and other intersubband lasers, another important physical aspect of optical gain must be considered: the nonequilibrium electron distribution functions $f_3(\varepsilon)$ and $f_2(\varepsilon)$ in the subbands. We have already seen that the electron in-plane energy distribution strongly affects their scattering rates and lineshape function. Since the electron distributions are a result of electron transport as well as the interaction of electrons and the photon field in the QCL, a proper treatment may have to invoke many-body formalisms. Simplified models of the electron distribution functions typically use quasi-equilibrium distributions as approximations for ideal cases where electron concentration is either very high or very low. For the purpose of optical gain estimation, we use the Boltzmann distribution to approximate the cases of a moderate or high level injection, as in Ref. 24. In this approximation, the density of states in the two-dimensional subbands is constant and the distribution functions are normalized according to

$$\frac{m_i}{\pi\hbar} = \int_0^\infty f_i(\varepsilon_i)\, d\varepsilon_i = n_i, \quad i = 2, 3\,, \tag{31}$$

and the distribution functions are then reduced to

$$f_i(\varepsilon_i) = \frac{\pi\hbar^2 n_i}{m_i K_B T} e^{-(\varepsilon_i)/(K_B T)}\,. \tag{32}$$

Although it is a common practice to use electron quasi-equilibrium distribution functions to approximate the nonequilibrium distributions, and it may be adequate for estimation purposes, further work should derive the electron distributions based on a proper transport model which includes many-body effects.

3.8. *Optical gain*

We have thus far considered the major physical aspects of optical gain in QCLs: the electron quasi-bound states, the electron relaxation rates among the subbands, the population inversion, the transition matrix element in the dipole moment approximation, the lineshape function, and the approximate electron distribution functions, all of them lead to an estimate of optical gain of the QCL. The general formulation of optical gain of lasers has been introduced and explained in Eq. (25). Following the procedure of the gain derivation in Refs. 23 and 25, the gain coefficient $g(\hbar\omega)$ is given by

$$g(\hbar\omega) = \frac{e^2|x_{23}|^2 m_3 \hbar\omega}{\epsilon_0 \hbar^2 dc\sqrt{\kappa_\infty}} \int_0^\infty L[\hbar\omega - \hbar\omega(\varepsilon)]\,[f_3(\varepsilon) - f_3(\varepsilon_2)]\, d\varepsilon\,, \tag{33}$$

where ε_0 is the permittivity of vacuum, d the width of the quantum well, c the speed of light, m_3 is the electron effective mass in subband 3, and x_{23} the transition matrix element. The spectral gain can then be calculated as a function of photon energy $\hbar\omega$, with parameter variations. We can examine Eq. (33) to identify the physical parameters we discussed in the previous sections, such as the ratio of electron population inversion between the lasing subbands, temperature, intra-subband or intersubband scattering rates, carrier concentration in the subbands, parameters of the nonparabolicity, as well as the material and structure parameters. Figure 9 shows the calculated optical gain at $T = 300$ K for various ratios of electron population inversion in the model QCL, and Fig. 10 the gain at various temperatures with an electron population inversion ratio of 5. Although the gain evaluation here is regarded to be approximate, we can make observations of some basic aspects of the gain behavior:

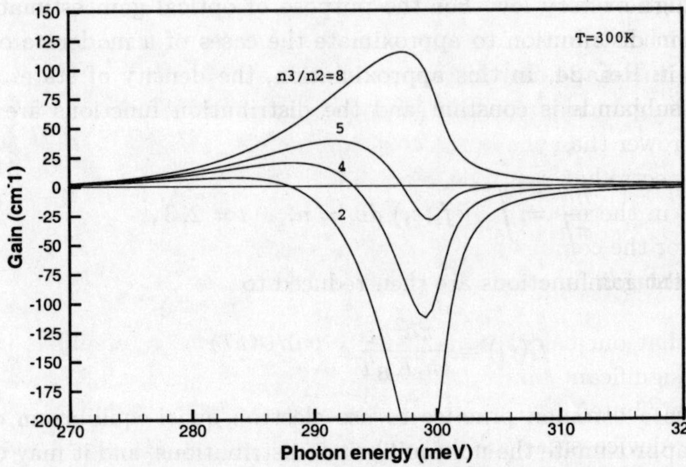

Fig. 9. Calculated optical gain at $T = 300$ K for various ratios of electron population inversion in the model QCL.

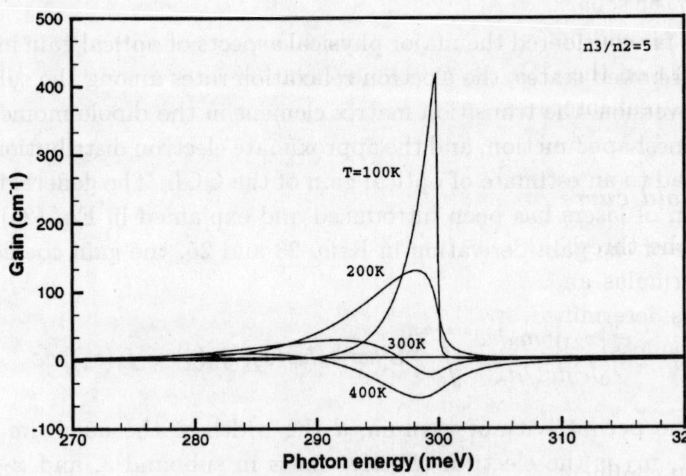

Fig. 10. Calculated optical gain with an electron population ratio of 5 at various temperatures in the model QCL.

1. The optical gain spectra of the intersubband lasers being studied are sensitive functions of the carrier population inversion between the subbands, temperature, and particularly, the phonon scattering rates, which impose their influence on the gain through the lineshape function.
2. The effects of the population inversion and temperature (through the distribution functions) are obvious as expected: a higher population inversion enhances the gain whereas a higher temperature decreases the gain magnitude.

3. In general the effects of the lineshape function lead to a reduction of the gain magnitude and a smoothness in gain spectra around the gain peak. The Lorentzian lineshape function typically used may either underestimate or overestimate the gain magnitude. The lineshape of the QCL as calculated is significantly narrower than the Lorentzian lineshape.
4. The gain spectra have a strong dependence on the phonon scattering mechanisms considered in the model calculation. The inclusion of the intrasubband phonon scattering or the combined intra- and inter-subband scattering rate will generate very different gain spectra for some laser structures.

We note that our calculations on step quantum well intersubband lasers have shown that a significant gain suppression will occur due to the intersubband phonon scattering contributed by the interface phonon mode if the lasing subband separation in the device structure is resonant to the interface phonon energy. This is expected since the large intersubband scattering rate at the interface phonon frequency will greatly reduce the lifetime of the lasing electron state in the upper subband. This effect should be considered in realistic intersubband laser design typically having three or more subbands; the interface phonon energy may be utilized to be equal to the separation of the lower two subbands for the benefit of population inversion, but should be avoided for the lasing subbands. The influence of confined phonon spectra on the QCL intersubband gain has been studied by Kisin *et al.*[26] and is still of great interest.

3.9. *Threshold current density and output power*

To estimate the threshold current density and optical power output, we can use the simple formulas and the related data given in Ref. 2. The threshold current density, J_{th}, is determined by the threshold condition, i.e., when the optical model gain equals the loss,

$$g_{\text{th}}\Gamma = \alpha_M + \alpha_I\,, \tag{34}$$

where Γ is the confinement factor, g_{th} the peak material gain at threshold, α_I the intrinsic loss due to electron absorption inside the active and cladding layers. We thus have

$$\alpha_I = \alpha_a(1-\Gamma) + \alpha_c\Gamma\,, \tag{35}$$

and α_M is the mirror loss which accounts for the useful transmitted optical output power at the two end mirrors of the laser device. The confinement factor represents a reduction of optical gain by the ratio of the power in the active region to the power across the entire laser structure, since the laser emission wavelength is much longer than the width of the active region. The mode confinement in multiple quantum well structures can be analyzed by solving the electromagnetic wave equation separately for each of the active and cladding layers. The procedure becomes quite tedious when a large number of layers is involved. For an estimate we use

$$\Gamma \simeq 2\pi^2(n_1^2 - n_2^2)d^2/\lambda^2 \,, \tag{36}$$

where n_1 and n_2 are the refractive indices of the active and cladding layers, respectively, d the active layer thickness, and λ is the free-space wavelength. Assuming the index of 3.56 for InGaAs and 3.15 for QCL cladding layers, $\Gamma \simeq 0.43$. As the material gain is proportional to the injection current density, we have $g_{\rm th} = g_N J_{\rm th}$, where g_N is the normalized gain coefficient and was measured to be 9 $\mathrm{kA^{-1}\ cm^{-3}}$ in Ref. 2. Eq. (34) is then recast as

$$J_{\rm th} = \frac{\alpha_M + \alpha_I}{g_N \Gamma} \,. \tag{37}$$

This yields $J_{\rm th} \simeq 7.75\ \mathrm{kA/cm^2}$ for typical values of $\alpha_M \simeq 20\ \mathrm{cm^{-1}}$, $\alpha_I \simeq 10\ \mathrm{cm^{-1}}$, and the estimated Γ of 0.43. The threshold current density is temperature dependent, and generally increases with increasing temperatures.

The optical power output of the QCL can be estimated using

$$P_o = A(J - J_{\rm th})\left(\frac{\eta_i h\nu}{e}\right)\frac{\alpha_M}{\alpha_M + \alpha_I} \,, \tag{38}$$

where A is the device area and η_i the internal quantum efficiency. Equation (38) means that the number of photons generated in the device per second is equal to the number of electrons injected per second but reduced by the factor of the internal quantum efficiency, and that only a fraction of the generated optical power is coupled out for useful output, while the rest is consumed by the intrinsic loss. Using the data presented in Ref. 2 we can estimate the device area to be of the order of $10^{-4}\ \mathrm{cm^2}$. With the previously obtained values of α_M, α_I, and $J_{\rm th}$, Eq. (38) yields the output power of tens to hundreds of milliwatts, proportional to the difference between the injected and the threshold current densities.

Due to the complexity of realistic QCL structures and physical processes involved, it is difficult to accurately predict the QCL terminal properties of practical QCL devices. With further efforts in laser modeling calculations and experimental precision, it is hopeful to have reasonably good agreement between theoretical expectations and actual device specifications, and to provide theoretical tools for device design optimization in the future.

4. Other Intersubband Lasers

We now turn to discuss other intersubband lasers that have been proposed or demonstrated. Our intention here is to compare and find features of these novel laser schemes and not for a detailed and exclusive coverage.

We have seen many good features of the QCLs, as highlighted in the previous sections. However, in addition to the requirement of sophisticated design capability and state-of-the-art epitaxy technology, their extension to even longer wavelengths is problematic. With decreased transition energies, the intersubband transitions tend to have much smaller radiative efficiency due to the enhanced phonon emission rate. Furthermore, the intrasubband free carrier absorption is roughly proportional to the squared wavelength above about 11 μm. This will translate into larger optical losses in the device at longer wavelengths. These factors may result in prohibitive high threshold current densities as well as temperature and power limitations for QCLs. For these reasons, optically pumped intersubband lasers have been pursued and they seem to be very attractive for operation at FIR and THz frequencies. We compare optical and electrical pumping next.

4.1. *Optical vs. electrical pumping*

Both electrical and optical pumping can create a source of electrons for the required population inversion, each with its advantages and disadvantages. Electrical pumping is generally preferred since it is convenient, economical and compact as compared to optical pumping for which another laser is needed. However, in electrically pumped intersubband lasers, the electrons are typically injected through a barrier by resonant tunneling into an upper subband, the same mechanism is used to depopulate the electrons from a lower subband to aid the population inversion, as in the case of the QCLs. Since the lasing wavelengths of interest are much longer than the width of the active region of these lasers, it leads to a small confinement factor and low output power. Therefore, multiple quantum wells or superlattices are often employed in the device structures. This situation places a more stringent requirement for an optimal intersubband engineering. As discussed previously, the transport and optical processes are many-body interactive in nature and accurate device modeling and optimization are a formidable task especially when a large number of layers is involved. Another related disadvantage is that for efficient electron injection, heavily doped contacts are usually required which tend to increase the cavity losses and reduce the output power.

In optical pumping, on the other hand, the carriers are excited to higher subbands or directly into the upper lasing subband with no tunneling involved for population or depopulation processes. In addition, heavily doped contact layers may not be necessary. Therefore, optically pumped intersubband lasers are simpler in design and they would be very attractive particularly for intersubband lasers aimed at FIR and THz frequencies. Optical pumping, however, is less efficient because the pumped photons usually have much higher energy than the target radiation energy, and all the carriers, not necessarily those in the lowest subband, may absorb the pumping energy. In any case, optical pumping remains a good practical option.

4.2. *Step quantum well intersubband lasers*

In this section, we outline two intersubband laser schemes based on step quantum well structures which we have proposed for laser emission at FIR or THz frequencies, as shown in Fig. 11(a) for an optically pumped structure, and (b) for an electrically pumped structure. We have investigated how a step quantum well structure modifies the intersubband relaxation rates, allowing population inversion between subbands to be achieved. We have also measured the intersubband relaxation rates in the step quantum wells with femtosecond differential transmission spectroscopy and experimentally observed population inversion.[27,28]

The basic idea of the step quantum well intersubband lasers is to design the structure so that it behaves as a four-level laser system, with population inversion between subband 3 and subband 2. For the optically pumped structure, a pump laser (e.g. CO_2 laser) is used to excite a doped quantum well, pumping carriers from subband 1 to subband 4 or higher subbands, where carriers can relax to subband 3 very fast. By tailoring the well width, barrier width and the material composition of the step region, the energy separation between subband 2 and subband 1 is designed to be larger than the optical phonon energy, while the separation between subbands 3 and 2 is less than the optical phonon energy. The carriers in subband 2 will be depopulated very fast to subband 1 through optical phonon emission. However, the relaxation rate from subband 3 to subband 2 is significantly reduced. When the relaxation rate from subband 2 to subband 1 is faster than that from subband 3 to subband 2, population inversion occurs to facilitate lasing action.

The step quantum well structure as shown in Fig. 11(a) was grown by MBE and it consists of 20 periods of 10 nm GaAs wells which are surrounded by 15 nm $Al_{.15}Ga_{.85}As$ step layers and 10 nm $Al_{.25}Ga_{.75}As$ cladding layers. The calculated energy separation between subbands 2 and 1 is 68 meV, which is larger than the optical phonon energy and the separation between subbands 3 and 2 is 28 meV, smaller than the optical phonon energy, corresponding to an emission in the THz frequency range.

The intersubband laser structure shown in Fig. 11(b) is based on the population inversion in asymmetric step quantum wells. We have proposed and investigated an electrically pumped intersubband laser structure at 10 μm wavelength.[30] This device structure has 10 stages and is comprised of multiple active regions connected by graded gap superlattices. The energy band profile of the structure indicates that the carriers are injected into the step quantum well through a thin barrier via tunneling, and stimulated emission takes place between the population-inverted third and second subbands. Carriers in the second subband can be removed efficiently to the first subband by tunneling and escape from the active region through the barrier. They are then transferred to the next active region through the graded gap superlattice to continue lasing action. The inner and outer cladding layers confine the optical wave to the active regions. Preliminary calculations indicate that this structure compare favorably with the QCL: the relaxation times are $\tau_{32} \simeq 14$ ps, and $\tau_{21} \simeq 2.1$ ps; the threshold gain for this particular structure is estimated to be 187 cm^{-1}, at a threshold current density of 2 kA/cm^2.[29] The realization of intersubband lasing at room temperature therefore appears feasible.[30]

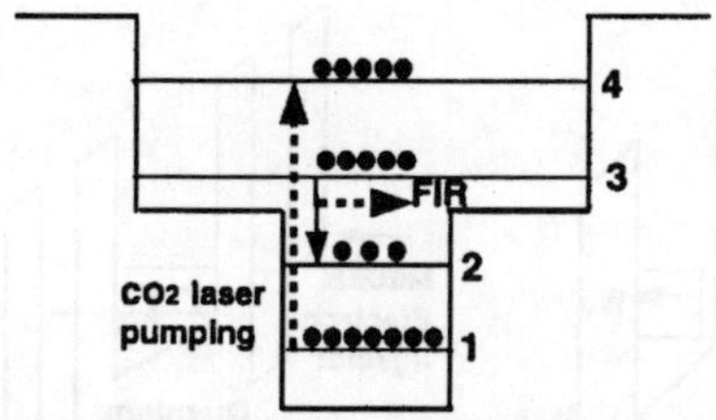

(a) A period of the active layer of an optically pumped FIR laser.

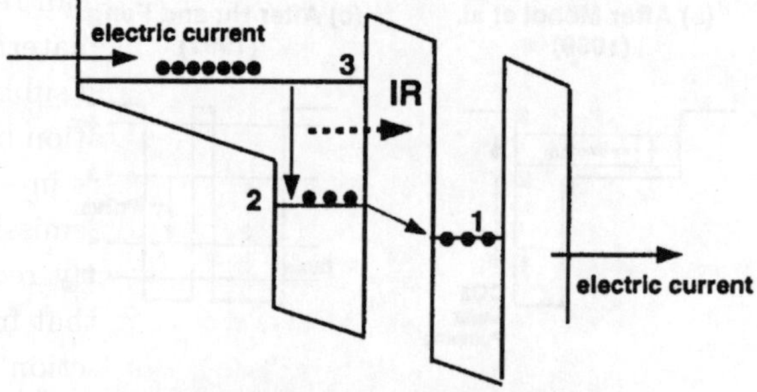

(b) A period of the active layer of an electrically pumped IR laser.

Fig. 11. (a) Optically pumped step quantum well intersubband laser structure and (b) electrically pumped step quantum well intersubband laser structure.

4.3. *Other intersubband laser schemes*

Let us look into some example schemes of intersubband laser structures that were proposed or demonstrated in the past to summarize essential features that have to be well controlled for device realization of intersubband lasers. The resonant tunneling diode laser in Fig. 12(a) was proposed by our group to use electron resonant tunneling as electrical pumping and subband transition between the resonant and ground states for lasing, making resonant tunneling diodes into lasers.[31] In scheme (b), large oscillator strengths were observed in asymmetrical step quantum wells[32] for which the rules of wavefunction parity and transition selection are relaxed. Carrier population inversion in coupled quantum wells are studied in scheme (c) to include intrawell and interwell effects for further optimization of resonant tunneling in multiple quantum wells for lasing.[33] In the intersubband laser scheme (d), design optimization and experimental work lead to the demonstration of the optically pumped laser action at the long wavelength (15.5 μm).[34,35] Keeping the QCL scheme in mind, we can see that these intersubband laser schemes have in one way or another explored similar physical aspects as in the QCLs: (1) resonant tunneling or an optical source for intersubband pumping; (2) tailoring the quantum barrier/well dimensions for the subband energies and relaxation rates; (3) use of asymmetrical

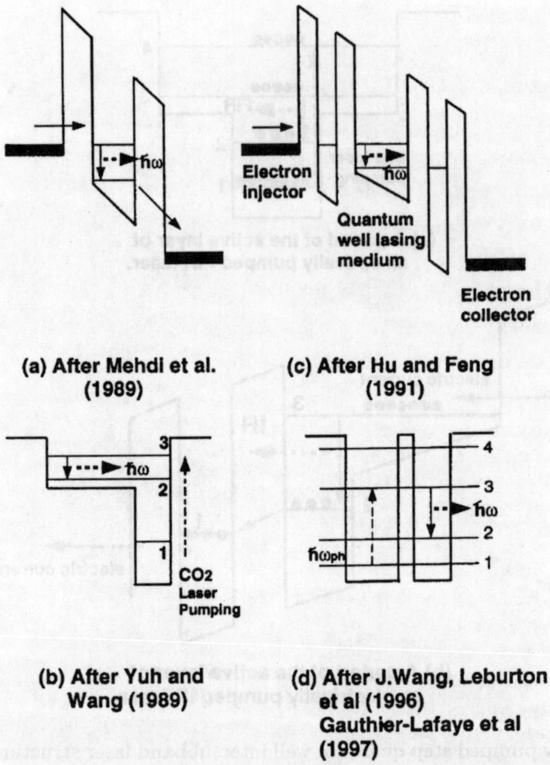

Fig. 12. Various intersubband laser schemes: (a) resonant tunneling diode laser, (b) optically pumped step quantum well laser, (c) electrically pumped coupled quantum well laser, and (d) optically pumped coupled quantum well laser.

structures to increase the stimulated transition strength, (4) a large number of the active regions is needed to enhance the optical power. For the demonstrated laser scheme in Fig. 12(d), one hundred asymmetric coupled quantum wells were grown for the waveguide core of the laser device to give an output power of 0.4 W at 77 K. From our discussions on the physical aspects and study of these laser schemes, it can be concluded that for a successful intersubband laser to work, it is essential to have a good device scheme that incorporates a careful design of the electron source (pumping), electronic or photonic to photonic energy conversion (the active region), and the addition/extraction of optical power (a large number of lasing elements and the cavity design). Furthermore, for these three design aspects to be implemented, it is indispensable to have capabilities of theoretical calculations for intersubband engineering and precision control of material growth.

5. Conclusions

Quantum cascade lasers and other intersubband lasers are a fascinating research field that stands at the forefront of semiconductor science and engineering. They represent a fine combination of quantum physics and the state-of-the-art material growth technology. Their development will fill the gap of presently underdeveloped mid- to far-infrared and terahertz sources, and therefore also result in important practical applications.

In this chapter/paper we have discussed the basic operational principles of the QCLs. A simplified model calculation is used to gain some insight into the major physical processes and provide estimates of the critical parameters in the device active region. Some of the estimated values are considered to be semi-quantitative, but it is our hope that the simplified model calculation can help to understand the device design considerations and show the importance of intersubband engineering. We have also briefly reviewed other intersubband laser schemes and their common design features. It is our belief that with ingenuous device schemes, further advances in quantum device modeling, and the availability of excellent material growth technology, a variety of intersubband lasers will be realized in the near future.

References

1. R. F. Kazarinov and R. A. Suris, "Possibility of amplification of electromagnetic waves in a semiconductor with a superlattice", *Sov. Phys. Semicond.* **5** (1971) 707–709.
2. J. Faist, F. Capasso, D. L. Sivco, C. Sirtori, A. L. Hutchinoson, and A. Y. Cho, "Quantum cascade laser", *Science* **264** (1994) 553–556.
3. F. Capasso, "Quantum cascade lasers: From band structure engineering to applications", *Proc. 1998 IEEE 16th Int. Semiconductor Laser Conf.*, 4–8 October 1998, Nara, Japan, pp. 223–224.
4. A. Tredicucci, C. Gmachl, F. Capasso, D. L. Sivco, A. L. Hutchinson, and A. Y. Cho, "Long wavelength superlattice quantum cascade laser at $\lambda \simeq 17$ μm", *Appl. Phys. Lett.* **74** (1999) 638–640.
5. G. Scamarcio, F. Capasso, C. Sirtori, J. Faist, A. L. Hutchinson, D. L. Sivco, and A. Y. Cho, "High power interminiband superlattice infrared lasers", *Conf. Laser Electro-Optics, 1997 Opt. Soc. Am. Tech. Dig. Ser. (CLEO'97)* **11** (1997) 423.
6. C. Sirtori, F. Capasso, J. Faist, A. L. Hutchison, D. L. Sivco, and A. Y. Cho, "Dual wavelength intersubband emitters", *CLEO'97* **11** (1997) 425–426.
7. F. Capasso, C. Gmachl, A. L. Hutchinson, A. Tredicucci, J. Faist, D. L. Sivco, J. N. Baillargeon, and A. Y. Cho, "High-performance, widely tunable, single-mode, mid-infrared distributed feedback quantum cascade lasers", *CLEO'98* **12** (1998) 339–340.
8. C. Gmachl, F. Capasso, E. Narimanov, J. Noeckel, A. D. Stone, J. Faist, D. L. Sivco, and A. Y. Cho, "High-power directional emission from microlasers with chaotic resonators", *Science* **280** (1998) 1556–1564.
9. S. Slivken, A. Matlis, A. Rybaltowski, Z. Wu, and M. Razeghi, "Low-threshold 7.3 μm quantum cascade lasers grown by gas source molecular beam epitaxy", *Appl. Phys. Lett.* **74** (1999) 2758–2760.
10. W. R. Frensley, "Boundary conditions for open quantum systems driven far from equilibrium", *Rev. Mod. Phys.* **62** (1990) 745–791.
11. C. S. Lent and D. J. Kirkner, "The quantum transmitting boundary method", *J. Appl. Phys.* **67** (1990) 6353.
12. J. P. Sun, R. K. Mains, K. Yang, and G. I. Haddad, "A self-consistent model of Γ-X mixing in GaAs/AlAs/GaAs quantum well structures using the quantum transmitting boundary method", *J. Appl. Phys.* **74** (1993) 5053–5060.

13. A. A. Lucas, E. Kartheuser, and R. G. Badro, "Electron-phonon interaction in dielectric films: Application to electron energy loss and gain spectra", *Phys. Rev.* **B2** (1970) 2488–2499.
14. J. J. Licari and Roger Evrard, "Electron-phonon interaction in a dielectric slab: Effect of electronic polarizability", *Phys. Rev.* **B15** (1977) 2254–2264.
15. H. B. Teng, J. P. Sun, G. I. Haddad, M. A. Stroscio, SeGi Yu, and K. W. Kim, "Phonon assisted intersubband transitions in step quantum well structures", *J. Appl. Phys.* **84** (1998) 2155–2164.
16. M. A. Stroscio, "Tailoring acoustic and optical phonon modes in mesoscopic and nanoscale semiconductor structures: Engineering scattering rates", *Phys. Low-Dim. Struct.* **11/12** (1997) 145–154.
17. J. P. Sun, G. I. Haddad, P. Mazumder, and J. N. Schulman, "Resonant tunneling diodes: models and properties", *Proc. IEEE* **86** (1998) 641–661.
18. R. K. Mains and G. I. Haddad, "Time-dependent modeling of resonant tunneling diodes from direct solution of the Schrödinger equation", *J. Appl. Phys.* **64** (1988) 3564–3569.
19. C. Sirtori, F. Capasso, J. Faist, A. L. Hutchinson, D. L. Sivco, and A. Y. Cho, "Resonant tunneling electron injection in quantum cascade lasers", *IEEE J. Quantum Electron.* **34** (1998) 1722–1729.
20. J. Faist, F. Capasso, C. Sirtori, D. L. Sivco, A. L. Hutchinson, and A. Y. Cho, "Laser action by tuning the oscillator strength", *Nature* **387** (1997) 777–782.
21. C. Sirtori, F. Capasso, J. Faist, and S. Scandolo, "Nonparabolicity and sum rule associated with bound-to-bound and bound-to-continuum intersubband transitions in quantum wells", *Phys. Rev.* **B50** (1994) 8663–8674.
22. M. Asada, "Intraband relaxation effect on optical spectra", in *Quantum Well Lasers*, ed. P. S. Zory, Jr., Academic Press, 1993.
23. B. Gelmont, V. B. Gorfinkel, and S. Luryi, "Theory of the spectral line shape and gain in quantum wells with intersubband transitions", *Appl. Phys. Lett.* **68** (1996) 2171–2173.
24. V. B. Gorfinkel, S. Luryi, and B. Gelmont, "Theory of gain spectra for quantum cascade lasers and temperature dependence of their characteristics at low and moderate carrier concentrations", *IEEE J. Quantum Electron.* **32** (1996) 1995–2003.
25. A. Yariv, *Optical Electronics*, 4th ed., Saunders Publishing, 1991.
26. M. V. Kisin, V. B. Gorfinkel, M. A. Stroscio, G. Belenky, and S. Luryi, "Influence of complex phonon spectra on intersubband optical gain", *J. Appl. Phys.* **82** (1997) 2031–2038.
27. X. Zhang, G. I. Haddad, J. P. Sun, A. Afzali-Kushaa, and C. Y. Sung, "Population inversion in step quantum wells at 10 μm wavelength", *1995 53rd Ann. Device Res. Conf. Dig.* IEEE, New York, NY, USA, 1995, pp. 118–119.
28. C. Y. Sung, T. B. Norris, A. Afzali-Kushaa, and G. I. Haddad, "Femtosecond intersubband relaxation and population inversion in stepped quantum well", *Appl. Phys. Lett.* **68** (1996) 435–437.
29. X. Zhang, C. Y. Sung, T. B. Norris, and G. I. Haddad, "Population inversion in asymmetric quantum wells and infrared intersubband lasers", *Proc. SPIE* **2694** (1996) 19–27.
30. X. Zhang and G. I. Haddad, "Room temperature 10 μm intersubband lasers based on carrier capture processes in step quantum wells", *Proc. 9th Int. Conf. Hot Carriers in Semiconductors*, July 31–August 4, Chicago, 1995, pp. 619–622.
31. I. Mehdi, G. I. Haddad, and R. K. Mains, "Novel use of resonant tunneling structures for optical and IR modulator", *Superlattices Microstruct.* **5** (1989) 443–449.

32. Y. J. Mii, K. L. Wang, R. P. G. Karunasiri, and P. F. Yuh, "Observation of large oscillator strengths for both $1 \to 2$ and $1 \to 3$ intersubband transitions of step quantum wells", *Appl. Phys. Lett.* **56** (1990) 1046–1048.
33. Q. Hu and S. Feng, "Feasibility of far-infrared lasers using multiple semiconductor quantum wells", *Appl. Phys. Lett.* **59** (1991) 2923–2925.
34. J. Wang, J. P. Leburton, F. H. Julien, and A. Sa'ar, "Design and performance optimization of optically pumped mid-infrared intersubband semiconductor lasers", *IEEE Photon. Technol. Lett.* **8** (1996) 1001–1003.
35. O. Gauthier-Lafaye, P. Boucaud, F. H. Julien, S. Sauvage, S. Cabaret, J. M. Lourtioz, V. Thierry-Mieg, and R. Planel, "Long-wavelength (≈ 15.5 μm) unipolar semiconductor laser in GaAs quantum wells", *Appl. Phys. Lett.* **71** (1997) 3619–3621.

International Journal of High Speed Electronics and Systems, Vol. 9, No. 4 (1998) 901–940

ADVANCES IN MEASUREMENTS OF PHYSICAL PARAMETERS OF SEMICONDUCTOR LASERS

G. E. SHTENGEL and R. F. KAZARINOV
Lucent Technologies Bell Laboratories, 9999 Hamilton Blvd., Breinigsville, PA 18031, USA

G. L. BELENKY
Department of Electrical Engineering, SUNY at Stony Brook, Stony Brook, NY 11794, USA
and
Lucent Technologies Bell Laboratories, 700 Mountain Ave., Murray Hill, NJ 07974, USA

M. S. HYBERTSEN and D. A. ACKERMAN
Lucent Technologies Bell Laboratories, 700 Mountain Ave., Murray Hill, NJ 07974, USA

We present a summary of the advances in characterization techniques allowing comprehensive study of physical processes in semiconductor lasers. The studies of the electrical characteristics and optical emission below threshold allow to measure the optical gain, linewidth enhancement factor, transparency wavelength, optical loss and carrier lifetime. Some other parameters, such as leakage current and wavelength chirp, can only be deduced from the above threshold measurements. Measurements of the carrier temperature and carrier heating in semiconductor lasers allow to obtain important information about the device performance at high injection current densities. Taken together, all these measurements provide critical experimental feedback in the laser design process. They also furnish essential information to guide our understanding of the microscopic physical processes determining the laser performance and our efforts to simulate those processes.

1. Introduction

The fast growing use of semiconductor lasers in various fields including fiber communication systems, optical data storage, remote sensing etc. places very stringent requirements on device performance. This requires a detailed understanding of physical processes governing the behavior of laser diodes. A semiconductor laser fundamentally operates by electrical transport of carriers to a central active layer of the device where they recombine to produce optical gain and support stimulated emission of photons. However, in modern, optimized devices, a non-trivial cascade of microscopic processes is involved.

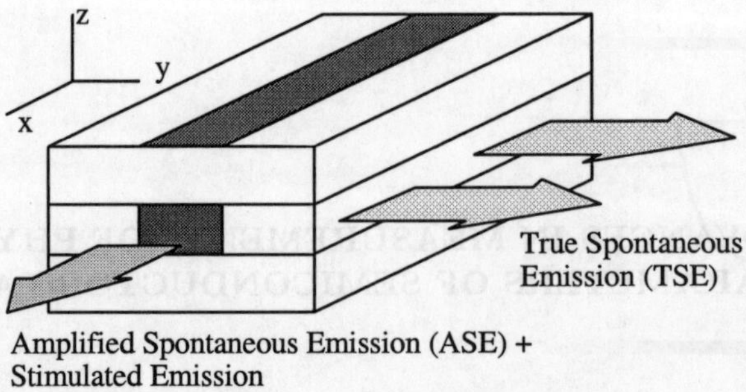

Fig. 1. Simplified scheme of the laser diode.

Most of the experimental techniques presented here are developed in relation to telecommunication lasers. These lasers are usually designed so the output radiation can be coupled into the single-mode optical fiber. Therefore the lasers themselves usually employ single mode waveguide design. Two common approaches are illustrated in Fig. 1. The active layer, which may be either bulk material or multiple quantum wells, is confined in the z-direction by the cladding layers of lower refractive index, forming an optical waveguide in this direction. In the y-direction the confinement is obtained by either etch and regrowth with material of lower index and lower conductivity (index guided buried heterostructure (BH)), or partial etch down to the cladding layer, forming a ridge waveguide laser. The laser resonator is usually obtained by cleaving the structure along the crystal axes. Cleaved facets form two mirrors of the laser cavity along the x-direction. Light emerging from these two facets travels along the laser cavity and is therefore affected by the presence of the optical gain or loss. At currents higher than threshold it is predominantly stimulated emission, below threshold it is amplified spontaneous emission (ASE). The light emerging from the side of a chip (y- or z- direction) experiences almost no amplification or absorption, because it travels only a very short distance inside the active layer media, and there is neither waveguide nor resonator in this direction. This is true spontaneous emission (TSE), almost completely unaffected by the geometry of the laser structure.

Study of this light emission offers tremendous insight to the operation of the laser, including such fundamental aspects as the optical gain and loss. However, optimization of the properties of the laser requires understanding the connection between the optical emission and the carrier transport processes in the device. Figure 2 illustrates this point for a multiple quantum well buried heterostructure laser. The active layer at the core of the diode is also at the center of a diode, which is operated in forward bias resulting in the injection of electrons and holes. The regions of material regrown along the edges of the active layer must be designed to minimize current flow around the active layer. Details of the carrier transport can have significant impact on steady state characteristics such as efficiency as well as the modulation response of the device. Within the active layer itself, electrons and

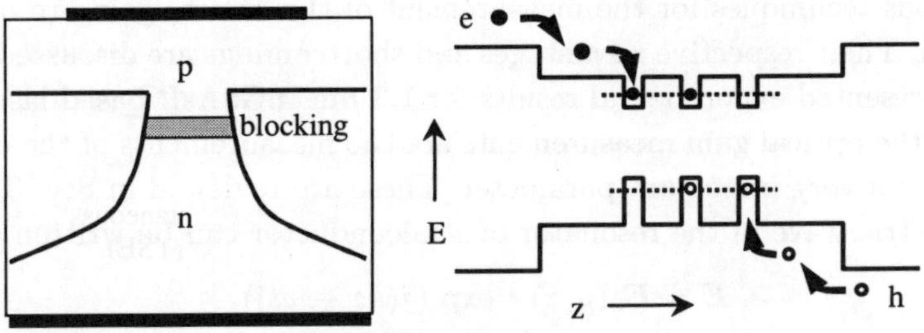

Fig. 2. Schematic cross section of a BH laser (left) and an illustration of some of the transport processes in the multiple quantum well active region (right).

holes are injected with higher energy due to the band offset at the heterojunction. Then the carriers must be captured into the quantum wells. There are different time scales associated to the carrier thermalization, cooling and capture. Furthermore, with multiple wells, the transport of carriers to populate wells further along the stack can involve several steps of capture and reemission. In general, the population of carriers above the barrier will be out of equilibrium with the populations bound in the wells. As the net stimulated emission rate rises, these populations will shift to support the required current flow. The light emission from the active region is determined by the arrangement of the carriers at each bias of the laser diode. The other carrier recombination processes, such as Auger, also depend on the details of the carrier distribution.

A full experimental characterization of all of the processes outlined would be very challenging indeed. In this review, a broad set of electrical and optical techniques is described which give complimentary information on the operation of the laser diode. In general, the carrier distribution in the active layer strongly depends on drive current below threshold, but not above threshold. Physical processes below threshold are critical in determining the operating point of the laser. Therefore studying the electrical characteristics and optical emission below threshold is often more informative in the process of understanding the device performance. Some other parameters, such as leakage current or wavelength chirp can only be deduced from above threshold measurements. Taken together, these measurements provide critical experimental feedback in the process of laser diode optimization. They are also essential information to guide our understanding of the microscopic physical processes described above and efforts to simulate those processes.

2. Measurements of the Optical Gain

One of the most important parameters relating the physical properties of the semiconductor structure to output characteristics of the laser diode is the optical gain. Optical gain and its dependence on the operating conditions determine not only the basic output characteristics, such as threshold current, but also the temperature dependence of the output characteristics, as well high-speed performance of the laser. Various techniques for the measurement of the optical gain are described in this section. Their respective advantages and shortcomings are discussed on the basis of the presented experimental results for 1.3 μm InGaAsP based lasers. Closely coupled to the optical gain measurements are the measurements of the optical loss, which is also a very important parameter. These are reviewed in Sec. 3.

The electric wave in the resonator of semiconductor can be written as:

$$\boldsymbol{E} = \boldsymbol{E}_0(y, z) \cdot \exp\left(j(kx - \omega t)\right) \tag{1}$$

where $\boldsymbol{E}_0(y, z)$ is the magnitude of the field, and the complex propagation constant k is:

$$k(\lambda) = k' + jk'' = \frac{2\pi\, N_{\text{eff}}(\lambda)}{\lambda} - \frac{j}{2} g(\lambda) \tag{2}$$

N_{eff} is an effective index of refraction for the optical mode, and g is the modal optical gain. The factor of 2 comes in because the optical gain is usually defined with respect to optical power, not optical field intensity. Modal optical gain $g(\lambda)$ is related to the material optical gain $G(\lambda)$:

$$g(\lambda) = \Gamma G(\lambda) - \alpha_{\text{tot}}\,, \tag{3}$$

where Γ is optical confinement factor (the fraction of the transverse optical mode overlapping the active layer and therefore experiencing optical gain).[1] The total optical loss α_{tot} consists of mirror loss and internal loss, usually attributed to free-carrier absorption and scattering from waveguide imperfections:

$$\alpha_{\text{tot}} = \alpha_{\text{mirr}} + \alpha_{\text{int}} = \frac{1}{L} \ln\left(\frac{1}{R}\right) + \alpha_{\text{int}}\,. \tag{4}$$

The representation of the mirror loss as $1/L \ln(1/R)$ is valid only Fabry–Perot Lasers, $R = \sqrt{R_1 \cdot R_2}$ being the geometrical mean of the mirror reflectivities, L the laser resonator length.

It should be pointed out that the optical gain is defined in Eq. (2) as (in terms of variations):

$$\delta g = -2\delta k'' = -\frac{4\pi}{\lambda} \Gamma \delta N''. \tag{5}$$

In Refs. 2 (Chapter 5.3.3, p. 209), and 3 (Chapter 15.6, p. 582–583), the authors use the opposite sign in the Eq. (1) which leads to the opposite sign in the relation between the gain and the imaginary part of the refractive index. It will also lead to an opposite sign in the definition of the linewidth enhancement factor, as it will be discussed later. In our treatment we follow definitions of C. Henry.[4] This difference is, of course, only in definitions and not in final results.

Optical gain can be extracted from both ASE and TSE, as well as using some other methods, such as direct transmission measurement or variable stripe length (VSL) technique.[5] Direct transmission measurement may be advantageous when both facets are anti-reflection AR coated (mirror loss is very high) and Fabry–Perot contrast is very small. VSL technique has an advantage of not being influenced by the resonator quality. However it requires a special sample preparation and cannot be used for a single device. Only the methods based on ASE and TSE measurements will be discussed here.

2.1. *Determination of the optical gain from the amplified spontaneous emission*

In this section the gain measurements based on the analysis of the ASE spectra will be discussed. Figure 3 shows the spectra of ASE in TE polarization for the 1.3 μm BH laser[6] with a bulk active region and uncoated mirror facets. There can be seen a very clean interference pattern of the Fabry–Perot resonator formed by the cleaved mirrors. This interference pattern is used to analyze the optical gain.

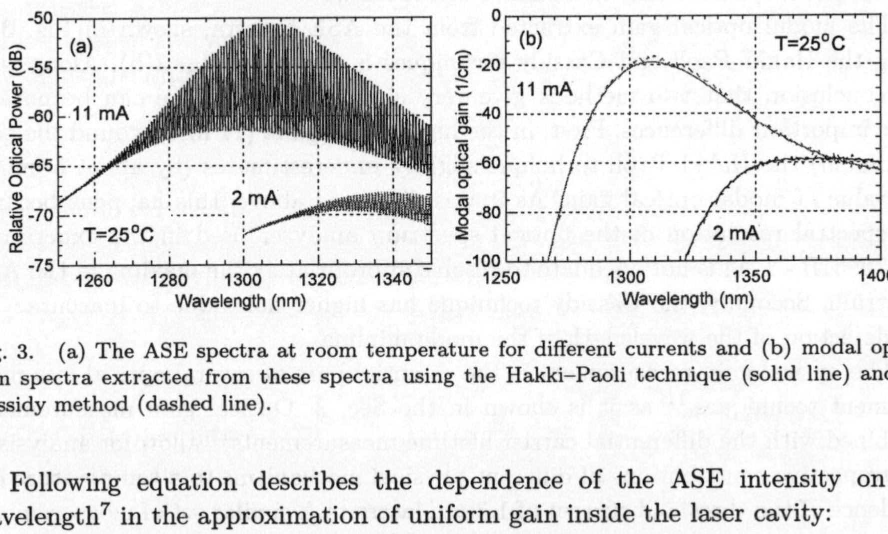

Fig. 3. (a) The ASE spectra at room temperature for different currents and (b) modal optical gain spectra extracted from these spectra using the Hakki–Paoli technique (solid line) and the Cassidy method (dashed line).

Following equation describes the dependence of the ASE intensity on the wavelength[7] in the approximation of uniform gain inside the laser cavity:

$$I(\lambda) = \frac{B(1 + R \cdot e^{(\Gamma G - \alpha_{\text{int}})L})(1 - R)}{(1 + R \cdot e^{(\Gamma G - \alpha_{\text{int}})L})^2 - 4R \cdot e^{(\Gamma G - \alpha_{\text{int}})L} \cdot \sin^2(\frac{2\pi L n}{\lambda})}, \tag{6}$$

where B is the proportionality coefficient equal to the total amount of ASE coupled into the lasing mode.

B. Hakki and T. Paoli proposed to determine the modal optical gain from the contrast of the ASE spectra.[8] Equation (6) is used to obtain:

$$g(\lambda) = \frac{1}{L} \ln \frac{\sqrt{r(\lambda)} - 1}{\sqrt{r(\lambda)} + 1} \tag{7}$$

where $r(\lambda)$ is the peak-to-valley ratio:

$$r(\lambda) = \frac{I_{\max}(\lambda)}{I_{\min}(\lambda)}. \tag{8}$$

D. Cassidy[9] proposed another way to analyze ASE in order to determine the material gain values. This technique, also called the mode-sum method, is based on a slightly different analysis of the same basic Eq. (6). E. Gordon demonstrated[7] that the optical gain in the oscillator system governed by Eq. (6) is:

$$g(\lambda) = \frac{1}{L} \ln \frac{p(\lambda) - 1}{p(\lambda) + 1}, \tag{9}$$

where p is the ratio of the integral over one mode to the mode minimum:

$$p(\lambda) = \frac{\int_{\lambda_1}^{\lambda_2} I(\lambda') d\lambda'}{I_{\min}(\lambda) \cdot (\lambda_2 - \lambda_1)}. \tag{10}$$

D. Cassidy[9] and later L. Lam Sin Cho[10] have shown theoretically and experimentally that the mode-sum method is less sensitive to the spectral bandwidth of the measurement system than the Hakki–Paoli technique.

The modal optical gain extracted from the ASE spectra, shown in Fig. 3(a) using the Hakki–Paoli and Cassidy techniques is shown in Fig. 3(b). Along with the conclusion that two methods give very similar results, there can be noticed some important differences. First, in the high gain regime (11 mA, around the gain maximum) the Hakki–Paoli technique slightly underestimates (by about 5 cm^{-1}) the value of modal optical gain. As it was discussed above, this happens because the spectral resolution of the optical spectrum analyzer used in the experiment (HP70951B - 1 Å) is not adequate to resolve appropriately the maxima in the ASE spectrum. Secondly, the Cassidy technique has higher noise due to inaccuracy in the detection of the wavelength of the mode minima.

The optical gain measurement is the essential part of several optical loss measurement techniques,[11] as it is shown in the Sec. 3. Optical gain measurements combined with the differential carrier lifetime measurements[12] allow for analysis of the respective contributions of different physical mechanisms to a temperature dependence of the threshold current of 1.3 μm lasers.[13] A similar set of measurements was also very useful for optimization of the high-speed performance of 1.55 μm lasers,[14] which will be discussed in the Sec. 7. Combinations of Hakki–Paoli and Cassidy methods were used to study the optical gain in lasers based on various material systems and designs in the entire wavelength range from visible[15] to far IR.[16]

2.2. *Determination of the optical gain from the true spontaneous emission*

An alternative way of determining the optical gain was first proposed by V. Gribkovskii and V. Samoilyukovich[17] and has been extended for many applications by C. Henry *et al.*[18] It was later used by P. Blood *et al.*,[19] as a part of simultaneous analysis of optical gain and recombination based on TSE measurements.

This method is based on the general relations between the rates of spontaneous emission, stimulated emission, and optical absorption. It was shown in Ref. 18, that if carriers have Fermi-like distribution functions, the material optical gain is related to the absorption coefficient in the following way:

$$G(\varepsilon,\mu,T)=\alpha(\mu,\varepsilon)\cdot\left\{\exp\left(\frac{\mu-\varepsilon}{kT}\right)-1\right\}, \tag{11}$$

where ε is photon energy, μ the separation between the quasi-Fermi levels of electrons and holes, and $\alpha(\mu,\varepsilon)$ the absorption coefficient of the material of the active layer. Generally, $\alpha(\mu,\varepsilon)$ itself depends on the band filling and, therefore, on μ. Another important equation relates the intensity of spontaneous emission $I_{\text{sp}}(\varepsilon)$ to the absorption coefficient:

$$I_{\text{sp}}\propto\varepsilon^2\cdot\alpha(\varepsilon,\mu)\cdot\exp\left(\frac{\mu-\varepsilon}{kT}\right), \tag{12}$$

where the proportionality sign allows for some constant factors to be omitted. They are of no importance because it is practically impossible to quantify the absolute measurements of the spontaneous emission. Combining Eqs. (11) and (12):

$$G(\varepsilon,\mu,T)\propto\frac{I_{\text{sp}}(\varepsilon)}{\varepsilon^2}\cdot\left\{1-\exp\left(\frac{\varepsilon-\mu}{kT}\right)\right\}. \tag{13}$$

In order to obtain the value of the material optical gain using Eq. (13) the value of the quasi-Fermi level separation, μ, should be known. In Ref. 18, the authors have proposed utilizing the property of the Fabry–Perot lasers to lase at the wavelength where the material gain is maximum. Therefore its first derivative with respect to wavelength is zero, allowing a determination of the value of the quasi-Fermi level separation at threshold. In order to determine this value at lower currents, the authors of Ref. 18 have proposed to use the fact that at the high energies the absorption coefficient $\alpha(\mu,\varepsilon)$ is independent of the injection level. This allows for determination of the value of the quasi-Fermi level separation below threshold. P. Rees and P. Blood have pointed out that this procedure can possibly lead to a serious error, since the Fermi-level separation has to be determined with very high accuracy in order for accurate calculation of the optical gain from the TSE spectra.[19,20] They have also pointed out that the carrier temperature enters the Eq. (13), which should be taken into account, since in the case of carrier heating the carrier temperature is not equal to the lattice temperature.

One more condition is required in order to use the Eq. (13) to determine quantitatively the gain spectra from the TSE spectra. Eq. (13) gives the value of the material optical gain in arbitrary units and a proper scaling factor needs to be determined in order to find the material or modal gain. C. Henry used the fact that a FP laser at threshold has a maximum gain equal to the total loss. If the laser has uncoated facets, the mirror loss can be easily calculated. Then the value of total loss can be estimated from L–I curve slope with an assumption that the internal efficiency is 100%. This approximation is not accurate in many cases, however, and this procedure cannot be used for the lasers with coated mirrors or lasers with wavelength-selective feedback, such as DFB or DBR lasers. P. Blood has proposed to use the known value of absorption coefficient above the absorption edge.[19]

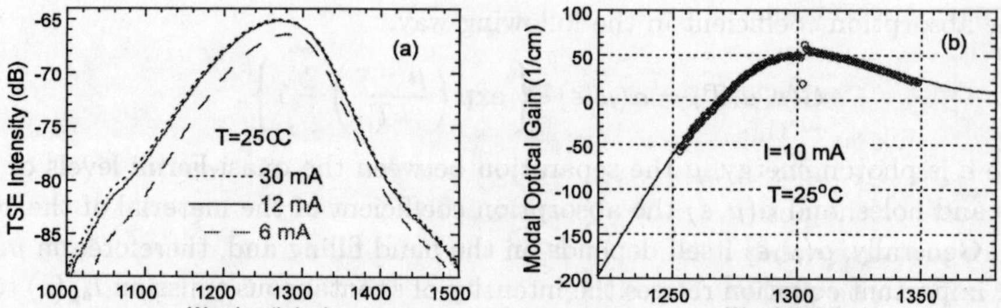

Fig. 4. The TSE spectra at 25°C for different currents (a) and modal optical gain spectrum (b) extracted from TSE spectra using the Henry technique (solid line). The gain spectrum extracted from the ASE spectrum taken at the same conditions is also shown (circles).

The values of the total loss and quasi-Fermi level separation can also be determined using the transparency measurements described in the following section. The TSE spectra, recorded from the side of an uncoated DFB laser at different currents are presented in Fig. 4(a). Noticeably, there are no Fabry–Perot ripples in the TSE spectra. The TSE spectrum is also much broader and extends much further into high energies (lower wavelengths) than the ASE, because the TSE is not affected by reabsorbtion in the active layer. The modal optical gain was determined from these spectra using the values of the total loss and the transparency energy, obtained using the zero-contrast method, described in Sec. 3. The modal optical gain spectra are plotted in Fig. 4(b) together with the spectra extracted from the ASE as described in Sec. 1. The gain spectrum extracted from the TSE is much wider than shown in Fig. 4(b), where the narrow wavelength range is chosen for the sake of comparing it to the gain spectrum extracted from the ASE.

The ASE spectrum of a DFB laser is distorted in the vicinity of the DFB wavelength, and so is the gain spectrum determined from it. The modal optical gain curves extracted from the TSE and the ASE lie very close to each other outside the vicinity of DFB wavelength. This confirms the validity of both techniques. Some discrepancy may arise partly from the fact that the total loss is a weak function of both energy and carrier concentration but was considered constant.

The significant advantage of the methods allowing for extraction of the modal optical gain from the ASE is that they give the absolute value of the gain, and, together with the transparency measurements, the total optical loss. These measurements are fairly easy and fast. The disadvantage is that the modal optical gain is measured in fairly small intervals of both wavelength and gain. When the modal optical gain becomes too low (typically below 150 cm^{-1}), the ripple contrast in the ASE spectrum is usually too small to be measured and analyzed accurately. Another disadvantage of these measurements is that the experimental samples must have a Fabry–Perot resonator.

The Henry technique uses the TSE spectra, which is not affected by the value or the spectral dependence of the mirror loss or grating. Therefore it can provide the information about the optical gain in cases when techniques using the ASE are not suitable. Another advantage is the much wider spectral range of measurements. This allows for applications such as analysis of the change of the real part of the refractive index,[4,21] that has many important applications. Change of the real part of the refractive index is proportional to the change of the effective refractive index, which defines the position of the lasing wavelength of the laser. For these studies only the relative value of change of the optical gain is important, so the Henry technique can be used directly.

2.3. *Determination of the gain in broad-area lasers*

The optical gain spectra can be determined using optical gain spectroscopy techniques.[22–24] This method utilizes the amplification of the spontaneous emission in an optically pumped active region of variable length. The data are fitted to a calculated linewidth produced by a model of the energy bands in the material to determine the gain as a function of carrier density. These techniques are useful before the device has been processed and provide information on the intrinsic properties of the material.

In Ref. 18, the gain spectra were determined using the fundamental relationships between the rates of spontaneous and stimulated emission and optical absorption. Gain was determined for bulk GaAlAs laser structure by measurements of the spontaneous emission observed transverse to the laser cavity for various levels of injected current, supplemented by measurements of the differential quantum efficiency and laser line energy at threshold. The gain value at maximum gain spectra was defined based on the fact that, at the lasing wavelength, the gain at threshold is a maximum and equal to the losses in the cavity. This method required knowledge of the internal efficiency value. In Ref. 25, this method was applied to a broad-area InGaAs/InGaAsP quantum-well laser.

The more direct method of the determination of the gain in broad area lasers was described by D. J. Bossert and D. Gallant.[26] Based upon far-field spatial filtering technique, the modal gain change with injection current was determined from below-threshold ASE spectra. Single lateral mode was separated by spatial far field pattern, and then Hakki–Paoli approach was used to determine value of the modal gain from the fringe contrast in InGaAs/GaAs[27] and midinfrared laser.[28]

3. Measurement of the Optical Loss

Internal optical loss is a fundamental characteristic of a semiconductor laser. The value of internal loss affects threshold current and external slope efficiency, and is, therefore, an important parameter. The optical loss has a number of different contributions; most important among them are the free carrier (intraband) absorption [29] and scattering loss on the waveguide nonuniformities.

An early (and still widely used) technique for measurement of internal optical loss was proposed by J. R. Biard *et al.*,[30] and extended by H. Casey *et al.*[31] It requires a set of lasers, varying in length but otherwise equivalent, in order to estimate the average value of loss as well as the internal quantum efficiency. This method does not take into account the systematic variation of the threshold condition due to variation in length and random variation between lasers. Also injection efficiency might not be a constant value across the set of different samples due to incomplete Fermi level pinning resulting from the carrier leakage.[32] Therefore this method could be used for rough evaluations of optical loss, but for the detailed studies a more accurate procedure is desirable.

Recently several new techniques have been proposed for optical loss measurements.[11,33–35] These are all based on the Eqs. (3) and (4) relating the modal optical gain, the material optical gain and the total loss in the laser. From Eq. (3) it is clear that the modal optical gain is equal to the total loss (with opposite sign) if the material optical gain is zero. This condition holds to good approximation for the energies well below the bandgap energy and holds rigorously at the transparency energy, i.e., at the transition point between absorption and gain.[34] Thus, finding the modal gain at these points provides the value of the total loss, which, in turn, provides internal loss if the mirror loss is known.

A technique based on below-bandgap measurements[34,35] is made difficult by the low intensity of ASE in this region, which limits the accuracy of the gain measurements. For the gain spectra shown in Fig. 3(b) the accuracy is no better than 2–5 cm^{-1}. Below bandgap loss measurements are thus adequate for rough estimates of loss, but become unacceptable for more demanding purposes.

Other methods reduce the problem of measuring internal loss to the task of determining the material transparency condition. E. Avrutin *et al.*,[34] proposed to determine transparency at a given current is to find the intersection of the gain curves in TE and TM polarizations under the assumption that the optical gain in different polarizations is equal when the material gain is zero. Figure 5 shows the curves of TE- and TM-polarized-gain measured for a typical MQW laser. There is one intersection point at high energy (short wavelength) as well as an indication that the two curves will converge at below-bandgap energies as the material optical gain tends toward zero. The possible source of an error of this method is the assumption

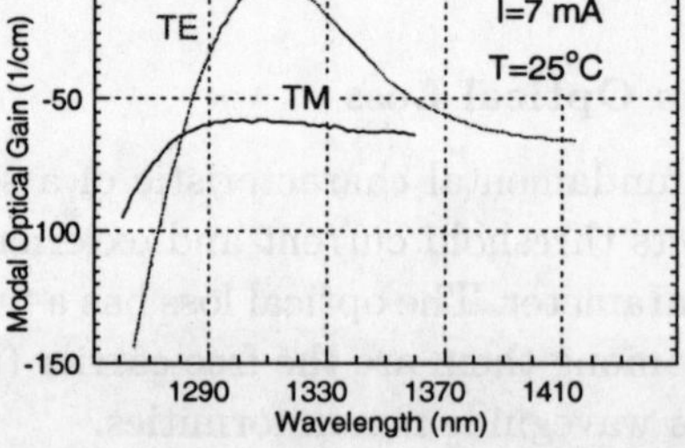

Fig. 5. Modal Optical Gain extracted from ASE spectra of MQW laser in TE and TM polarizations.

that the total optical loss is equal in the TE and TM polarizations. This may lead to an error, in this case of about 2 cm^{-1}.

P. Andrekson *et al.*,[33] demonstrated the use of a modulated external probe laser to find the transparency condition, using the optical detection properties of the laser medium.[36] In this method, an AC voltage induced on the terminals of the laser under test changes polarity when the DC bias current passes through a value corresponding to transparency at the wavelength of the probe laser.

Equivalently, the wavelength of the probe laser can be swept through transparency at a fixed DC bias current. Figure 6(a) shows a typical plot of the dependence of the induced voltage on the probe wavelength. Also plotted (dotted line) is the ASE spectrum of the laser (this is the same laser sample as in Fig. 5). Corresponding features appear in both spectra due to the Fabry–Perot cavity resonances of the laser under test, which enhance the detection properties of the laser at the wavelengths of Fabry–Perot resonances.

It was demonstrated that the external probe measurements are very sensitive to the spectral purity of the probe signal.[11] Some commercially available tunable external cavity lasers (ECL) do not have sufficient signal-to-noise ratio (SNR), which is usually degraded by the presence of the parasitic ASE in the ECL signal. This may result in an error in determination of the transparency wavelength and, therefore, optical loss. To avoid this parasitic effect it was suggested to use the fact that the amplitude of the resonances in the induced voltage spectrum depends on the value of the material optical gain and equals zero at the material transparency. In the absence of any parasitic ASE component in the probe signal, as in Fig. 6(a), the resonances in the voltage signal vanish at the same probe wavelength at which the induced AC voltage changes polarity. The effect of low spectral SNR is illustrated in Fig. 6(b). In this case, the ECL, operating at the edge of its range, provided insufficient suppression of its own ASE. Figure 6(b) shows an offset of $\sim$ 2 nm between the probe wavelengths for which the induced AC voltage equals zero and the wavelengths at which voltage resonances vanish. This difference translates to about 7 cm^{-1} difference on a gain/loss scale.

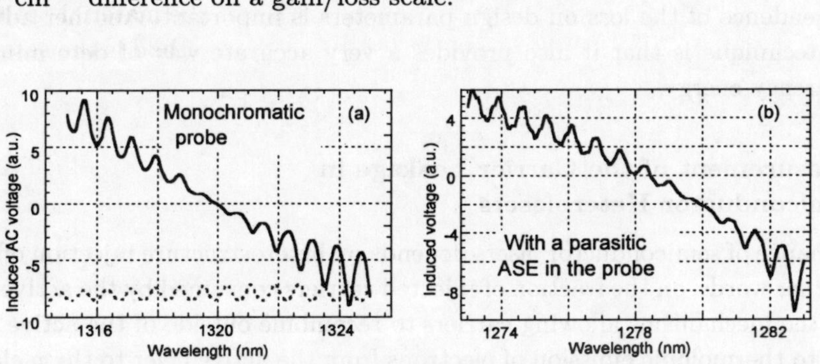

Fig. 6. Dependence of the induced AC voltage on the laser terminals on the wavelength of the external probe light in the absence (a) and presence (b) of the parasitic signal. Corresponding part of the ASE spectrum is also shown (dashed line).

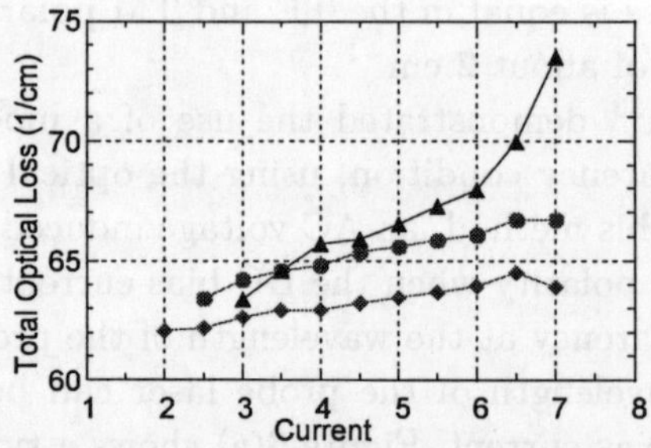

Fig. 7. The total loss determined with different methods: using the "zero-voltage" technique (triangles), using the "zero-contrast" rather than "zero-voltage" criterion with the same probe source (circles), and using the TE-TM crossing method (diamonds).

Figure 7 shows the total loss data measured in three ways: using the "zero-voltage" technique (triangles), using a "zero-contrast" rather than "zero-voltage" criterion with the same probe source (circles), and using the TE-TM crossing method (diamonds) discussed above. The effect of the inadequate and varying spectral purity of the probe source gives a serious error in magnitude of total loss, and its dependence on bias current as seen at the lowest and highest currents for the "zero-voltage" set of data. The TE-TM crossing technique gives somewhat better results, underestimating the value of the loss due to the reason explained above, but allowing data to be obtained over a slightly wider range. Smoothly varying and reproducible loss data were observed when using the "zero-contrast" method. The estimated accuracy of this measurement is about 0.5–1 cm^{-1}.

This method made it possible to carry out the study of the loss dependence on temperature and carrier concentration.[13] It was shown that optical loss is a weak function of both temperature and carrier concentration and does not play an important role in determining the dependence of the laser output characteristics (such as threshold current and output efficiency) on temperature. This method is also useful in a process of optimizing the laser structure, when understanding the dependence of the loss on design parameters is important. Another advantage of this technique is that it also provides a very accurate way of determining the transparency energy.

4. Measurement of the Carrier Leakage in Semiconductor Heterolasers

Performance of semiconductor lasers depends on heterostructure injection efficiency or, in other words, on the fraction of injected carriers consumed by the active region. One of the mechanisms allowing carriers to recombine outside of the active layer is related to thermoionic emission of electrons from the active layer to the *p*-cladding. This phenomenon is usually referred to as heterobarrier leakage. Another mechanism of carrier escaping from active region prior their recombination is lateral transport of carriers through the blocking structure (in the case of CMBH type devices) or the effect of carrier spreading for broad area lasers. Carrier leakage can affect the laser slope efficiency resulting in its reduction with increase in current and temperature and can contribute to the temperature dependence of the threshold current. The carrier overflow into the barriers and the separate confinement heterostructure has received a great deal of attention due to the influence of carrier concentration in SCH layer on the static and dynamic properties of MQW lasers.[37–39] The most frequent objects for studies of carrier leakage are InGaAsP/InP heterostructures due to the importance of their high temperature applications and due to the low value of the conduction band discontinuity between the InP cladding layer and the conduction band of the quaternary InGaAsP materials of the waveguide. The effect of heterobarrier leakage on the performance of blue-green and midinfrared lasers was also studied.[28,40,41] In this section we will discuss the experimental approaches to studying carrier leakage phenomena in semiconductor lasers focusing our attention on studies of heterobarrier leakage.

4.1. *Optical technique of studying the carrier leakage*

The obvious method of studying the efficiency of carrier leakage under different conditions is to register the light resulting from recombination of carriers outside the active region.[42–46] Electro-luminescence image of InGaAsP/InP CMBH laser at 100 mA forward bias, recorded by E. J. Flynn and D. A. Ackerman is shown in Fig. 8(a) for a leaky laser and in Fig. 8(b) for a device with low leakage. The detected radiation is InP band to band recombination at $\lambda = 0.97$ μm. Low and high acceptor levels in their p-InP cladding layers characterize these high and low leakage devices respectively.

One of the earliest experiments in which the intensity of thermionic flow of carriers out of the active region in InGaAsP/InP heterostructure was measured under different conditions is described by E. J. Flynn.[46] In a 1.3 μm LED structure, electrons leaking from the active layer into the p-InP cladding were confined in a subsidiary quaternary layer (leak detection layer). This quaternary layer has a different composition from the active region and the recombination emission

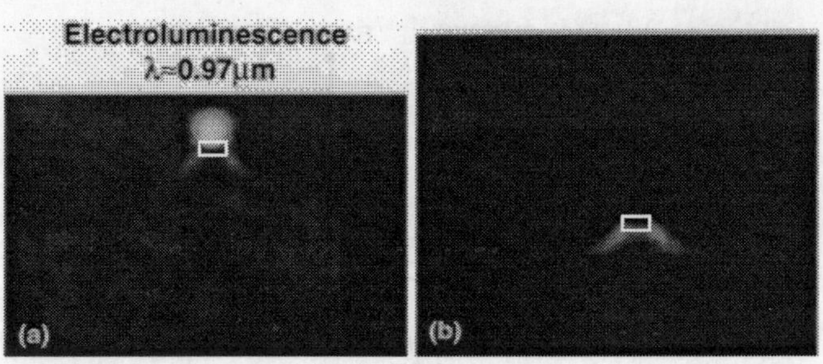

Fig. 8. Electrolumininescence images at 100 mA forward bias: (a) leaky laser and (b) very low leakage laser. Boxes approximately define the position of the active layer. The detected radiation is InP recombination light (Courtesy of E. J. Flynn).[46]

caused by carrier leakage ($\lambda = 1.05$ μm) was optically detected. Experimental results[42–46] showed that significant electron leakage can occur in InGaAsP/InP double heterostructures constituting one of the mechanisms of sublinearity of the light-current characteristics and also of the temperature dependence of the threshold current in laser diodes.

The method of spectrally resolved luminescence was used to study heterobarrier carrier leakage in CMBH lattice matched and compressively strained 1.3 μm InGaAsP/InP lasers with an active layer of the order of 1 μm in width.[44] The waveguide forming a step graded SCH consisted of two quaternary materials with slightly a different composition corresponding to the band gaps of 1.1 μm and 1.15 μm respectively. The intensity of the corresponding emission lines allows conclusions about carrier concentration in different parts of the waveguide. Spontaneous emission spectra were measured in the range of injection currents 1–100 mA and heat-sink temperatures 20–80°C through the window opening in the contact metallization. It was shown that barrier emission intensity continues to increase even above threshold. Measurements of the temperature dependence of the slope efficiency and the intensity of the emission caused by electron leakage at different temperatures and different pumping currents above the laser threshold were performed.[44] The results indicate that the increase of the carrier leakage in MQW 1.3 μm InGaAsP/InP lasers correlates quantitatively with decreased high temperature slope efficiency. Laser structures with low carrier overflow into barriers and SCH layer exhibit better high-temperature performance. Experiments[44] show that higher values of compressive strain would have to be applied in order to reduce the carrier overflow into the barriers and into the separate confinement region.

4.2. *Electrical technique of studying the carrier leakage*

Observation and measurements of the leakage current in 1.3 μm InGaAsP/InP lasers with the bulk active region by an electrical technique was carried out for the first time by T. R. Chen[47] employing a laser-bipolar-transient-type device. A schematic representation of the device used in this work is shown in Fig. 9.

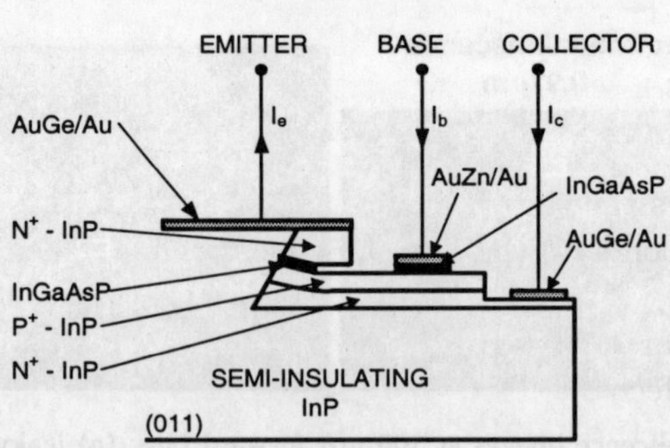

Fig. 9. Schematic representation of the laser-bipolar-transistor structure.[47]

The emitter base junction functions as in an ordinary laser diode. The collector-base junction is reverse biased, as in a bipolar transistor, to collect the leakage current. The laser part of the transistor was not different from that of a terrace mesa double heterostructure laser. However, an extra N^+–InP layer was added in order to collect the heterobarrier electron leakage current. Under forward bias condition, the emitter-base junction acts as a laser diode. The electrons are injected from the InP emitter into the quaternary region. In this structure, the electrons that overcome the heterobarrier A and arrive at the base collector junction B are swept out by the electrical field in the reverse-biased junction. A more complex structure to measure the heterobarrier leakage by an electrical method was implemented in Ref. 48. In addition to conventional InGaAsP/InP laser diode layers, the device contains a small area collector, whose layer structure is shown in the inset to Fig. 10.

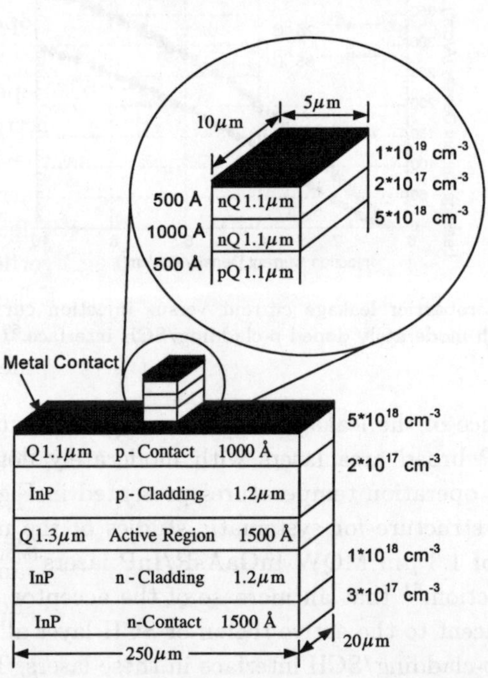

Fig. 10. Schematic cross-section of the laser with leakage collector.[48]

Figure 11 displays the energy band diagram in the cross-section of the device. The diagram corresponds to the operating bias regime and illustrates the idea of the experiment.

Since the collector area was only 1% of the diode area, the presence of the collector does not affect the laser performance. The crucial fabrication step in the processing of the structure in Fig. 9 was evaporation of a self-aligned metal contact to the *p*-quaternary layer, which had to provide an ohmic contact to the laser diode without degrading the collector *p*–*n* junction.

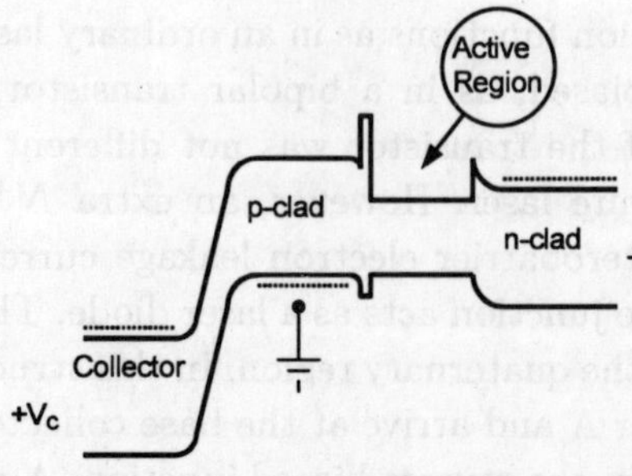

Fig. 11. Structure band diagram of the device shown on Fig. 9 under bias.[48]

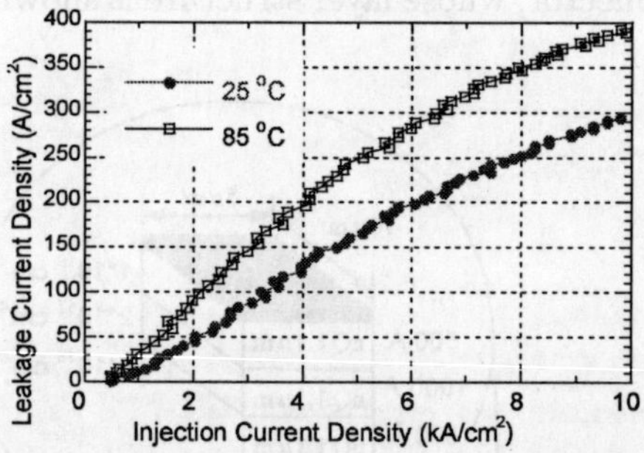

Fig. 12. Measured heterobarrier leakage current versus injection current for 1.3 μm MQW InGaAsP/InP lasers with moderately doped p-cladding/SCH interface.[50]

Typical dependence of the leakage current on injection current for the 1.3 μm MQW InGaAsP/InP broad area lasers with moderately doped p-cladding/SCH interface at different operation temperatures presented in Fig. 12.[50] The use of a three terminal laser structure for systematic studies of the nature of the performance degradation of 1.3 μm MQW InGaAsP/InP lasers[49,50] allowed to confirm the theoretical prediction,[51] that an increase of the acceptor concentration in the p-cladding layer adjacent to the active region or SCH layer allows to minimize the voltage drop at the p-cladding/SCH interface in these lasers. The reduction of this voltage drop allows to avoid a substantial reduction of the barrier for thermionic emission of electrons from the active region.

5. Electrical and Optical Measurements of RF Modulation Response Below Threshold

Most of the experimental methods aimed at determining the carrier concentration are based on a model using the balance equations of carriers and photons in the laser. These equations, also known as "Rate Equations", can be derived from the Maxwell equations (see Statz and deMars[52] and Kleinman[53]), and in a simplified single mode case can be written as:

$$\frac{dn}{dt} = \eta_{\text{int}} \frac{I}{eV_{\text{act}}} - R - \frac{c}{N_{\text{eff}}} \frac{\frac{dG}{dn}(N - N_{\text{tr}})}{1 + \in S} S \,,$$

$$\frac{dS}{dt} = \frac{c}{N_{\text{eff}}} \Gamma \frac{\frac{dG}{dn}(N - N_{\text{tr}})}{1 + \in S} S - \frac{S}{\tau_p} - \beta \Gamma R_{\text{sp}} \,, \tag{14}$$

where n and S are the carrier and photon concentrations respectively, I is the pumping current, e is the electron charge, V_{act} is the active laser volume, R is the total recombination rate (without stimulated emission), R_{sp} the spontaneous emission rate, c the light velocity in vacuum, N_{eff} the effective refractive index, G the material optical gain in the active layer, Γ the optical confinement factor, and τ_p is the photon lifetime. Internal efficiency η_{int} accounts for an imperfect current injection efficiency into the active layer. The term $(1+ \in S)$ describes the gain saturation with power, $\in$ being the phenomenological gain compression parameter. It was first introduced by T. Koch and R. Linke[54] in order to characterize the spectral hole burning. In general this parameter may be used to describe the reduction of the optical gain above threshold due to any other process, such as spatial hole burning, carrier heating etc. In Eq. (14), $\in$ is a composite value, which allows characterizing the reduction of the optical gain due to all the above effects in a linear approximation. Since the rate equations in the form of Eq. (14) treat the laser as a medium with single spectral and spatial mode, only the value of the gain at the lasing wavelength is important. Therefore the wavelength-independent phenomenological parameter $\in$ can be used. It should be noticed, however, that the use of this parameter is not valid when considering the effects that involve the gain spectra, since different effects distort the gain spectra differently and nonuniformly.

The first equation in the set (Eq. (14)) describes the balance of the carrier plasma. The carriers are introduced by the injection current (first term), and then recombine spontaneously or nonradiatively (second term), as well as through stimulated emission of radiation (last term). The second equation describes the balance of photons inside the laser cavity. In this set of equations two important processes are ignored: carrier transport through the SCH and MQW layers[55] and carrier capture into quantum wells (in the case of QW lasers).[56,57]

The stimulated emission term in Eq. (14) can be neglected below threshold. Then, expressing the recombination in terms of the carrier concentration and carrier lifetime τ_s:

$$\frac{dn}{dt} = \eta_{\text{int}} \frac{I}{eV_{\text{act}}} - \frac{n}{\tau_s} \,. \tag{15}$$

For small-signal analysis R. Olshansky *et al.*, suggested using a small-signal pulse[58] or sinusoidal[59] excitation. Equation (15) can be modified for this case:

$$\frac{d\delta n}{dt} = \eta_{\text{int}} \frac{\delta I}{eV_{\text{act}}} - \frac{\delta n}{\tau_d} \,. \tag{16}$$

The total number of carriers can be found by integrating the differential carrier lifetime over the bias current:[60]

$$n(I) = \eta_{\text{int}} \cdot \int_0^I \tau_d(I')dI' . \tag{17}$$

Various techniques, based on Eqs. (15)–(17) were used to measure the carrier concentration and the recombination rates in semiconductor lasers. A number of authors studied these parameters through carrier lifetime measurements.[58–62] These measurements use different approaches but all give as a result the value of the carrier lifetime. This value and its dependence on the current characterize various recombination mechanisms in the active layer of semiconductor laser. The major mechanism is the spontaneous radiative recombination — a standard bimolecular recombination. Non-radiative processes such as trap or interface and Auger recombination have also been addressed.

5.1. *Determination of the differential carrier lifetime from the device impedance*

The equivalent circuit of a semiconductor laser diode in the small-signal modulation regime below threshold is shown in Fig. 13. It can be derived from the rate equations discussed above.[63,64] The active layer can be represented as a RC circuit with a characteristic time equal to the differential carrier lifetime.

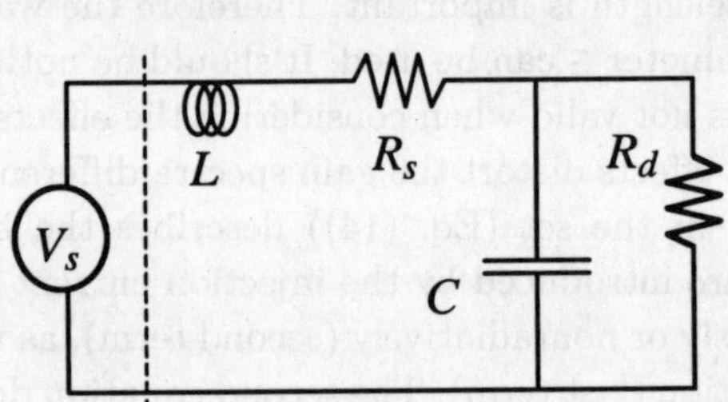

Fig. 13. Simple equivalent circuit of laser diode below threshold.

Taking into account a series resistance, R_s, introduced by the contacts and cladding layers, and a bonding wire inductance L, we have a circuit with the total impedance of

$$Z(\omega) = j\omega L + R_s + \frac{R_d}{1 + j\omega\tau_d} , \tag{18}$$

where $\tau_d = R_d C$, and R_d is the static differential resistance of the p-n junction. This model does not take into account leakage paths and a blocking structure capacitance. It also considers that transport and capture-escape times are much faster than the differential carrier lifetime and may then be neglected.

Equation (18) has two important consequences. First, the laser impedance below threshold is frequency dependent and therefore the amplitude of the current modulation is not constant over a broad frequency range if a standard signal generator (equivalent to a voltage source) is used in the small-signal response experiments.[12,65] Secondly, the differential carrier lifetime can be extracted directly from electrical measurements of the laser impedance.[66] The laser impedance as function of frequency can be measured using a network analyzer. The real and imaginary parts of the laser impedance are described by:

$$\mathrm{Re}\ (Z(\omega)) = R_s + \frac{R_d}{1 + (\omega\tau_d)^2}, \tag{19}$$

$$\mathrm{Im}\ (Z(\omega)) = \omega L - \frac{\omega\tau_d R_d}{}.1 + (\omega\tau_d)^2 \tag{20}$$

Plotted in Fig. 14 are the measured real (a) and imaginary (b) parts of the impedance of a laser with the bulk active layer at room temperature and current of 1 mA. The solid lines are fits to Eqs. (19) and (20) with the parameters shown. Fitting the real and imaginary parts of the laser impedance gives very close values for all model parameters.

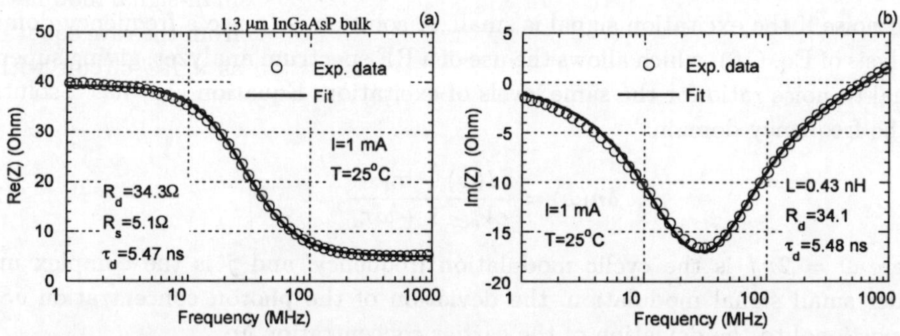

Fig. 14. Measured real (a) and imaginary (b) parts of the laser diode impedance (circles) and fit to Eqs. (19) and (20) (solid lines).

The measured laser impedance shows a frequency-independent series component, R_s, of approximately 5 Ω and a frequency- and bias-dependent component, R_d, associated with the differential impedance of a diode *p-n* junction.[63,64,67] The magnitude of both components can also be measured as a function of current in the CW regime: R_s by extrapolating the above-threshold value of dV/dI to threshold and R_d by measurement of the sub-threshold value of $dV/dI(I) - R_s$.[68] These values were found to be very close to those extracted from the RF measurements.[12]

This measurement technique is based on a simple purely electrical measurement. It is useful when optical detection is difficult, (e.g. at low currents and in the wavelength ranges where no fast detectors are available). On the other hand, this method is not very accurate at high currents, where the value of the differential resistance, R_d becomes too small. This model in general is correct only as long as the transport effects including the capture-escape process[55–69] can be neglected. In the case of highly doped MQW lasers this model is not accurately correct and more complicated model is required.[56,57]

One more important limitation of this technique is that the laser diode itself might have parasitic capacitance formed by the contact pads, blocking layers etc. This capacitance is effectively connected in parallel with the laser *p-n* junction capacitance, C, in the equivalent circuit of Fig. 13. It is usually thought to be small compared with the junction capacitance, but the latter decreases as the current through the laser decreases, and at some current the junction capacitance may become comparable to the parasitic one. Then the equivalent circuit of Fig. 13 is no longer valid.

5.2. *Determination of the differential carrier lifetime from the optical response measurements*

The original technique for determination of differential carrier lifetime proposed to use a small-signal current step excitation.[58] The optical response curve was fitted to an exponential form based on Eq. (16). This technique has a disadvantage of high noise if the excitation signal is small. A solution is to use a frequency domain analysis of Eq. (16), which allows the use of a RF spectrum analyzer, giving superior signal-to-noise ratio for the same levels of excitation. Equation (16) has a solution in the frequency domain:

$$\delta n(\omega) = \frac{\delta I(\omega)}{eV_{\text{act}}} \frac{\tau_d}{1 + \omega\tau_d}, \tag{21}$$

where $\omega = 2\pi f$ is the cyclic modulation frequency, and j is the complex unit. Under small signal modulation, the deviation of the photon concentration δS is proportional to the deviation of the carrier concentration δn.

Originally, many authors used Eq. (21) considering that the current amplitude is frequency independent. However, most of the commercially available signal generators are power sources, which in the experiment are loaded on the laser diode. This connection has an equivalent circuit of a voltage source loaded on an $r = 50\ \Omega$ resistor plus a laser impedance $Z(\omega)$. The frequency dependence of the impedance results in the frequency dependence of the amplitude of current modulation, therefore it should be taken into account in order to extract a correct value of the differential lifetime.[12,65] Equation (21) can be modified to yield:

$$\delta n(\omega) = \frac{1}{eV_{\text{act}}} \cdot \frac{\delta V(\omega)}{r + Z(\omega)} \cdot \frac{\tau_d}{1 + j\omega\tau_d}. \tag{22}$$

The voltage modulation amplitude $\delta V(\omega)$ can be considered to be constant over the frequency range. Therefore, knowledge of $Z(\omega)$ can be used to correct the measured optical modulation, in order to use the single-pole fitting procedure proposed in Ref. 59. Using Eq. (22), we can write:

$$F(\omega) = |\delta S(\omega)| \cdot |r + Z(\omega)| \propto \frac{1}{\sqrt{1 + (\omega \tau_d)^2}}, \tag{23}$$

where the corrected response function, $F(\omega)$, can be used to extract the differential lifetime. The measured optical response curve (circles in Fig. 15) was corrected using Eq. (23) (squares in Fig. 15), and then fit to a single pole roll-off from which the differential carrier lifetime was extracted.

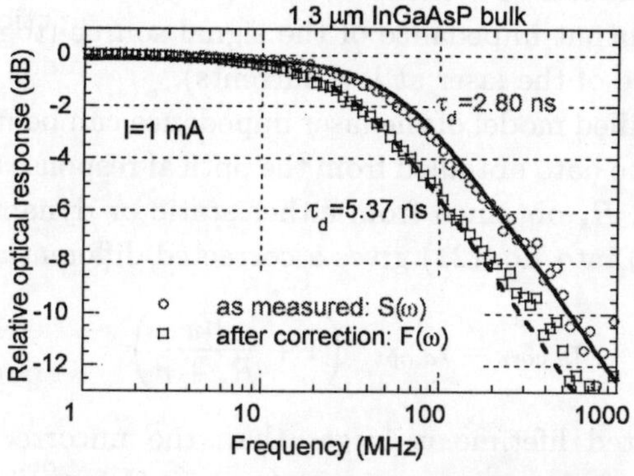

Fig. 15. Optical response curves: as measured (circles), and corrected using Eq. (23) (squares). Lines are single-pole fits.

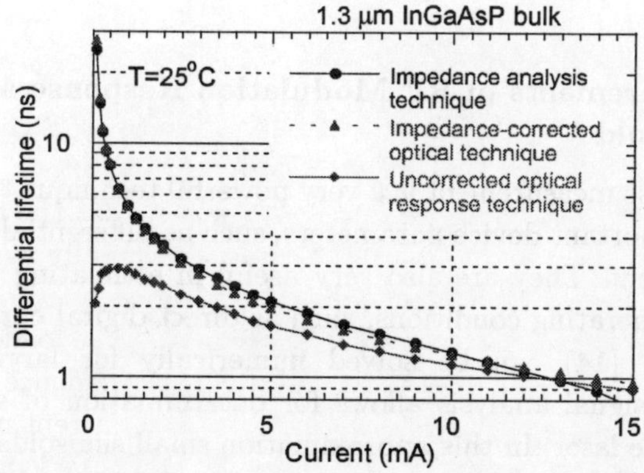

Fig. 16. Dependence of differential carrier lifetime (determined using different techniques) on current.

The results of the measurements of the differential carrier lifetime using all three of the above techniques are plotted in Fig. 16. When the differential carrier lifetime is measured using the uncorrected optical response technique (rhombs), it shows noticeable saturation and decrease at low injection currents. Other authors[59] have observed this behavior. Additional recombination mechanisms were introduced to account for this effect. Comparison of the differential carrier lifetimes obtained using the impedance analysis technique (triangles), and the impedance-corrected optical technique (circles) to the data obtained using uncorrected optical response illustrates the large effect of the rapidly varying laser impedance, especially at low bias current. The corrected differential lifetimes give very close values within the entire current range and no longer show the saturation and decrease with decreasing current. There is no need to invoke an extra mechanism of current consumption as described in Ref. 59. It can be concluded, that the apparent saturation in lifetimes is due to the finite output impedance of the signal source (together with the high differential impedance of the laser at low currents).

Finally, the simplified model of the laser impedance can be used for correction of the differential lifetime data obtained from the optical response measurements using the values of R_d and R_s obtained from either static or dynamic measurements.[65] Substituting Eq. (18) into Eq. (21) gives a corrected differential carrier lifetime:

$$\tau_{d,\mathrm{corr}} = \tau_{d,\mathrm{opt}} \cdot \left(1 + \frac{R_d}{R_s + r}\right) \tag{24}$$

in which the corrected lifetime is larger than the uncorrected, optically measured lifetime by the factor in parentheses ($r = 50\ \Omega$ is the signal source output impedance). For example, this correction factor equals 2 when R_d is about 55 Ω, which occurs at approximately 0.7 mA. Use of this simple correction gives the same value of the differential lifetime as the technique based on impedance analysis and impedance corrected optical measurement.

6. Optical Measurements of RF Modulation Response and RIN Above Threshold

Small-signal response measurement is a very powerful technique that allows analyzing a number of important device parameters, such as differential gain and the gain suppression coefficient. They are also very useful in estimating the device performance under real operating conditions, such as direct digital current modulation.

The set of Eq. (14) can be solved numerically for large signal analysis. Also, simple small-signal analysis allows for determination of some fundamental characteristics of the laser. In this approximation small sinusoidal current modulation around the static value results in sinusoidal modulation of the carrier concentration and photon density:[70,71]

$$\begin{aligned} I(t) &= I_0 + I_1 \exp(j\omega t)\,, \\ n(t) &= n_0 + n_1(\omega) \exp(j\omega t)\,, \\ S(t) &= S_0 + S_1(\omega) \exp(j\omega t)\,, \end{aligned} \tag{25}$$

Following Refs. 70 and 71, the solution for optical power response can be written as

$$\frac{P(\omega)}{I(\omega)} = \frac{hv}{q}\eta H(\omega)\,, \tag{26}$$

where the two-parameter modulation transfer function $H(\omega)$ is defined as in Ref. 2, (p. 200):

$$H(\omega) = \frac{\omega_R^2}{\omega_R^2 - \omega^2 + j\omega\gamma}\,. \tag{27}$$

The two parameters in Eq. (27) are the electron-photon resonance frequency ω_R and the damping factor γ. Expressions for these values are derived in the course of the same analysis:

$$\omega_R \approx \sqrt{\frac{\frac{c}{N_{\text{eff}}}\frac{dG}{dn}S_0}{\tau_p(1+\in S_0)}} \tag{28}$$

and

$$\gamma = \frac{1}{\tau_s} + \frac{c}{N_{\text{eff}}}\frac{dG}{dn}\frac{S_0}{1+\in S_0} + \frac{1}{\tau_p}\frac{\in S_0}{1+\in S_0}\,, \tag{29}$$

where photon lifetime τ_p is proportional to the total optical loss:

$$\frac{1}{\tau_p} = \frac{c}{N_{\text{eff}}}\alpha_{\text{tot}}\,. \tag{30}$$

Shown in Fig. 17 are the amplitude (a) and phase (b) of the laser optical modulation response at different DC biases, measured using HP8510 Network Analyzer with HP83440D high-speed *p-i-n* detector. The response curves show typical resonant behavior described by Eq. (27). Additional drooping of the response curves, seen in the Fig. 17(a) has been attributed to carrier transport through the SCH layers[55] and to the carrier capture and thermoionic emission processes in QW lasers.[72]

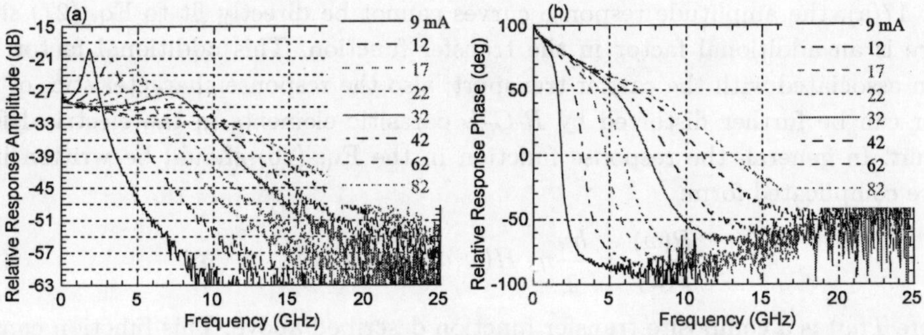

Fig. 17. Relative Amplitude (a) and Phase (b) of the laser response to small-signal direct current modulation, measured using a *p-i-n* detector and HP8510 Network Analyzer (at different DC bias currents).

As it can be seen from Eqs. (28) and (29), the resonance frequency and damping factor contain important information on the material parameters: differential gain and the gain compression coefficient. If these values were known as functions of the injection current (or optical power), the above parameters could be determined.

The damping factor γ increases proportionally to the square of the resonance frequency ω_R^2. R. Olshanslky *et al.*,[73] have defined the proportionality coefficient between the γ and ω_R^2 and as the K-factor:

$$\gamma = K f_R^2 + \gamma_0 \,, \qquad K \approx 4\pi^2 \left(\tau_p + \frac{\in}{\frac{c}{N_{\text{eff}}} \frac{dG}{dn}} \right) . \tag{31}$$

The resonant frequency increases with current and so does the small-signal modulation bandwidth (the frequency at which the response amplitude drops by 3dB from its DC value), but only up to a certain level. As the DC injection current increases, the response flattens (as it can be seen in the Fig. 17(a)). At some point the damping factor becomes large enough that response amplitude drops by 3dB at the frequency below ω_R. It was shown[73] that maximum possible bandwidth is approximately:

$$f_{3dB\,\max} \approx \sqrt{2} \frac{2\pi}{K} \,. \tag{32}$$

6.1. *Determination of the resonant frequency and damping factor using carrier subtraction technique. Determination of the differential gain and the gain compression coefficient*

The differential gain and the gain saturation parameters can be determined from the dependencies of the resonance frequency and damping factor on the optical power (DC drive current) using Eqs. (28)–(31). These dependencies can both be determined from the small-signal response measurements described in the previous section. However an accurate procedure is not trivial. As it can be seen from Fig. 17(a), the amplitude response curves cannot be directly fit to Eq. (27) since there is an additional factor in the transfer function. This additional factor has been associated with the carrier transport; also the response characteristics of the laser can be further distorted by *R-C-L* parasitic elements in the electrical feed circuit. In general, the response function in the Eq. (26) should be written in a more complicated form:

$$\frac{P(\omega)}{I(\omega)} = \frac{hv}{q} \eta \cdot H(\omega, \omega_R, \gamma) \cdot T(\omega) \,, \tag{33}$$

where $T(\omega)$ is a composite transfer function described above. This function cannot be written analytically in the general case.

P. Morton has proposed an elegant way to determine the resonant frequency and damping factor from the response measurements. He demonstrated that in the assumption that the transfer function $T(\omega)$ does not depend on the DC bias and the modulation efficiency $I(\omega)$ is constant above threshold, the ratio of two response amplitudes does not depend on any factors other than respective resonant frequencies and damping factors. Since the relative response amplitude curves are typically measured in the logarithmic (dB) units, it is convenient to express the difference between the two responses in this way:

$$\begin{aligned}\text{Response}_1(f) - \text{Response}_2(f) &= 20\log_{10}(P_1(f)) - 20\log_{10}(P_2(f)) \\ &= 10\log_{10}\left(\frac{(f_{R2}^2 - f^2)^2 - (f\gamma_2/2\pi)^2}{(f_{R1}^2 - f^2)^2 - (f\gamma_1/2\pi)^2}\right) + \text{const}\,.\end{aligned} \tag{34}$$

Shown in Fig. 18(a) are the relative response amplitudes measured using *p-i-n* detector and HP71400 RF spectrum analyzer. The data was taken for the same device as in the Fig. 17. The signal noise at low amplitude levels and high modulation frequencies is considerably lower than in the case when the measurements were done using the Network Analyzer, because the RF spectrum analyzer has considerably higher dynamic range. This is very useful because the low noise at high frequencies is important for successful fitting of the difference curves to Eq. (34). The response curves in Fig. 18(a) are not as smooth as in Fig. 17(a) because of some calibration inaccuracy (in the particular case of the data shown in Fig. 18(a) this inaccuracy is not very high). But the calibration data is not needed since the calibration correction enters as an additional factor into the function $T(\omega)$. P. Morton had demonstrated[74] that even high calibration inaccuracy is completely eliminated by the subtraction procedure. This feature makes the subtraction technique very valuable for the lasers having high parasitic RC parameters or packaged for DC or low-frequency operation.

The results of the subtraction procedure are presented in Fig. 18(b). The differential curves obtained by subtraction of the response amplitudes at 17 mA from

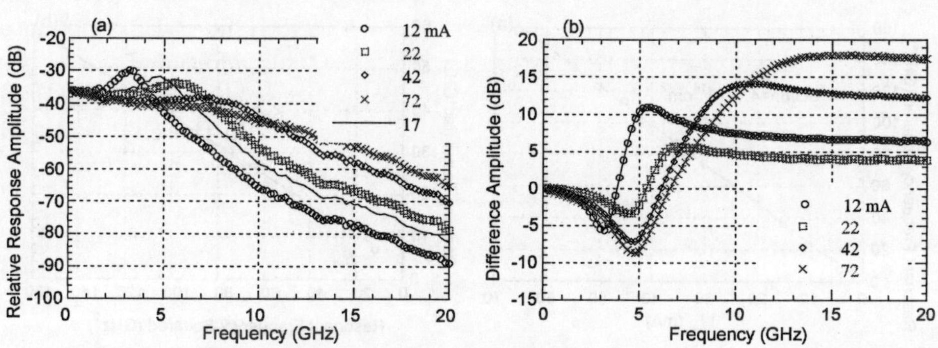

Fig. 18. Relative response amplitudes (a) measured using *p-i-n* detector and RF spectrum analyzer (for the same laser as in Fig. 17), and the results of subtraction of the reference curve (solid line in (a)) from the response curves.[74] Solid lines in (b) are fits to the Eq. (34).

the response amplitudes at each current are shown with the same symbols, as the original curves in the Fig. 18(a). These differential curves are very smooth and can easily be fitted to Eq. (34) (solid lines). It turns out that even though the curves are fitted to a five-parameter expression, the fitting procedure converges very well and gives very accurate data for the resonant frequencies and damping factors.

The results of the curve fitting performed on the differential curves are presented in Fig. 19. Solid curves are linear fits. The same procedure was performed with the differential curves obtained by subtracting the response at different currents to check the accuracy of the procedure. The values of the K-factor and differential gain were within 5%, which confirms the robustness of the method.

Knowledge of the dependence of the resonant frequency on current allows determining the differential gain in the material of the active layer. Using Eq. (28) we can write:

$$\frac{dG}{dn} \approx \frac{4\pi^2 f_R^2}{\frac{c}{N_{\mathrm{eff}}} S_0}\tau_p = \frac{4\pi^2 eV}{\Gamma \eta_i \frac{c}{N_{\mathrm{eff}}}} \frac{f_R^2}{(I - I_{\mathrm{th}})}, \tag{35}$$

where I_{th} is the threshold current, and we used the relation between the photon density and the output power:

$$S_0 \approx \frac{\Gamma}{V} \cdot \frac{P_{\mathrm{out}}}{hv} \frac{1}{\frac{c}{N_{\mathrm{eff}}}\alpha_{\mathrm{mirr}}} = \frac{\Gamma}{V} \cdot \frac{\eta_i}{e}\tau_p (I - I_{\mathrm{th}}), \tag{36}$$

where the mirror loss α_{mirr} is given by the first term of Eq. (4) for an uncoated device.

In the current treatment the effects of the carrier transport and capture into the QW are ignored. More accurate phenomenological treatment yields another factor χ in the right side of Eq. (35). ($\chi = 1 + \tau_d/\tau_e$, where τ_d is the carrier capture time including diffusion across SCH and τ_e the thermionic emission lifetime).[72]

Differential gain was estimated from the data presented in Fig. 19(a) using Eq. (35). Ignoring the transport effects ($\chi = 1$) and assuming the internal efficiency $\eta_i = 1$, we obtain $dG/dn = 4.1 \times 10^{-16}$ cm^2.

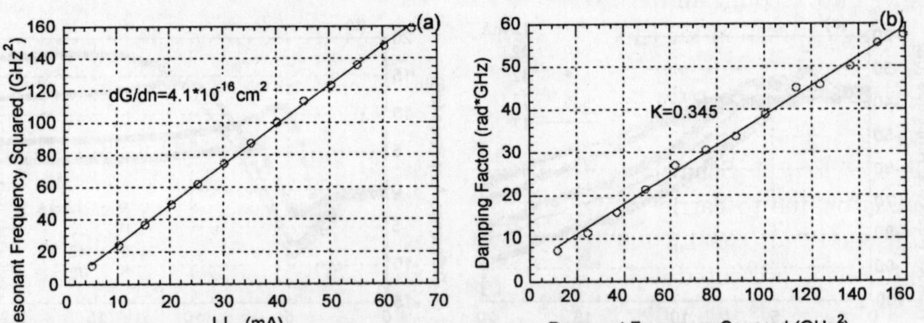

Fig. 19. Results of the curve fitting of the differential curves presented in the Fig. 18(b): (a) Resonant Frequency Squared as a function of the current deviation above threshold, (b) Damping Factor as a function of Resonant Frequency Squared.

6.2. *Determination of the resonant frequency and differential gain from the RIN measurements*

Relative intensity noise (RIN) measurements represent the alternative technique used for studies of high-speed dynamics in QW lasers. Like the modulation response technique described above, this approach allows for determining of dG/dn and K-factor through numerical fitting procedure. However, the signal measured is not the response to the external modulation of the bias current, but the noise spectra of the laser itself. This noise power is associated with the fluctuation of the concentrations in photon and electron systems caused by spontaneous light emission events. Deviation of the electron and photon densities from equilibrium values leads to their damped oscillations with the frequency of electron-phonon resonance f_R and damping factor γ. Hence, the measurements of the laser noise power spectra at different bias currents provide enough information to determine such important laser parameters as differential gain and K-factor.

Relative Intensity Noise (RIN) is defined (see p. 221–225 in Ref. 2) as the mean square value of the noise power related to the mean value of the output power squared:

$$RIN \equiv \frac{\langle \delta P(t)^2 \rangle}{P_0^2}. \tag{37}$$

It is more convenient to define RIN per unit bandwidth because the measurement bandwidth can vary under different experimental conditions. Then:

$$\frac{RIN}{\Delta f} \equiv \frac{2S_{\delta P}(\omega)}{P_0^2}, \tag{38}$$

where $S_{\delta P}(\omega)$ is double sided noise power spectral density (if the spectral density is defined as single-sided, then the factor 2 should be removed). As defined, RIN is measured in dB/Hz.

Small signal analysis applied to the rate equations similar to the Eq. (14) but taking into account an internal noise source and transport of carriers leads to the following expression for the spectral dependence of the RIN:

$$RIN = \frac{4}{\pi}\delta f_{st}\frac{f^2 + (\gamma^*/2\pi)^2}{(f_r^2 - f^2)^2 + f^2(\gamma/2\pi)^2}, \tag{39}$$

where — δf_{st} — Schawlow-Townes linewidth and γ^* — differs from γ in the denominator due to carrier transport through the active region of a MQW laser.[72] RIN spectra for the 1.3 μm InGaAsP/InP MQW DFB laser at the different bias conditions is presented in Fig. 20.[75] The resonance frequency, damping factor and Shawlow-Townes linewidth can be determined from fitting the experimental RIN spectra to Eq. (39).

The RIN measurement technique was utilized to study the high-speed characteristics of 1.3 μm and 1.55 μm InGaAsP/InGaAs[76–78] and 1.55 μm InGaAs/InGaAlAs[79] lasers.

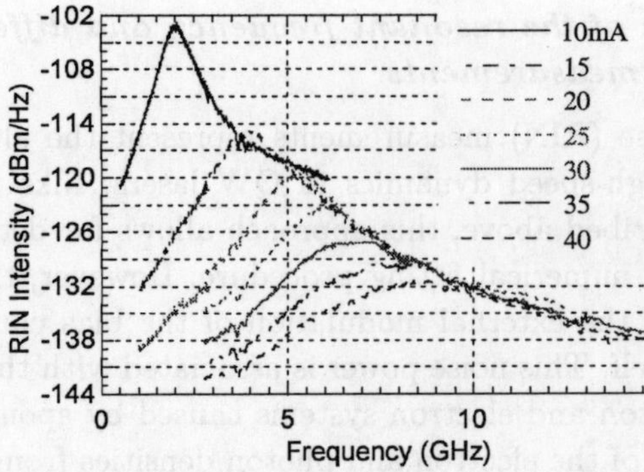

Fig. 20. Dependence of the RIN spectra for 1.3 μm CMBH InGaAsP/In MQW DFB laser at 25°C on bias current.

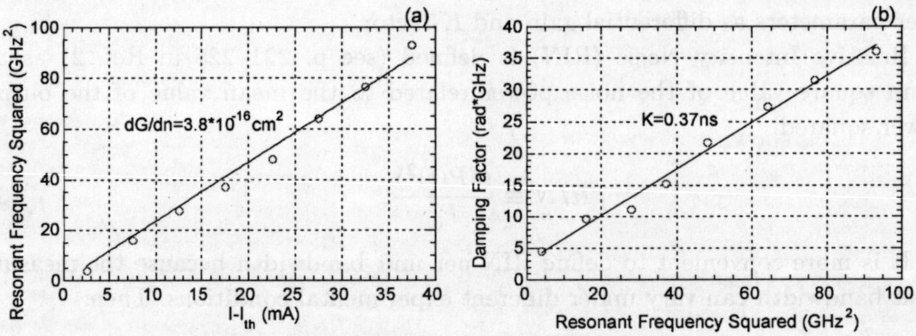

Fig. 21. Determination of the differential gain (a) and K-factor (b) from the slopes of the theoretically predicted linear dependencies.

As it was mentioned in the previous section, by taking account of the carrier transport properties of the structure one can obtain the dependence of the resonance frequency on current. This allows to determine the differential gain. Experimental results for the same laser as in Fig. 20 are presented in Fig. 21.[75] The values of the differential gain and K-factor are also shown.[75]

The main advantage of the RIN technique with respect to modulation response measurements is elimination of the problem with high frequency modulation experimental equipment. At the same time, the low level of the noise for high quality lasers makes it is necessary to use low noise preamplifiers. Also, there is no low frequency roll-off in the RIN spectra, which is related to the transport effects.

7. Measurements of the Linewidth Enhancement Factor

Fleming and Mooradian[80] first observed that the static linewidth of the semiconductor laser is much larger than predicted by classical Schawlow-Townes formula. C. Henry has introduced the linewidth enhancement factor,[81] which relates small variations of the real and imaginary parts of the refractive index due to a change of the carrier concentration:

$$\alpha_T = \frac{\left.\frac{\partial N'}{\partial n}\right|_T}{\left.\frac{\partial N''}{\partial n}\right|_T} = \left.\frac{\partial N'}{\partial N''}\right|_T . \tag{40}$$

C. Henry[81] had shown that spontaneous emission noise in the active layer of the semiconductor laser causes the fundamental broadening of the Lorentzian laser linewidth:

$$\Delta v = \frac{c}{N_{\text{eff}}} \frac{R_{\text{sp}}}{G} \frac{\alpha_{\text{mirr}}\, hv}{8\pi P}(1 + \alpha_T^2) \tag{41}$$

where, again, the P is output power, R_{sp} the spontaneous emission rate, and α_{mirr} the mirror loss. The linewidth enhancement factor is labeled α_T here to emphasize the fact that variations are due to a change of carrier concentration at constant carrier temperature.

Linewidth enhancement factor α_T is a critical parameter determining both the static linewidth of the semiconductor laser and instantaneous wavelength change of the semiconductor laser under direct modulation, also known as the wavelength chirp. It was shown that for most applications, low values of parameter α_T are desirable. Determination of the parameter α_T is therefore important for understanding and improving the laser design and performance. This can be done using various methods; some related to the gain measurements described in previous sections. C. Henry[4] has proposed to use the Kramers–Krönig relation to determine the α_T parameter from the optical gain data obtained from TSE measurements described in Sec. 2.

Shortly after C. Henry,[81] K. Vahala and A. Yariv developed a theory predicting the same linewidth enhancement factor.[82] C. Harder *et al.*, also proposed a method to determine the linewidth enhancement factor α_T from the high-frequency modulation experiment.[83] This method involves considerable effort since both FM and AM modulation indices have to be measured. FM modulation index measurement at high frequency requires a FP interferometer (supercavity detector), which is sometimes inconvenient.

F. Devaux *et al.*,[84] have proposed another method for measuring the linewidth enhancement factor using the above threshold modulation response measurements, very similar to those discussed in Sec. 6.1. They proposed to utilize the FM/AM conversion in a dispersive media such as standard optical fiber. The optical spectrum of the modulated laser (as well as any other source) contains modulation side-bands around the central optical frequency. These side-bands have different group velocities in the fiber, and therefore different delays after traveling through some distance of the fiber. This results in sharp drops (nulls) in the response curve at frequencies corresponding to the phase difference of π between these side-bands. The exact frequencies of these nulls are determined by both the linewidth enhancement factor and fiber dispersion and can be measured using a standard network analyzer. Therefore the linewidth enhancement factor can be determined very accurately.

The disadvantage of this method is that it can only be used for characterization of spectrally single-moded sources. For example it cannot be used for FP lasers.

7.1. *Measurements of linewidth enhancement factor α_T from ASE and TSE spectra*

I. Henning and J. Collins proposed a very simple technique for measuring the linewidth enhancement factor α_T based on the ASE measurements.[85] Equation (40) can be used with other equations relating the change of the real and imaginary parts of the refractive index to the values, which can be measured experimentally. Rewriting Eq. (5):

$$\delta N'' = -\frac{1}{\Gamma}\frac{\delta g}{4\pi}\lambda \tag{42}$$

and

$$\Gamma\frac{\delta N'}{N'} \approx \frac{\delta N_{\text{eff}}}{N_{\text{eff}}} = \frac{\delta\lambda}{\lambda}\,, \tag{43}$$

where N_{eff} is the effective group index. Substituting Eqs. (42) and (43) into Eq. (40) we get:

$$\alpha_T = -\frac{4\pi N_{\text{eff}}}{\lambda^2}\frac{\partial\lambda}{\partial g}\bigg|_T\,. \tag{44}$$

By recording the ASE at different currents below threshold and measuring the change of the gain and the wavelength shift of Fabry–Perot peaks with current one can determine the linewidth enhancement factor α_T and its dependence on the wavelength and current. Shown in the Fig. 22(a) is a set of α_T factor curves measured using the above procedure.

This method is now commonly used for measuring the linewidth enhancement factor α_T.[86–88] P. Morton has shown how these measurements together with the gain measurements and carrier lifetime measurements[12] allow for optimization of

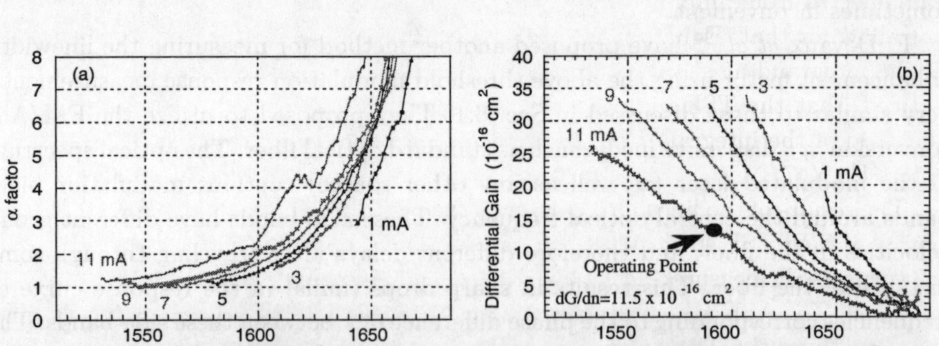

Fig. 22. The linewidth enhancement factor (a) and differential gain (b) versus wavelength at various bias currents.

the laser structure for high-speed operation.[14] Presented in Fig. 22 are the measured dependencies of linewidth enhancement factor and differential gain on the wavelength for various currents. Optimizing the DFB wavelength from the data shown in Fig. 22 allowed achieving the small-signal direct modulation bandwidth of 26 GHz and linewidth enhancement factor of close to 1.

Shown in Fig. 22(b) is the set of the differential gain spectra determined using the below threshold gain spectra combined with the carrier lifetime measurements discussed in Secs. 2.1 and 5.1 respectively. It should be pointed out that the value of the differential gain obtained using this method (indicated by solid circle in Fig. 22(b)) is in very good agreement with the value of the differential gain obtained in the same device from the response measurements described in Sec. 6.1.[14]

An alternative measurement technique for determination of α_T utilizes the optical gain spectra determined from TSE as described in Sec. 2.2. C. Henry[4] has used the Kramers–Krönig relation for small changes of refractive index:

$$\delta N' = \frac{2}{\pi} P \int_0^\infty \frac{\delta N'' \cdot \varepsilon'}{\varepsilon'^2 - \varepsilon^2} \cdot d\varepsilon' \,, \tag{45}$$

where P indicates taking the principal part integral. Equation (42) can be used to relate δG and $\delta N''$:

$$\delta N' = -\frac{hc}{2\pi^2} P \int_0^\infty \frac{\delta G}{\varepsilon'^2 - \varepsilon^2} \cdot d\varepsilon' \,. \tag{46}$$

The coefficient α_T can be determined by substituting Eq. (46) into Eq. (40). After some simple algebra:

$$\alpha_T(\varepsilon) = \frac{\left.\frac{\partial N'}{\partial \mu}\right|_T}{\left.\frac{\partial N''}{\partial \mu}\right|_T} = \frac{2}{\pi} \frac{\varepsilon \cdot P \int_0^\infty \delta G(\varepsilon')\big|_T \frac{d\varepsilon'}{\varepsilon'^2 - \varepsilon^2}}{\delta G(\varepsilon)\big|_T} \,. \tag{47}$$

Equation (47) allows for determination of α_T from the gain spectrum. In order to perform this procedure, the gain spectra should be obtained in a very broad energy range. However, the commonly used Hakki–Paoli technique[8] for extraction of the gain spectra from amplified spontaneous emission (ASE) from the laser facet does not allow for that. Equation (47) requires some caution in treating the singularity in the integrand at $\varepsilon' = \varepsilon$, a procedure described by C. Henry.[4] It should also be noticed, that the TSE intensity can be measured in arbitrary units because only the ratio of the integral of the spontaneous emission to its value at certain energy enters Eq. (47).

The results of the linewidth enhancement factor measurements in the DFB using both ASE and TSE spectra were reported in Ref. 21. Obtained values of $\alpha_T \approx 2.0 \pm 0.2$ (using ASE measurements) and $\alpha_T \approx 2.1 \pm 0.2$ (using TSE measurements) agree very well. The ASE measurements are usually easier to perform and more accurate, but sometimes the TSE type of the measurements are the only available solution.

Both methods described in this section are based on the below threshold measurements, and therefore can give only an asymptotic value of the parameters close to threshold. However they have an advantage, since they can be used to study the dependence of the linewidth enhancement factor on carrier concentration and other parameters that cannot be varied in the real operating conditions above threshold. On the other hand the results of the above threshold measurements[82,84] are more accurate since they can be obtained in the exact operating conditions of the device.

8. Measurements of the Carrier Temperature and Carrier Heating in Semiconductor Lasers

Typical semiconductor lasers operate at the injection current densities of an order of 10 kA/cm^2 or higher. Under these conditions one can expect a significant difference between the lattice temperature T_L and the temperature of the electron-hole-plasma in the active region T_{e-h}. This difference is determined by the energy acquired by carriers in the active layer and their energy relaxation time. Both quantities have been disputed in the literature (see Ref. 89 and references there).

Since the modal optical gain, wavelength chirp and carrier leakage over the heterobarrier are sensitive to the carrier temperature T_{e-h}, the study of carrier heating is important for understanding the laser physics and improving device design. Experimental studies of the heating in heterostructures using optical excitation and time resolved spectroscopy were carried out by different groups (see for example the review[90]). In semiconductor lasers and light-emitting diodes the carrier temperature is often estimated from the high-energy tail of the spontaneous emission spectra.[91–94] Carrier temperatures up to 400 K at room temperature were reported in InGaAsP LED's.[91] The authors of Ref. 92 studied InGaAsP/InP buried heterostructure lasers with a bulk active layer. They did not observe carrier heating at room temperature for current densities up to 10 kA/cm^2 (within an experimental accuracy of 10 K). The authors of Ref. 93 studied photoexcieted InGaAsP/InP broad area quantum-well structures and found considerable heating at low temperatures. In AlInAs/GaInAs/InP bulk double heterostructure light-emitting transistors, significant carrier heating at cryogenic temperatures was determined from the spontaneous emission profile[94] as well as using an analysis of thermionic emission.[95]

The problem with the estimation of carrier temperature from the high energy tail of the spontaneous emission arises from the fact that the density of states as well as matrix element of the optical transition are not well known functions of energy. This makes it difficult to analyze the spontaneous emission (see for example Fig. 4(a)). Study of the thermionic emission is also nontrivial since it requires a special sample geometry and is generally unsuitable for real laser structures.

8.1. *Determination of carrier heating from wavelength chirp and TSE measurements*

Recently a new experimental technique was proposed, which allows to measure the rate of change of the carrier plasma temperature with pumping current above threshold.[21] This method is based on the relationship between the rate of carrier heating and the wavelength chirp. Wavelength chirp results from the dependence of the real part of refractive index on current. The real and imaginary parts of refractive index are related through the Kramers–Krönig relation. Under lasing conditions the distortion of spectral profile of the optical gain (and therefore the imaginary part of refractive index) is caused by spectral hole burning[54,96,97] and carrier heating.[97,98] However, the experimental studies of fast dynamics of the optical gain in semiconductor laser structures have shown that the electron-hole plasma thermalization time is much shorter than the time needed for thermal equilibrium between the plasma and crystal lattice.[99] Therefore in the treatment of Ref. 21, the carrier heating was considered to be dominant and the spectral hole burning effect was neglected. In this case the energy distribution of electrons and holes is described by Fermi functions with the same temperature. This temperature differs from the lattice temperature; electrons and holes also have different quasi-Fermi level energies when voltage is applied to a laser diode. Sources of carrier heating above threshold are injection of energetic carriers from heterobarriers into active layer and free-carrier absorption. The first effect depends on the injection current and the second on the optical field. Therefore to first order approximation, both effects are proportional to the difference between operating and threshold currents.

In a static case above threshold, the optical gain is equal to the optical loss. At the lasing wavelength, the reduction of the optical gain due to carrier heating is compensated by of the increase of the carrier concentration. Expressing this in terms of variations of the imaginary part of the refractive index yields:

$$\left.\frac{\partial N''}{\partial \mu}\right|_T d\mu + \left.\frac{\partial N''}{\partial T}\right|_\mu dT = 0\,. \tag{48}$$

However, pinning of the optical gain does not occur at all wavelengths. The real part of the refractive index N' is related to the imaginary part N'' (and optical gain) via the Kramers–Krönig relation. This results in the dependence of the refractive index N' on current above threshold and in the wavelength chirp. The full change of the real part of the index N' due to the same small variations $d\mu$ and dT, can be expressed in the same terms as in Eq. (48):

$$\left.\frac{\partial N'}{\partial \mu}\right|_T d\mu + \left.\frac{\partial N'}{\partial T}\right|_\mu dT = dN'\,. \tag{49}$$

The linewidth enhancement factor as defined in Eq. (40) can also be defined as a derivative with respect to the quasi-Fermi level separation because only two variables out of three (n, μ and T) are independent for Fermi-distributed carriers:

$$\alpha_T = \frac{\left.\dfrac{\partial N'}{\partial \mu}\right|_T}{\left.\dfrac{\partial N''}{\partial \mu}\right|_T}\,. \tag{50}$$

This definition allows for easier measurement, because quasi-Fermi level separation μ is easier to determine than the carrier concentration.

The Kramers–Krönig relation suggests that there is an analogous parameter relating small variations of the real and imaginary parts of the refractive index due to the change of the carrier temperature at a fixed value of the quasi-Fermi level separation:

$$\alpha_\mu = \frac{\left.\frac{\partial N'}{\partial T}\right|_\mu}{\left.\frac{\partial N''}{\partial T}\right|_\mu} = \left.\frac{\partial N'}{\partial N''}\right|_\mu . \tag{51}$$

This definition is very close to one by C. Hultgren and E. Ippen.[100] It should be noted that unlike α_T, the parameter α_μ could be determined only from TSE measurements using a formula similar to Eq. (47). Determination of this parameter from ASE measurements is impossible because the shift of Fabry–Perot peaks with ambient temperature is determined mainly by the change of the effective index due to the change of the crystal lattice temperature, and not the carrier plasma temperature.

Combining Eqs. (48)–(51), one can relate the change of the real part of the refractive index and change of the carrier temperature:

$$dN' = -(\alpha_T - \alpha_\mu) \cdot \left.\frac{\partial N''}{\partial T}\right|_\mu \cdot dT . \tag{52}$$

Equation (52) may be rewritten in terms of laser parameters that can be measured experimentally. Equations (42) and (43) can be used to express the change of the lasing wavelength due to the change of the carrier temperature in the active layer:

$$d\lambda = \frac{\lambda^2}{4\pi \cdot N_{\text{eff}}} (\alpha_T - \alpha_\mu) \cdot \left.\frac{\partial g}{\partial T}\right|_\mu \cdot dT . \tag{53}$$

Carrier heating lifts the pinned carrier concentration (and quasi-Fermi level separation) above threshold and results in wavelength chirp. The carrier heating affects the lasing wavelength in two ways, described by the two terms in brackets $(\alpha_T - \alpha_\mu)$ in Eq. (53). The first term corresponds to the change of the real part of the refractive index due to change of the quasi-Fermi level separation. The second term describes the change in real part of the refractive index due to the change of the gain profile produced by the carrier heating.

To determine accurately the change of wavelength with current in a small-signal approach, one can use a Fabry–Perot etalon method or a gated delayed self-homodyne technique.[101] Rewriting Eq. (53) for frequency chirp due to current modulation:

$$\frac{dT}{dI} = -\frac{4\pi \cdot N_{\text{eff}}}{(\alpha_T - \alpha_\mu) \cdot c} \frac{1}{\left.\frac{\partial g}{\partial T}\right|_\mu} \beta \tag{54}$$

where $\beta = \frac{dv}{dI}$ is the chirp parameter.

Equation (54) establishes the relation between the rate of carrier heating above threshold and wavelength chirp. Coefficients $\frac{\partial g}{\partial T}\Big|_{\mu}$, α_T and α_μ can be determined from the gain measurements described in previous sections. Their values were:

$$\frac{\partial g}{\partial T_{e-h}}\Big|_{\mu} \approx -0.45 \pm 0.05 \frac{\text{cm}^{-1}}{K}\,, \ \alpha_T \approx 2.1 \pm 0.2\,, \ \alpha_\mu \approx -1.4 \pm 0.3\,,$$

and $\beta = 156$ MHz/mA.[21] Using these values, the rate of change of the carrier temperature with current was estimated to be approximately 0.13 K/mA. The accuracy of the estimation is about 25%. This allows for an estimation of the carrier temperature difference between a threshold of 15 mA and a typical operating current of 50 mA to be 4.7 K, which is consistent with the results of C. Henry *et al.*[92] This is considerably less than estimated by other authors (e.g. Ref. 102) where the energy relaxation time of the electron-hole plasma was considered to be few picoseconds. The results of Ref. 21 suggest a subpicosecond energy relaxation time, which is consistent with the results of pump-probe experiments.[99]

Acknowledgments

The authors acknowledge C. L. Reynolds and L. Shterengas for useful discussion, and T. Aherne for correcting errors. G. L. Belenky thanks ARO for support under grant DAAG55-97-1-0009.

References

1. H. Casey and M. Panish, *Heterostructure Lasers*, Academic Press, New York, 1978.
2. L. C. Coldren and S. W. Corzine, "Diode lasers and photonic integrated circuits", in *Wiley Series in Microwave and Optical Engineering*, John Wiley and Sons, New York, 1995.
3. A. Yariv, *Optical Electronics*, 4th Edition, Saunders College Publishing, New York, 1991.
4. C. H. Henry, R. A. Logan, and K. A. Bartness, "Spectral dependence of the change in refractive index due to carrier injection in GaAs lasers", *J. Appl. Phys.* **52** (1981) 4457–4461.
5. A. Oster, G. Erbert, and H. Wenzel, "Gain spectra measurements by a variable stripe length method with current injection", *Electron. Lett.* **33** (1997) 864–886.
6. J. L. Zilko, L. J. P. Ketelsen, Y. Twu, D. P. Wilt, S. G. Napholtz, J. P. Blaha, K. E. Strege, V. G. Riggs, D. V. Vanharen, S. Y. Leung, P. M. Nitzsche, J. A. Long, G. Przyblek, J. Lopata, M. W. Focht, and L. A. Koszi, "Growth and characterization of high yield, reliable, high power, high-speed InP/InGaAsP capped mesa buried heterostructure distributed feedback lasers", *IEEE J. Quantum Electron.* **25** (1989) 2091–2095.
7. E. I. Gordon, "Optical master oscillators and noise", *Bell Syst. Tech. J.* (1964) 507–539.
8. B. W. Hakki and T. L. Paoli, "Gain spectra in GaAs double-heterojunction injection lasers", *J. Appl. Phys.* **46** (1975) 1299–1306.
9. D. T. Cassidy, "Technique for measurement of the gain spectra of semiconductor lasers", *J. Appl. Phys.* **56** (1984) 3096–3099.

10. L. A. Lam Sin Cho, P. M. Smowton, and B. Thomas, "Spectral gain measurements for semiconductor laser diodes", *IEE Proc. Part J.* **137** (1990) 64–68.
11. G. E. Shtengel and D. A. Ackerman, "Internal optical loss measurements in 1.3 μm InGaAsP lasers", *Electron. Lett.* **31** (1995) 1157–1159.
12. G. E. Shtengel, D. A. Ackerman, P. A. Morton, E. J. Flynn, and M. S. Hybertsen, "Impedance-corrected carrier lifetime measurements in semiconductor lasers", *Appl. Phys. Lett.* **67** (1995) 1506–1508.
13. D. A. Ackerman, G. E. Shtengel, M. S. Hybertsen, P. A. Morton, R. F. Kazarinov, T. Tanbun-Ek, and R. A. Logan, "Analysis of gain in determining T_0 in 1.3 μm semiconductor lasers", *IEEE J. Selected Topics in Quantum Electron.* **1** (1995) 250–263.
14. P. A. Morton, D. A. Ackerman, G. E. Shtengel, R. F. Kazarinov, M. S. Hybertsen, T. Tabun-Ek, R. A. Logan, and A. M. Sergent, "Material characteristics of 1.55 μm high-speed MQW lasers", *IEEE Photon. Tech. Lett.* **7** (1995) 833–835.
15. V. Kozlov, A. Salokatve, A. V. Nurmikko, D. C. Grillo, Li He, J. Han, Y. Fan, M. Ringle, and R. L. Gunshor, "Gain characteristics of blue/green II-VI quantum well diode lasers", *Appl. Phys. Lett.* **65** (1994) 1863–1864.
16. C. Sirtori, J. Faist, F. Capasso, D. L. Sivco, A. L. Hutchinson, and A. Y. Cho, "Pulsed and continuous-wave operation of long wavelength infrared ($\lambda = 9.3$ μm) quantum cascade lasers", *IEEE J. Quantum Electron.* **33** (1997) 89–93.
17. V. P. Gribkovskii and V. A. Samoilyukovich, "Determination of the amplification coefficient, luminescence power, and separation between Fermi quasilevels in injection lasers", *Zhurnal Prikladnoi Spektroskopii* **11** (1969) 170–172.
18. C. H. Henry, R. A. Logan, and F. R. Merrit, "Measurement of gain and absorption spectra in AlGaAs buried heterostructure lasers", *J. Appl. Phys.* **51** (1980) 3042–3050.
19. P. Blood, A. I. Kucharska, J. P. Jacobs, and K. Griffiths, "Measurement and calculation of spontaneous recombination current and optical gain in GaAs-AlGaAs quantum-well structures", *J. Appl. Phys.* **70** (1991) 1144–1156.
20. P. Rees and P. Blood, "Derivation of gain spectra of laser diodes from spontaneous emission measurements", *IEEE J. Quantum Electron.* **31** (1995) 1047–1050.
21. G. E. Shtengel, R. F. Kazarinov, G. L. Belenky, and C. L. Reynolds, Jr., "Wavelength chirp and dependence of carrier temperature on current in MQW InGaAsP lasers", *IEEE J. Quantum Electron.* **33** (1997) 1396–1402.
22. K. L. Shaklee and R. F. Leheny, "Direct determination of optical gain in semiconductors", *Appl. Phys. Lett.* **18** (1971) 475–477.
23. P. S. Gross and W. G. Oldham, "Theory of optical gain measurements", *IEEE J. Quantum Electron.* **11** (1975) 190–197.
24. O. Hildebrand, E. Goebel, and K. Loehnert, "Investigations on unsaturated optical gain spectra of GaAs/GaAlIAs DHS lasers", *J. Appl. Phys.* **15** (1978) 149–152.
25. R. O. Miles, M. A. Dupertuis, F. K. Reinhart, and P. M. Brosson, "Gain measurements in InGaAs/InGaAsP multiquantum-well broad area lasers", *IEE Proc. Part J.* **139** (1992) 33–38.
26. D. J. Bossert and D. Gallant, "Improved method for gain/index measurements of semiconductor lasers", *Electron. Lett.* **32** (1996) 338–339.
27. D. J. Bossert and D. Gallant, "Gain, refractive index and α-parameter in InGaAs-GaAs SQW broad-area lasers", *IEEE Photon. Tech. Lett.* **8** (1996) 322–324.
28. D. V. Donetsky, G. L. Belenky, D. Z. Garbuzov, H. Lee, R. U. Martinelly, G. Taylor, S. Luryi, and J. C. Connolly, "Direct measurements of heterobarrier leakage current and modal gain in 2.3 μm double QW p-substrate InGaAsSb/AlGaAsSb broad area lasers", *IEEE Electron. Lett.* **35** (1999) 298–299.

29. C. H. Henty, R. A. Logan, F. R. Merritt, and J. P. Luongo, "The effect of the intervalence band absorption on the thermal behavior of InGaAsP lasers", *IEEE J. Quantum Electron.* **19** (1983) 947–952.
30. J. R. Biard, W. N. Carr, and B. S. Reed, "Analysis of a GaAs laser", *Trans. Metallurg. Soc. AIME* **230** (1964) 286–290.
31. H. C. Casey, Jr., M. B. Panish, W. O. Schlosser, and T. L. Paoli, "GaAs-AlxGa1-xAs heterostructure laser with separate optical and carrier confinement", *J. Appl. Phys.* **45** (1974) 322–333.
32. P. M. Smowton and P. Blood, "On the determination of internal optical mode loss of semiconductor lasers", *Appl. Phys. Lett.* **70** (1997) 2365–2367.
33. P. A. Andrekson, N. A. Olsson, T. Tanbun-Ek, R. A. Logan, D. L. Coblentz, and H. Temkin, "Novel technique for determining internal loss of individual semiconductor laser", *Electron. Lett.* **28** (1992) 171–172.
34. E. A. Avrutin, I. E. Chebunina, I. A. Eliashevitch, S. A. Gurevich, and G. E. Shtengel, "TE and TM optical gains in AlGaAs/GaAs single-quantum-well lasers", *Semicond. Sci. Technol.* **8** (1993) 80–87.
35. L. J. P. Ketelsen, "Simple technique for measuring cavity loss in semiconductor lasers", *Electron. Lett.* **30** (1994) 1422–1424.
36. V. V. Luc, P. G. Eliseev, M. A. Manko, and G. T. Mikaelian, "Electrical diagnostics of the amplifier operation and a feasibility of signal registration on the basis of the voltage saturation effect in junction laser diodes", *IEEE J. Quantum Electron.* **19** (1983) 1080–1083.
37. A. G. P. Agrawal and N. K. Dutta, *Semiconductor Lasers*, Van Nostrand Reinhold, New York, 1993, p. 127.
38. H. Yamazaki M. Yamaguchi, and M. Kitamura, "Spectral linewidth rebroadening in MQW-DFB LDs caused by spontaneous emission noise in SCH/barrier layers", *IEEE Photon. Tech. Lett.* **6** (1994) 341–343.
39. H. Ishikawa and I. Suemune, "Analysis of temperature dependent optical gain of strained quantum well taking account of carriers in SCH layer", *IEEE Photon. Tech. Lett.* **6** (1994) 344–347.
40. M. Buijs, K. Shahzad, S. Flamholtz, K. Haberen, and J. Gaines, "Carrier leakage in blue-green II-VI semiconductor lasers", *Appl. Phys. Lett.* **67** (1995) 1987–1989.
41. B. Lane, D. Wu, H. J. Yi, J. Diaz, A. Rybaltowski, S. Kim, M. Erdtmann, H. Jeon, and M. Razeghi, "Study on the effect of minority carrier leakage in InAsSb/InPAsSb double heterostructure", *Appl. Phys. Lett.* **70** (1997) 1447–1449.
42. Liang-Hui Chen, J. C. V. Mattos, F. C. Prince, and N. B. Patel, "Spatially resolved observation of carrier leakage in 1.3-μm $In_{1-x}Ga_xAs_yP_{1-y}$ lasers", *Appl. Phys. Lett.* **44** (1984) 520–522.
43. S. Yamakoshi, T. Sanada, O. Wada, I. Umebu, and T. Sakurai, "Direct observation of electron leakage in InGaAsP/InP double heterostructure", *Appl. Phys. Lett.* **40** (1982) 144–146.
44. A. A. Bernussi, H. Temkin, D. L. Coblentz, and R. A. Logan, "Effect of barrier recombination on the high temperature performance of quaternary multiquantum well lasers", *Appl. Phys. Lett.* **66** (1995) 67–69.
45. D. Garbuzov, G.-J. Shiau, V. Bulovic, M. Boroditsky, C.-P. Chao, and S. R. Forrest, "Photoluminescence study of excess carrier spillover in 1.3 μm wavelength strained multi-quantum-well InGaAsP/InP laser structures", *Appl. Phys. Lett.* **66** (1995) 1307–1309.
46. E. J. Flynn and D. A. Ackerman, "Electron leakage and the excess voltage at *p-P* heterojunctions in InGaAsP/InP lasers", *IEDM 1997*, 15.3.1, p. 389.

47. T. R. Chen, S. Margalit, U. Koren, K. L. Yu, L. C. Chiu, A. Hasson, and A. Yariv, "Direct measurement of the carrier leakage in an InGaAsP/InP laser", *Appl. Phys. Lett.* **42** (1983) 1000–1002.
48. G. L. Belenky, R. F. Kazarinov, J. Lopata, S. Luryi, T. Tanbun-Elk, and P. A. Garbinski. "Direct measurement of the carrier leakage out of the active region in InGaAsP/InP laser heterostructures", *IEEE Trans. Electron. Devices* **42** (1995) 215–218.
49. G. L. Belenky, C. L. Reynolds, D. V. Donetsky, G. E. Shtengel, M. Hybertsen, M. A. Alam, G. A. Baraff, R. K. Smith, R. F. Kazarinov, J. Winn, and L. E. Smith, "Role of *p*-doping profile and regrowth on the static characteristics of 1.3 μm MQW InGaAsP/InP lasers. Experiment and modeling", *IEEE J. Quantum Electron.* **35** (1999) 1515–1520.
50. G. L. Belenky, D. V. Donetsky, C. L. Reynolds, Jr., R. F. Kazarinov, G. E. Shtengel, S. Luryi, and J. Lopata, "Temperature performance of 1.3 μm InGaAsP-InP lasers with different profile of *p*-doping", *IEEE Photon. Tech. Lett.* **9** (1997) 1558–1560.
51. R. F. Kazarinov and M. R. Pinto, "Carrier transport in laser heterostructures", *IEEE J. Quantum Electron.* **30** (1994) 49–53.
52. H. Statz and G. deMars, in *Quantum Electronics*, ed. C. H. Towns, Columbia Univ. Press, New York, 1960, p. 235.
53. D. A. Kleinman, *Bell. Syst. Tech. J.* **43** (1964) 1505.
54. T. L. Koch and R. A. Linke, "Effect of nonlinear gain reduction on semiconductor laser wavelength chirping", *Appl. Phys. Lett.* **48** (1986) 613–615.
55. R. Nagarajan, M. Ishikawa, T. Fukushima, R. S. Geels, and J. E. Bowers, "High speed quantum-well lasers and carrier transport effects", *IEEE J. Quantum Electron.* **28** (1992) 1990–2008.
56. S. Weisser, I. Esquivias, P. J. Tasker, J. D. Ralston, B. Romero, and J. Rosezweig, "Impedance characteristics of quantum-well lasers", *IEEE Photon. Tech. Lett.* **6** (1994) 1421–1423.
57. I. Esquivias, S. Weisser, B. Romero, J. D. Ralston, and J. Rosezweig, "Carrier capture and escape times in $In_{0.35}Ga_{0.65}As$-GaAs multiquantum-well lasers determined from high-frequency electrical impedance measurements", *IEEE Photon. Tech. Lett.* **8** (1996) 1294–1296.
58. R. Olshansky, C. B. Su, J. Manning, and W. Powazinik, "Measurement of radiative and nonradiative recombination rates in InGaAsP and AlGaAs light sources", *IEEE J. Quantum Electron.* **20** (1984) 838–854.
59. R. Olshansky, J. LaCourse, T. Chow, and W. Powazinik, "Measurement of radiative, auger, and nonradiative currents in 1.3 μm InGaAsP buried heterostructure lasers", *Appl. Phys. Lett.* **50** (1987) 310–312.
60. G. H. B. Thompson, "Analysis of radiative and nonradiative recombination law in lightly doped InGaAsP Lasers", *Electron. Lett.* **19** (1983) 154–155.
61. K. Konnerth and C. Lanza, "Delay between current pulse and light emission of a gallium arsenide injection laser", *Appl. Phys. Lett.* **4** (1964) 120–121.
62. Y. Zou, J. S. Osinski, P. Godzinski, P. D. Dapkus, W. C. Rideout, W. F. Sharfin, J. Schlafer, and F. D. Crawford, "Experimental study of 1.55 mm compressively strained semiconductor lasers", *IEEE J. Quantum Electron.* **29** (1993) 1565–1575.
63. J. Katz, S. Margalit, C. Harder, D. Wilt and A. Yariv, "The intrinsic equivalent circuit of a laser diode", *IEEE J. Quantum Electron.* **17** (1981) 4–7.
64. R. S. Tucker, "Circuit model of double-heterojunction laser below threshold", *IEE Proc. Part I.* **128** (1981) 101–106.

65. E. J. Flynn, "Carrier-density-independent radiative constant in 1.3 μm buried heterostructure lasers", *J. Appl. Phys.* **78** (1995) 4046–4052.
66. G. E. Shtengel, D. A. Ackerman, and P. A. Morton, "True carrier lifetime measurements of semiconductor laser", *Electron. Lett.* **31** (1995) 1747–1748.
67. M. Morishita, T. Ohmi, and J. Nishizawa, "Impedance characteristics of double-heterostructure laser diodes", *Solid-State Electron.* **22** (1979) 951–962.
68. W. B. Joyce and R. W. Dixon, "Electrical characterization of heterostructure lasers", *J. Appl. Phys.* **49** (1978) 3719–3728.
69. S. C. Kan and K. Y. Lau, "Intrinsic equivalent circuit of quantum-well lasers", *IEEE Photon. Tech. Lett.* **4** (1992) 528–530.
70. R. S. Tucker, "High-speed modulation of semiconductor lasers", *J. Lightwave Technol.* **3** (1985) 1180–1192.
71. J. E. Bowers, "High-speed semiconductor laser design and performance", *Solid State Electron.* **30** (1987) 1–11.
72. T. Fukushima, R. Nagarajan, M. Ishikawa, and J. E. Bowers, "High-speed dynamics in InP based multiple quantum well lasers", *Jpn. J. Appl. Phys.* **32** (1993) 70–83.
73. R. Olshansky, P. Hill, V. Lanisera, and W. Powazinik, "Frequency response of 1.3 μm InGaAsP, high speed semiconductor lasers", *IEEE J. Quantum Electron.* **23** (1987) 1410–1418.
74. P. A. Morton, T. Tanbun-Ek, R. A. Logan, A. M. Sergent, P. F. Sciortino, Jr., and D. L. Coblentz, "Frequency response substraction for simple measurement of intrinsic laser dynamic properties", *IEEE Photon. Tech. Lett.* **4** (1992) 133–136.
75. G. L. Belenky, C. L. Reynolds, L. Shterengas, D. V. Donetsky, G. E. Shtengel, M. Hybertsen, R. F. Kazarinov, and S. Luryi, "High-speed dynamics in 1.3 μm InGaAsP/InP multiple quantum well DFB lasers with different *p*-doping profiles", unpublished.
76. L. D. Westbrook, N. C. Fletcher, D. M. Cooper, M. Stevenson, and P. C. Spurdens, "Intensity noise in 1.55 μm GaInAs quantum well buried heterostructure lasers", *Electron. Lett.* **25** (1989) 1183–1184.
77. M. C. Tatham, I. F. Lealman, C. P. Seltzer, L. D. Westbrook, and D. M. Cooper, "Resonance frequency, damping and differential gain in 1.55 μm multiple quantum-well lasers", *IEEE J. Quantum Electron.* **28** (1992) 408–414.
78. H. Lu, G. P. Li, and T. Makino, "High-speed performance of partly gain-coupled 1.55-μm strained layer multiple-quantum-well DFB lasers", *IEEE Photon. Tech. Lett.* **8** (1993) 861–863.
79. A. Wakatsuki, Y. Kawamura, Y. Noguchi, and H. Iwamura, "Effect of conduction-band discontinuity on lasing characteristics of 1.5 μm InGaAs/In(Ga)AlAs MQW-FP lasers", *IEEE Photon. Tech. Lett.* **4** (1993) 383–386.
80. M. W. Fleming and A. Mooradian, "Fundamental Line broadening of single mode GaAlAs diode lasers", *Appl. Phys. Lett.* **38** (1981) 511–513.
81. C. H. Henry, "Theory of the linewidth of semiconductor lasers", *IEEE J. Quantum Electron.* **18** (1982) 259–264.
82. K. Vahala and A. Yariv, "Semiclassical theory of noise in semiconductor lasers", *IEEE J. Quantum. Electron.* **19** (1983) 1096–1109.
83. C. Harder, K. Vahala, and A. Yariv, "Measurement of the linewidth enhancement factor α of semiconductor lasers", *Appl. Phys. Lett.* **42** (1983) 328–330.
84. F. Devaux, Y. Sorel, and J. F. Kerdilis, "Simple measurement of fiber dispersion and of chirp parameter of intensity modulated light emitter", *J. Lightwave Technol.* **11** (1993) 1937–1940.

85. I. D. Henning and J. V. Collins, "Measurements of the semiconductor laser linewidth broadening factor", *Electron. Lett.* **19** (1983) 927–929.
86. L. D. Westbrook and B. Eng, "Measurements of dg/dN and dn/dN and their dependence on photon energy in $\lambda = 1.55$ μm InGaAsP laser diodes", *IEE Proc. Part J.* **133** (1986) 135–142.
87. W. Rideout, B. Yu, J. LaCourse, P. K. York, K. J. Beernink, and J. J. Coleman, "Measurement of the carrier dependence of differential gain, refractive index, and linewidth enhancement factor in strained-layer quantum well lasers", *Appl. Phys. Lett.* **56** (1990) 706–708.
88. A. S. Schönfelder, S. Weisser, J. D. Ralston, and J. Rosenzweig, "Differential gain, refractive index, and linewidth enhancement factor in high-speed GaAs-based MQW lasers: Influence of strain and *p*-doping", *IEEE Photon. Tech. Lett.* **6** (1994) 891–893.
89. V. I. Tolstikhin and M. Willander, "Carrier heating effects in dynamic-single-frequency GaInAsP-InP laser diodes", *IEEE J. Quantum Electron.* **31** (1995) 814–833.
90. J. Shah, "Photoexcieted hot carriers: From CW to 6fs in 20 years", *Solid-State Electron.* **32** (1989) 1051–1056.
91. J. Shah, R. F. Leheny, and R. E. Nahory, "Hot-carrier effects in 1.3 μm $In_{1-x}Ga_xAs_yP_{y-1}$ light emitting diodes", *Appl. Phys. Lett.* **39** (1981) 618–620.
92. C. H. Henry, R. A. Logan, H. Temkin, and F. R. Merrit, "Absorption, emission, and gain spectra in 1.3 μm InGaAsP quaternary lasers", *IEEE J. Quantum Electron.* **19** (1983) 941–946.
93. D. Z. Garbuzov, V. P. Chalyi, A. E. Svelokuzov, V. B. Khalfin, and A. L. Ter-Martirosyan, "Auger recombination and heating of carriers at high rates of photoexcitation of InGaAsP/InP ($\lambda = 1.3$ μm) and InGaAsP/GaAs ($\lambda = 0.85$ μm) quantum-well heterostructures", *Sov. Phys. Semicond.* **22** (1988) 410–413.
94. G. L. Belenky, P. A. Garbinski, S. Luryi, M. Mastrapasqua, A. Y. Cho, R. A. Hamm, T. R. Hayes, E. J. Laskowski, D. L. Sivco, and P. R. Smith, "Collector-up-light-emitting charge injection transistors in *n*-InGaAs/InAlAs/*p*-InGaAs and *n*-InGaAs/*p*-InGaAs heterostructures", *J. Appl. Phys.* **73** (1993) 8618–8627.
95. G. L. Belenky, A. Kastalsky, S. Luryi, P. A. Garbinski, A. Y. Cho, and D. L. Sivco, "Measurement of the effective temperature of majority carriers under injection of hot minority carriers in heterostructures", *Appl. Phys. Lett.* **64** (1994) 2247–2249.
96. R. F. Kazarinov, C. H. Henry, and R. A. Logan, "Longitudinal mode self-stabilization in semiconductor lasers", *J. Appl. Phys.* **53** (1982) 4631–4644.
97. K. L. Hall, J. Mark, E. P. Ippen, and G. Eisenstein, "Femtosecond gain dynamics in InGaAsP optical amplifiers", *Appl. Phys. Lett.* **56** (1990) 1740–1742.
98. M. S. Stix, M. P. Kesler, and E. P. Ippen, "Observation of subpicosecond dynamics in GaAlAs laser diodes", *Appl. Phys. Lett.* **48** (1986) 1722–1724.
99. K. L. Hall, G. Lenz, A. M. Darwish, and E. P. Ippen, "Subpicosecond gain and index nonlinearities in InGaAsP diode lasers", *Opt. Commun.* **111** (1994) 589–612.
100. C. T. Hultgren and E. P. Ippen, "Ultrafast refractive index dynamics in AlGaAs diode laser amplifiers", *Appl. Phys. Lett.* **59** (1991) 635–637.
101. D. M. Baney and W. V. Sorin, "Measurement of a modulated DFB laser spectrum using gated delayed self-homodyne technique", *Electron. Lett.* **24** (1988) 669–670.
102. V. D. Pishchalko and V. I. Tolstikhin, "Carrier heating effects in InGaAsP/InP injection heterolasers", *Sov. Phys. Semicond.* **24** (1990) 288–294.

International Journal of High Speed Electronics and Systems, Vol. 9, No. 4 (1998) 941–978

LATERAL INJECTION LASERS

EDWARD (TED) H. SARGENT and J. M. (JIMMY) XU
Department of Electrical and Computer Engineering,
University of Toronto, Toronto, Ontario, Canada M5S 1A1

Since its invention in the 60's, the semiconductor laser has relied on vertical injection of electrons and holes in order to pump the active region to inversion. While many aspects of semiconductor laser development have seen tremendous innovation and marked progress, this vertical injection scheme has remained unchanged. In contrast, the electronic transistor has evolved from the early dominance of vertical injection to today's largely lateral injection CMOS platform, which yielded new performance, functionality, and integrability and enabled the astonishingly successful integrated circuit revolution.

The lateral current injection laser offers advantages which go far beyond optoelectronic integrability: it enables new functionalities, post-fabrication processing and engineering, and in some aspects of normal operation it can lead to greatly improved performances. To realize its full potential, this new class of semiconductor lasers deserves the same attention and the intensive iterations of efforts in theory, design, fabrication, and experimental probing as its vertical cousin has been received.

We introduce a first-principles, combined theoretical-experimental approach to the exploration of the LCI laser. It reveals the role of lateral ambipolar drift-diffusion of carriers in either hindering or, potentially, in enhancing the provision of gain to the desired optical mode; of two-dimensional bandstructure engineering of the injection path and active region to achieve low differential resistance and efficient modal gain provision; of selective formation and lateral positioning of heterojunction in enabling high-efficiency device operation; and of adiabatic injection of electrons and holes across diffusively-graded heterojunctions in facilitating carrier capture into 2D quantum well states.

Our results point to the tremendous promise of lasers based on lateral injection of current in enabling vertical cavity lasers with vastly increased performance and functional diversity; multi-terminal devices such as lasers with capacitive gain and wavelength tunability; high-speed directly-modulated lasers with reduced chirp; and functional devices with directly integrated high-speed longitudinal modulation.

1. Carrier Injection in Semiconductor Lasers: A Perspective

Electronic device technology underwent long ago an evolution from the vertical injection paradigm of the bipolar junction transistor to the lateral approach of field effect transistor technology. Semiconductor lasers have, in contrast, remained over thirty years dominated, and to a large extent constrained, by vertical injection technology. Countless cycles of complementary theoretical and experimental investigations have vastly improved the performance of vertical injection lasers and increased the sophistication of designs in the vertical and longitudinal dimensions.

In fact, the vertical injection paradigm is inherently flawed from the point of view of integration: as one example, an integrated heterodyne receiver recently demonstrated required 150 processing steps.[1] The exploration of the lateral degree of freedom has remained dormant, and essentially unchanged, for the past two decades. As a result, the lateral current injection (LCI) laser has yet to become a serious contender in a field dominated by its greatly perfected vertical injection cousin.

This situation could change dramatically as demand becomes more pressing for optoelectronic integrated circuits (OEICs), an application area in which LCI lasers are natural candidates.

Ease of integration is only one of many instances of the great potential of the LCI laser which results from the release of a new degree of freedom. For example, under lateral current injection, electrical and optical confinement may be decoupled, enabling the use of non-conducting cladding layers and a semi-insulating substrate. This decoupling, and the allowance for the use of non-conductive cladding layers, enable the possibilities of modulation/tuning of gain, power, and wavelength via the use of a gate.

Post-fabrication processing via deposition and/or etching from the top is now not only permitted but also readily doable, which allows us to add or extract different functionalities such as a dielectric DFB laser, a DBR-VCSEL, laser-transistor, and photonic wiring.

If one may extrapolate from the rate and richness of progress in the more constrained paradigm of vertical injection lasers, the exploration of an additional degree of freedom holds great promise for both improved performance in normal laser operation and new device possibilities.

In this work, we begin by motivating in further detail the promise of lateral injection of current. We employ arguments both general (e.g. the new design possibilities resulting from the release of a new spatial degree of freedom) and specific (e.g. the possibility of improved carrier capture via an adiabatic energetic transition from cladding to quantum wells, enabling improved high-speed laser performance). We then proceed to discuss new issues in semiconductor laser theory which arise under lateral injection of current. This is a vast and ever-shifting territory: every answer to an important question in LCI laser theory and design bring up ten further questions of still greater interest and promise.

We discuss new fabrication process challenges which arise in seeking to fabricate LCI lasers — and, in particular, in developing a process toolkit which could form a basic for monolithic, integrated optoelectronics. We report new experimental results of LCI lasers which we have fabricated, and discuss these in terms of emerging LCI laser theory. Finally, we return to some of our original hypothesis, motivations, and questions — and summarize with a sampling of promising directions for future work.

2. Motivation for Lateral Injection of Current in Semiconductor Lasers

In present-day lasers, current is injected predominantly in the vertical direction (Fig. 1(a)). Injecting current laterally (Fig. 1(b)) opens up a new spatial degree of freedom for exploration and, as presented below, contributes to enhancing device performance in conventional roles; enabling monolithic optoelectronic integrated circuits; and enabling novel functional devices.

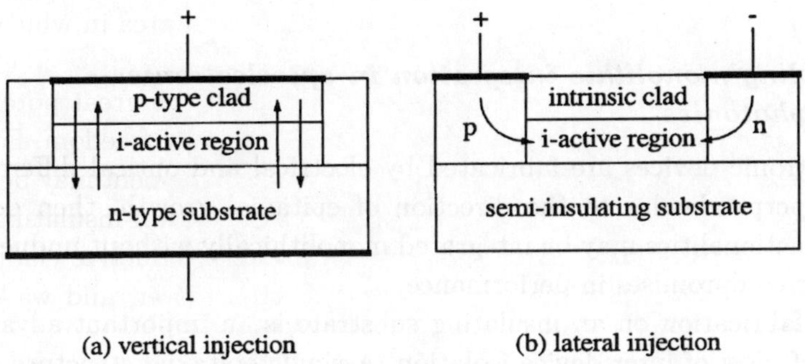

Fig. 1. Vertical and lateral injection concepts.

2.1. *Improving device operation in conventional roles*

Lateral injection of current opens up new possibilities for designing devices for use in traditional applications.

The high intrinsic parasitic capacitance of vertical lasers results from heavy doping throughout the epitaxial layer structure and limits the speed of direct modulation of device output. By enabling a decoupling of vertical and lateral directions, lateral current injection permits dopants to be added only at the level and locations required.

This decoupling has other consequences. In the lateral injection paradigm, the vertical structure of a device may be designed so as to optimize confinement in that direction — for example, the vertical heterostructure may be chosen with a view to achieving a circular mode profile or to devising quantum-confined electrical properties which are particularly conducive to high-speed operation. In vertical injection lasers it is necessary to design the multi-quantum well heterostructure with a view to ensuring an even distribution of carriers among the wells and to minimise loss and heating. In contrast, the design of the electrical and optical structure of an LCI laser is much less constrained, since the lateral direction assumes primary responsibility for carrier injection.

In LCI lasers based on multi-quantum well active regions, increased differential gain and high modulation bandwidth may be obtained without compromising injection uniformity. Nonlinear modulation characteristics in vertical lasers with large numbers of quantum wells result, in large part, from difficulty in injecting holes evenly into the different parts of the structure. Clearly, it is advantageous to use the vertical structure for optical and electronic confinement and the lateral direction for smooth injection of electrons and holes parallel to the quantum wells.

The potential for the use of lateral current injection lasers in new explorations of laser operation is another excellent prospect. The lack of obstruction (such as a metallic contact) above the active region allows for direct optical access to the heart of the device. Light may be injected or extracted in a precise, spatially selective manner along a new direction.

2.2. *Enabling monolithic integration in optoelectronics and photonics*

If optoelectronic devices are fabricated by electrical and optical differentiation in the plane perpendicular to the direction of epitaxial growth, then devices with differing functionalities may be integrated monolithically without undue processing demands or compromises in performance.

Device fabrication on an insulating substrate is an important advantage from the point of view of inter-device isolation. A single epitaxial structure compatible with both laser and transistor operation may then be used to form optoelectronic integrated circuits — aggregates of initially independent, then selectively interconnected, electronic, and optical devices. Planar integration is, in turn, compatible with the formation of microwave stripline interconnections on the surface of the substrate.

The potential and demand for integration extends beyond the range of electronics and optoelectronics into the realm of photonic integration. As an illustration, conventional semiconductor lasers typically rely on the use of a vertical facet between semiconductor and air in order to create high-reflectivity mirrors for optical feedback and lasing. In a photonic integrated circuit, the use of such facets would necessitate sending light from the optical source into free space, only to be recaptured into the waveguide at a later point. This cannot typically be done without significant loss of optical power.

Ideally, photonic interconnections — analogous to the role played by wires in an electronic circuit — would be used to link monolithically integrated photonic devices. The LCI laser offers renewed opportunities for realizing this goal because it does not suffer from the constraint present in vertical injection lasers that the cladding material participate in both optical confinement and carrier injection. Instead, dielectric optical waveguides could be deposited and modified following the elaboration of the electrical structure on the planar substrate, as depicted in Fig. 2.

A generic lateral injection *p–n* or *p–i–n* laser structure might be coupled with a resonant cavity to form a Fabry–Perot laser; a longitudinal grating to provide wavelength-selective feedback distributed along the length of the cavity (a distributed feedback, or DFB, laser); or a vertical distributed Bragg reflector (DBR) to form a surface-emitting laser. A longitudinal photodetector could be

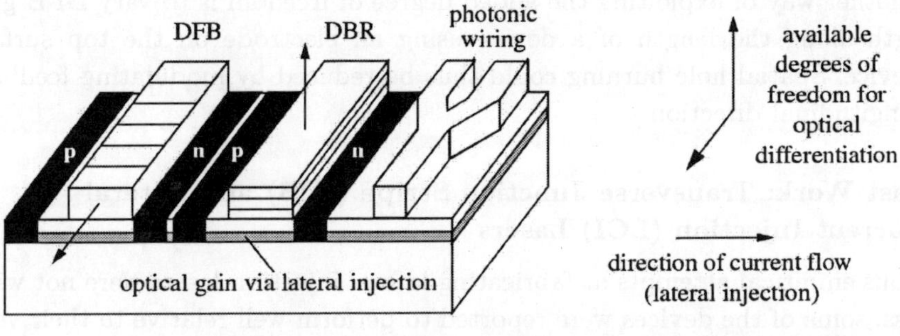

Fig. 2. Photonic integrated circuit built atop the lateral injection paradigm.

created by coupling light into a waveguide with an anti-reflection coated facet. These photonic devices, formed using the same basic electronic structure, would be coupled together using "photonic wires", in analogy with electronic wiring.

2.3. *Enabling functional devices*

Lateral injection offers new possibilities for the functional photonic devices. A hybrid of a number of functional device ideas is illustrated in Fig. 3.

Since lateral injection of current releases an additional spatial degree of freedom, the vertical dimension may be used for capacitive modulation of the gain in the active region while the lateral direction is used for current injection. Were capacitive modulation simply a mechanism for controlling the injection of carriers into the active region, it could be viewed as an elegant approach to integrating a laser with a transistor, in itself a desirable prospect.

In fact, injection modulation is only one of a number of possibilities enabled by lateral injection. The electronic properties of the active region may be modified by establishing an electric field across the active region, allowing the shape of the quantum wells and of their confined electronic states to be tailored. By modifying the optical gain spectrum in this manner, capacitive tuning of the lasing wavelength could be achieved.

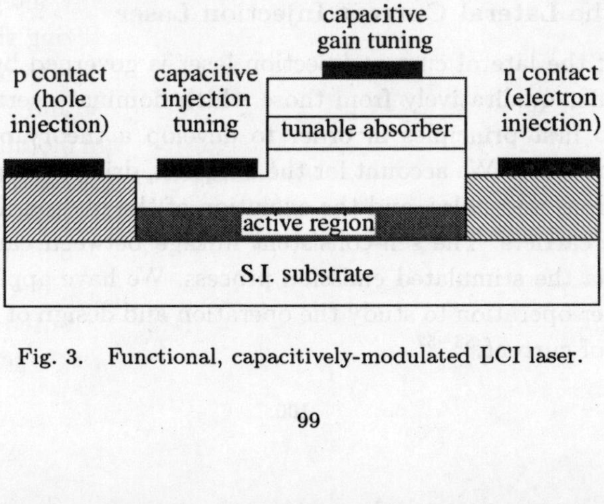

Fig. 3. Functional, capacitively-modulated LCI laser.

Another way of exploiting the added degree of freedom is to vary DFB grating strength along the length of a device using an electrode on the top surface of the device. Spatial hole burning could thus be reduced by modulating feedback in the longitudinal direction.

3. Past Work: Transverse Junction Stripe (TJS) and Lateral Current Injection (LCI) Lasers

Previous empirical attempts at fabricating lateral injection lasers were not without success: some of the devices were reported to perform well relative to their vertical injection counterparts. On the other hand, a number of fabrication trials yielded devices whose performance was worse than anticipated and was not explained on physical grounds.[2–52]

There existed prior to this work no systematic attempt at understanding the nature and importance of the physical mechanisms which underlie the operation of the lateral current injection laser. The issues of nonuniform gain in these lasers is one example. Researchers at Fujitsu speculated and at University of Toronto showed that since injecting *p*- and *n*- contacts are separated by a much greater distance in LCI lasers than in conventional vertical injection lasers, the lateral carrier density distribution may be nonuniform, and this may impede device performance. Neither provided a quantitative estimate of the contact separation at which carrier density becomes significantly nonuniform. Neither sought to establish a clear relationship describing interactions between the optical mode and carrier density profiles. Neither clarified the role of either gain nonuniformity or gain asymmetry in limiting performance, either below or above the lasing threshold.

Without a clear understanding of the importance of various mechanisms, the only method of designing improved devices would be one of trial-and-error, either in the form of further empirical attempts or additional, unguided numerical simulations. Without access to generalized, physically-motivated relationships between structure and performance, fabrication trials would provide little information as to the underlying mechanisms at work, little insight would be accrued, and little guidance would be made available for future reference.

4. Theory of the Lateral Current Injection Laser

The operation of the lateral current injection laser is governed by physical mechanisms which differ qualitatively from those which dominate vertical devices. We have returned to first principles in order to develop a theory of lateral current injection laser operation. We account for the diffusion, drift, generation and recombination of electrons and holes, and the evolution of the photon field in a medium containing these carriers. The self-consistent linkage between carrier and photon density occurs via the stimulated emission process. We have applied this physical model of LCI laser operation to study the operation and design of devices based on lateral injection of current.[53–57]

Interband lateral injection lasers, like their vertical injection counterparts, rely on forward biasing of a bipolar junction in order to generate optical gain. Population inversion, as described by the condition of Bernard and Duraffourg, may naturally be expected to be a requirement for net stimulated emission and lasing. However, the simple form of this condition will not exist if there exists a variation in the degree of inversion over the lateral extent of the active region. The conditions for, and consequences of, a nonuniform gain distribution demand a quantitative analysis.

On a similar note, purely intuitive arguments may not satisfactorily resolve the question of the need for built-in lateral carrier confinement in the form of heterojunctions. Upon first consideration, one might suppose that if heterojunctions were necessary for room temperature lasing of vertical injection lasers, so too would they be needed for room-temperature operation of lateral injection lasers. On the other hand, the much greater spatial separation of electron and hole injectors might, on first consideration, suggest that carriers would have ample opportunity to recombine radiatively before leaving the active area. Through development of first-principles theory, we provide an intuitive explanation for the correct answer — an answer which is, in our view, not self-evident without proper fundamental and self-consistent treatment.

4.1. *Theory of ambipolar drift and diffusion*

When a forward bias is applied across a *p–n* or *p–i–n* (hetero) junction, holes are injected from the *p*-type material and electrons from the *n*-type material into the ambipolar active region. In a *p–i–n* structure, an accumulation of holes near the *p–i* interface and an accumulation of electrons at the *n–i* interface would set up an internal field which would induce carriers to drift towards their respective far interfaces. Self-consistent account must be taken of the establishment of any internal electric field. Recombination processes represent a further consideration: if electrons and holes are present in the same spatial region, they may interact through various mechanisms in such a way as to recombine, the momentum and energy being taken away by photons and phonons. In the case of stimulated emission, there is a mutual coupling between the densities of carriers and photons. Electron and hole flow in the presence of concentration gradients (diffusion) and electric fields (drift) must be considered self-consistently with carrier recombination; with the role of a photon field built up by stimulated recombination and optical feedback; and with the development of an internal electric field, as described by Poisson's equation. One such fully self-consistent solution for a particular structure is depicted for illustrative purposes in Fig. 4.[58]

The complete, exact system of equations is not solvable analytically. In the interest of obtaining analytical solutions (helpful in allowing clarification of basic physical issues, though not always quantitatively exact), the assumption of quasi-charge neutrality, also known as the electrical neutrality condition or charge balance

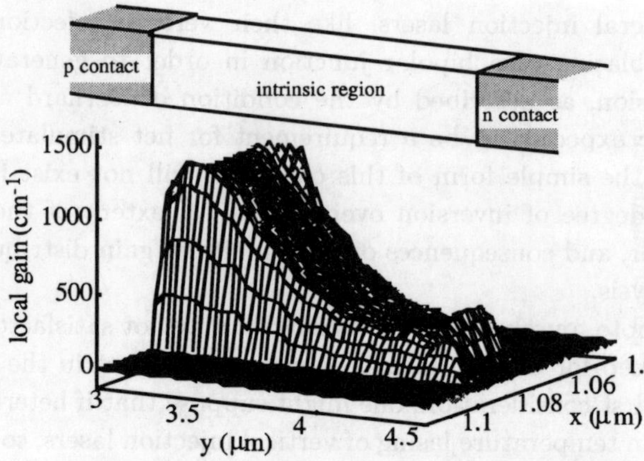

Fig. 4. Lateral carrier density cross-section.

assumption, is introduced. It is not to say that no internal electric field is established. In fact, an electric field does indeed arise from the difference in carrier mean free paths: it tends to keep the excess carrier concentration of electrons and holes approximately equal. The quasi-charge neutrality assumption is valid as long as the the local differences between electron and hole densities required to maintain the internal electric field are not large compared to the total excess carrier densities.

Within the quasi-charge neutrality assumption, the explicit dependence on electric field may be eliminated, and a differential equation obtained in which only the carrier density appears explicitly. The result takes the form of a simple diffusion equation:

$$\frac{d^2n}{dx^2} = \frac{\Re}{D}$$

where the ambipolar diffusivity D is related to electron and hole diffusivies and the electron-to-hole mobility ratio b according to:

$$D = \frac{D_n + bD_p}{b+1}.$$

As long as the excess carrier densities significantly exceed the equilibrium carrier densities, the assumptions made and equations developed to this point are justified. These approximations are consistent with modest doping levels and high injection conditions inside a typical semiconductor laser. We illustrate typical solutions and their deviations from ideal quasi-charge neutrality in Fig. 5.

In order to explore the roles of electrons and holes in contributing to the ambipolar diffusivity, we rewrite this quantity:

$$D = \frac{2}{b+1}D_n = \frac{2}{1+1/b}D_p.$$

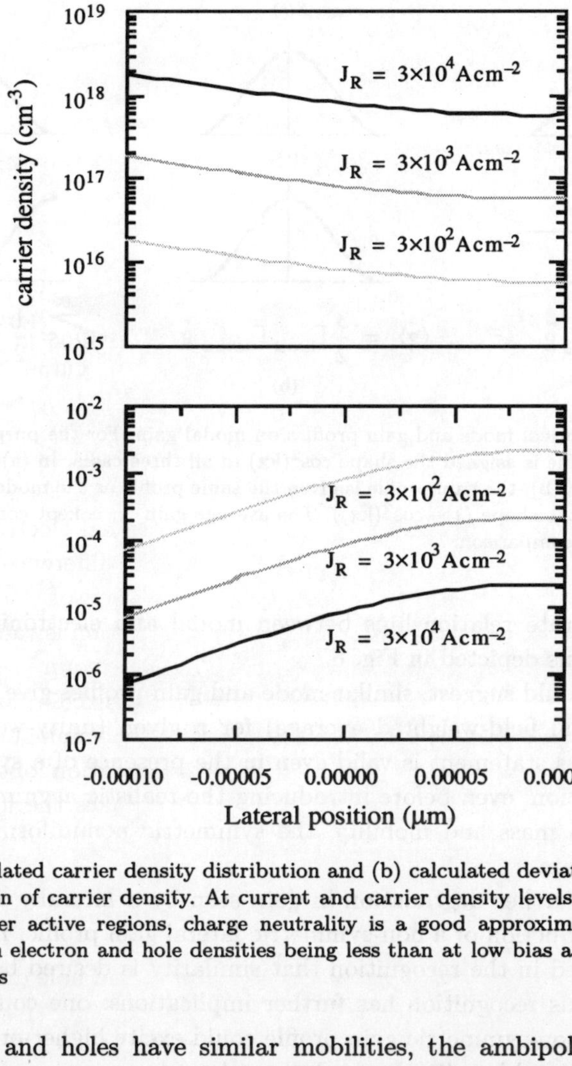

Fig. 5. (a) Calculated carrier density distribution and (b) calculated deviation from charge neutrality as a fraction of carrier density. At current and carrier density levels of interest in typical semiconductor laser active regions, charge neutrality is a good approximation, with the fractional difference in electron and hole densities being less than at low bias and less than near the threshold-level bias

If electrons and holes have similar mobilities, the ambipolar diffusivity will equal their common diffusivity. If holes are much less mobile than electrons, then $D = 2D_p$. The heavier carrier — the limiting carrier — will govern the process of ambipolar drift-diffusion. Electrons, typically much more mobile than holes, will flow to wherever they are needed to pair up with holes and satisfy quasi-charge neutrality. Holes will linger near to their point of injection.

4.2. *Modal gain in the presence of a laterally nonuniform carrier density profile*

In an active region with laterally nonuniform carrier density and gain distribution, the notion of modal gain will hold, but the notion of lateral confinement factor is not immediately useful in describing modal gain. We consider, as an extreme example

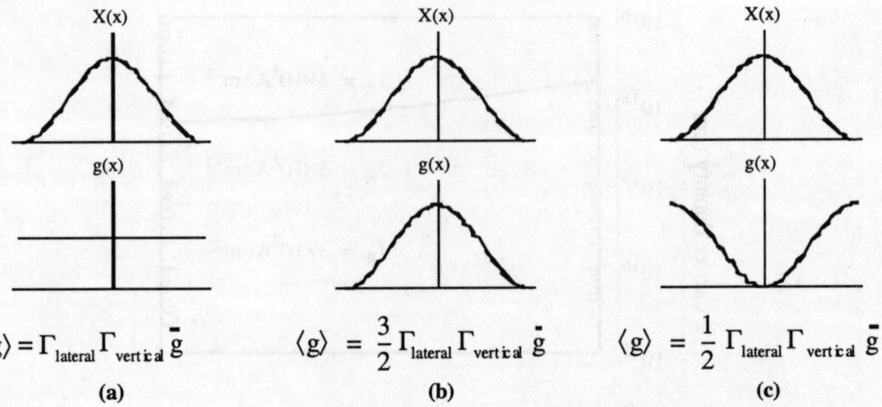

Fig. 6. Impact of optical mode and gain profiles on modal gain. For the purpose of calculation, the lateral mode profile is asigned the shape $\cos^2(kx)$ in all three cases. In (a), the gain profile is assumed to be flat. In (b), the gain profile is given the same profile as the mode profile. In (c), the gain profile is given the shape $(1 - \cos^2(kx))$. The average gain, g, is kept constant from (a)–(c) in order to facilitate comparison.

designed to illustrate relationships between modal and electronic structure, the three configurations depicted in Fig. 6.

As intuition would suggest, similar mode and gain profiles give rise to a greater modal gain (modal field-weighted average) for a given (unity-weighted) average material gain. This statement is valid even in the presence of a symmetric carrier and gain distribution: even before introducing the realistic asymmetry in electron and hole effective mass and mobility, the symmetric nonuniformity of Fig. 6(c) degrades modal gain.

As expected, further degradation in gain seen by a laterally symmetric mode results from introduction of a non-symmetric lateral gain profile. In fact, this realization is embodied in the recognition that similarity is desired between gain and mode profiles. This recognition has further implications: one could imagine that a given laterally non-symmetric gain profile could excite higher-order modes more efficiently than it would excite the fundamental mode.

It is worth noting that, in fact, the former scenario is *not* a consequence of the antisymmetric distribution of the lateral field. The optical gain profile which gives rise to stimulated emission interacts with the photon density distribution rather than with the field distribution (Fig. 7).

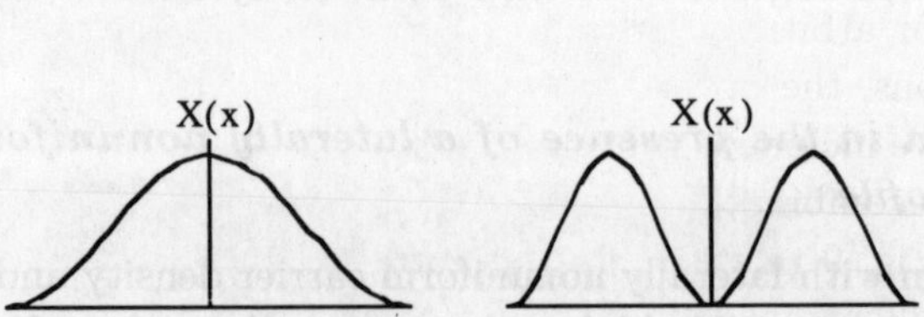

Fig. 7. Intensity profiles of the fundamental and first order lateral modes of a waveguide.

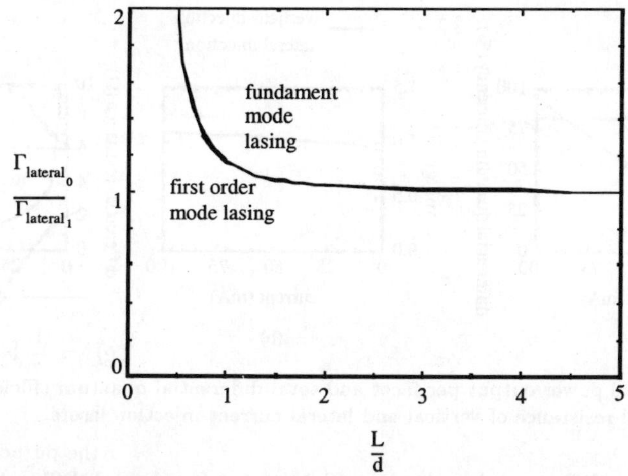

Fig. 8. Ratio of fundamental to first-order mode lateral confinement factor required for the two modes to experience the same net modal gain.

In Fig. 8 we illustrate calculated conditions under which fundamental- and first-order modes will lase in a device described by a single ambipolar diffusion length L_a and active region width d.

We have shown more recently[59] that LCI lasers which exploit the effects of lateral nonuniformities produced by ambipolar drift-diffusion are capable of exhibiting novel optoelectronic functionality while preserving a high lasing efficiency.

4.3. *Performance implications of lateral gain nonuniformity*

The implications of reduced overlap between lasing mode and the gain-providing carrier profile are qualitatively clear: threshold currents will be increased, and efficiencies reduced. This is not necessarily to say, however, that the quantitative impact of this effect will be important — even measurable — in any and all lateral injection lasers.

In order to highlight the practical implications of lateral gain uniformity on performance, we have made a detailed study of comparable lateral and vertical injection structures.[60] In order to include all key quantitatively important mechanisms, we simulated two closely comparable devices using our fully self-consistent two-dimensional finite-element simulator.[61] This permits us to solve self-consistently for arbitrary structures Poisson's equation, the electron and hole continuity equations, the optical wave equation, and the photon rate equation in two dimensions. We include in our model spontaneous, stimulated, Shockley–Read–Hall, and Auger recombination.

We report in Fig. 9 relationships between current, voltage, and output optical power for a vertical current injection (VCI) and LCI laser, both emitting at 1.3 μm and having active region width 1.1 μm. Given in the figure are the calculated optical

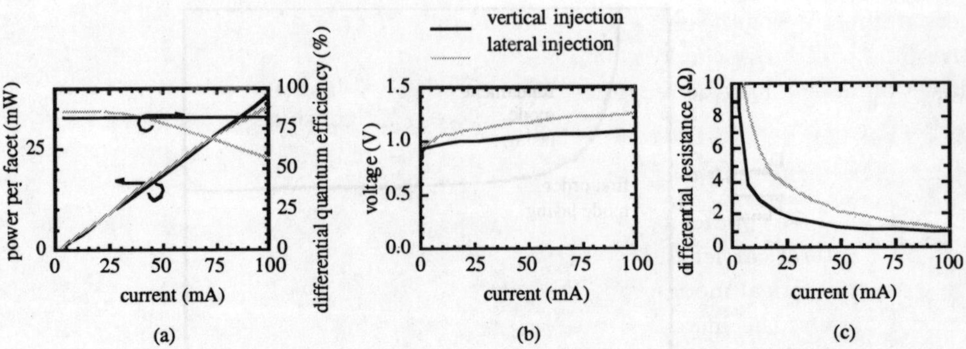

Fig. 9. (a) Optical power output per facet and total differential quantum efficiency, (b) voltage, and (c) differential resistance of vertical and lateral current injection lasers.

output per facet, differential quantum efficiency, voltage, and differential resistance, all as a function of current, for the vertical and lateral devices. The reported results raise a number of issues meriting in-depth consideration and intuitive explanation.

4.3.1. *Higher initial efficiency in lateral current injection laser*

The higher initial differential quantum efficiency in the LCI laser is a consequence of a difference in modal free carrier absorption between the lateral and vertical devices. The fraction of the optical mode which overlaps with the doped contact/cladding regions is ten times greater in the vertical injection case than in the lateral case. This arises because the vertical dimension of the high index region is usually substantially less than an optical half-wavelength inside the medium, whereas the lateral dimension is generally much greater. In the structures considered, the component of modal free carrier absorption due to the contact regions is ten times higher as a result.

Irrespective of the quantitative importance of this initial efficiency improvement, the preceding results highlight one important general feature of the LCI family of devices. When a new spatial degree of freedom is released, a number of design features which are mutually constraining in the vertical injection case become decoupled in the lateral case. In the vertical injection paradigm, the direction of current injection and of tightest optical confinement are the same and the optical mode therefore overlaps vertically with regions of heavily-doped cladding material. In the lateral injection case, the direction of injection and of tightest confinement are orthogonal, and the resulting design flexibility affords the possibility of creating a device with a greater initial differential quantum efficiency and lower capacitance.

A number of similar benefits arise when consideration is extended to more advanced laser devices. For example, the separate confinement heterostructure (SCH) of an LCI multi-quantum well laser may engineered to be essentially free of carriers, affording reduced free-carrier absorption and associated wavelength chirp under direct modulation. In addition, since carriers in the LCI laser are injected directly from heavily-doped contacts into the quantum wells, the frequency-damping effects associated with ambipolar drift-diffusion in the SCH of a vertical injection laser[75] may be reduced.

4.3.2. *Lower differential efficiency under high bias*

LCI lasers whose contact separations exceed their carrier ambipolar diffusion length exhibit a lateral carrier distribution asymmetry which leads to a reduced overlap between the optical mode profile and the local material gain distribution. At high optical powers, the increased rate of stimulated emission gives rise to a further decrease in the ambipolar diffusion length, reduced mode-gain overlap, and a further deterioration in differential quantum efficiency.

A number of physical effects which could conspire to reduce the differential quantum efficiency in lateral injection lasers may be envisaged. These are described below and their relative importance considered. These effects are traceable to the lack of carrier density (distribution) pinning above the lasing threshold.

4.3.2.1. Reduced lateral gain-mode overlap

As the recombination time decreases with increasing optical power, the degree of lateral carrier density asymmetry increases. Because holes are generally much heavier than electrons in unstrained III–V semiconductor materials, the peak of the lateral carrier distribution is at the *p–i* junction (Fig. 10). In contrast, the maxi-

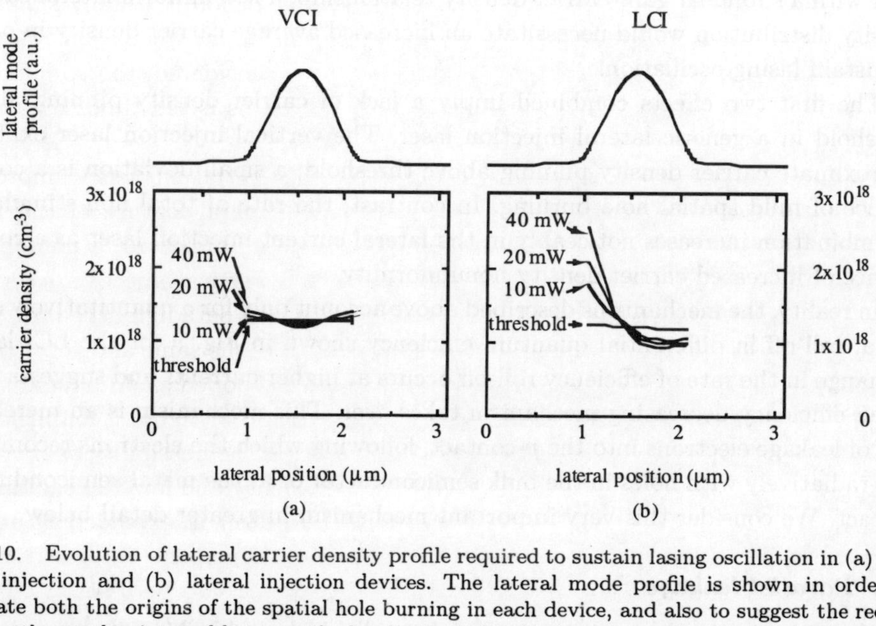

Fig. 10. Evolution of lateral carrier density profile required to sustain lasing oscillation in (a) vertical injection and (b) lateral injection devices. The lateral mode profile is shown in order to indicate both the origins of the spatial hole burning in each device, and also to suggest the reduction in the overlap integral between the local gain and optical mode profile which constitutes the model gain.

mum of the lateral optical mode distribution is in the lateral middle of the intrinsic active region. Whereas the carrier and gain distributions change with increasing current, the optical mode profile remains fixed in strongly index-guided devices, and the overlap deteriorates. In view of this degradation, the carrier density *average* over the lateral extent of the active region must grow in order to preserve a constant modal gain.

4.3.2.2. Superlinear dependence of recombination rate on carrier density

Because the asymmetry of the carrier density distribution grows with increasing optical power, the higher moments of the carrier density distribution would increase even if the *average* carrier density were to remain constant. As a result, recombination mechanisms characterized by superlinear dependencies on the local carrier density — mechanisms which include bimolecular (spontaneous) and Auger recombination — contribute to higher "parasitic" recombination currents in the device.

4.3.2.3. Sublinear gain-carrier density dependence

A third factor arises in devices whose active regions exhibit a sublinear dependence of gain on carrier density. In a bulk active region laser with an approximately linear gain-carrier density relationship, if the optical intensity were approximately constant over the width of the active region, then as the lateral carrier density nonuniformity increased above threshold, a constant average carrier density would be sufficient to maintain zero net round-trip gain. However, in a quantum-well laser with a sublinear gain-carrier density relationship, a less uniform lateral carrier density distribution would necessitate an increased average carrier density in order to sustain lasing oscillation.

The first two effects combined imply a lack of carrier density pinning above threshold in a generic lateral injection laser. The vertical injection laser exhibits approximate carrier density pinning above threshold; a small deviation is a consequence of mild spatial hole burning. In contrast, the rate of total non-stimulated recombination increases noticeably in the lateral current injection laser as a consequence of increased carrier density nonuniformity.

In reality, the mechanisms described above account only for a quantitatively slow initial roll-off in differential quantum efficiency shown in Fig. 9 for the LCI laser. A change in the rate of efficiency roll-off occurs at higher currents and suggests that a *new* efficiency-decreasing mechanism takes over. This mechanism is an increased flux of leakage electrons into the *p*-contact, following which the electrons recombine non-radiatively with holes in the bulk semiconductor or at the metal-semiconductor contact. We consider this very important mechanism in greater detail below.

4.4. *Parallel leakage*

In a lateral injection laser without special parallel leakage-blocking structures, it is at least imaginable that ambipolar flow of electrons and holes could occur through the high-bandgap cladding material above and below the active region. In conventional buried heterostructure lasers, semi-insulating material or *p–n–p–n* blocking structures are used to prevent analogous parallel leakage. In further contrast, the differential resistance of the active region of an LCI laser could be significantly higher, in view of its much greater carrier path/injection cross-section aspect ratio, than that of a vertical injection device. This implies a more rapid buildup of potential across the combined structure, giving rise to an earlier onset of parallel (non-injected) leakage current.

To illustrate the impact of this effect, we provide in Fig. 11 calculated band diagrams for the same pair of vertical and lateral injection lasers analyzed in the previous section.

The diagrams are for zero bias, threshold, and 40 mW/facet output power for both lasers. An appreciable portion of the differential bias applied in the evolution from threshold to 40 mW/facet output power lies across the intrinsic active region of the device. This effect is much smaller in the VCI case.

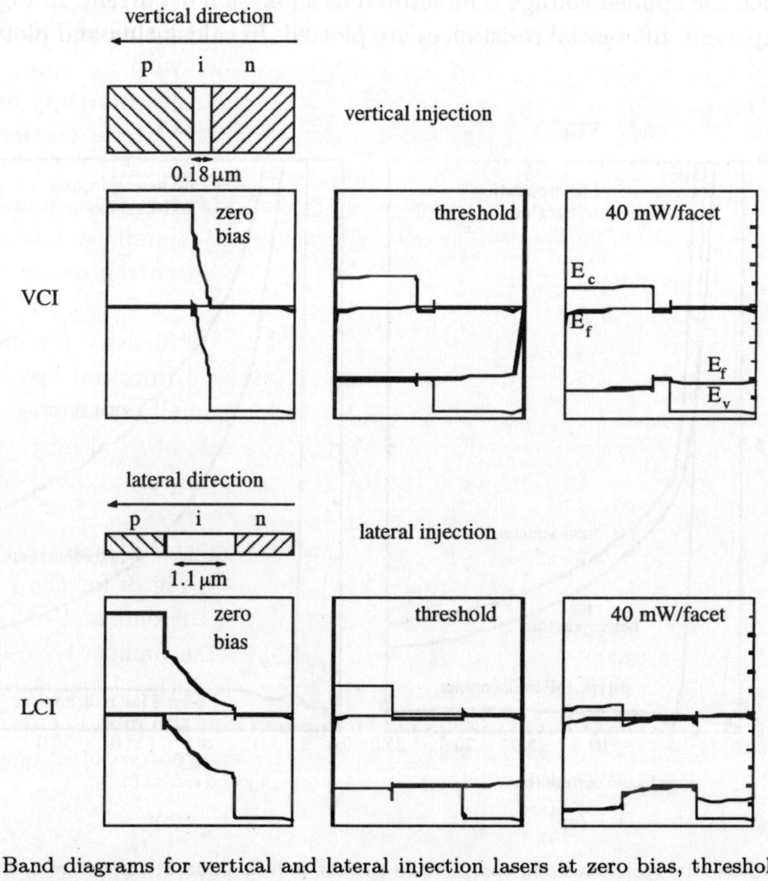

Fig. 11. Band diagrams for vertical and lateral injection lasers at zero bias, threshold current, and at 40 mW/facet output power.

Also evident in the band diagrams is a difference in the rate of increase of the potential across the depleted p–i heterojunction (a vertical junction in the VCI case and a lateral one in the LCI case) under increasing bias current. As the density of current passing over a heterojunction rises, the voltage drop across the depletion layer also increases. This voltage accumulation, which occurs principally at the p-side barrier, increases as the logarithm of current density for large current densities (see, for example, Ref. 62). For the same current injected into the active region, the current density in the LCI laser active region considered is about six times greater than in the VCI laser because of the sixfold decrease in the cross-sectional area available for injection in the LCI case. In view of the preceding logarithmic relation, the differential resistance associated with this voltage accumulation is expected to be somewhat less than twice that of the vertical injection laser. A further source of differential resistance lies in the buildup of voltage across bulk semiconductor contacts, or tubs.

The question of which of these effects has the greatest impact on the operation of real devices remains. To quantify these effects, each laser is divided into segments across which the applied voltage is monitored as a function of current. In Fig. 12 the series component differential resistances are plotted. In calculating and plotting the

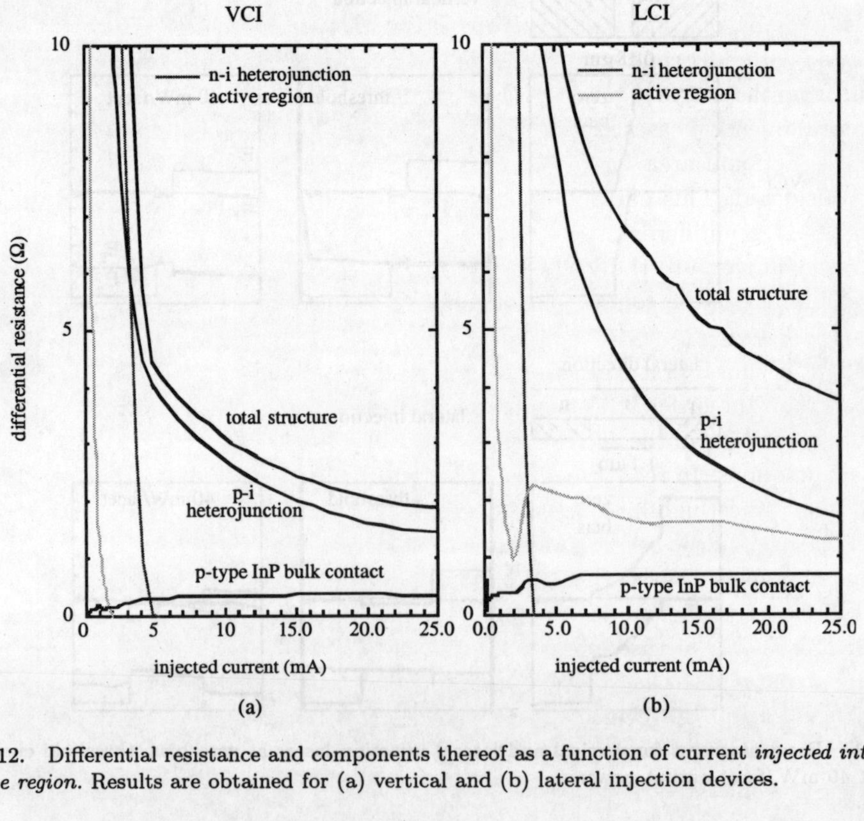

Fig. 12. Differential resistance and components thereof as a function of current *injected into the active region.* Results are obtained for (a) vertical and (b) lateral injection devices.

component differential resistances, it is not the total current, but instead the current injected into the active region, that was used. In this way voltage and leakage effects are isolated. The differential resistances treated herein may therefore be considered to appertain to the active region in isolation from its surrounding, potentially leaky, cladding layers.

For both vertical and lateral lasers, the voltage across the n–i heterojunction is constant above threshold. In contrast, the voltage across the p–i heterojunction rises much more rapidly in the LCI laser case than the vertical injection case. The voltage drop across the active region of the VCI laser is very nearly pinned above threshold, while it continues to grow above threshold in the LCI case. The continued development of the p–i heterojunction potential dominates the differential resistance of the VCI laser, while the combination of this potential with that across the active region is responsible for the higher differential resistance in the LCI laser. As expected, the LCI laser p–i heterojunction differential resistance is typically less than twice that of the vertical injection case.

The gradual decrease in the resistance of the active region itself is a consequence of the increase in the average carrier density in the active region with increasing bias. As the bias increases, gain-mode overlap deteriorates, a higher average is needed to sustain laser oscillation, and the differential resistance of the LCI laser decreases gradually.

Also worthy of comparison are the quantitative differential resistance values associated with bulk conduction through the p-contact. Close to the metallic contact and far from the active region, the cross-sectional p-contact area seen in the vertical and lateral injection cases are similar. On the other hand, at the active region itself, the cross-sectional area for injection is approximately six times less for the LCI laser considered. This current constriction effect accounts for the approximately two times larger differential resistance associated with the p-contact in the LCI laser case compared to the VCI case.

We summarize the conclusions of this analysis as follows:

- It might seem on first consideration that with injecting contacts separated by a distance on the order of an ambipolar diffusion length, most electrons and holes would recombine in the active region even without the benefit of carrier-confining heterojunctions. In reality, self-consistent treatment of the ambipolar regime reveals that since high-mobility electrons readily reach the opposite contact, electron leakage will be severe unless strong carrier confinement is introduced.
- It might appear that with high-bandgap intrinsic material in parallel with the LCI laser active region, parallel leakage through cladding layers should be minimal. This is true at threshold, where the voltage applied across the active region and parasitic paths is slightly greater than that corresponding to the active region bandgap. However, because of higher current densities in LCI lasers, the above-threshold differential resistance of the active region is typically higher than in vertical lasers. For this reason, voltage may rapidly accumulate across the par-

asitic path above the lasing threshold so that parallel leakage causes the lasing efficiency to drop drastically at low or moderate powers.

- Since *p–n–p–n* current blocking structures are not readily realized in the lateral injection scheme, other solutions to the leakage problem must be found. One straightforward but powerful solution to the problem of early leakage onset is to increase the contact doping. At first this might seem ineffective, for it increases access to both carrier paths (both active region injection as well as parallel leakage) equally. In fact, since increasing the contact doping lowers the heterojunction differential resistance under high injection, the total active region differential resistance is reduced and the onset of substantial parallel leakage postponed.
- It might appear that the strategy frequently adopted in vertical injection laser design of grading doping downward from high to moderate levels as the active region is approached could usefully be applied in LCI lasers. In fact, this is not only unnecessary, but also does not aid in resolving the leakage problem. It is not necessary because heavy doping right at the heterojunction between the *p*-contact and the active region will not give rise to intolerable modal free-carrier absorption in view of the minimal confinement of the lateral mode to the lateral contacts. It is not helpful because, for the differential resistance of the heterojunction to be usefully reduced, heavy doping must exist within a Debye length of the junction.

4.5. *Hybrid injection lasers*

Oe *et al.*,[63] reported a novel LCI laser with very strong performance relative to many previously-reported attempts at realizing LCI lasers. Using a simple bulk active region, a device was obtained which lased at 1.5 μm continuous wave at room temperature with a threshold of 10 mA and optical power per facet in excess of 10 mW. The essential structural features of this device are illustrated in Fig. 13.

Current guides — doped layers of intermediate composition — are placed above and below the active region. We have previously demonstrated two-dimensional fully self-consistent numerical simulation that current guides help to smooth out

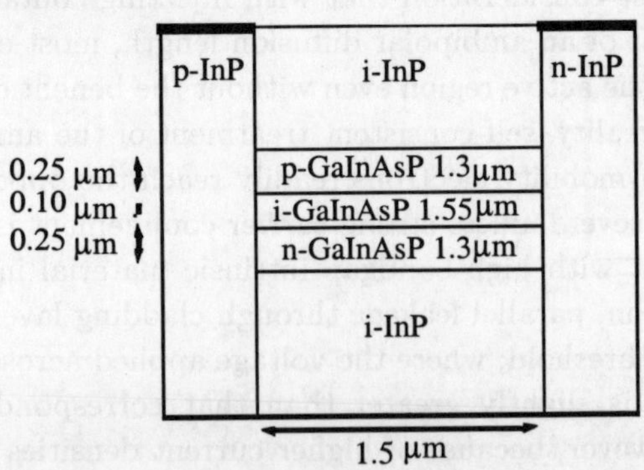

Fig. 13. Key structural features of the LCI laser of Oe *et al.*

the lateral material gain profile in the device,[64] permitting efficient pumping of the fundamental optical mode.

However, another effect of current guides is to reduce the series resistance of the active region by increasing the area available for the injection of a given current. Not only is resistive heating decreased, facilitating continuous wave room temperature operation; but also, as voltage accumulates more slowly across leakage paths in parallel with the active region, the onset of significant leakage current occurs at higher optical powers.

We reported a combined theoretical-experimental study in which we assessed the importance of the improvements offered by current guides.[65] In order to assess the impact of the use of current guides, the fully self-consistent two-dimensional light emitter simulator reported in Ref. 61 was used to model a number of structures with and without current guides. The three structures illustrated in Fig. 14 were considered first. One incorporates current guides similar to those employed by Oe *et al.*; one uses no current guides; and the final, extreme case uses not only current guides but also employs doped cladding for additional hybrid injection and further reduced series resistance.

The L–I and V–L characteristics of these three devices are shown in Fig. 15. (A V–L characteristic was produced instead of a V–I characteristic, since the optical power L provides a better measure of the current injected into the active region.) The L–I characteristics in Fig. 15(a) show a pronounced roll-off in lasing efficiency in the case of the device without current guides, while the deterioration comes much later in the case of the devices with current guides. Because it had less

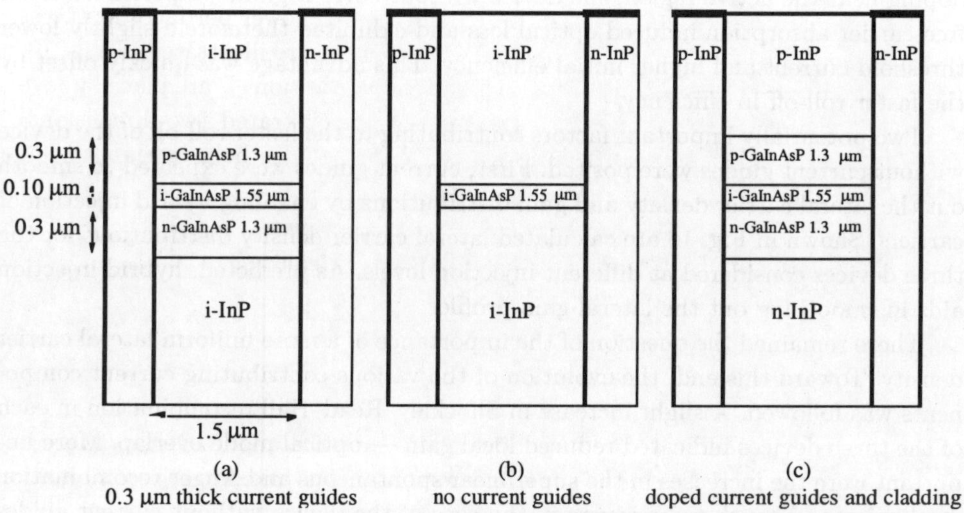

Fig. 14. Structures simulated to reveal the impact of current guides and associated hybrid injection. While not indicated explicitly in the figure, the structures simulated were assumed to have identical vertical refractive index profiles in order to achieve the same modal profiles and confinement factors, enabling thereby fair comparison of device performance.

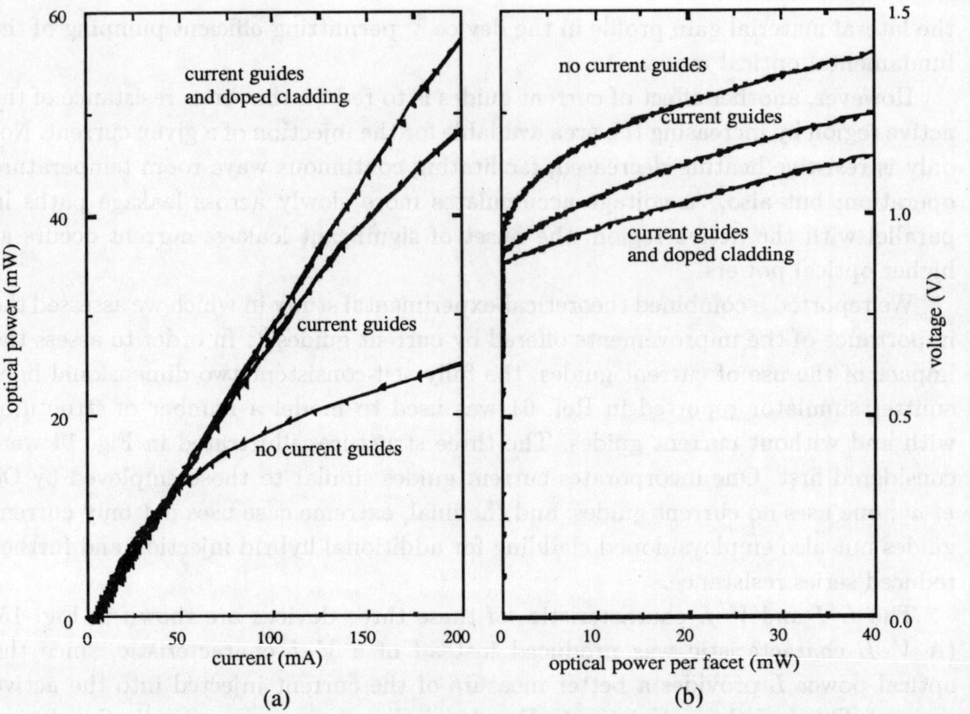

Fig. 15. (Simulation). The effect of hybrid injection on the performance of an LCI laser.

doping near the active region, the device without current guides experienced lower free-carrier absorption induced optical loss and exhibited therefore a slightly lower threshold current and higher initial efficiency. This advantage was quickly offset by the faster roll-off in efficiency.

Two potentially important factors contributing to the faster roll-off of the device without current guides were posited. First, current guides were expected to smooth out the lateral carrier density and gain distributions by enabling hybrid injection of carriers. Shown in Fig. 16 are calculated lateral carrier density distributions for the three devices considered at different injection levels. As predicted, hybrid injection aids in smoothing out the lateral gain profile.

There remained the question of the importance of a more uniform lateral carrier density. Toward this end, the evolution of the various contributing current components was followed. A slight increase in Shockley–Read–Hall recombination in each of the three devices indicated reduced local gain — optical mode overlap. More important were the increases in the superlinear spontaneous and Auger recombination mechanisms — much more severe in the case of the device without current guides than of those with.

Much more quantitatively important, however, is leakage through parasitic paths: this overwhelms any of the preceding effects and is primarily responsible

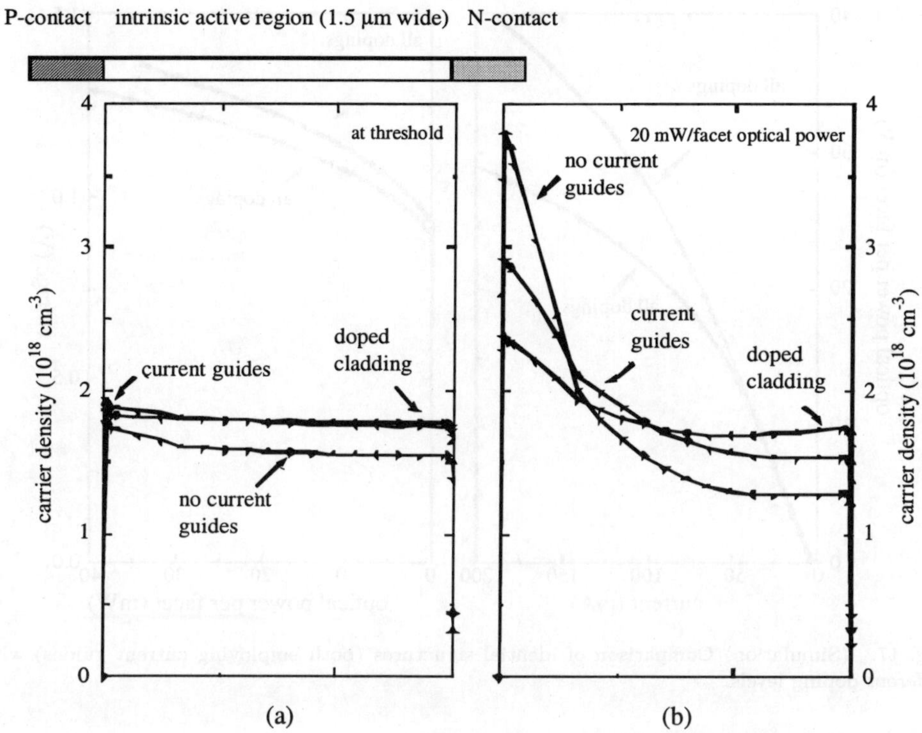

Fig. 16. (Simulation) Lateral carrier density profiles for the devices of Fig. 14, considered at (a) threshold and (b) 20 mW/facet optical power.

for the efficiency roll-off seen at different points on the L–I characteristics of the three devices. The larger voltage required to inject a given current into the active region of the device without current guides is accompanied by increased leakage through parallel parasitic paths which see this same larger voltage difference.

It is at this point worth considering the origins of series resistance in a lateral current injection laser in order to find other ways in which to reduce its importance and enable high-power, low-leakage operation. It was shown in the previous section that above the lasing threshold, by far the largest component of the differential series resistance above threshold in a typical LCI laser comes from the increase in voltage applied across heterojunction between the p-type contact and the active region in order to inject more current into the active region. (This phenomenon and its implications have also been studied in the vertical injection case.)[62] The voltage drop across the p–i heterojunction increases approximately as the logarithm of the current density and decreases as $\ln(\mu_p N_A)$ (where μ_p is the hole mobility and N_A the contact acceptor concentration).

The effect of p-contact doping on series resistance and leakage effects was considered by simulating a device identical to that of Fig. 14(b) but with all dopings reduced from 2×10^{18} cm^{-3} to 1×10^{18} cm^{-3}. The resulting L–I and V–L (Fig. 17)

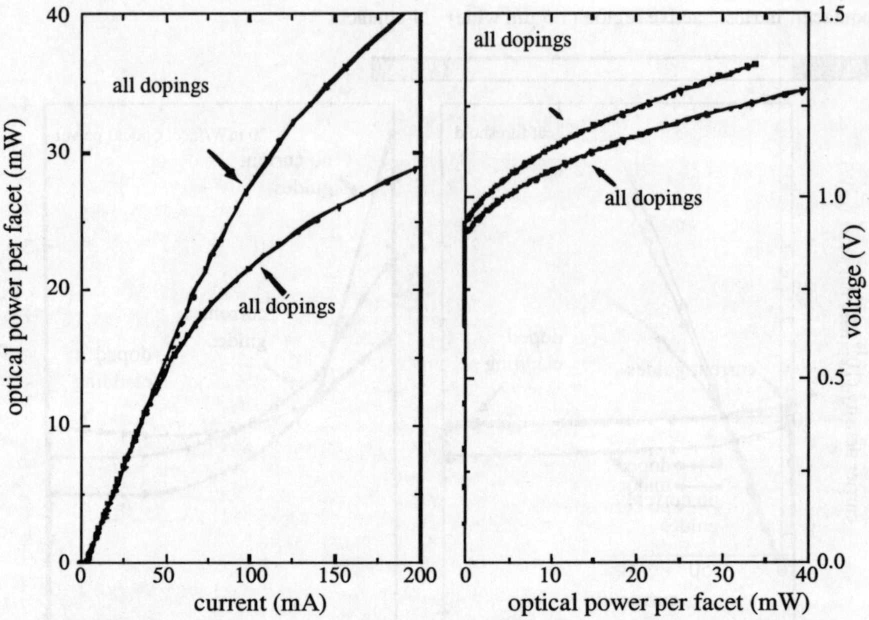

Fig. 17. (Simulation) Comparison of idential structures (both employing current guides) with different doping levels.

reveal that this moderate reduction in doping leads to a much earlier roll-off in the L–I characteristic. LCI laser performance is a sensitive function of p-contact doping in the range of technological feasibility.

While the resistance of the heterojunction between the high-bandgap p-contact and the active region is determined by the doping of the p-contact, an SCH region with intermediate composition and no intentional doping could exhibit ambipolar conduction at moderate biases and consequent hybrid injection. This undoped layer could therefore serve to lower the series resistance of the active region. In order to test this hypothesis, the performance of another set of three structures was compared: one had doped current guides $\lambda = 1.3$ μm), one no current guides, and one an *undoped* SCH region ($\lambda = 1.3$ μm). All doped regions were doped to a level of 2×10^{18} cm^{-3}. The result L–I and V–I characteristics, shown in Fig. 18, indicate that the device with an undoped SCH in fact performs better than that with doped current guides, and much better than the device without current guides. At the lasing threshold, an appreciable population of electrons and holes co-exist in the SCH and contribute to a moderation ambipolar conductance. The reason for the somewhat superior performance of the undoped current guides over the doped current guides is that both guides, top and bottom, participate in *hole* injection in the undoped case. Since the dominant series resistance term to be reduced is associated with hole injection, the above-threshold differential resistance of the active region is best reduced by maximizing the area available for hole injections.

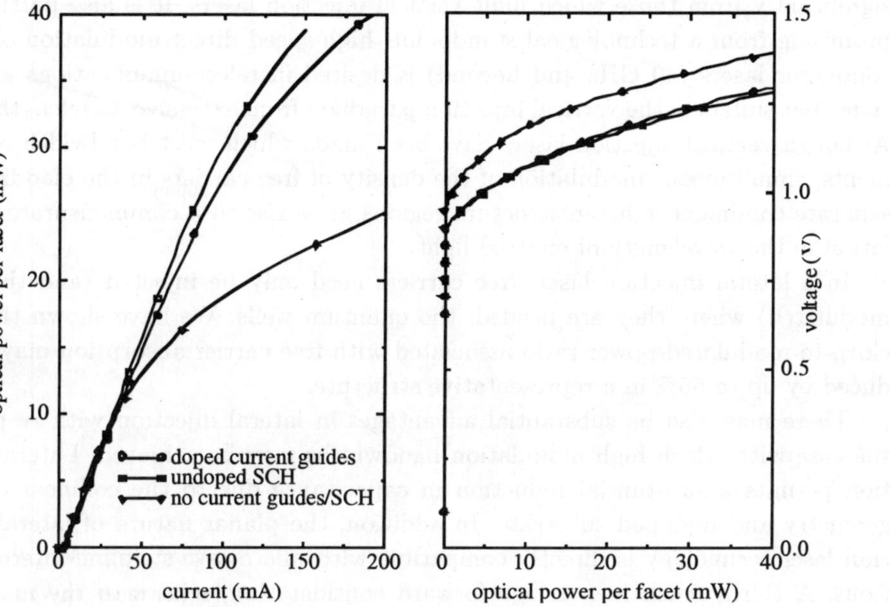

Fig. 18. (Simulation) Performance of LCI lasers with doped versus undoped currents guides (device with no current guides at all also shown as reference).

4.6. *Capture, escape, and dynamic performance*

A matter of great theoretical and practical interest is the nature of carrier injection from three-dimensional states in bulk contacts to lower-dimensional states in quantum-confined structures such as wells and wires. This problem of carrier capture has long been of interest in vertical injection structures, for it is one of the mechanisms responsible for limiting the speed of direct laser modulation. In the vertical geometry, carriers must lose substantial energy and change momentum in their direction of propagation. In the lateral geometry, carriers may preserve their momentum in the direction of travel. In either lateral or vertical injection, injected carriers must relax into energetic states from which they may contribute to optical gain at the lasing wavelength. In vertical injection devices, carriers experience large, sudden changes in band energies as they move along their vertical trajectory. The lateral injection paradigm offers the possibility of preserving an abrupt heterostructure in the vertical direction in order to maximize quantum confinement, while simultaneously permitting a more gradual energetic transition along the lateral path of carriers. Such an adiabatic lateral transition from 3D injector states into 2D active region states arises naturally by virtue of a lateral grading of the band structure in LCI lasers fabricated using quantum well intermixing.

The preceding discussion points to a substantial new area of investigation: the dynamic operation of the lateral current injection laser. The LCI laser merits study on this front for reasons of general interest, since qualitative mechanisms may differ significantly from those which limit vertical injection lasers. It is also particularly promising from a technological standpoint: high-speed direct modulation of semiconductor lasers (40 GHz and beyond) is desired in telecommunications applications, but suffers in the vertical injection paradigm from excessive wavelength chirp. Although vertical injection lasers have been made which meet bandwidth requirements, simultaneous modulation of the density of free carriers in the cladding and separate confinement heterostructure regions gives rise to a commensurate modulation in the wavelength of emitted light.

In a lateral injection laser, free carriers need only be injected (and therefore modulated) where they are needed: the quantum wells. We have shown that the chirp-to-modulated-power ratio associated with free carrier absorption may be reduced by up to 50% in a representative structure.

There may also be substantial advantages in lateral injection with respect to the ease with which high modulation bandwidths may be achieved. Lateral injection permits a substantial reduction in capacitance due to the collinear contact geometry and undoped substrate. In addition, the planar nature of lateral injection laser technology is directly compatible with microwave stripline interconnections. A third, much less straightforward consideration, relates to the matter of carrier transport. In the vertical injection laser, the effective travel time is increased due to the multiple quantum well carrier capture and escape processes. The LCI approach has the advantage of reducing the capture-escape problem, as carriers travel along the quantum wells rather than across them. Moreover, the two-dimensional bipolar lateral transport along the quantum wells might be faster due to the reduced dimensionality of the system. There are thus indications that the speed limit of LCI lasers can actually exceed that of the vertical injection lasers, despite a seemingly longer carrier travel path in the active region of the LCI structures.

This issue of the ultimate speed limits of vertical and lateral lasers illustrates the need for a fundamental approach to the study of dynamic operation. In contemplating transit time and small- and large-signal modulation in both devices, qualitative considerations are seemingly insufficient to resolve the issue of the relative modulation bandwidth in the two device families. A time-dependent model which accounts for in-plane capture delays and which is free from the usual charge neutrality assumption must be adopted. This photon-assisted bipolar 2D lateral injection/transport problem is, on its own, of considerable scientific merit but has yet to be assessed in the literature.

5. Laser Fabrication

We have pursued two paths for lateral injection laser fabrication. The first involves ion implantation doping and intermixing; the second relies on etch and regrowth buried heterostructure LCI lasers. A third available method is to use diffusion for contact formation — a method reported in a number of publications on lateral injection and transverse junction devices.

5.1. *Ion implanted lateral injection lasers*

We fabricated a set of LCI lasers in order to explore the mechanisms newly revealed in the preceding discussions, and to enable comparison of theory and experiment.[66] We employed ion implantation, an attractive technique for spatially selective doping and/or intermixing of semiconductor heterostructures. By promoting the alloying of adjacent heteroepitaxial layers (e.g. quantum wells and barriers), vacancies introduced during the course of ion implantation enable regrowth-free formation of the lateral heterobarriers required in LCI lasers for adequate current confinement.[67]

The epitaxial structure was grown by organo-metallic vapor phase epitaxy on InP:Fe. The bottom cladding consisted of 1 μm InP:Fe. The active region, with photoluminescence peak at 1.55 μm, was made up of ten 95 A 1% compressively strained undoped wells separated by 190 A 0.5% tensile strained $Ga_{0.18}In_{0.82}As_{0.25}P_{0.75}$ ($Q = 1.1$ μm) p-type 5×10^{17} cm^{-3} barriers. Above and below this multi-quantum well (MQW) structure were 500 A thick layers of unstrained GaInAsP ($Q = 1.07$ μm) doped p-type 5×10^{17} cm^{-3} and 1×10^{18} cm^{-3}, respectively. The top cladding consisted of 0.5 μm undoped InP, on top of which was grown a 0.2 μm thick contact layer of undoped $Ga_{0.47}In_{0.53}As$. Iron doping was used in the substrate and lower cladding to reduce parallel leakage current. Compressively strained quantum wells were used to minimize Auger recombination and break the heavy hole — light hole degeneracy, thereby reducing the heavy hole effective mass and increasing the hole mobility (and therefore the ambipolar diffusion length). Current guiding layers above and below the MQW, along with the thick p-doped barriers, served to reduce the series resistance of the active region by promoting hybrid injection of holes.

The processing sequence carried out is depicted in Fig. 19. The GaInAs contact layer was removed using a selective wet etch in the vicinity of the future ridge (Fig. 19(a)) in order to avoid optical absorption of the lasing mode. SiO_2 was deposited and patterned; ridges were formed using a selective wet etch, stopping on the upper GaInAsP current guide (Fig. 19.(b)). The SiO_2 served as a self-aligned mask for selective implantation of phosphorus into the quantum wells with a dose of 2×10^{14} cm^{-2}. The dielectric mask was then stripped and the samples subjected to rapid thermal annealing in the proximity geometry in order to intermix the wells (Fig. 19(c)). Samples were masked to reveal the intended n-contact regions (Fig. 19(d)) and implanted with silicon, then masked for implantation of beryllium in order to form the p-contacts (Fig. 19(e)). Another anneal was performed, also using an InP proximity cap, in order to activate the implanted dopant ions. Ti/Pt/Au was evaporated for the p-contact while Ni/Au/Ge/Au was evaporated for the n-contact. Samples were cleaved into 500 μm bars and mounted on aluminum nitride carriers.

The devices lased only at low temperatures. While the ultimate objective of LCI laser fabrication is to realize room temperature, continuous wave, low threshold, high efficiency lasing, this experimental trial permitted us nevertheless to study the physical mechanisms which underlie LCI laser operation. Our temperature-

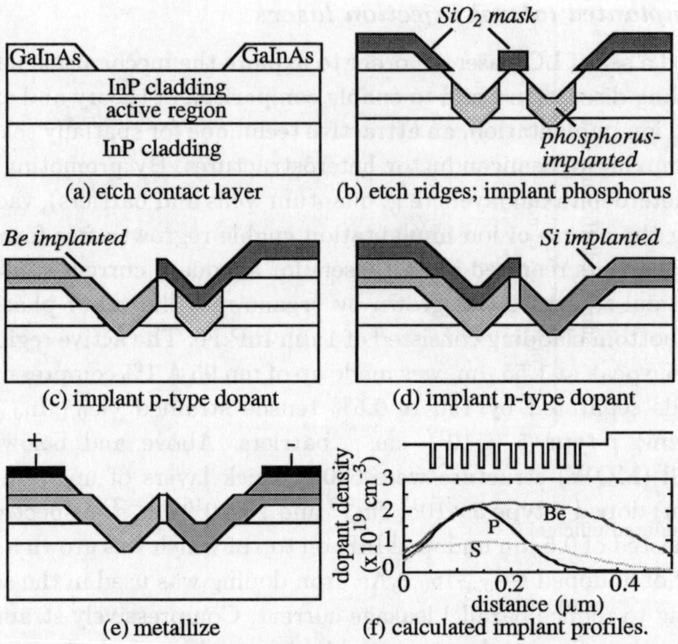

Fig. 19. Process sequence for ion-implanted LCI lasers.

dependent probing of the performance of the LCI lasers fabricated permitted us to reveal which mechanisms prevented successful room temperature lasing of these devices and to propose technologically feasible design and processing improvements.

We first studied the dependence of measured output parameters on pulse width and duty cycle and identified a region over which the output did not vary with pulsed excitation conditions; we concluded that device heating was not significant in this regime, and that the device lattice temperature was thus well approximated by the heatsink temperature. We show in Fig. 20 the measured evolution of the stimulated emission efficiency of this laser as a function of heatsink temperature. We fit this to a model of stimulated efficiency which accounts for the temperature dependence of the various recombination mechanisms and of leakage over the heterobarriers. Using typical values for material, we estimated the degree of lateral heterobarrier formation obtained by implanting and annealing to have been around 30 meV. This is in reasonable agreement with the measured room-temperature photoluminescence spectral shift of 38 meV for implanted versus unimplanted materials annealed under the same conditions (inset of Fig. 20). Thus, one main reason for non-lasing at room temperature was the insufficient degree of quantum well intermixing: for room temperature operation, our $4kT$ rule of thumb would imply that more than 100 meV of effective bandgap shift in the quantum well region is required. Significantly greater intermixing (preferably up to 200 meV) would enable improved efficiency and better high-power operation. Ion implantation and

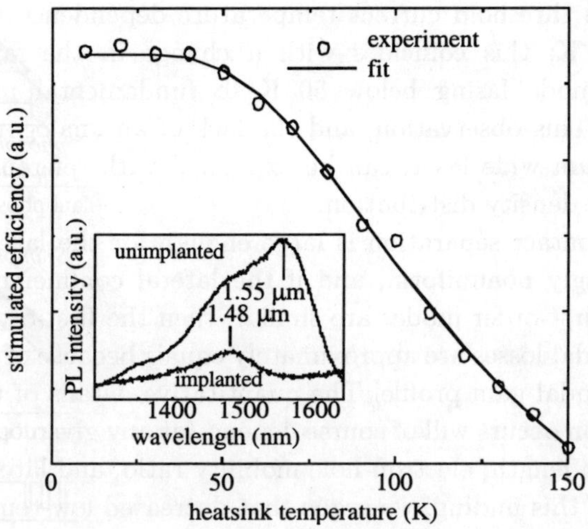

Fig. 20. Stimulated efficiency versus heatsink temperature for a device with a 3 μm wide ridge.

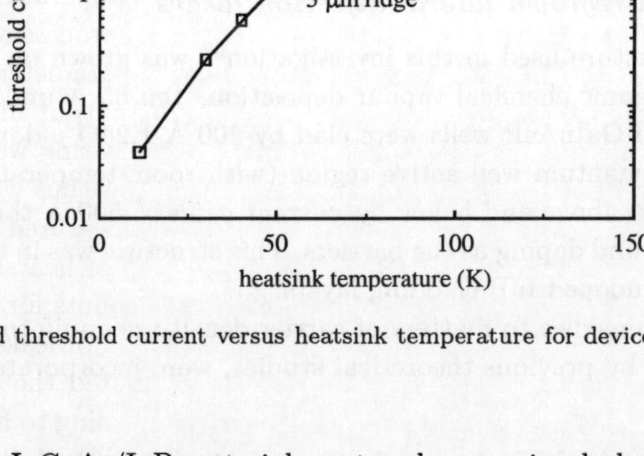

Fig. 21. Measured threshold current versus heatsink temperature for devices with ridge widths of 1 μm and 3 μm.

annealing in the InGaAs/InP materials system has previously been shown experimentally to allow bandgap shifts as large as 200 meV.[68] The intermixing of thinner wells and thick, high-bandgap barriers would enable the needed amount of carrier confinement.

We compare in Fig. 21 the temperature dependence of threshold current for lasers with 3 μm and 1 μm wide ridges. The fact that the threshold current is appreciably lower for the wider-ridge devices would suggest that the 1 μm wide device experiences a much larger scattering loss. There is a discontinuous change in the slope of the threshold current temperature dependence for the 3 μm wide device around 50 K; this coincides with a change in the far field from first-order transverse mode lasing below 50 K to fundamental mode lasing above this temperature. This observation, and the lack of an analogous discontinuity in the case of the 1 μm wide laser, can be explained with reference to the laterally nonuniform carrier density distribution.

If the lateral contact separation is large enough for the lateral carrier density profile to be strongly nonuniform, and if the lateral confinement factors for the fundamental and first-order modes are similar, then the first-order mode may lase first (assuming modal losses are approximately equal) because of its greater overlap with the lateral modal gain profile. The quantitative details of the ridge width at which this transition occurs will of course depend (at any given temperature) on the ambipolar diffusion length, electron-hole mobility ratio, and lateral modal profiles.

On the basis of this finding, we argue that decreased low-temperature mobility arising as a result of increased ionized impurity scattering below $\sim$ 50 K gives rise to a reduced ambipolar diffusion length. This decrease in the ambipolar diffusion length enables first-order mode lasing in the 3 μm wide ridge device prior to fundamental mode lasing. In the 1 μm wide ridge device, there is no first-order lateral mode admitted (or its confinement factor is very small), so lasing always occurs first on the fundamental mode. However, as the ambipolar diffusion length drops at low temperatures, the threshold current drops more slowly as gain-mode overlap worsens with increasing lateral carrier density nonuniformity.

5.2. *Etched-and-regrown lateral injection lasers*

The epitaxial structure used in this investigation[69] was grown on Fe:InP via low-pressure metal organic chemical vapour deposition. Ten 55 Å undoped 1.5% compressively strained GaInAsP wells were clad by 200 Å 1.25Q *p*-doped (cm^{-3}) barriers. This multi-quantum well active region (with room-temperature PL peak at 1585 nm) was clad above and below by *current guides*,[7] 500 Å thick layers of the same composition and doping as the barriers. This structure was in turn sandwiched between 0.5 μm undoped InP cladding layers.

A number of remedies to matters of carrier density nonuniformity and current leakage, informed by previous theoretical studies, were incorporated into this epitaxial structure:

- compressively strained quantum wells were used to reduce the heavy-hole effective mass and increase thereby the mobility of the limiting carrier;
- current guiding layers[7] and doped barriers were introduced for "hybrid" injection of carriers — predominantly lateral injection aided by some vertical carrier flow — into the active region. *p*-type guides were used to enhance injection of holes, the limiting carrier;
- compressively strained quantum wells were also used in aid of reducing parasitic recombination effects (e.g. Auger recombination);

- many quantum wells were used for a high confinement factor low threshold carrier density. Parasitic effects (e.g. Auger recombination and carrier leakage) were thereby minimized.

The processing sequence is depicted in Fig. 22. Strong lateral carrier confinement, shown to be essential for high-performance operation, is achieved by etching followed by regrowth of a higher-bandgap material. Etch-and-regrowth trials on bulk InP material were used to achieve a planar morphology.

The dependence of threshold current, spontaneous efficiency, and stimulated efficiency are given in Fig. 23 for devices with 1 μm and 2 μm wide ridges. At heatsink temperatures of 50 K and below, the 2 μm device lased in a first-order lateral mode rather than the fundamental mode (observed by measuring the lateral far-field profile). The transition from fundamental- to first-order mode lasing is also manifest in a commensurate jump in threshold current.

First-order mode lasing may be favoured over fundamental mode lasing in an LCI laser if the lateral gain profile is strongly nonuniform. For this effect to account for the characteristic of Fig. 23, the threshold ambipolar diffusion length would have to drop to below its room temperature value at temperatures of 50 K and below. The low-temperature hole mobility in GaInAsP drops below its room temperature value as ionized-impurity scattering becomes dominant at low temperatures. The ambipolar diffusion length is also expected to drop below its room temperature value, so that the lateral gain nonuniformity becomes more severe, and lasing on the first-order mode is anticipated. The 1 μm wide device exhibits consistent fundamental mode operation from 10 K – 300 K, with a continuous decrease in stimulated

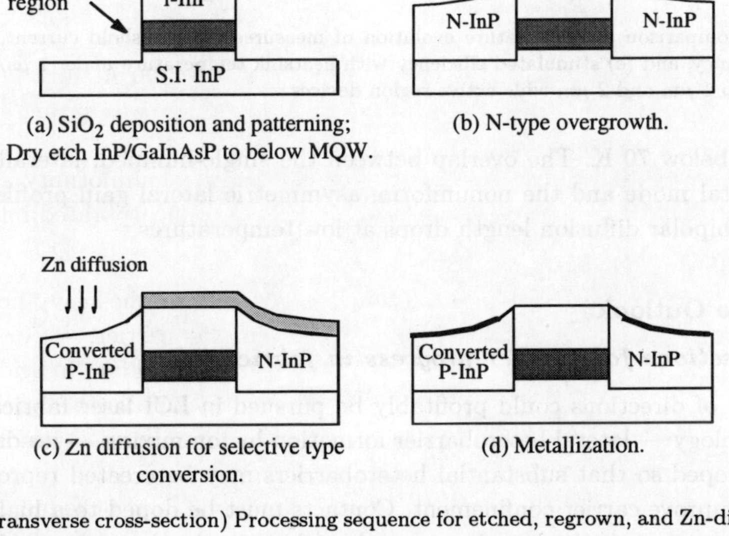

Fig. 22. (Transverse cross-section) Processing sequence for etched, regrown, and Zn-diffused type-converted LCI laser.

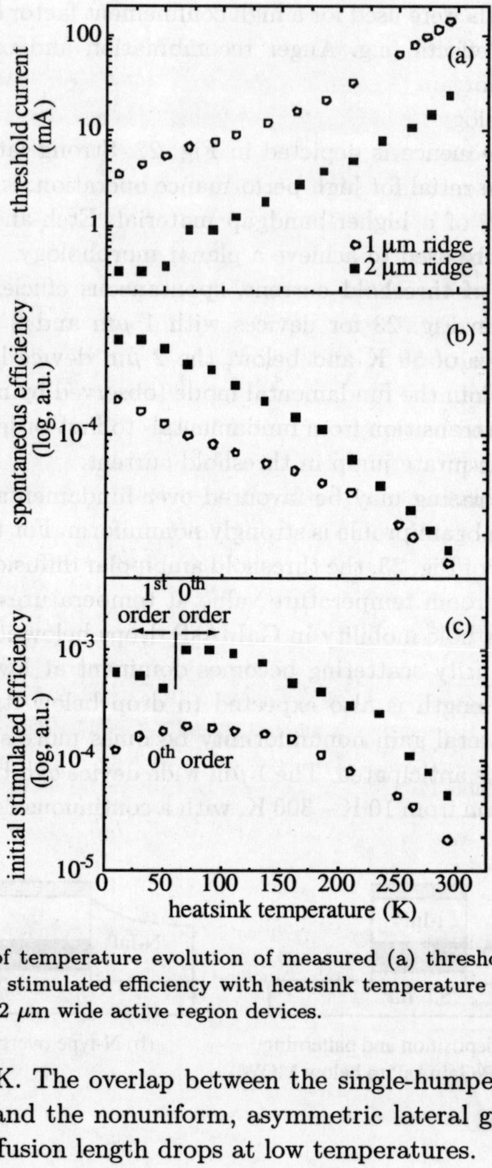

Fig. 23. Comparison of temperature evolution of measured (a) threshold current, (b) spontaneous efficiency, and (c) stimulated efficiency with heatsink temperature under 1 μs/0.1% pulsed excitation in 1 μm and 2 μm wide active region devices.

efficiency below 70 K. The overlap between the single-humped, laterally centered fundamental mode and the nonuniform, asymmetric lateral gain profile decreases as the ambipolar diffusion length drops at low temperatures.

6. Future Outlook

6.1. *Directions for further progress in fabrication*

A number of directions could profitably be pursued in LCI laser fabrication. One key technology — lateral heterobarrier formation by ion mixing — needs to be further developed so that substantial heterobarriers may be created reproducibly in order to improve carrier confinement. Contacts must be doped to a high level and residual damage must be largely removed — thus, ion implantation and annealing (without undue surface decomposition) also require further study. The series resistance associated with metal-semiconductor contacts and bulk semiconducting regions is also an important experimental feature needing further study and control.

Regrowth technology also requires further improvement. The morphology achieved in MOCVD and particularly in MBE deserves further attention, as do issues of nonradiative recombination at the growth interface and elimination of an n-type doping spike.

Even once these base technologies have been further honed, there exists the potential for further technological development in reducing process complexity and the number of mask levels. A scheme for simultaneous formation of n- and p-contact using a single metallization step is an example of one such simplifying process.

The development of an etch-free regrown-LCI laser process also deserves exploration. Damage created in a semiconductor during dry etching is a serious concern for etch-and-regrowth processing. Selective-area growth (SAG) permits planar differentiation without recourse to damaging dry etching. This process would begin with selective growth of the laser active region on a dielectric-masked wafer surface.

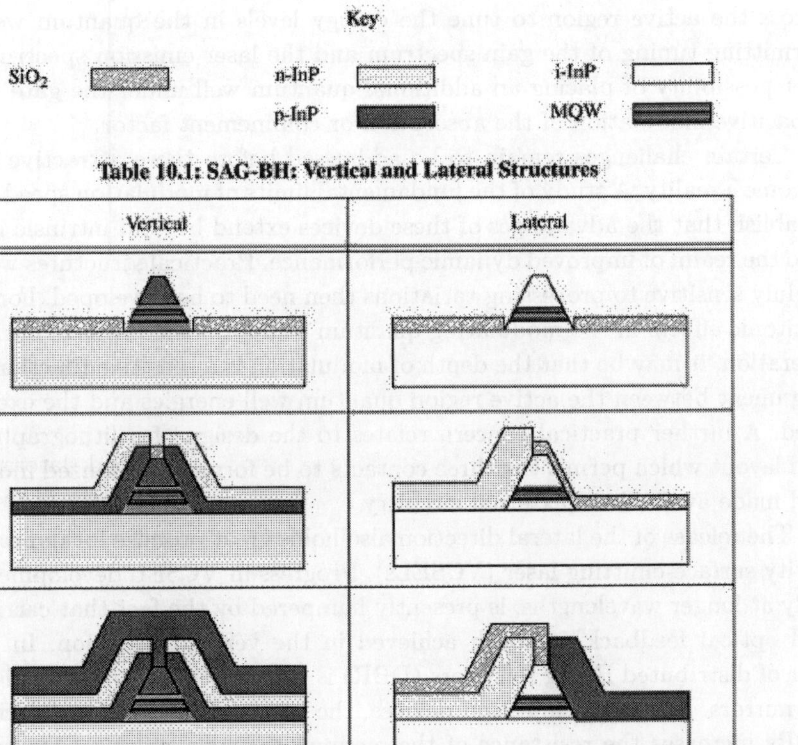

Fig. 24. Selective-area regrown LCI laser.

The mask would be removed, and a buried heterostructure formed by an additional growth atop this structure. One useful feature of this approach is that the properties of the dielectric mask pattern influence the rate of growth of the active region; lasers with controlled differences in wavelength may therefore be grown simultaneously. It is seen in Fig. 24 that many of the same process steps could be used for vertical and lateral SAG lasers, so that exploratory SAG-LCI trials could readily be combined with ongoing vertical injection SAG efforts.

7. Device Opportunities

One of the most exciting opportunities in the study of the lateral direction is the potential to apply the theoretical insights, design improvements, and technological advances to the fabrication of novel functional devices.

LCI lasers permit the use of an insulating vertical heterostructure and therefore lend themselves to control via the application of an electric field using a third terminal. This electrical field could be used to modulate the injection of current by controlling the conductivity of the channel. Such a structure might be thought of as an integrated laser-transistor. Alternatively, an electric field could be applied across the active region to tune the energy levels in the quantum wells, thereby permitting tuning of the gain spectrum and the laser emission spectrum. The further possibility of placing an additional quantum well under the gate would allow capacitive modulation of the absorption or confinement factor.

Certain challenges remain to be addressed before these attractive possibilities become a reality. A study of the fundamental limits of modulation speed is needed to establish that the advantages of these devices extend beyond intrinsic integrability into the realm of improved dynamic performance. Practical structures which are not unduly sensitive to processing variations then need to be developed. For example, if excitonic effects in the modulating quantum well form the physical basis for device operation, it may be that the depth of modulation is a sensitive function of spectral alignment between the active region quantum well energies and the exciton energy used. A further practical concern relates to the design of a lithographic sequence and layout which permit the three contacts to be formed, optimized independently, and made available to external circuitry.

The release of the lateral direction also holds great promise for improved vertical cavity surface-emitting laser (VCSELs). Progress in VCSEL development, particularly at longer wavelengths, is presently hampered by the fact that carrier injection and optical feedback are both achieved in the vertical direction. In VCSELs, a pair of distributed Bragg reflectors (DBR) is typically used to create high reflectivity mirrors. In vertical injection devices, the heterostructure associated with these DBRs increases the resistance of the medium through which current must be injected in order to reach the active region. Doping in the DBRs must be kept to a moderate level if free-carrier absorption is not to be unduly severe. The need for this compromise highlights the constraining nature of the purely vertical approach.

In VCSELs intended for long-wavelength operation, vertical injection suffers another drawback. A recent breakthrough in long-wavelength VCSEL technology has been the introduction of wafer fusion: mirrors are grown in the high-index-contrast AlGaAs/GaAs system and are fused to a GaInAsP-based active region wafer.[70] This wafer-fused interface is acceptable from an optical standpoint, but results in tremendous increases in threshold voltage.

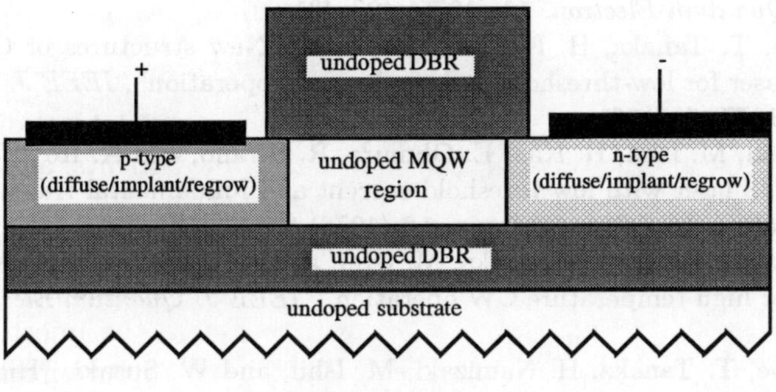

Fig. 25. Lateral injection VCL.

The lateral injection VCSEL portrayed in Fig. 25 would address many of these issues in a natural and convenient manner. In particular, it would allow:

- the use of completely intrinsic DBR mirrors for very low free-carrier absorption;
- the use of heavily-doped injectors;
- current to be injected along a path which would circumvent the fused interface entirely;
- multi-quantum well structures to be placed at several standing wave peaks of the vertically-confined optical mode. Because of restrictions on active region thickness in conventional VCSELs, gain regions are typically placed at only one standing wave peak. As a result, the threshold carrier density is high and L–I characteristics usually degrade at low optical powers.

If the planar dimension can be better employed in lateral injection lasers, opportunities in the longitudinal direction merit exploration as well. A more uniform distribution of carriers and photons could be achieved by varying the coupling strength of a DFB grating along its length. In vertical injection structures, it is difficult to modify the grating strength without simultaneously modifying the resistance. As argued previously, lateral injection of current enables a decoupling of these two factors so that feedback strength and injection strength could be controlled independently. Further exploration along the longitudinal dimension is also promising for the development of split electrode devices which would not suffer from undesired electrical coupling as they do in the vertical injection paradigm.

References

1. F. Fidorra, H. Heidrich, R. Kaiser, and D. Trommer, "Monolithic integrated heterodyne receivers", *Lasers and Electro-Optics Society Annual Meeting*, 1995, Vol. 1, pp. 17–18.
2. H. Namizaki, H. Kan, M. Ishii, and A. Ito, "Transverse-junction-stripe-geometry double-heterostructure lasers with very low threshold current", *J. Appl. Phys.* **45** (1974) 2785–2786.
3. H. Namizaki, "Transverse-junction-stripe lasers with a GaAs *p–n* homojunction", *IEEE J. Quantum Electron.* **11** (1975) 427–431.
4. W. Susaki, T. Tanaka, H. Kan, and M. Ishii, "New structures of GaAlAs lateral-injection laser for low-threshold and single-mode operation", *IEEE J. Quantum Electron.* **13** (1977) 587–591.
5. H. Namizaki, M. Ishii, H. Kan, E. Ohmura, R. Hirano, and A. Ito, "Long-life GaAs-GaAlAs TJS laser with low threshold current and fundamental transverse and single longitudinal mode", *Opt. Commun.* **18** (1976) 39–40.
6. E. Oomura, R. Hirano, T. Tanaka, M. Ishii, and W. Susaki, "TJS laser with buried *p*-region for high temperature CW operation", *IEEE J. Quantum Electron.* **14** (1978) 460–461.
7. H. Kumabe, T. Tanaka, H. Namizaki, M. Ishii, and W. Susaki, "High temperature single-mode CW operation with a junction-up TJS laser", *Appl. Phys. Lett.* **33** (1978) 38–39.
8. K. Isshiki, N. Kaneno, H. Kumabe, H. Namizaki, K. Ikeda, and W. Susaki, "Ten-thousand-hour operation of crank transverse-junction-stripe lasers grown by metal-organic chemical vapor deposition", *IEEE J. Lightwave Technol.* **4** (1986) 1475–1481.
9. K. Mitsunaga, K. Fujiwara, M. Nunoshita, and T. Nakayama, "Si-doped GaAs/AlGaAs TJS laser by MBE", *J. Vac. Sci. Technol.* **B2** (1984) 256–258.
10. S. Nita, H. Namizaki, S. Takamiya, and W. Susaki, "Single-mode junction-up TJS lasers with estimated lifetime of 10^6 hours", *IEEE J. Quantum Electron.* **15** (1979) 1208–1209.
11. Y. Kadota, K. Chino, Y. Onodera, H. Namizaki, and S. Takamiya, "Agingl behavior and surge endurance of 870–900 nm AlGaAs lasers with nonabsorbing mirrors", *IEEE J. Quantum Electron.* **20** (1984) 1247–1251.
12. H. Kumabe, S. Nita, Y. Seiwa, T. Tanaka, T. Sogo, S. Horiuchi, and S. Takamiya, "Continuous-wave 15-mW operation test of single-mode high-power crank structure TJS laser diodes", *Conference on Lasers and Electro-Optics (CLEO '81)*, pp. 152–153.
13. H. Namizaki, H. Kumabe, and W. Susaki, "Spontaneous emission behavior in AlGaAs TJS lasers", *IEEE J. Quantum Electron.* **17** (1981) 799–803.
14. S. Nagai, S. Nita, K. Yamashita, T. Sogo, and S. Takamiya, "Noise characteristics of multi-longitudinal-mode transverse junction stripe laser diodes", *9th European Conference on Optical Communication*, 1983, pp. 143–146.
15. H. Kumabe, K. Isshiki, H. Namizaki, T. Nishioka, H. Koyama, and S. Takamiya, "Mechanical stress suppressed AlGaAs high power TJS laser diodes", *15th Conference on Solid State Devices and Materials*, 1983, pp. 329–332.
16. Y. J. Yang, Y. C. Lo, G. S. Lee, K. Y. Hsieh, and R. M. Kolbas, "Transverse junction stripe laser with a lateral heterobarrier by diffusion enhanced alloy disordering", *Appl. Phys. Lett.* **49** (1986) 835–837.
17. Y. Sin, K. Y. Hsieh, J. H. Lee, and R. M. Kolbas, "Surface and bulk leakage currents in transverse junction stripe lasers", *J. Appl. Phys.* **69** (1991) 1081–1090.

18. K. L. Yu, U. Koren, T. R. Chen, and A. Yariv, "A groove GaInAsP laser on semi-insulating InP using a laterally diffused junction", *IEEE J. Quantum Electron.* **18** (1982) 817–819.
19. T. R. Chen, U. Koren, K. L. Yu, K. Y. Lau, L. C. Chiu, A. Hasson, S. Margalit, and A. Yariv, "High-power, single-mode operation of an InGaAsP/InP laser with a grooved transverse junction using gain stabilization", *Appl. Phys. Lett.* **41** (1982) 225–228.
20. L. B. Chang and L. Z. Shia, "Room-temperature continuous wave operation of a visible AlGaAs/InGaP transverse junction stripe laser grown by liquid phase epitaxy", *Appl. Phys. Lett.* **60** (1992) 1090–1092.
21. L. B. Chang and L. Z. Shai, "Fabrication methods for an AlGaAs/InGaP/AlGaAs visible laser with transverse junction stripe structure grown by liquid phase epitaxy", *Solid State Electron.* **36** (1993) 1049–1054.
22. M. Ishii, K. Kamon, M. Shimazu, M. Mihara, H. Kumabe, and K. Isshiki, "Planar TJS lasers fabricated in semi-insulating GaAs substrates for optoelectronic integrated circuits", *Electron. Lett.* **23** (1987) 179–181.
23. J. Ohta, K. Kuroda, K. Mitsunaga, K. Kyuma, K. Hamanaka, and T. Nakayama, "Buried transverse-junction stripe laser for optoelectronic-integrated circuits", *J. Appl. Phys.* **61** (1987) 4933–4935.
24. J. Ohta, K. Kuroda, K. Mitsunaga, K. Kyuma, K. Hamanaka, and T. Nakayama, "Monolithic integration of a transverse-junction stripe laser and metal-semiconductor field-effect transistors on a semi-insulating GaAs substrate", *Electron. Lett.* **23** (1987) 509–510.
25. F. Brillouet, E. V. K. Rao, J. Beerens, and Y. Gao, "A new low capacitance transverse junction stripe AlGaAs/GaAs laser for planar laser-MESFET integration", *Proceedings of the Fourteenth International Symposium on Gallium Arsenide and Related Compounds*, 1987, pp. 789–792.
26. F. Brillouet, E. V. K. Rao, and J. Beerens, "New low capacitance transverse junction stripe AlGaAs/GaAs laser for planar laser-MESFET integration", *Electron. Lett.* **24** (1988) 97–99.
27. Y. Honda, I. Suemune, and M. Yamanishi, "Monolithic integration of a lateral current injection laser and a JFET based on the same modulation doped structure", *22nd Conference on Solid State Devices and Materials*, 1990, pp. 1179–1180.
28. N. Yasuhira, I. Suemune, Y. Kan, and M. Yamanishi, "Selectively doped double-heterojunction lateral current injection ridge waveguide AlGaAs/GaAs laser", *Appl. Phys. Lett.* **56** (1990) 1391–1393.
29. Y. Honda, I. Suemune, N. Yasuhira, and M. Yamanishi, "A new optoelectronic device based on modulation-doped heterostructure: demonstration of functions as both lateral current injection laser and junction field effect transistor", *IEEE Photon. Technol. Lett.* **2** (1990) 881–883.
30. Y. Honda, I. Suemune, N. Yasuhira, and M. Yamanishi, "Continuous-wave operation of lateral current injection ridge waveguide AlGaAs/GaAs laser with a selectively-doped heterostructure", *Jpn. J. Appl. Phys. (Part 1)* **30** (1991) 990–991.
31. P. A. Evaldsson, G. W. Taylor, P. Cooke, P. R. Claisse, C. A. Burrus, and B. Tell, "Small-signal and continuous wave operation of the lateral current injection heterostructure field-effect laser", *Appl. Phys. Lett.* **60** (1992) 1697–1699.
32. S. K. Sargood, G. W. Taylor, P. R. Claisse, T. Vang, P. Cooke, D. P. Docter, P. A. Kiely, and C. A. Burrus, Jr., "A quantum-well inversion channel heterostruc-

ture as a multifunctional component for optoelectronic integrated circuits", *IEEE J. Quantum Electron.* **29** (1993) 136–149.
33. G. A. Vawter, J. L. Merz, and L. A. Coldren, "Monolithically integrated transverse-junction-stripe laser with an external waveguide in GaAs/AlGaAs", *IEEE J. Quantum Electron.* **25** (1989) 154–162.
34. H. Kawanishi, M. Hafich, B. Lenz, and P. Petersen, "AlGaAs transverse junction stripe laser with distributed Bragg reflector", *Electron. Lett.* **16** (1980) 738–740.
35. H. Kawanishi, M. J. Hafich, R. A. Skogman, B. S. Lenz, and P. E. Petersen, "GaAs/AlGaAs distributed feedback-transverse junction stripe laser using a hybrid liquid phase epitaxy/metal-organic chemical vapor deposition growth technique", *J. Appl. Phys.* **52** (1981) 4447–4449.
36. M. Kameya, S. Tai, K. Mitsunaga, K. Kyuma, K. Hamanaka, and T. Nakayama, "Spectral linewidth of an AlGaAs/GaAs DFB-TJS external-cavity laser with optical phase control loop", *Electron. Lett.* **22** (1986) 1231–1232.
37. K. Mitsunaga, S. Noda, K. Kojima-K, M. Kameya, K. Kyuma, K. Hamanaka, and T. Nakayama, "GaAs/AlGaAs distributed feedback transverse junction stripe laser grown by molecular beam epitaxy", *Appl. Phys. Lett.* **50** (1987) 1622–1624.
38. W. Hsin, M. Ogura, J. P. Weber, S. C. Wang, J. J. Yang, M. C. Wu, S. Wang, and J. R. Whinnery, "GaAs/AlGaAs surface emitting laser diode with vertical distributed feedback optical cavity and transverse junction buried heterostructure", *IEDM Tech. Dig.* (1987) 792–795.
39. H. Yamada, H. Watanabe, H. Ito, and H. Inaba, "Surface-emitting GaAs light-emitting diode/laser diode with modified coaxial transverse junction (CTJ) structure", *Electron. Lett.* **24** (1988) 77–78.
40. L. Mitsunaga, M. Kameya, K. Kojima, S. Noda, K. Kyuma, K. Hamanaka, and T. Nakayama, "CW surface-emitting grating-coupled GaAs/AlGaAs distributed feedback laser with very narrow beam divergence", *Appl. Phys. Lett.* **50** (1987) 1788–1790.
41. T. Takamori, L. A. Coldren, and J. L. Merz, "Folded-cavity transverse junction stripe surface-emitting laser", *Appl. Phys. Lett.* **55** (1989) 1053–1055.
42. W. Yang, A. Gopinath, and M. Hibbs-Brenner, "Planar GaAs-AlGaAs MQW transverse junction ridge waveguide lasers using shallow zinc diffusion", *IEEE Photon. Technol. Lett.* **7** (1995) 848–850.
43. A. Furuya, M. Makiuchi, O. Wada, T. Fujii, and H. Nobuhara, "AlGaAs/GaAs lateral current injection (LCI)-MQW laser using impurity-induced disordering", *Jpn. J. Appl. Phys.* **26** (1987) L134–135.
44. A. Furuya, M. Makiuchi, and O. Wada, "Continuous-wave operation of lateral current injection multiquantum-well laser", *Electron. Lett.* **24** (1988) 1282–1283.
45. A. Furuya, M. Makiuchi, O. Wada, and T. Fujii, "AlGaAs/GaAs lateral current injection multiquantum well (LCI-MQW) laser using impurity-induced disordering", *IEEE J. Quantum Electron.* **24** (1988) 2448–2453.
46. O. Wada, A. Furuya, and M. Makiuchi, "Planar, compatible OEIC's based on multiquantum well structures", *IEEE Photon. Technol. Lett.* **1** (1989) 16–18.
47. Y. Suzuki, S. Mukai, H. Yajima, and T. Sato, "Transverse junction buried heterostructure (TJ-BH) AlGaAs diode laser", *Electron. Lett.* **23** (1987) 384–386.
48. K. Shimoyama, M. Katoh, M. Noguchi, Y. Inoue, H. Gotoh, Y. Suzuki, and T. Satoh, "Transverse junction buried heterostructure (TH-BH) laser diode grown by MOVPE", *J. Crystal Growth* **93** (1988) 803–808.

49. K. Shimoyama, M. Katoh, Y. Suzuki, T. Satoh, Y. Inoue, S. Nagao, and H. Gotoh, "CW operation and extremely low capacitance of TJ-BH MQW laser diodes fabricated by entire MOVPE", *Jpn. J. Appl. Phys.* **27** (1988) L2417–L2419.
50. K. Shimoyama, M. Katoh, Y. Suzuki, T. Satoh, Y. Inoue, S. Nagao, and H. Gotoh, "CW operation and extremely low capacitance of TJ-BH laser diodes having MQW active layers fabricated by entire MOVPE", *20th Conference on Solid State Devices and Materials*, 1988 pp. 343–346.
51. Y. Kawamura, Y. Noguchi, and H. Iwamura, "Lateral current injection InGaAs/InAlAs MQW lasers grown by GSMBE/LPE hybrid method", *Electron. Lett.* **29** (1993) 102–103.
52. K. Oe, Y. Noguchi, and C. Caneau, "GaInAsP lateral current injection lasers on semi-insulating substrates", *IEEE Photon. Technol. Lett.* **6** (1994) 479–481.
53. E. H. Sargent, G. L. Tan, and J. M. Xu, "Physical model of OEIC-compatible lateral current injection lasers", *IEEE Selected Topics Quantum Electron.* **3** (1997) 507–512.
54. E. H. Sargent, C. Caneau, C. Zah, and J. M. Xu, "OEIC-enabling lateral current injection lasers: Underlying physics mechanisms and design for improved high-power efficiency", *International Topical Workshop on Contemporary Photonic Technologies*, 1998.
55. E. H. Sargent and J. M. Xu, "Integration-compatible lateral current injection lasers: Design of 2-D heterostructure devices", *Conference on Lasers and Electro-Optics/Pacific Rim*, 1997.
56. E. H. Sargent, G. L. Tan, and J. M. Xu, "Investigations of lateral current injection lasers: Internal operating mechanisms and doping profile engineering", *Compound Semiconductors 1996*, 1997, pp. 651–4.
57. E. H. Sargent and J. M. Xu, "Investigations of the physical mechanisms governing the performance of OEIC-compatible *p–i–n* active region lateral current injection lasers", *LEOS '96 Annual Meeting*, 1996, pp. 314–315.
58. E. H. Sargent, G. L. Tan, D. A. Suda, and J. M. Xu, "Internal operating mechanisms of OEIC-compatible lateral injection lasers: Intrinsic differences from the vertical injection paradigm", *15th IEEE International Semiconductor Laser Conference (ISLC)*, 1996.
59. E. H. Sargent and F. Meshkati, "OEIC-enabling lateral current injection (LCI) lasers: Efficient functional devices using lateral beam-steering", *European Conference on Integrated Optics*, 1999.
60. E. H. Sargent, G. Tan, and J. M. Xu, "Lateral current injection lasers: Underlying mechanisms and design for improved high-power efficiency", *IEEE Lightwave Technol.* **16** (1998) 1854–1864.
61. G.-L. Tan, N. Bewtra, K. Lee, and J. M. Xu, "A two-dimensional nonisothermal finite element simulation of laser diodes", *IEEE J. Quantum Electron.* **29** (1993) 822–835.
62. R. F. Kazarinov and M. R. Pinto, "Carrier transport in laser heterostructures", *IEEE J. Quantum Electron.* **30** (1994) 49–53.
63. K. Oe, Y. Noguchi, and C. Caneau, "GaInAsP lateral current injection lasers on semi-insulating substrates", *IEEE Photon. Technol. Lett.* **6** (1994) 479–481.
64. D. A. Suda, H. Lu, T. Makino, and J. M. Xu, "An investigation of lateral current injection laser internal operation mechanisms", *IEEE Photon. Technol. Lett.* **7** (1995) 1122–1124.
65. E. H. Sargent, K. Oe, C. Caneau, and J. M. Xu, "OEIC-enabling LCI lasers with current guides: Combined theoretical-experimental investigation of internal operating mechanisms", *IEEE J. Quantum Electron.* **34** (1998) 1280–1287.

66. E. H. Sargent, J. M. Xu, C. Caneau, and C. Zah, "Experimental investigation of physical mechanisms underlying lateral current injection laser operation", *Appl. Phys. Lett.* **73** (1998) 285–287.
67. E. H. Sargent and J. M. Xu, *Proc. CLEO/Pacific Rim* (1997).
68. J. Z. Wan, J. G. Simmons, and D. A. Thompson, *J. Appl. Phys.* **81** (1997) 765.
69. E. H. Sargent, D. A. Suda, A. Margittai, F. R. Shepherd, M. Cleroux, G. Knight, N. Puetz, T. Makino, A. J. SpringThorpe, G. Chik, and J. M. Xu, "Experimental study of LCI lasers fabricated by single MOCVD overgrowth followed by selective dopant diffusion", *IEEE Photon. Technol. Lett.* **10** (1998) 1536–1538.
70. N. M. Margalit, D. I. Babic, K. Streubel, R. P. Mirin, R. L. Naone, J. E. Bowers, and E. L. Hu, "Submilliamp long wavelength vertical cavity lasers", *Electron. Lett.* **32** (1996) 1675–1677.

International Journal of High Speed Electronics and Systems, Vol. 9, No. 4 (1998) 979–1005

EXTENDED WAVELENGTH (1.0 TO 1.3 μm) InGaAs/GaAs QUANTUM DOT GaAs-BASED VERTICAL-CAVITY SURFACE-EMITTING AND LATERAL-CAVITY EDGE-EMITTING LASERS

D. G. DEPPE and D. L. HUFFAKER
Microelectronics Research Center,
Department of Electrical and Computer Engineering,
The University of Texas at Austin, Texas 78712-1084, USA

An important advance in InGaAs/GaAs quantum dot lasers has been the demonstration of lasing at wavelengths significantly longer than that possible using InGaAs strained quantum wells, extending beyond 1.3 μm. These fully GaAs-based active regions are compatible with commercial vertical-cavity surface-emitting laser (VCSEL) technology based on selective oxidation, and offer novel performance due to their three-dimensional electronic confinement. This chapter reviews the status of long wavelength quantum dot edge-emitting lasers and VCSELs, and presents some of the new physical principles needed to understand their novel device characteristics.

1. Introduction

Semiconductor quantum dot (QD) lasers were studied theoretically in the late 1980's,[1–3] and subsequently demonstrated in 1994 and thereafter.[4–10] Key to the QD laser demonstration has been the finding that self-organization during growth of highly strained InGaAs can form three-dimensional islands with a crystal lattice coherent to the GaAs substrate.[11–13] These self-organized QDs exhibit good luminescence efficiency, and as compared to planar quantum well lasers, also exhibit novel device characteristics. The lasing threshold is temperature insensitive in the absence of nonradiative recombination, and the lasing wavelength has been extended to 1.33 μm for GaAs-based QD lasers.[14–19] The longer wavelength QDs with emission between 1.0 and 1.3 μm are useful for GaAs-based vertical-cavity surface-emitting lasers (VCSELs), and QD VCSELs have also been demonstrated at wavelengths as long as 1.15 μm.[20–25]

In this chapter, we describe recent advances and the relevant device physics for longer wavelength GaAs-based QD lasers. We focus on the electronic structure of inhomogenously broadened QD ensembles, their gain characteristics, dynamic response, and the influence of nonradiative recombination. Recent studies for both edge-emitters and VCSELs are reviewed. The performance of QD lasers continue to rapidly improve, and the emphasis here is placed on understanding how the device physics changes for QDs as compared to the extensively studied planar quantum well system. In Sec. 2, we review the molecular beam epitaxy (MBE) of InGaAs QDs grown on GaAs. In Sec. 3, we describe the electronic structure of the QDs, and present nonlinear rate equations that describe electron or hole capture in the QD ground state. The laser threshold conditions in terms of the cavity parameters and QD characteristics are given for edge-emitters and VCSELs. In Sec. 4, we present the latest results for 1.3 μm QD edge-emitting lasers, and in Sec. 5, do the same for QD VCSELs. Section 6 gives a brief summary.

2. Epitaxial Growth and Electroluminescence Characteristics of Quantum Dots

Quantum dots can be formed by the self-organization of highly strained epitaxial material that leads to three-dimensional islanding.[11–13] So far, other approaches for forming small volume three-dimensionally confined III–V semiconductor structures have inevitably suffered from surface-problems. Etching in particular leads to a large surface recombination velocity, and the high surface-to-volume ratio for QDs kills the luminescence. The self-organization can lead to high quality heterointerfaces through *in-situ* formation under the same growth conditions used for high quality epitaxy. An important aspect of self-organized QDs is that the three-dimensional surfaces can accommodate a much greater indium content than planar growth before dislocations form due to strain, and longer emission wavelength is possible as compared to InGaAs quantum wells coherently strained to GaAs. The precise emission wavelength, however, depends on the QD dimensions. Typically, a QD's area is greater than its height (height taken as normal to the growth plane). The high In content and large InGaAs QD size have extended the lasing wavelength to beyond 1.3 μm,[16] while planar InGaAs quantum well lasers suffer serious performance problems for wavelengths much beyond 1.0 μm. The technological importance of the 1.3 μm for fiber optic interconnects has generated a great deal of interest in applying the InGaAs QD technology to VCSELs and low power edge-emitting lasers, and we focus on these long wavelength QDs below.

During the initial 1994 demonstrations of QD lasers, 1.3 μm light emission from InGaAs/GaAs QDs was also first demonstrated.[26] These $In_{0.50}Ga_{0.50}As$ average composition QDs were grown using metal organic vapor phase epitaxy by alternating layers of InAs and GaAs. In 1995, 1.3 μm photoluminescence was achieved by molecular beam epitaxy (MBE) from $In_{0.30}Ga_{0.70}As$ average composition QDs grown using alternating layers of column III and column IV material.[27] However, the crystal quality in these early experiments remains unclear. Although 1.3 μm emission has also been demonstrated for planar InGaAs quantum wells, the material quality to date is far inferior to that needed to make lasers, and early studies on the 1.3 μm InGaAs QDs did not include enough information to accurately assess the luminescence efficiency.

In 1998, 1.3 μm light emitting diodes (LEDs) were reported along with a careful study of growth conditions and electroluminescence efficiency.[28] These dots were grown by MBE using cycled, sub-monolayer deposition of In, Ga, and As, a technique also used in earlier studies. A wide range of growth parameters were studied to show that the luminescence efficiency depends strongly on the mount of In, Ga, and As supplied in each cycle to the growth surface. The best luminescence efficiency is achieved by alternately supplying 0.25 ML of In and 0.25 ML of Ga in each column III deposition cycle, and this same growth recipe led to the first demonstration of a 1.3 μm GaAs-based QD laser.[16] Continuous-wave operation of 1.3 μm edge emitting lasers has also been reported using MBE with either multiple stack or a single stack $In_{0.50}Ga_{0.50}As$ QD active region.[18,19] So far, the slow growth rate inherent to MBE has been required to reproducibly grow high quality QDs, especially for long wavelengths.

Several studies show that the radiative efficiency can be improved by surrounding the InGaAs QDs with $In_xGa_{1-x}As$, where $x \leq 0.2$ and x and the QD deposition are chosen to avoid dislocations.[29–32] The detailed reasons for the material improvement are still being investigated, but some important trends in the change of the QD growth are apparent in initial data from atomic force microscope (AFM) studies and *in-situ* characterization using reflection high energy electron diffraction (RHEED). Figure 1 shows AFM images of the 1.3 μm InGaAs QDs grown either without (a) or with (b) a thin $In_{0.15}Ga_{0.85}As$ buffer layer. The QD density increases by approximately a factor of two over when the InGaAs buffer layer is absent, which also increases the maximum optical gain from the 1.3 μm QD ensemble.

Covering the InGaAs QDs with $In_xGa_{1-x}As$ where $x \leq 0.2$ also improves the luminescence efficiency. Nonradiative recombination in the QDs is a serious problem, and the QD lasing threshold can suffer thermal run-away due to the temperature dependence of nonradiative recombination. The nonradiative recombination also has an excitation level dependence, with the luminescence efficiency decreasing for even moderately increasing excitation levels. The excitation level dependence comes from the high nonradiative recombination rates in the wetting layer and higher energy QD levels. This nonradiative recombination problem seems to be

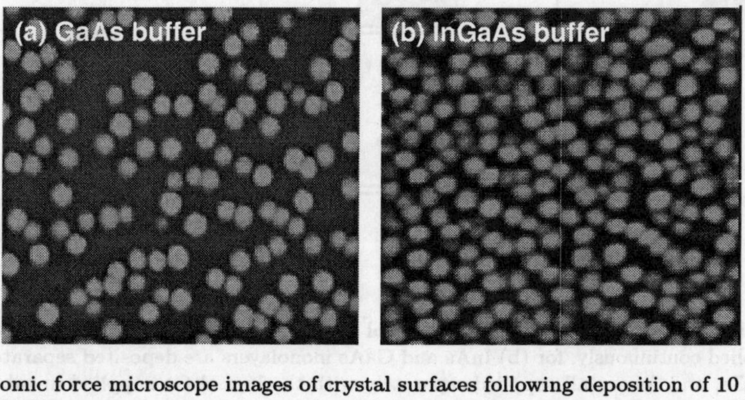

Fig. 1. Atomic force microscope images of crystal surfaces following deposition of 10 monolayers of InGaAs (50% average In content) on either (a) GaAs or (b) InGaAs buffer of ~10% In.

reduced when the QDs are covered with $In_xGa_{1-x}As$ where $x \leq 0.2$. Because the crystal strain can also be viewed as a pressure effect, one might speculate that inter diffusion and even melting occur when the heterointerfaces become highly strained by covering with GaAs. These problems are reduced for covering with $In_xGa_{1-x}As$, perhaps because the QD strain is accommodated over a larger volume. Improved $In_{0.50}Ga_{0.50}As$ QD laser characteristics at 1.3 μm have been achieved by growing the QDs on an $In_{0.10}Ga_{0.90}As$ buffer layer. Continuous-wave lasing has been obtained with threshold currents as low as $\sim$ 4 mA at room temperature, and threshold current density as low as 40 A/cm^2 (Ref. 19). These QD laser results are discussed in Sec. 4.

The impact of different deposition techniques on the QD electroluminescence efficiency has been studied for different deposition conditions while keeping the average deposited In content and the total number of deposited MLs constant. The different techniques are illustrated in Fig. 2 and involve controlling the sub-ML supply of both In and Ga, as well as the supply of As. Figure 3 shows AFM photographs of the crystal surfaces with the growths terminated directly following QD deposition. The total QD material deposition is 10 MLs with an average In composition of 50%, and the growth temperatures for each are 510°C. Figure 3(a) shows $In_{0.50}Ga_{0.50}As$ QDs formed by a constant supply of In, Ga, and As to the crystal surface. The as-formed QDs (without overgrowth) have an average dot size of $\sim$ 350 Å and dot density of $\sim 10^{11}$ dots/cm^2. However, the luminescence efficiency is poor for the QDs of Fig. 3(a). The large clusters that can be seen are presumably relaxed by dislocations. For the constant supply of In, Ga, and As the QD luminescence efficiency drops significantly for depositions greater than 7 MLs.

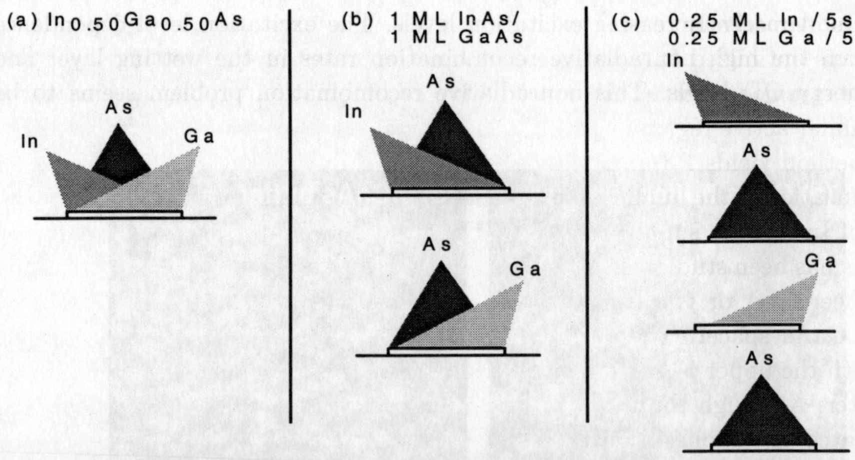

Fig. 2. Different deposition techniques to control size and density of QDs. For (a) the In, Ga, and As are supplied continuously, for (b) InAs and GaAs monolayers are deposited separately, and for (c) the In, Ga, and As are all supplied separately with the In and Ga supplied in sub-monolayer depostions.

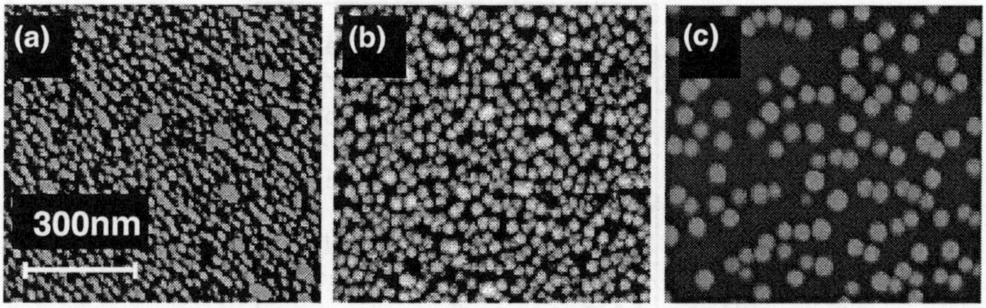

Fig. 3. Atomic force microscope images of the epitaxial crystal surfaces after depositions of 10 monolayers of InGaAs with 50% average In content, but with the different deposition techniques. (a), (b), and (c) correspond to depositions shown in Fig. 2(a), (b), and (c).

Figure 3(b) shows QDs formed by 5 repetitions of a 1 ML InAs/1 ML GaAs cycle (again giving 10 total MLs of average 50% In content). The measured average QD size is ~ 400 Å and density is $\sim 5 \times 10^{10}$ dots/cm^2. Electroluminescence from this active region has a center wavelength of 1.22 μm, but the luminescence efficiency is also decreased over QDs grown under similar conditions but with fewer MLs. Although a 14 ML deposition of 1 ML InAs/1 ML GaAs exhibits electroluminescence peaked at ~ 1.3 μm, its luminescence efficiency is further decreased as compared to the 10 ML deposition, due to dislocation formation.

Figure 3(c) shows QDs formed by 20 cycles of 0.25 ML In, 5 sec. As, 0.25 ML Ga, 5 sec. As. The dot density is reduced to $\sim 10^{10}$/cm^2 and the dot diameter is much larger (~ 500 Å) compared to Figs. 3(a) and (b). This difference in the dot density and size indicates that the In and Ga growth in the absence of As allows a larger surface migration of the column III atoms during QD formation, and favors larger QDs. Apparently the easy migration allows the QDs to form with larger size and perhaps optimized In grading within the QDs themselves to minimize defects due to the strained material. The room temperature electroluminescence measured from a similar active region in an LED is 1.32 μm. It is significant that the sub-ML growth method yields 1.3 μm emission with a total In content of 5 ML.

Optimization of the luminescence efficiency requires careful control of the precise amount of In and Ga supplied in a sub-ML during growth of the long wavelength QDs. This has been studied using LEDs.[28] Either a single layer of QDs in a 0.19 μm GaAs spacer layer or two layers (two-stack) separated by 0.19 μm of GaAs in a 0.38 μm GaAs spacer. The GaAs spacers are surrounded by AlGaAs confining layers, with the upper *p*-AlGaAs layer oxidized to form oxide apertures of ~ 10 μm in diameter. Although the sub-ML deposition cycles use equal amounts of In and Ga separated by supplies of 5 sec. As flux and the total deposited QD material is held fixed at 10 MLs, the precise amount of In and Ga deposited are varied between 0.50 ML/0.50 ML, 0.25 ML/0.25 ML and 0.15 ML/0.15 ML.

The electroluminescence light output power versus current curves for five different QD LEDs are measured and converted to output efficiency per QD layer as

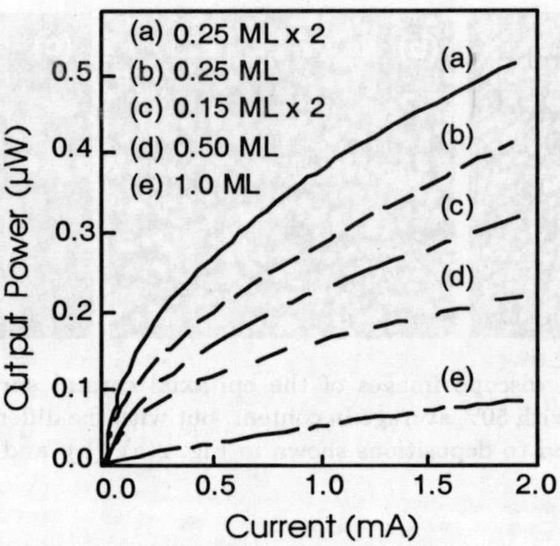

Fig. 4. Collected output power versus current for different QD growth depositions.

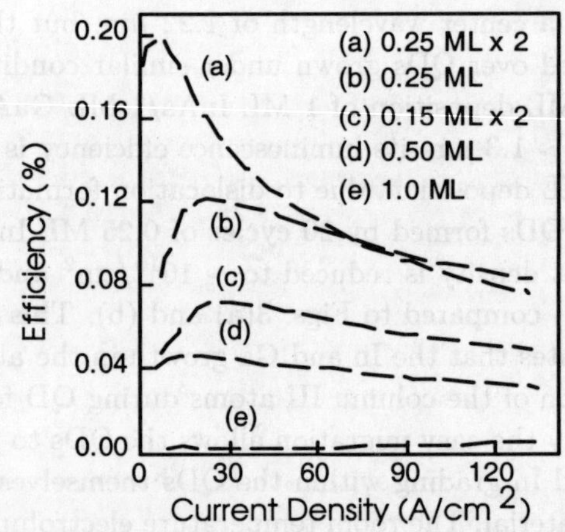

Fig. 5. Computed efficiency based on curves of Fig. 4 and estimated collection efficiency.

a function of current density. This data is plotted in Figs. 4(a)–(e) and 5(a)–(e). In both figures, curves (a)–(d) represent data from 10 ML total QD material grown using different sub-monolayer thicknesses and curve (e) shows data from a 10 ML growth of 1 ML InAs/1 ML GaAs. As shown in Fig. 2, the maximum output power of 0.5 μW is achieved at 2 mA from the two-stack 0.25 ML/.25 ML QD LED. In Fig. 3, the single layer 0.25 ML/0.25 ML QD LED has a peak efficiency 0.2% at 8 A/cm^2 which is comparable to that of good $In_{0.20}Ga_{0.80}As$ quantum wells emitting at shorter wavelengths (0.98 μm). However, the efficiency rapidly decreases with increasing current density due to a higher nonradiative recombination rate due to the increasing number of electrons and holes contained in each QD. The decreased efficiency of the two-stack 0.25 ML/0.25 ML sample compared to the single stack is most likely due to inefficient carrier injection into the two QD layers and nonradiative recombination at heterointerfaces outside the QD regions. All of the sub-monolayer QD LEDs have significantly higher efficiency than the 1 ML InAs/1 ML GaAs LED.

In general, larger QDs show narrower linewidths. Mukai *et al.*, have reported a 37 meV linewidth at 1.33 μm and Mirin *et al.*, report 28 meV at 1.32 μm.[26,27] However, a dependence of the linewidth on the sub-ML supply is also found. The electroluminescence spectra from the two-stack 0.25 ML/0.25 ML LED for several current densities are shown in Figs. 6 and 7. For low current density, when only the ground state (at 1.32 μm) is excited, the full-width at half maximum (FWHM) of

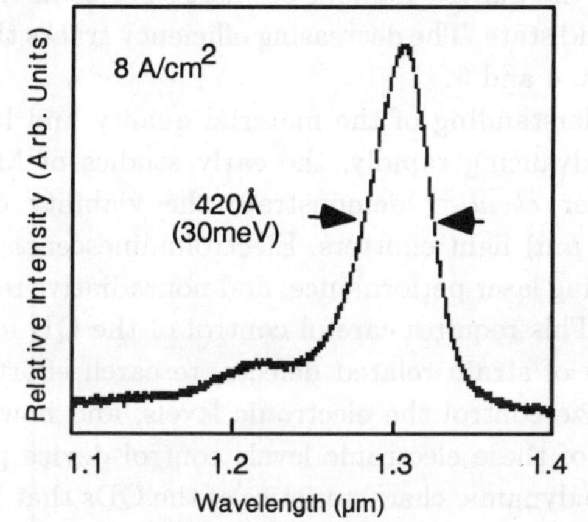

Fig. 6. Electroluminescence spectrum from the ground state emission of 1.3 μm InGaAs QDs grown on GaAs.

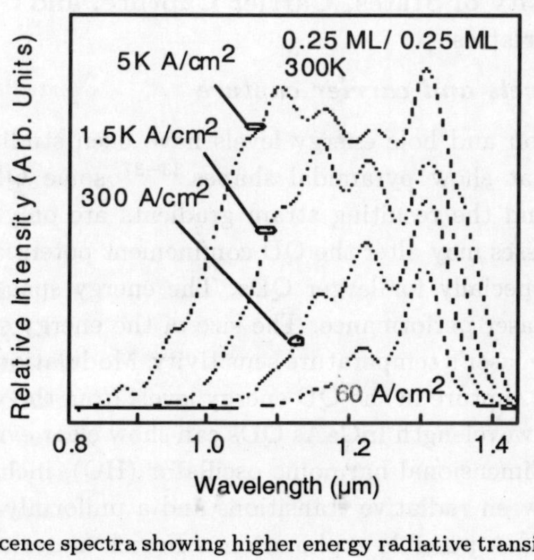

Fig. 7. Electroluminescence spectra showing higher energy radiative transitions from 1.3 μm QDs.

30 meV indicates good QD size uniformity. The single layer 0.25 ML/0.25 ML QD LED exhibits ground state emission at 1.32 μm with a FWHM of 40 meV, the single layer 0.5 ML/0.5 ML QD LED exhibits ground state emission at 1.31 μm with a FWHM of 41 meV, and the two stack 0.15 ML/0.15 ML QD LED exhibits ground state emission at 1.32 μm with a FWHM of 50 meV. From Fig. 6, only the ground state transitions are excited for the current density of 8 A/cm^2. As the current density is increased from 60 A/cm^2 to 5×10^3 A/cm^2, a total of four well-resolved higher energy states (1.22 μm, 1.17 μm, 1.14 μm and 1.12 μm) and a wetting layer at 0.95 μm are also excited. The ground state emission at 1.32 μm dominates the spectrum throughout this range of current densities due to the combined effects of heating and carrier loss from the higher energy states. The strong level filling that occurs is due to the limited number of energy states in the QD active region, especially in the ground state. The decreasing efficiency tracks the increasing current density shown in Figs. 4 and 5.

Although the understanding of the material quality and luminescence characteristics of QDs is advancing rapidly, the early studies of Mukai *et al.*,[26] Mirin *et al.*,[27] and Huffaker *et al.*,[28] demonstrate the viability of InGaAs QDs for long wavelength (1.3 μm) light emitters. Electroluminescence efficiency is one key parameter in controlling laser performance, and nonradiative recombination centers must be eliminated. This requires careful control of the QD formation conditions. With the elimination of strain related defects, research efforts can focus on how the QD shape and size control the electronic levels, and how the energy spacing and spectral density of these electronic levels control device performance. This is especially true for the dynamic characteristics of the QDs that include the radiative spontaneous emission rate and the rate of electron and hole capture and scattering among the discrete QD energy levels.

3. Quantum Density of States, Carrier Capture, and Gain Characteristics

3.1. *Electronic levels and carrier capture*

Although the electron and hole energy levels have been studied theoretically for InAs/GaAs QDs that show pyramidal shapes,[33–37] some QD features such as interface diffusion and the resulting strain gradients are only now being characterized. These processes may alter the QD confinement potential and therefore the energy spectrum, especially for larger QDs. The energy spectra of the QDs has a major impact on laser performance. The size of the energy separations between levels determines the laser's temperature sensitivity. Modulation characteristics can be limited by carrier capture in the QD energy levels from the wetting layer.[38] Experimentally, longer wavelength InGaAs QDs can show clear evidence of the energy spectrum of a two-dimensional harmonic oscillator (HO), including both uniform energy spacings between radiative transitions and a uniformly stepped saturation of these transitions.[14] A possible explanation is In rich material at the QD center, leading to a gradient in the In concentration. Since the height of the QD is typically much less than its diameter, the energy spacing due to the QD height is much larger than due to its diameter. For longer wavelength QDs there are six radiative transitions observed that are separated by nearly constant energies of ~ 65 meV.[28] Also, larger QDs in general appear to have shorter radiative lifetimes than smaller QDs, and this supports the idea that QD emission is based on superradiance.[39–41]

Describing the QD energy levels as that of a two-dimensional HO agrees with the available experimental data for the 1.3 μm QDs,[14,15,42,43] and provides a useful mathematical framework from which to calculate the electronic structure and the QD dynamic response. However, the reader is cautioned that more experiments are required before the long wavelength QD electronic structure is really understood. Figure 8 illustrates the energy levels given equal electron and hole energy level separations and equal effective masses. The spontaneous luminescence spectra suggest that selection rules govern the radiative electron-hole transitions within the QDs, as illustrated in Fig. 8(a) so that each electron level is radiatively coupled to only one hole level. The reduced density of states (density of radiative transitions) based on the level degeneracies of the 2-D HO, assuming a Gaussian distribution of energy levels, is given by[15,43]

$$\rho_{\mathrm{red}}(\omega) = 4\sqrt{\frac{\ln(2)}{\pi}}\frac{n_{\mathrm{QD}}}{\Delta\omega}\sum_{m=0}^{M}(m+1)e^{-\ln(2)\frac{(\omega_g+m\omega_0-\omega)^2}{(\Delta\omega/2)^2}} \tag{1}$$

where n_{QD} is the quantum dot density, $\Delta\omega$ is the full-width at half-maximum of the emission linewidth, ω_g is the ground state emission frequency at line center, and m labels a collection of degenerate energy levels with degeneracy of $m+1$. The reduced density of states for QDs, along with the spontaneous lifetime, establish the gain characteristics from a QD ensemble.[15]

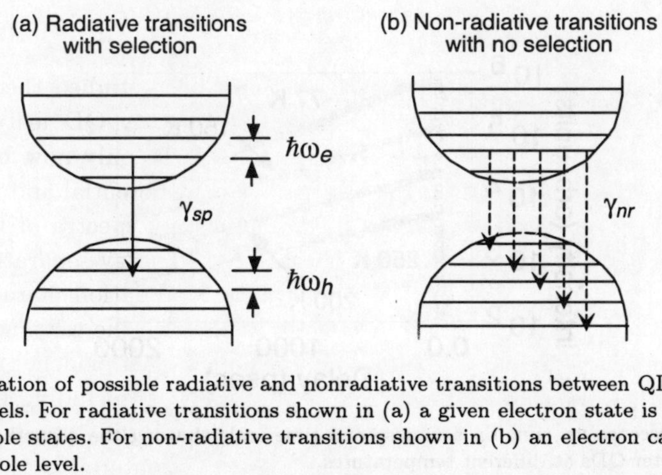

Fig. 8. Illustration of possible radiative and nonradiative transitions between QD's electron and hole energy levels. For radiative transitions shown in (a) a given electron state is coupled to only one or a few hole states. For non-radiative transitions shown in (b) an electron can recombine to any available hole level.

Assuming that nonradiative recombination occurs through defects, we do not expect selection rules to hold for nonradiative recombination. In other words, any QD electron can nonradiatively recombine with any hole in the same QD, as illustrated in Fig. 8(b). As discussed in Sec. 2, it is found experimentally that although nonradiative recombination may be negligible when only the ground state is occupied, the 1.3 μm QDs can presumably hold up to 42 electron-hole pairs.[43] The radiative current density required to fill the QDs to this level is ~ 250 A/cm^2, assuming a radiative recombination lifetime of ~ 400 psec.[39,44] If the nonradiative electron-hole recombination rate were, for example, 1 nsec. for only ground state occupation, the nonradiative lifetime would shorten to 240 picoseconds if the QD were totally filled with 42 electrons-hole pairs, requiring a nonradiative current density of 420 A/cm^2. The thermal distribution of carriers can also increase the nonradiative current density further by lengthening the radiative lifetime. The above picture is consistent with the measured trends in the efficiency versus current density and the efficiency versus temperature. Both increasing current density and increasing temperature decrease the QD luminescence efficiency.

Time resolved photoluminescence has been used to characterize the spontaneous emission rate from a 1.3 μm QD ensemble over a range of temperatures.[44] The luminescence decay data measured for ground state emission at different temperatures is shown in Fig. 9. A spectral filter is used to insure that only ground state emission is collected. As the temperature is increased, lengthening of the spontaneous emission time occurs due to thermal excitations of electrons and holes to higher energy levels. This thermal excitation reduces the probability of simultaneous electron and hole occupancy in radiatively coupled quantum states.

The dynamic response of the 1.3 μm QD ensemble has also been characterized using time resolved photoluminescence, and carrier relaxation in QDs has been an important topic sometimes of debate.[45–48] For the 1.3 μm QDs, the measured rise time for the ground state luminescence depends weakly on temperature and excitation level, but is less than 5 psecs. for all excitation conditions at temperatures

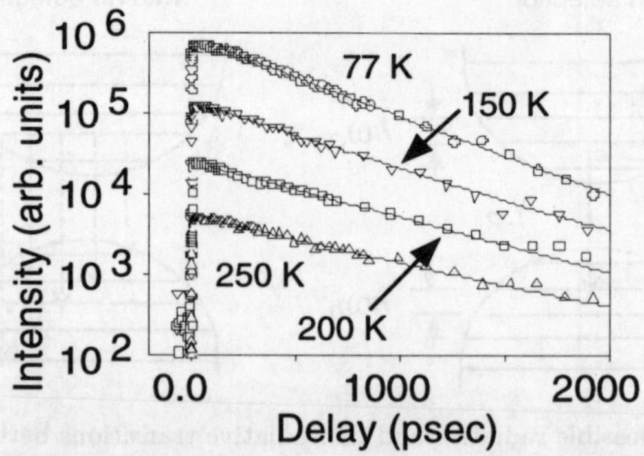

Fig. 9. Luminescence vs. time on a nanosecond time scale showing the ground state decay for the 1.3 micrometer QDs at different temperatures.

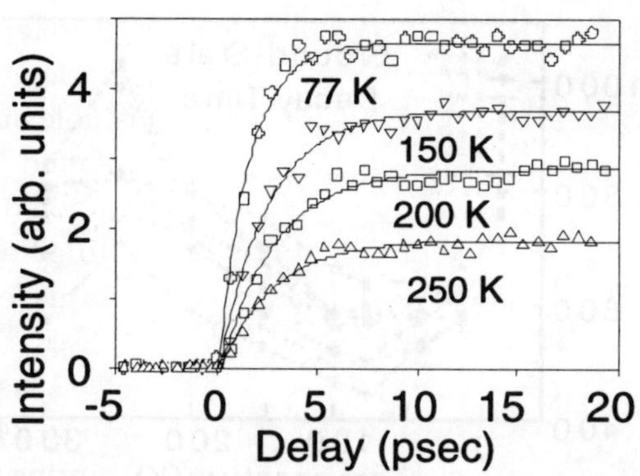

Fig. 10. Luminescence vs. time on a picosecond time scale showing the ground state rise for the 1.3 micrometer QDs at different temperatures.

$\leq$ 300 K. The measured rise times for different temperatures are shown in Fig. 10. In measuring the rise times, care must be taken to insure that diffusion effects due not limit to carrier capture from a bulk GaAs region. This can be done by growing heteroconfining barriers around the QD region, and keeping the total absorption region less than a few thousand angstroms in thickness. Then, excitation in either the wetting layer or bulk absorbing regions will result in different temporal responses, and reveal what bulk diffusion effects may limit the capture time. Under these conditions, the temperature dependence of the rise time due to carrier relaxation can be identified.

Figures 11 and 12 show the recombination times and rise times versus temperatures for a range of pump levels. A minimum in both the recombination times and rise times occurs at $\sim$ 77 K. At temperatures greater than $\sim$ 150 K we see that the spontaneous lifetime begins to lengthen, presumably due to excitation of carriers to

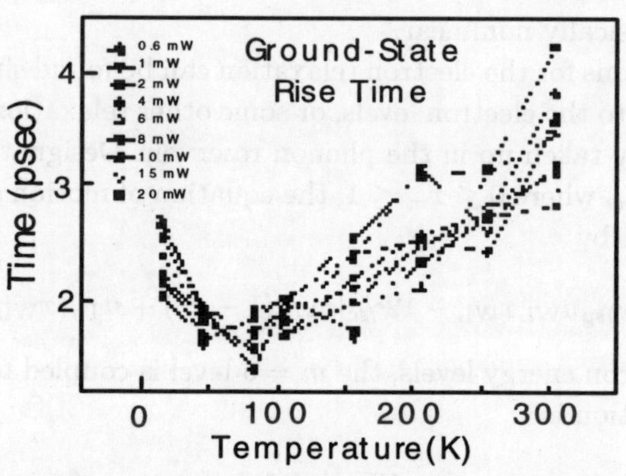

Fig. 11. Ground state rise time of 1.3 μm QDs measured versus temperature for several pump powers.

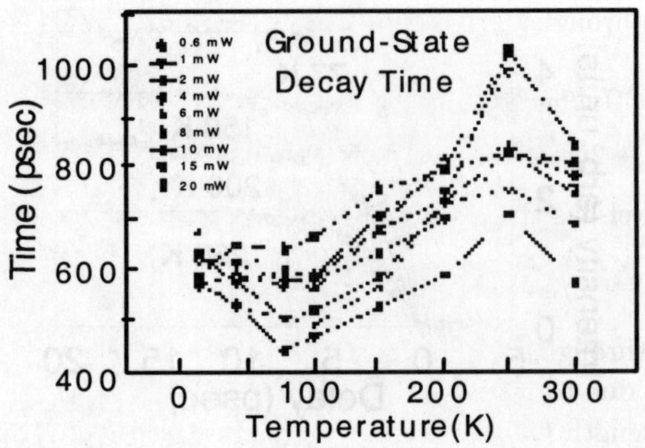

Fig. 12. Ground state decay time of 1.3 μm QDs measured versus temperature for several pump powers.

higher QD energy levels. The increase of the spontaneous emission rate with temperature is consistent with the assumption of selection rules between electron and hole energy levels for radiative transitions, similar to the case for planar quantum wells. However, the increase in recombination times and rise times for temperatures below ~ 77 K is not clearly understood. The low temperature increase in the rise time is consistent with the carrier capture from the GaAs regions as limited by carrier diffusion, since the diffusion coefficient will decrease for decreasing temperatures.

The QD dynamic response can be modeled using the HO model using the two-dimensional level degeneracies, and assuming that the QDs are coupled to a wetting layer with one-dimensional confinement. We assume that hole levels are filled with much greater rates than electron levels, so that the response is limited by electrons. We also assume that the wetting layer is filled with a pump and that the electrons relax to the QD ground state from the wetting layer. Each ground state electron must then relax through the five QD electron levels, with transition rates set in part by the level degeneracies. Since the Pauli exclusion principle holds, the electron relaxation is intrinsically nonlinear.

The rate equations for the electron relaxation can be found simply from coupling a phonon reservoir to the electron levels, or some other relaxation mechanism whose energy is eventually taken up in the phonon reservoir. Designating each level as m with occupation x_m, where $0 \leq x_m \leq 1$, the equation of motion of the wetting layer occupation is given by

$$\frac{dx_{\rm WL}}{dt} = -\gamma_{\rm sp} u_{\rm WL} x_{\rm WL} - 12\gamma_{\rm sc}[x_{\rm WL}(1 - x_5) + n_{\rm Th}(x_{\rm WL} - x_5)] \,. \tag{2}$$

Assuming six electron energy levels, the $m = 5$ level is coupled to the wetting layer. Its equation of motion is

$$\begin{aligned} \frac{dx_5}{dt} &= -\gamma_{\rm sp} u_5 x_5 - 10\gamma_{\rm sc}[x_5(1 - x_4) + n_{\rm Th}(x_5 - x_4)] \\ &\quad + N_{\rm WL}\gamma_{\rm sc}[x_{\rm WL}(1 - x_5) + n_{\rm Th}(x_{\rm WL} - x_5)] \,. \end{aligned} \tag{3}$$

The levels $m = 1$ through 4 have the equation of motion given by

$$\frac{dx_m}{dt} = -\gamma_{\rm sp} u_m x_m - 2m\gamma_{\rm sc}[x_m(1 - x_{m-1}) + n_{\rm Th}(x_m - x_{m-1})]$$
$$+ 2(m+2)\gamma_{\rm sc}[x_{m+1}(1 - x_m) + n_{\rm Th}(x_{m+1} - x_m)]\,. \quad (4)$$

And the $m = 0$ level has the equation of motion given by

$$\frac{dx_0}{dt} = -\gamma_{\rm sp} u_0 x_0 + 6\gamma_{\rm sc}[x_1(1 - x_0) + n_{\rm Th}(x_1 - x_0)]\,. \quad (5)$$

Equations (2) through (5) allow both the steady state spectral emission and the dynamic response due to intraband transitions be calculated for the 1.3 μm QDs, and shorter wavelength QDs as well. We assume that $n_{\rm Th}$ is due to the thermal reservoir of phonon modes and thus has the Bose–Einstein temperature dependence given by

$$n_{\rm Th} = \frac{1}{e^{\hbar\omega_e/(KT)} - 1} \quad (6)$$

where $\hbar\omega_e$ is the electron energy level separation, which we assume is greater that half the energy separation of the radiative transitions. The inclusion of $n_{\rm Th}$ in the rate equations allows temperature effects to be studied for comparison with experimental data.

Equations (2) through (6) allow the dynamic response of electrons and holes to be calculated as they scatter from one discrete energy level to another. Nonlinear effects arise in the scattering and give rise to level saturation. Figure 13 shows the calculated rise times using Eqs. (2) through (6) for either $T = 300$ K or $T = 77$ K. For these curves we assume that level populations are all zero at time $t = 0$ except for $x_{\rm WL}$. We assume that the $t = 0$ value of $x_{\rm WL}$ is set by the temporally short incident pump pulse that we assume is a delta function in time. One can see that the nonlinear rate equations give the correct qualitative trends in the 300 K calculation in that the rise time decreases as the pump level is increased. Note also that the calculated rise time for 77 K is much faster than for 300 K. Both trends stem from

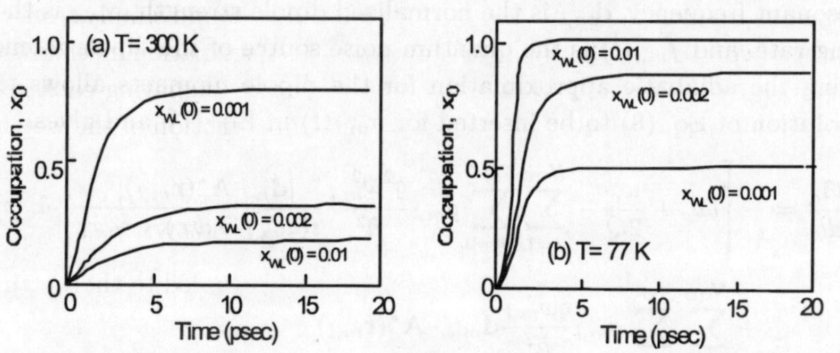

Fig. 13. Calculated time dependence of the ground state occupation rise time for different initial wetting layer populations for either (a) $T = 300$ K or (b) $T = 77$ K. The initial wetting layer population is assumed to be created by a short excitation pulse.

the large value of wetting layer states as compared to the number of QD levels that capture electrons. Because of the large difference in state numbers, for an electron captured in the upper level of a QD there is a high probability that it will be thermally re-excited back into the wetting layer. This re-excitation process slows down the capture, and thus the rise time of the ground state emission. At high pump levels the re-excitation back to the wetting layer is less likely due to the increased wetting layer population. At low temperature re-excitation back to the wetting layer is less likely because of the smaller thermal phonon population. Diffusion effects that may limit carrier capture from the GaAs region are not included in the calculated results of Fig. 13.

3.2. *Lasing threshold conditions*

The QD laser has been treated theoretically in several papers.[43,49–51] Here we use formulae that are also given elsewhere.[43] The field coupling of the lasing mode to the QD dipole moments can be derived from the laser Hamiltonian[52] and is described by the two equations of motion

$$\frac{d\hat{a}_c}{dt} = -\left(i\omega_c + \frac{\omega_c}{2Q}\right)\hat{a}_c(t) + \sum_{l=1}^{N_{\mathrm{QD}}}\sum_{m=0}^{\infty} g_{m,l}\frac{q\omega_{m,l}}{\hbar}\mathbf{d}_{m,l}\cdot\mathbf{A}_c^*(\mathbf{r}_l)\hat{\sigma}_{m,l}(t) + \hat{f}_c(t) \quad (7)$$

and

$$\frac{d\hat{\sigma}_{m,l}}{dt} = -(i\omega_{m,l} + \gamma_{m,l})\,\hat{\sigma}_{m,l} + \frac{q\omega_{m,l}}{\hbar}\mathbf{d}_{m,l}^*\cdot\mathbf{A}_c(\mathbf{r}_l)\hat{a}_c(t) + \hat{f}_{m,l}(t)\,, \quad (8)$$

where l is used to label the different QDs of the QD ensemble, m is used as above to label the energy levels of QD l with $g_{m,l}$ giving the degeneracy of each level, $\hat{a}_c(t)$ is the quantum field amplitude of the field mode, ω_c is the passive cavity oscillation frequency of the field mode, $\omega_c/(2Q)$ is one-half the photon loss rate of the passive cavity, $\mathbf{A}_c(\mathbf{r}_n)$ is the normalized vector field amplitude of the cavity mode at the QD position $\mathbf{r}_l$, q is the electronic charge, $\hat{f}_c(t)$ is the quantum noise operator of the cavity mode, and for QD l, $\hat{\sigma}_{m,l}(t)$ is the quantum dipole moment amplitude, $\omega_{m,l}$ is the resonant frequency, $\mathbf{d}_{m,l}$ is the normalized dipole strength, $\gamma_{m,l}$ is the dipole dephasing rate, and $\hat{f}_{m,l}(t)$ is the quantum noise source of the dipole moment.

Making the adiabatic approximation for the dipole moments allows the integrated solution of Eq. (8) to be inserted for $\hat{\sigma}_{m,l}(t)$ in Eq. (7), and gives

$$\begin{aligned}\frac{d\hat{a}_c}{dt} = &-\left[i\omega_c + \frac{\omega_c}{2Q} - \sum_{l=1}^{N_{\mathrm{QD}}}\sum_{m=0}^{\infty} g_{m,l}\frac{q^2\omega_{m,l}^2}{\hbar^2}\frac{|\mathbf{d}_n\cdot\mathbf{A}_c^*(\mathbf{r}_{m,l})|^2}{i(\omega_{m,l}-\omega_L)+\gamma_{m,l}}\right]\hat{a}_c(t)\\ &+\sum_{l=1}^{N_{\mathrm{QD}}}\sum_{m=0}^{\infty} g_{m,l}\frac{q\omega_{m,l}}{\hbar^2}\mathbf{d}_{m,l}\cdot\mathbf{A}_c^*(\mathbf{r}_{m,l})\\ &\times\int_{-\infty}^{t} dt'\, e^{-(i\omega_{m,l}+\gamma_{m,l})(t-t')}\,\hat{f}_{m,l}(t') + \hat{f}_c(t)\,, \end{aligned} \quad (9)$$

where ω_L is the actual lasing frequency. In Eq. (9), the first summation over l leads to stimulated emission and frequency pulling of the lasing mode, while the second summation over the dipole noise operators leads to spontaneous emission. Threshold is achieved when the real part of the stimulated emission term approximately equals the photon loss term, or for

$$\frac{\omega_c}{Q} = 2\sum_{l=1}^{N_{\mathrm{QD}}}\sum_{m=0}^{\infty} g_{m,l}\frac{q^2\omega_{m,l}^2}{\hbar^2}\frac{\gamma_{m,l}|\mathbf{d}_{m,l}\cdot\mathbf{A}_c^*(\mathbf{r}_{m,l})|^2}{(\omega_{m,l}-\omega_L)^2+\gamma_{m,l}}. \tag{10}$$

Equation (10) gives the general form for lasing threshold that can be used to describe different QD lasers. As shown below, it includes the electronic density of states given in Eq. (1) only if inhomogeneous broadening dominates the emission linewidth, and also includes the optical confinement factor through the field normalization of $\mathbf{A}_c^*(\mathbf{r}_{m,l})$. Note that if inhomogeneous broadening is eliminated from the QD ensemble, the threshold will be set by the QD dephasing rate. For this case the ground state frequency $\omega_{o,l}$ will be the same for all QDs, so that cavity tuning and the dephasing play a major role in establishing threshold. The inhomogeneous broadening in QD ensembles greatly exceeds the homogeneous linewidth. However, there have been some estimates of the homogeneous linewidth,[53] which is temperature dependent due to coupling to a phonon reservoir. For low temperature measured linewidths for single QDs appear to be $\sim$ 100 μeV.[54] This dephasing rate sets a maximum limit for the low temperature characteristic scattering time between energy levels of $\sim$ 7 psec, but this scattering rate is expected to increase significantly at room temperature.[43]

For the inhomogeneously broadened system the summation in Eq. (10) can be replaced with a reduced density of states of electronic transitions at the frequency. Replacing ω_L with ω and assuming that the ensemble contains enough QDs with a Gaussian distribution in energy levels so that the summations can be approximated by integrals, we use

$$\sum_{l=1}^{N_{\mathrm{QD}}}\sum_{m=0}^{\infty} g_{m,l}\omega_{m,l}^2\frac{\gamma_{m,l}}{(\omega_{m,l}-\omega)^2+\gamma_{m,l}^2} = \pi\omega^2\int_{-\infty}^{\infty}dx_l\int_{-\infty}^{\infty}dy_l\,\rho_{\mathrm{red}}(\omega) \tag{11}$$

where the reduced density of states accounting for inhomogeneous broadening in the QD ensemble is then given by Eq. (1). Equilibration between QDs within the ensemble that leads to a negative temperature characteristic can also be treated,[55] but is not included here.

The threshold for an edge-emitting laser geometry can be expressed in terms of the spontaneous emission rate, cavity loss rate, ω_c/Q, and the effective density of states (Eq. (1)). The magnitude squared of the dipole moment in Eqs. (9) or (10) is directly related to the spontaneous emission rate.[52] The summation over the emitter positions that includes the normalization of the lasing mode's field strength mathematically gives rise to the optical confinement factor in edge-emitting lasers.

Using subscripts to explicitly account for occupation by either electrons or holes, the threshold condition in Eq. (10) is given as[15,43]

$$\frac{\omega_o}{Q} = \nu_g \left[\frac{1}{L} \ln\left(\frac{1}{R}\right) + \alpha \right] = \gamma_{\text{sp,o}} \frac{\pi^2 c^3 \Gamma(x_{c,0} - x_{\nu,0})}{\Delta z n^3 \omega_0^2} \rho_{\text{red}}(\omega_o) \tag{12}$$

where $n_{c,0}$ is the average occupancy of the QD ground state by the electron with the subscript $m = 0$, $n_{\nu,0}$ is the average occupancy of the ground state from the valence band, n is the refractive index, c is the speed of light in open space, ν_g is the optical group velocity in the laser cavity, L is the cavity length, the mirror reflectivity is R, ω_c is the cavity resonance, α is the waveguide loss, and $\Gamma/\Delta z$ is the waveguide confinement factor normalized by the active region thickness. $\Gamma/\Delta z$ come from the field normalization and also sets the spontaneous emission rate into the lasing mode. The temperature dependence of threshold comes due to the spontaneous and nonradiative currents required to maintain a given population inversion $x_{c,0} - x_{\nu,0}$ in Eq. (12) for different temperatures. From our assumption of equal effective masses for electrons and holes we have $x_{c,0} - x_{\nu,0} = 2x_0 - 1$. Optical transparency occurs for $x_0 = 0.5$, with net absorption occurring for $x_0 < 0.5$, and net stimulated emission occurring for $x_0 > 0.5$.

For VCSELs, the summation in Eq. (10) results in a different form due to the different cavity geometry. The threshold condition can be put directly in terms of the mirror reflectivities, assuming a plane wave mode approximation, and the summation becomes

$$\ln\left(\frac{1}{\sqrt{R_T R_B}}\right) = \sqrt{\frac{\ln(2)}{\pi}} \frac{8\pi^2 c^2}{n^2 \omega_0^2 \tau_{\text{sp}}} \frac{n_{\text{QD}}}{\Delta\omega}. \tag{13}$$

Equations (12) and (13) can be used to determine just what QD ensemble characteristics are required to achieve lasing in a particular cavity system.

4. 1.3 μm Quantum Dot Edge-Emitting Laser

The 1.3 μm InGaAs/GaAs QD lasers are potentially important for optical interconnects, and also exhibit novel lasing behavior. As mentioned in the introduction, the deep potential wells and narrow gain spectrum lead to room temperature ground state operation approaching that of ideal QD lasers. The first 1.3 μm QD laser was demonstrated by Huffaker *et al.*[16] It was approximately one year later that CW operation at room temperature was reported for a 1.3 μm QD laser that uses a three-layered vertically-stacked active region by Mukai *et al.*[18] Park *et al.*, then showed that CW lasing operation is also possible for a single layered 1.3 μm QD laser.[19] Park *et al.*'s 1.3 μm QD laser threshold currents are impressive. They report a CW threshold current of ~ 4 mA at a current density of 45 A/cm^2, 25 A/cm^2 for broad-area operation, and cryogenic lasing operation at 4 K shows a threshold current density of only 6 A/cm^2. Despite these and the successes of many other researchers in developing QD lasers, nonradiative recombination still limits performance.

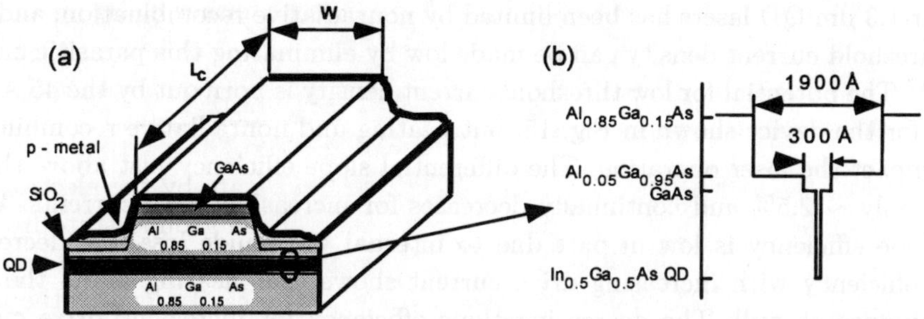

Fig. 14. Schematic illustrations of (a) the edge-emitting QD laser heterostructure and (b) the conduction band offsets in the absence of electic fields.

The 1.3 μm QD lasers reported so far have been grown using MBE,[16,17,19] and a schematic of one of the laser heterostructures is shown in Fig. 14. The temperature of the growth is 610°C, except for the quantum dot region. The single layer QD active region is grown at 510°C at the center of a waveguide clad by p and n $Al_{0.85}Ga_{0.15}As$ layers (inset shown in Fig. 14). A p^+ GaAs contact layer is grown in the final epitaxy. The undoped waveguide consists of $Al_{0.05}Ga_{0.95}As$ and is 0.194 μm thick. Short period superlattices (200 Å thick) composed of alternating layers of $Al_{0.85}Ga_{0.15}As$ and GaAs to gradually change aluminum composition from 85% to 5% are grown at the interfaces between the waveguide and cladding layers to smooth the heterointerfaces. The Si and Be doping is decreased to $\sim 10^{16}$ cm^{-3} within 0.5 μm of the $Al_{0.85}Ga_{0.15}As$ layers surrounding the waveguide, and at $\sim 2.10^{17}$ cm^{-3} in the remaining $Al_{0.85}Ga_{0.15}As$. The QD layer immediately follows depositions of 150 Å of GaAs and 25 Å of $In_{0.09}Ga_{0.91}As$, and are formed using the cycled sub-monolayer epitaxy sequence.[28] The 12 monolayer deposition has an average composition of $In_{0.5}Ga_{0.5}As$, and the QDs are covered with 150 Å of GaAs. Atomic force microscope images of the crystal surface with the growth halted just after the 12 monolayer deposition shows a QD density of $n_{QD} \sim 1.5 \times 10^{10}$ cm^{-2} with lateral sizes of $\sim$ 500 Å similar to that shown in Fig. 1(b). The QD ground states radiate with an ensemble full-width-at-half-maximum energy spread of $\hbar\Delta\omega \approx 30$ meV.

Lasers with stripes that range from 4 to 59 μm wide are fabricated by wet etching shallow ridge waveguides through most of the upper cladding layer. Cavity lengths that range from $\sim$ 1 mm to $\sim$ 5 mm are formed by cleaving, and high reflectivity (HR) facet coatings consisting of 4 pairs of ZnSe/MgF_2 are deposited on some of the lasers. The plane wave reflectivity of this mirror stack is > 0.95. The room temperature lasing characteristics of an edge-emitting laser with a stripe width of 9 μm and cavity length of 1.02 mm that has both end facets HR coated are shown in Fig. 15. With p-up mounting the CW threshold current is 4.1 mA, which is the lowest value yet reported at the time of this writing for CW room-temperature operation of a QD laser. The inset shows the ground state lasing wavelength of 1.32 μm just above the threshold. It has been suggested that the room temperature operation of similar 1.3 μm QD lasers has been limited by nonradiative recombination, and that the threshold current density can be made low by eliminating this parasitic current path.[17] The potential for low threshold current density is born out by the 45 A/cm^2 value for the device shown in Fig. 15, but heating and nonradiative recombination still impact the laser operation. The differential slope efficiency just above threshold is only $\sim$ 2.5% and continually decreases for increasing drive currents. While the slope efficiency is low in part due to internal waveguide loss, the decreasing slope efficiency with increasing drive current shows that heating limits the laser performance as well. The decreasing slope efficiency for increasing drive current can be explained by a strongly temperature sensitive threshold. Figure 16 shows that a remarkably low threshold current density of 6 A/cm^2 is obtained at 4 K for

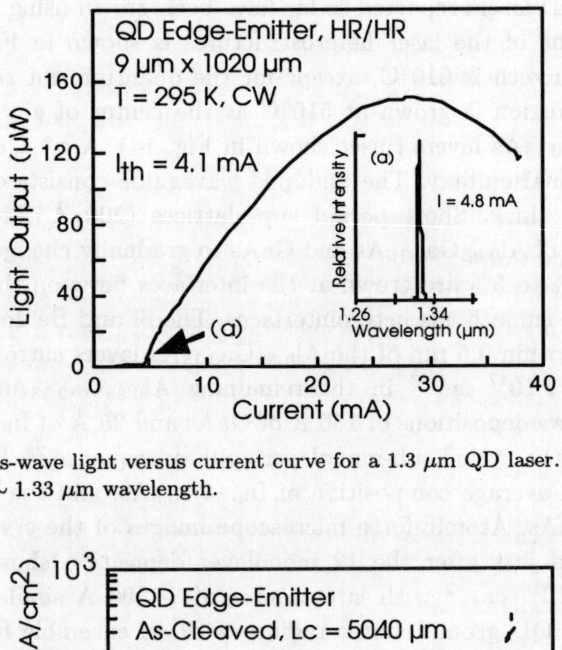

Fig. 15. Continuous-wave light versus current curve for a 1.3 μm QD laser. The inset shows the lasing spectrum at $\sim$ 1.33 μm wavelength.

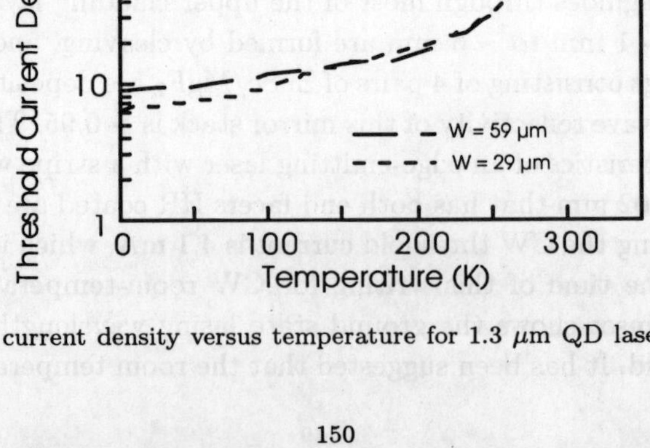

Fig. 16. Threshold current density versus temperature for 1.3 μm QD lasers of different stripe widths.

a 29 μm wide stripe laser. In general, the threshold current density increases for increasing temperatures. However, at $\sim$ 150 K the threshold increases less steeply. This is the temperature for which thermal QD coupling through the wetting layer occurs.[17,43] Above $\sim$ 250 K the threshold increases more steeply due to increasing nonradiative recombination that accompanies significant population of the QD upper energy levels and wetting layer.

Estimates suggest that even the room temperature threshold current density of the 1.3 μm QD laser can be in the range of 10 to 20 A/cm^2, despite some thermal population of the upper energy levels.[17,43] However, since the nonradiative recombination rate is quite sensitive to the current density, the radiative component of the current density can be difficult to separate from the nonradiative components. For 4 K and low excitation we assume that nonradiative recombination is negligible and arrive at a gain estimate for the QD ensemble. For the QD density of 2×10^{10} cm^{-2}, a degeneracy of two for the ground state levels, and a recombination time that we estimate to be $\sim$ 640 psec, the transparency current is $J_{\text{trans}} = qn_{\text{QD}}/\tau_{\text{rec}} \approx 5$ A/cm^2, with the maximum ground state gain obtained for a current density twice that at 10 A/cm^2 (in the absence of gain clamping due to stimulated emission). The reduced density of states at the ground state peak energy is given by $\rho_{\text{red}}(\omega_g) = 4\sqrt{\frac{\ln(2)}{\pi}}\frac{n_{\text{QD}}}{\Delta\omega} \approx 8.3 \times 10^{-4}\frac{\text{states}}{\text{cm}^2(\text{rad/sec})}$. The laser parameters for the threshold condition in Eq. (12) are as follows: the refractive index $n = 3.4$, c is the speed of light in open space, v_g is the optical group velocity (approximated to be $\approx 6.7 \times 10^7$ m/sec), L is the cavity length, R is mirror reflectivity product $(= \sqrt{R_1R_2})$, λ_c is the free space wavelength of the cavity resonance, α is the waveguide loss, and $\Gamma/\Delta z$ is the waveguide confinement factor normalized by the active region thickness. For our waveguide we calculate $\Gamma/\Delta z = 3.57 \times 10^6$ m^{-1}. The maximum modal gain from the QD ensemble ground state is then estimated from Eq. (1) to be $\sim$ 23 cm^{-1}. The threshold gain of 6 A/cm^2 would correspond to a modal gain of $\sim$ 5 cm^{-1}, so that the end loss of $\frac{1}{L}\ln(\frac{1}{R}) \approx 2.5$ cm^{-1} would require that $\alpha \approx 2.5$ cm^{-1} as well. With HR coatings of ≥ 0.95 reflectivity, the end loss is readily made negligible as compared to the internal waveguide loss. Along with heating, this leads to the low external differential slope efficiency shown in Fig. 15.

5. Quantum Dot Vertical-Cavity Surface-Emitting Laser

While an advantage for VCSELs is that the InGaAs/GaAs QDs are compatible with AlAs/GaAs distributed Bragg reflectors (DBRs) and selective oxidation,[21–25,56] the optical gain from a single layer QD ensemble is smaller than that of a quantum well for present inhomogeneous linewidths. Since a certain minimum gain bandwidth is believed necessary to obtain lasing over a useful temperature range, the gain spectrum should overlap the VCSEL's cavity resonance for minimum threshold. For InGaAs quantum well VCSELs, the gain peak shifts with temperature at $\sim$ 2.5 Å/K while the cavity resonance shifts at $\sim$ 0.5 Å/K.

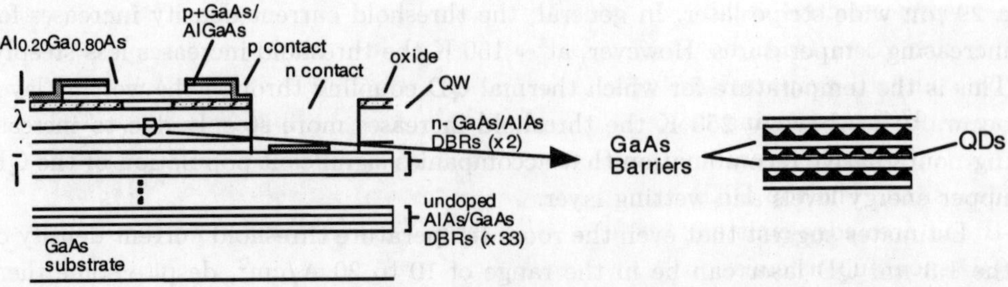

Fig. 17. Schematic illustration of a VCSEL that uses a three-stack QD active region. Intracavity contacting and undoped mirrors are used to obtain low optical loss.

The condition for the ground state lasing threshold is given by Eq. (13). Clear ground state lasing for a QD VCSEL with well-formed dots has been reported by Zou *et al.*[25] using the VCSEL structure illustrated in Fig. 17, and we analyze these results below. The QD layers have $n_{\mathrm{QD}} \approx 5 \times 10^{10}$ cm^{-2}, $\omega_o = 1.76 \times 10^{15}$ rad/sec, $\hbar\Delta\omega \approx 76$ meV, and τ_{sp} is measured under non-resonant excitation at low temperature to be ~ 800 psec. For a three-stack QD active layer, and assuming $\tau_{\mathrm{sp}} = 800$ psec and $c/n = 9 \times 10^7$ m/sec, it is calculated from Eq. (13) that $\sqrt{R_T R_B} \geq 0.9983$ must be achieved for lasing threshold. However, unknowns exist as to the actual values of τ_{sp} and $\sqrt{R_T R_B}$. Wang *et al.*, also reported a spontaneous lifetime of ~ 880 psec for nonresonantly pumped $In_{0.5}Ga_{0.5}As/GaAs$ QDs.[57] A spontaneous lifetime of ~ 380 psec has been reported for resonantly pumped $InP/In_{0.48}Ga_{0.52}P$ QDs.[39] Kurtenbach *et al.*, argue that the spontaneous decay from resonantly pumped emitters more accurately measures the Einstein coefficient for spontaneous emission. For a spontaneous lifetime of 400 psec the mirror reflectivity product required for threshold reduces to a more easily achieved 0.9966. Because the QD emission characteristics depend on the dot size and density, different ensembles give different gain and therefore different requirements of mirror reflectivity.

The semiconductor heterostructure is grown using molecular beam epitaxy on a semi-insulating GaAs substrate. The active region is centered within an $Al_{0.13}Ga_{0.87}As$ (formed from short period superlattice) full-wave cavity spacer. The bottom distributed Bragg reflector (DBR) consists of 35 AlAs/GaAs pairs, 33 of which are undoped. A single upper AlGaAs/GaAs DBR pair is formed from a quarter-wave of AlGaAs (with graded Al composition), $Al_{0.97}Ga_{0.03}As$, and a quarter-wave of GaAs. The cavity spacer is undoped, while the first two pairs of the lower AlAs/GaAs DBR are doped *n*-type, and the upper DBR layers are doped *p*-type. Heavily doped AlGaAs/GaAs contact layers are grown on top of the *p*-type quarter-wave layers, but selectively etched from the VCSEL cavity region. The QDs are formed from five monolayers of alternating InAs and GaAs. Each of the InGaAs QD layers are separated by 150 Å of GaAs with 300 Å of GaAs grown on either side of the three-stack QD active region.

After selective etching to form a trench stopping at the n-type GaAs layer of the first AlAs/GaAs bottom DBR pair, the n-type contact metalization is formed by standard evaporation/liftoff and alloying. Selective oxidation is used to form aperture sizes ranging from 10 μm to 2 μm in diameter.[21–25,56] After oxidation, the top contact layers are etched away from the optical cavity and p-contacts are deposited. The top DBR mirrors are completed with a six pair MgF/ZnSe DBR. Broad-area edge-emitters are also fabricated with 60 μm wide stripes to study the QD spontaneous emission at low excitation level.

Figure 18 shows the spontaneous electroluminescence spectrum from edge-emission at a low current of 5 A/cm^2. Two VCSEL resonances, one at 1.027 μm and a smaller one at 1.041 μm, can also be seen in the edge-emission. The QD ground-state emission is centered at 1.07 μm with an inhomogeneously broadened spectral width of 76 meV. Figure 19 shows the QD lasing characteristics for a 10 μm diameter oxide-apertured VCSEL for increasing temperatures. An inset shows the lasing spectrum above threshold consisting of three transverse modes. An uncalibrated Ge detector is used to the measure the output intensity. As the heat sink temperature is increased the lasing threshold increases from 703 μA (895 A/cm^2) at 300 K, to 833 μA at 310 K, to 1037 μA at 320 K, and 1080 μA at 327 K (our maximum possible measurement temperature). The peak lasing wavelength just above threshold changes from 10 672 Å at 300 K, to 10 679 Å at 310 K, to 10 685 Å at 320 K, to 10 687 Å at 327 K, giving the temperature dependence of the cavity resonance of $\sim$ 0.5 Å/K typical for VCSELs. The threshold voltage is $\sim$ 2.6 V and for a thermal resistance of $\sim$ 2.5 K/mW,[58] the threshold power at 300 K gives a temperature rise of only $\sim$ 5 K. Therefore, each of the three transverse modes are within the 76 meV ground state spectral width of the low level QD spontaneous emission of Fig. 18. Although lasing is also obtained in smaller apertures, the threshold does not monotonically change with aperture size due to variation in

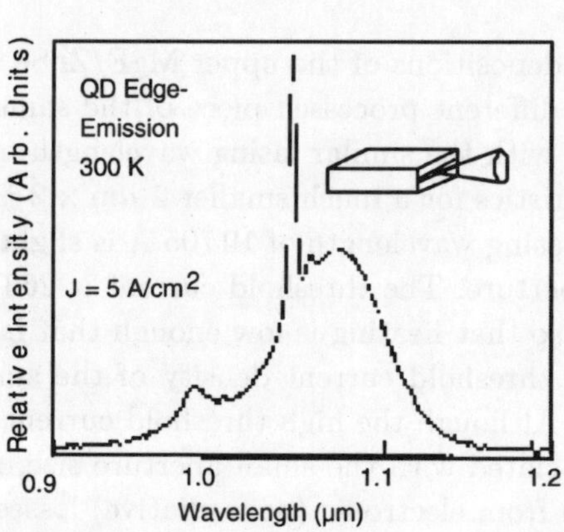

Fig. 18. Spontaneous emission spectra from the edge of a QD VCSEL showing the ground state emission. The current density is kept low at 5 A/cm^2 to prevent excitation of higher lying energy levels.

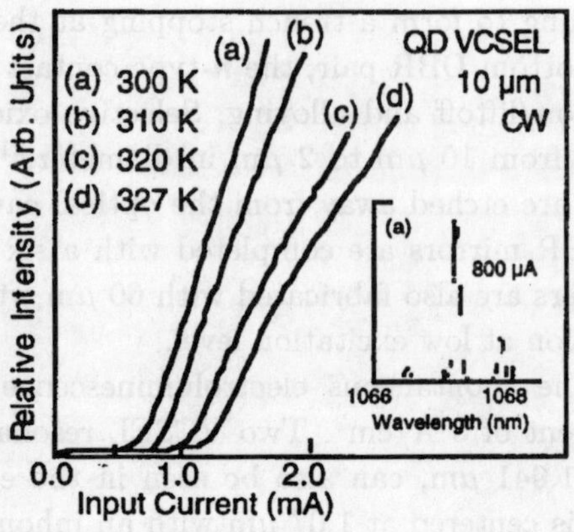

Fig. 19. Continuous-wave lasing operation for a 10 μm oxide-aperture VCSEL for temperatures around room temperature.

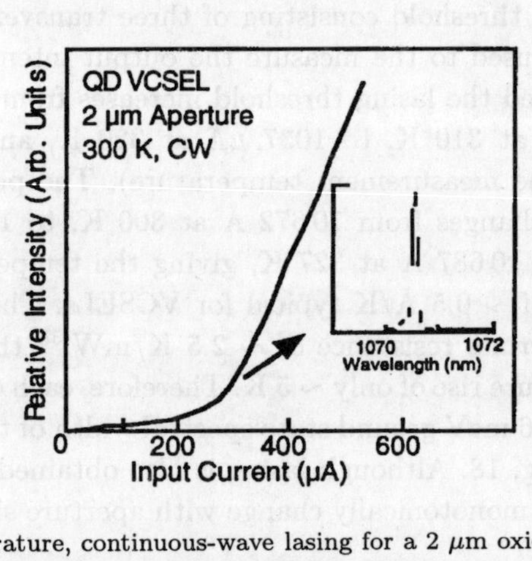

Fig. 20. Room temperature, continuous-wave lasing for a 2 μm oxide-aperture QD VCSEL.

the cavity Q (different depositions of the upper MgF/ZnSe DBR). Apertures with 6 μm diameter from a different processed piece of the same wafer yields $\sim$ 1 mA threshold currents, but with the similar lasing wavelength to that in Fig. 19.

The lasing characteristics for a much smaller 2 μm $\times$ 2 μm square aperture are shown in Fig. 20. The lasing wavelength of 10 705 Å is slightly longer than that for the 10 μm diameter aperture. The threshold current is 268 μA and the threshold voltage is again 2.6 V, so that heating is low enough that lasing still occurs at the QD ground state. The threshold current density of the smaller aperture is quite high at ~ 6.7 kA/cm^2. Although the high threshold current may in part be due to higher optical loss associated with the small aperture size, there is undoubtedly a significant contribution from electronic (nonradiative) losses discussed in Sec. 2.

Mirror reflectivities for the cavity structure of Fig. 17 are used to calculate cavity Q for comparison with measured below threshold spontaneous linewidths for the QD VCSELs. A plane wave mode approximation is used assuming free carrier absorption of 3 cm^{-1} in the n-type and 7 cm^{-1} in the p-type regions, and various values of distributed loss in the upper MgF/ZnSe DBR. The calculated cavity linewidths are compared to the measured linewidths for either the 10 μm, 6 μm, or 2 μm apertured VCSELs. We focus here on the 10 μm aperture VCSEL that most closely approximates a planar system, and that has the lowest threshold current density. The calculated mirror reflectivity of the lower 35 pair AlAs/GaAs DBR (including n-type absorption) is $R_B = 0.9997$, and that of the top p-type AlGaAs/GaAs pair combined with the 6 pair MgF/ZnSe DBR (with no absorption or scattering) is $R_T = 0.9987$. The calculated mirror reflectivity product is then $\sqrt{R_T R_B} = 0.9992$, well above that needed (> 0.9983) from the estimate of the maximum QD gain assuming $\tau_{sp} = 800$ psec. However, the calculated spectral full-width at half-maximum for this reflectivity is 0.57 Å which gives an extremely high quality factor for the cavity of $Q = \lambda/\Delta\lambda = 19\,000$. The calculated linewidth of 0.57 Å is much less than that measured for the cavity spontaneous emission at low current densities. The lowest order transverse modes consist of two closely spaced orthogonal polarizations separated by ~ 2.2 Å, but the longer wavelength polarization is more intense for each aperture size. We estimate the linewidth of the single lowest order mode between 30 A/cm^2 and 60 A/cm^2 to be ~ 1.9 Å for the 10 μm, which gives an experimental Q of ~ 5600.

The experimental Q may be limited by loss in the upper MgF/ZnSe DBRs, and we have calculated linewidths and mirror reflectivities based on this assumption. We also note, though, that while the lower AlAs/GaAs DBR may not be ideal, it is the mirror reflectivity product that is mainly responsible for setting the cavity linewidth and lasing threshold. Holding the bottom mirror reflectivity fixed at $R_B = 0.9997$, values of distributed loss in the upper MgF/ZnSe DBR of 100 cm^{-1}, 200 cm^{-1}, and 300 cm^{-1} give calculated cavity linewidths of 1.1 Å, 1.8 Å, and 2.4 Å, respectively, and upper DBR reflectivity values of $R_T = 0.9959$, 0.9931, and 0.9903, respectively. The distributed loss of 200 cm^{-1}, although it seems rather high, agrees not only with the measured linewidth but also the differential efficiency measured on quantum well VCSELs of the similar design.[59] The estimated experimental reflectivity product is then found to be $\sqrt{R_T R_B} \approx 0.9964$. From Eq. (12) we can estimate that for $\sqrt{R_T R_B} \approx 0.9964$ the spontaneous lifetime must be $\tau_{sp} \approx 350$ psec, and even $\sqrt{R_T R_B} \approx 0.9978$ (or $R_T = 0.9959$) requires $\tau_{sp} \approx 570$ psec for threshold. This range is within what Kurtenbach *et al.* measured for resonantly pumped $InP/In_{0.48}Ga_{0.52}P$ QDs and that reported for 1.3 μm InGaAs QDs.[39,44]

If the parameters in Eq. (13) are correct, the stimulated emission rate from the three-stack QD active region is significantly greater than that inferred from the nonresonantly pumped spontaneous lifetime. This may mean that a superradiance effect exists that, under heavy pumping, shortens the spontaneous lifetime and leads to lasing threshold. However, Eq. (12) may also be incorrect by a small error. The difference in the emission process between resonant and nonresonant excitation of the QDs lies in the coherence of the electronic wavefunctions in the

QD energy levels. This coherence may be established under high level excitation in which the QD level populations are maintained close to unity, while the stimulated emission is intrinsically a resonant excitation process for lasers.

6. Summary

QD lasers are rapidly advancing and have many advantages over planar quantum well lasers. This is partly due to the electronic density of states that results from the quantum confinement in three-dimensions. If the inhomogeneous broadening is reduced for the QD ensembles, homogeneous linewidths may be less than 5 meV at room temperature. For this narrow of a linewidth the gain of a single QD layer can exceed the gain of a planar quantum well, and at much lower current density. When this is obtained QD lasers can be expected to significantly outperform planar quantum well lasers. Note that this extends to spectroscopy applications as well, since the reduced density of states should also reduce the linewidth enhancement factor. The second interesting advantage of QDs is that in general they offer a wider range of lasing wavelengths for coherent lattice strain to a given substrate material. This allows the 1.3 μm InGaAs QD laser to be realized with epitaxial growth on GaAs substrates. These lasers can become quite important for fiber optics.

Acknowledgments

Research performed on QD lasers at The University of Texas and described in this book chapter has been supported by the National Science Foundation, the Army Research Office, the State of Texas Advanced Research Program, and the Defense Advanced Research Project Agency.

References

1. Y. Arakawa and H. Sakaki, "Multidimensional quantum well laser and temperature dependence of its threshold current", *Appl. Phys. Lett.* **40** (1982) 939–941.
2. M. Asada, Y. Miyamoto, and Y. Suematsu, "Gain and the threshold of three-dimensional quantum-box lasers", *IEEE J. Quantum Electron.* **22** (1986) 1915–1921.
3. K. Vahala, "Quantum box fabrication tolerance and size limits in semiconductors and their effect on optical gain", *IEEE J. Quantum Electron.* **24** (1988) 523–530.
4. N. Kirstaedter, N. N. Ledentsov, M. Grundmann, D. Bimberg, V. M. Ustinov, S. S. Ruvimov, M. V. Maximov, P. S. Kop'ev, Zh. I. Alferov, U. Richter, P. Werner, U. Gosele, and J. Heydenreich, "Low threshold, large T_o injection laser emission from InGaAs quantum dots", *Electron. Lett.* **30** (1994) 1416–1417.
5. H. Shoji, K. Mukai, N. Ohtsuka, M. Sugawara, T. Uchida, and H. Ischikawa, "Lasing at three-dimensionally quantum-confined sublevel of self-organized $In_{0.5}Ga_{0.5}As$ quantum dots by current injection", *IEEE Photon. Tech. Lett.* **7** (1995) 1385–1387.
6. N. N. Ledentsov, V. A. Shchukin, M. Grundmann, N. Kirstaedter, J. Bohrer, O. Schmidt, D. Bimberg, V. M. Ustinov, A. Y. Egorov, A. E. Zhukov, P. S. Kop'ev, S. V. Zaitsev, N. Yu. Gordeev, Zh. I. Alferov, A. I. Borovkov, A. O. Kosogov, S. S. Ruvimov, P. Werner, U. Gosele, and J. Heydenreich, "Direct formation of vertically coupled quantum dots in Stranski-Krastanow growth", *Phys. Rev.* **B54** (1996) 8743–8750.

7. K. Kamath, P. Bhattacharya, T. Sosnowski, T. Norris, and J. Phillips, "Room-temperature operation of $In_{0.4}Ga_{0.6}As/GaAs$ self-organized quantum dot lasers", *Electron. Lett.* **32** (1996) 1374–1375.
8. H. Shoji, Y. Nakata, K. Mukai, Y. Sugiyama, M. Suagawara, N. Yokoyama, and H. Ishikawa, "Room temperature CW operation at the ground state of self-formed quantum dot lasers with multi-stacked dot layer", *Electron. Lett.* **32** (1996) 2023.
9. Q. Xie, A. Kalburge, P. Chen, and A. Madhukar, "Observation of lasing from vertically self-organized InAs three-dimensional island quantum boxes on GaAs (001)", *IEEE Photon. Tech. Lett.* **8** (1996) 965–967.
10. R. Mirin, A. Gossard, and J. Bowers, "Room temperature lasing from InGaAs quantum dots", *Electron. Lett.* **32** (1966) 1732–1734.
11. P. R. Berger, K. Chang, P. Bhattacharya, J. Singh, K. K. Bajaj, "Role of strain and growth conditions on the growth front profile of $In_xGa_{1-x}As$ on GaAs during the pseudomorphic growth regime", *Appl. Phys. Lett.* **53** (1988) 684–686.
12. S. Guha, A. Madhukar, and K. C. Rajkumar, "Onset of incoherency and defect introduction in the initial stages of molecular beam epitaxy growth of highly strained $In_xGa_{1-x}As$ on GaAs (100)", *Appl. Phys. Lett.* **57** (1990) 2110–2112.
13. D. Leonard, M. Krishnamurthy, C. M. Reaves, S. P. Denbaars, and P. M. Petroff, "Direct formation of quantum-sized dots from uniform coherent islands of InGaAs on GaAs surfaces", *Appl. Phys. Lett.* **63** (1993) 3203–3205.
14. G. Park, O. B. Shchekin, D. L. Huffaker, and D. G. Deppe, "Lasing from InGaAs/GaAs quantum dots with extended wavelength and well-defined harmonic oscillator energy levels", *Appl. Phys. Lett.* **73** (1998) 3351–3353.
15. Z. Zou, O. B. Shchekin, G. Park, D. L. Huffaker, and D. G. Deppe, "Threshold temperature dependence of lateral cavity quantum dot lasers", *IEEE Photon. Tech. Lett.* **10** (1998) 1673–1675.
16. D. L. Huffaker, G. Park, Z. Zou, O. B. Shchekin, and D. G. Deppe, "1.3 μm room-temperature GaAs-based quantum-dot laser", *Appl. Phys. Lett.* **73** (1998) 2564–2466.
17. G. Park, D. L. Huffaker, Z. Zou, O. B. Shchekin, and D. G. Deppe, "Temperature dependence of lasing characteristics for long-wavelength (1.3 μm) GaAs-based quantum dot lasers", *IEEE Photon. Tech. Lett.* **11** (1999) 301–303.
18. K. Mukai, Y. Nakata, K. Otsubo, M. Sugawara, N. Yokoyama, and H. Ishikawa, "1.3 μm CW lasing of InGaAs/GaAs quantum dots at room temperature with a threshold current of 8 mA", *IEEE Photon. Tech. Lett.* **11** to be published.
19. G. Park, O. B. Shchekin, S. Csutak, and D. G. Deppe, "Room-temperature continuous-wave operation of a single-layered 1.3 μm quantum dot laser", *Appl. Phys. Lett.* to be published.
20. H. Saito, K. Nishi, I. Ogura, S. Sugov, and Y. Sugimoto, "Room-temperature lasing operation of a quantum-dot vertical-cavity surface-emitting laser", *Appl. Phys. Lett.* **69** (1996) 3140–3142.
21. D. L. Huffaker, O. Baklenov, L. A. Graham, B. G. Streetman, and D. G. Deppe, "Quantum dot vertical-cavity surface-emitting laser with a dielectric aperture", *Appl. Phys. Lett.* **70** (1997) 2356–2358.
22. D. L. Huffaker, L. A. Graham, and D. G. Deppe, "Low-threshold continuous-wave operation of an oxide-confined vertical-cavity surface-emitting laser based on a quantum dot active region and half-wave cavity", *Electron. Lett.* **33** (1997) 1225–1226.
23. J. A. Lott, N. N. Ledentsov, V. M. Ustinov, A. Yu. Egorov, A. E. Zhukov, P. S. Kop'ev, Zh. I. Alferov, and D. Bimberg, "Vertical cavity lasers based on vertically coupled quantum dots", *Electron. Lett.* **33** (1997) 1150–1151.
24. D. L. Huffaker, H. Deng, and D. G. Deppe, "1.15 μm wavelength oxide-confined

quantum dot vertical-cavity surface-emitting laser", *IEEE Photon. Tech. Lett.* **10** (1998) 185–187.
25. Z. Zou, D. L. Huffaker, S. Csutak, and D. G. Deppe, "Ground state lasing from a quantum dot oxide-confined vertical-cavity surface-emitting laser", *Appl. Phys. Lett.* **75** (1999) 22–25.
26. K. Mukai, O. Nobuyuki, S. Mitsuru, and S. Yamzaki, "Self-formed $In_{0.5}Ga_{0.5}As$ quantum dots on GaAs substrates emitting at 1.3 μm", *Jpn. J. Appl. Phys.* **33** (1994) L1710–L1712.
27. R. Mirin, J. Ibbetson, K. Nishi, A. Gossard, and J. Bowers, "1.3 μm photoluminescence from InGaAs quantum dots", *Appl. Phys. Lett.* **67** (1995) 3795–3797.
28. D. L. Huffaker and D. G. Deppe, "Electroluminescence efficiency of 1.3 μm wavelength InGaAs/GaAs quantum dots", *Appl. Phys. Lett.* **73** (1998) 520–522.
29. M.-E. Pistol, N. Carlsson, C. Persson, W. Seifert, and L. Samuelson, "Observation of strain effects in semiconductor dots depending on cap layer thickness", *Appl. Phys. Lett.* **67** (1995) 1438–1440.
30. L. F. Lester, A. Stintz, H. Li, T. C. Newell, E. A. Pease, B. A. Fuchs, and K. J. Malloy, "Optical characterisitcs of 1.24 μm InAs quantum-dot laser diodes", *IEEE Photon. Tech. Lett.* **11** (1999) 931–933.
31. K. Nishi, H. Saito, S. Sugou, J.-S. Lee, "A narrow photoluminescence linewidth of 21 meV at 1.35 μm from strain-reduced InAs quantum dots covered by $In_{0.2}Ga_{0.8}As$ grown on GaAs substrates", *Appl. Phys. Lett.* **74** (1999) 1224–1226.
32. V. M. Ustinov, N. A. Maleev, A. E. Zhukov, A. R. Kovsh, A. Yu. Egorov, A. V. Lunev, B. V. Volovik, I. L. Krestnikov, Yu. G. Musikhin, N. A. Bert, P. S. Kop'ev, Zh. I. Alferov, N. N. Ledentsov, and D. Bimberg, "InAs/GaAs quantum dot structures on GaAs substrates emitting at 1.3 μm", *Appl. Phys. Lett.* **74** (1999) 2815–2817.
33. H. Jiang and J. Singh, "Strain distribution and electronic spectra of InAs/GaAs self-assembled dots: An eight-band study", *Phys. Rev.* **B56** (1997) 4696–4701.
34. F. Adler, M. Geiger, A. Bauknecht, F. Scholz, H. Schweizer, M. H. Pilkuhn, B. Ohnesorge, and A. Forchel, "Optical transitions and carrier relaxation in self assembled InAs/GaAs quantum dots", *J. Appl. Phys.* **80** (1996) 4019–4026.
35. K. H. Schmidt, G. Medeiros-Ribeiro, M. Oestreich, and P. M. Petroff, "Carrier relaxation and electronic structure in InAs self-assembled quantum dots", *Phys. Rev.* **B54** (1996) 11346–11353.
36. M. Grundman, O. Stier, and D. Bimberg, "InAs/GaAs pyramidal quantum dots: Strain distribution, optical phonons, and electronic structure", *Phys. Rev.* **B52** (1999) 11969–11981.
37. M. Grundmann, N. N. Ledentsov, O. Stier, D. Bimberg, V. M. Ustinov, P. S. Kop'ev, and Zh. I. Alferov, "Excited states in self-organized InAs/GaAs quantum dots: Theory and experiment", *Appl. Phys. Lett.* **68** (1996) 979–981.
38. D. Klotzkin, K. Kamath, K. Vineberg, P. Bhattacharya, R. Murty, and J. Laskar, "Enhanced modulation bandwidth (20 GHz) of InGaAs-GaAs self-organized quantum-dot lasers at cryogenic temperatures: Role of carrier relaxation and differential gain", *IEEE Photon. Tech. Lett.* **10** (1998) 932–934.
39. A. Kurtenbach, W. W. Ruhle, and K. Eberl, "Intrinsic radiative lifetimes of $InP/In_{0.48}Ga_{0.52}P$ quantum dots", *Solid State Commun.* **96** (1995) 265–269.
40. E. Hanamura, "Rapid radiative decay and enhanced optical nonlinearity of excitons in a quantum well", *Phys. Rev.* **B38** (1988) 1228–1234.
41. M. Sugawara, "Theory of spontaneous-emission lifetime of Wannier excitons in mesoscopic semiconductor quantum disks", *Phys. Rev.* **B51** (1995) 10743–10754.
42. K. Mukai, N. Ohtsuka, H. Shoji, and M. Sugawara, "Emission from discrete levels

in self-formed InGaAs/GaAs quantum dots by electric carrier injection: Influence of phonon bottleneck", *Appl. Phys. Lett.* **68** (1996) 3013–3015.
43. D. G. Deppe, D. L. Huffaker, S. Csutak, Z. Zou, G. Park, and O. B. Shchekin, "Spontaneous emission and threshold characteristics of 1.3 μm InGaAs/GaAs quantum dot GaAs-based lasers", *IEEE J. Quantum Electron.* **35** (1999) 1238–1246.
44. D. G. Deppe, T. Boggess, L. Zhang, D. L. Huffaker, and O. Shchekin, unpublished.
45. K. H. Schmidt, G. Medeiros-Ribeiro, M. Oestreich, and P. M. Petroff, "Carrier relaxation and electronic structure in InAs self-assembled quantum dots", *Phys. Rev.* **B54** (1996) 11346–11353.
46. M. Grundman, O. Stier, and D. Bimberg, "InAs/GaAs pyramidal quantum dots: Strain distribution, optical phonons, and electronic structure", *Phys. Rev.* **B52** (1995) 11969–11981.
47. M. Grundmann, N. N. Ledentsov, O. Stier, D. Bimberg, V. M. Ustinov, P. S. Kop'ev, and Zh. I. Alferov, "Excited states in self-organized InAs/GaAs quantum dots: Theory and experiment", *Appl. Phys. Lett.* **68** (1996) 979–981.
48. H. Benisty, C. M. Sotomayor-Torres, and C. Weisbuch, "Intrinsic mechanism for the poor luminescence properties of quantum box systems", *Phys. Rev.* **B44** (1991) 10945–10948.
49. L. V. Asryan and R. A. Suris, "Inhomogeneous linewidth broadening and the threshold current density of a semiconductor quantum dot laser", *Semicond. Sci. Technol.* **11** (1996) 554–567.
50. D. Bimberg, N. Kirstaedter, N. N. Ledentsov, Zh. I. Alferov, P. S. Kop'ev, and V. M. Ustinov, "InGaAs-GaAs quantum dot lasers", *IEEE J. Select. Top. Quantum Electron.* **3** (1997) 196–205.
51. L. V. Asryan and R. A. Suris, "Temperature dependence of the threshold current density of a quantum dot laser", *IEEE J. Quantum Electron.* **34** (1998) 841–850.
52. M. Sargent, M. O. Scully, and W. E. Lamb, *Laser Physics*, Addison-Wesley, Reading, 1974.
53. A. V. Uskov, K. Nishi, and R. Lang, "Collisional broadening and shift of spectral lines in quantum dot lasers", *Appl. Phys. Lett.* **74** (1999) 3081–3083.
54. M. Grundman, J. Christen, N. N. Ledentsov, J. Bohrer, D. Bimberg, S. S. Ruvimov, P. Werner, U. Richter, U. Gosele, J. Heydenreich, V. M. Ustinov, A. Yu. Egorov, A. E. Zhukov, P. S. Kop'ev, and Zh. I. Alferov, "Ultranarrow luminescence lines from single quantum dots", *Phys. Rev. Lett.* **74** (1995) 4043–4046.
55. A. E. Zhukov, V. M. Ustinov, A. Yu. Egorov, A. R. Kovsh, A. F. Tsatsulnikov, N. N. Ledentsov, S. V. Zaitsev, N. Yu. Gordeev, P. S. Kopev, and Z. I. Alferov, "Negative characteristic temperature of InGaAs quantum dot injection laser", *Jpn. J. Appl. Phys.* **36** (1997) 4216–4218.
56. J. M. Dallesasse, N. Holonyak, Jr., A. R. Sugg, T. A. Richard, and N. El-Zein, "Hydrolization oxidation of AlGaAs-AlAs-GaAs quantum well heterostructures", *Appl. Phys. Lett.* **57** (1990) 2844–2846.
57. G. Wang, S. Fafard, D. Leonard, J. E. Bowers, J. L. Merz, and P. M. Petroff, "Time resolved optical characterization of InGaAs/GaAs quantum dots", *Appl. Phys. Lett.* **64** (1994) 2815–2817.
58. D. L. Huffaker and D. G. Deppe, "Multiwavelength, densely-packed 2×2 vertical-cavity surface-emitting laser array fabricated using selective oxidation", *IEEE Photon. Tech. Lett.* **8** (1996) 858–860.
59. D. L. Huffaker and D. G. Deppe, "Intracavity contacts for low-threshold oxide-confined vertical-cavity surface-emitting lasers", *IEEE Photon. Tech. Lett.* **11** (1999) 934–936.

International Journal of High Speed Electronics and Systems, Vol. 9, No. 4 (1998) 1007–1080

GaN-BASED LASER DIODES

MANIJEH RAZEGHI
Center for Quantum Devices,
Department of Electrical and Computer Engineering,
Northwestern University, Evanston, Illinois 60208, USA
E-mail: razeghi@ece.nwu.edu; http://cqd.ece.nwu.edu

We discuss optical properties of III-Nitride materials and structures. These properties are critical for the development of III-Nitride-based light-emitting diodes and laser diodes. Minority carrier diffusion length in GaN has been determined to be $\sim 0.1\ \mu$m. The properties of lasing in GaN have been studied using optical pumping. The red shift of emission peak observed in stimulated emission of GaN has been modeled and attributed to many-body interactions at high excitation. The correlation of photoluminescence and optical pumping has shown that band-to-band, or shallow donor-related bandtail to valence band transition is the necessary mechanism of lasing in GaN. This work showed that the thermal instability of InGaN at growth temperature is of main concern in the fabrication of InGaN-based MQW laser diode structures. Photoluminescence has shown that the InGaN composition is very sensitive to the growth temperature. Therefore InGaN growth temperature should be strictly controlled during InGaN-based MQW growth. This work discovered that proper annealing or Si-doping of InGaN/GaN MQW structures that are properly annealed could reduce the lasing threshold and improve the slope efficiency. Over-annealing of these MQWs can lead to thermal degradation of the active layer. Si-doping in over-annealed MQW structure further degrades its quality. The degradation has been attributed to the increase of defects and/or nonuniform local potential formation. *P*-type doping on the top of InGaN/GaN could also lead to the formation of compensation layer which also degrades laser diode performances. Optical confinement and carrier confinement in InGaN-based laser diode structures are evaluated for optimum laser diode design. The state-of-the-art and fundamental issues of InGaN-based light-emitting diodes and laser diodes are discussed.

1. Introduction

1.1. *Overview of the physical properties and issues related to GaN based materials*

Unlike more conventional semiconductors, such as silicon (Si) or gallium arsenide (GaAs) which have a diamond or zinc-blende structure with a cubic symmetry, GaN based materials crystals are wurtzite in their stable form with a hexagonal symmetry. They are polar crystals because they do not have a center of symmetry,[1] thus also exhibiting piezoelectricity, pyroelectricity[2] and second harmonic generation.[3,4] The large difference in electronegativity between the group III and group V elements (Al = 1.18, Ga = 1.13, In = 0.99, N = 3.0) results in very strong chemical bonds in the III-Nitride material system, which are the origin of most of the interesting III-Nitride physical properties, including their large bandgap energy.

GaN-based materials have a bandgap energy ranging from 6.2 to 1.9 eV, corresponding to a wavelength from 200 to 650 nm. This spectral range covers the visible spectrum (blue, green, yellow and red) as well as the near ultraviolet region in which the atmosphere transmits. The bandgap is direct which is most appropriate for optical devices. Lower intrinsic carrier density caused by large bandgap results in lower leakage and dark currents in devices, which is especially important in photodetectors and high temperature electronics. Another consequence of the strong chemical bonding is the physical (high melting points, mechanical strength) and chemical stability of these materials. These also enjoy high thermal conductivity. Their effective masses are higher than conventional semiconductors, thus leading to lower carrier mobilities, but this is made up for by the high saturated electron drift velocities predicted for this material system. The refractive indices of III-Nitrides are lower compared to narrower gap semiconductors, which results in a lower reflectivity at the interface. This is an advantage for photodetector efficiency, but a disadvantage when trying to achieve lasers with low threshold currents.

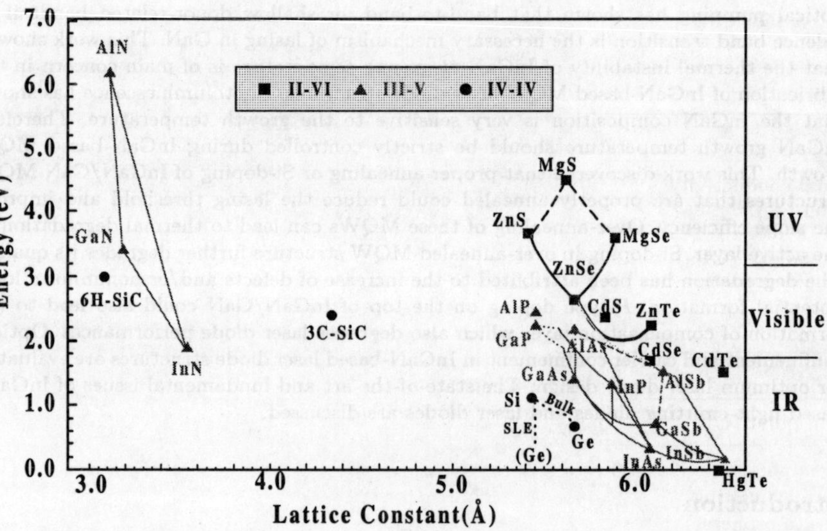

Fig. 1. Energy bandgap and lattice constants of common semiconductors.

All these properties yields these materials with a very high potential for operation in the visible-to-ultraviolet spectral region and in harsh environments (radiation, heat) such as those typical of space applications, high frequency, high power and high temperature electronics. Details of the physical properties of GaN materials are summarized in Table 1. A diagram comparing the lattice *a* constants and energy gaps of III-Nitrides and other semiconductors is shown in Fig. 1.

Table 1. Physical properties of III-Nitrides (from Refs. 6 and 7, except[a] from Ref. 8).

	GaN	AlN	InN
Room temperature energy gap (eV)	3.44	6.2	1.9
Lattice constant, a (Å)	3.189[a]	3.112	3.545
Lattice constant, c (Å)	5.186[a]	4.982	5.703
Thermal expansion coefficient, $\alpha_a (10^{-6}\ K^{-1})$	4.3 (17 – 477 C)[a]	5.27 (20 – 800 C)	5.6 (280 C)
Therma expansion coefficient, $\alpha_c (10^{-6}\ K^{-1})$	4.0 (17 – 477 C)[a]	4.15 (20 – 800 C)	3.8 (280 C)
Electron effective mass, $m_e(m_0)$	0.2		0.11
Hole effective mass, $m_h(m_0)$	0.8		0.5 (m_{hh} calculated) 0.17 (m_{1h} calculated)
refractive index, n	2.35 (1.0 μm) 2.60 (0.38 μm)	2.2 (0.60 μm) 2.5 (0.23 μm)	2.56 (1.0 μm) 3.12 (0.66 μm)
$\varepsilon(0)$	10.4 (E||c), 9.5 (E⊥c)	9.14	
$\varepsilon(\infty)$	5.8 (E||c), 5.4 (E⊥c)	4.84	9.3
Thermal conductivity, κ (W/cm K)	1.3	2.0	
Bulk modulus (GPa)	207[a]		
TO (meV)	69.3 (TO⊥), 66.1 (TO||)	82.6	59.3
LO (meV)	92.5 (LO⊥)	112.8	86.1
Melting point (C)	> 1700	3000	1100
Nitrogen partial pressure at melting point (atmospheres)	3000	100	$\gg 10^5$
ΔG^0 (kcal/mol)	−33.0	−68.2	−23.0
Heat capacity, C_p (cal mol^{-1} K^{-1})	9.7	7.6	10.0

Although GaN based devices are being rapidly demonstrated and commercialized, there are many barriers that must be overcome before the full potential of these materials can be realized as reliable devices. First, their high melting point and the extremely high nitrogen partial pressures near the melting point make their bulk growth very difficult. Therefore, high quality GaN substrates do not exist and the synthesis of nitride crystals has been carried in the form of thin films on a nonnative substrate. The dissimilarity between the substrate material and GaN based materials generally leads to poor structural quality as a result of the lattice and thermal mismatch. Lateral epitaxial overgrowth has proved successful in reducing the density of defects but has yet to prove to be a commercially viable alternative.[5] Moreover, nitride alloys with different compositions are also lattice mismatched, which leads to dislocations within heterostructures. High free electron and hole concentrations are often difficult to achieve because most dopant elements form deep levels in wide bandgap nitride semiconductors. The addition of more dopant source during the growth process frequently results in degradation of the structural and optical properties.[6–8] The metal contacts to the *p*-layers often have high contact resistances, thus leading to heating and premature device failures.

1.2. *Brief history and applications of GaN based lasers*

The driving force behind the exceptional interest in GaN based materials has been their potential for numerous significant device applications, in particular ultraviolet (UV)-blue-green lasers which are the focus of this chapter. However, these materials have not always received such attention.

The difficulty to achieve single crystalline materials, with controllable optical and electrical (both *n*- and *p*-type) properties made GaN based devices almost impossible to attain. Advances in the growth technology allowed the epitaxy of GaN based thin films with reasonable physical properties in the 1970s and early 1980s.[9,10] The first major breakthrough in this field was the use of a thin nucleation layer of AlN or GaN,[11,12] which facilitated the nearly two-dimensional growth of GaN, vastly improving and controlling the structural quality and as a result the electrical and optical quality. The second major breakthrough was the successful demonstration of *p*-type GaN through low-energy electron beam irradiation treatment or thermal annealing of magnesium doped GaN.[13,14] Before this advance, only *n*-type GaN could be achieved. These successes for GaN were then translated to ternary alloys, AlGaN and GaInN.[15,16]

Numerous GaN based devices were soon demonstrated. The first candela-class, blue light emitting diodes (LEDs) were commercialized in 1994 by Nichia Chemical Industries using GaInN/GaN double heterostructures.[17] High-brightness blue, green, and yellow GaInN quantum well LEDs were then reported in 1995.[18] Pulsed operation GaN based laser diodes were first announced by Nichia Chemical Industries in late 1995.[19] By the end of 1997, laser diodes with an estimated lifetime of 10,000 hours at room temperature in continuous mode were reported.[5]

The primary interest in shorter wavelength lasers is the ability to focus the beam to smaller spot sizes which scale as λ^{-2}, thus quadrupling the storage density of optical media by reducing the laser wavelength in half. Currently used AlGaAs based lasers for CD-ROM operate at a wavelength of 780 nm. The objective will be to achieve a digital video disk (DVD) system capable of storing 15 gigabytes by the year 2000. Such DVD-RAM system would require a laser diode operating in continuous wave (CW) at 60°C, with an output power of 30–40 mW, and an operating voltage of 3 V at 100 mA. The requirements for a DVD-ROM system would only be a 4–5 mW CW laser. In both cases, the laser should have a wavelength of 400–430 nm. It must not be too short to avoid transmission losses in air.

Blue-green laser diodes are also expected to complement existing red AlGaAs based lasers for use in projective displays. For example, a 55 inch large display needs a luminosity of 500 cd/m^2, which requires at least a 6.6 W, 1.8 W and 1.2 W red, green and blue lasers, respectively. Chemical analysis is another area of application for UV-blue-green lasers because the wavelength could be tuned to absorption lines of specific airborne chemicals which need to be detected. Other applications are undersea survey, communication and space communication. It comes out that sea water has the minimum absorption around 460–540 nm (minimum $\sim$ 490 nm),[20] which provides a good window for undersea survey and communication. Covert space-to-space communication needs to use the opaque nature of the atmosphere in the wavelength range of 240–280 nm. Finally, laser printing is also an important applications for short wavelength laser diodes. These would need to emit at a wavelength higher than 430 nm to avoid the decomposition of the toner components, with a single mode CW output power higher than 6 mW for fast printing.

1.3. *Fundamental issues of laser diode design*

Refractive indices (optical confinement)

One critical parameter for LED and LD design is a good knowledge of the refractive indices of different layers at the lasing wavelength. In the case of InGaN-based LEDs and LDs, the layers involved are InGaN, AlGaN, and GaN. Thus the refractive indices as a function of the alloys at the lasing wavelength should be known. Currently, there are no complete experimental measurements of these refractive indices. The best knowledge is shown in Ref. 21. They used the measured refractive indices of GaN as a function of wavelength to derive the refractive indices of AlGaN and InGaN. Figure 2 is their results for AlGaN, InGaN with different compositions.

Double heterostructure (DH) and single quantum well (SQW) structures have been used as the structures of InGaN-based LEDs. While InGaN-based LDs uses multiple quantum well structure. For three-layer symmetrical DH like the one shown in Fig. 3, the optical confinement factor for TE mode is calculated as:

$$\Gamma_T = \frac{\int_{-d/2}^{d/2} \phi^2(y)dy}{\int_{-\infty}^{+\infty} \phi^2(y)dy} \tag{1}$$

where $\phi(y)$ is the profile of the guided mode. To know $\phi(y)$, the eigenvalue equation in the waveguide has to be solve numerically. However, for the fundamental

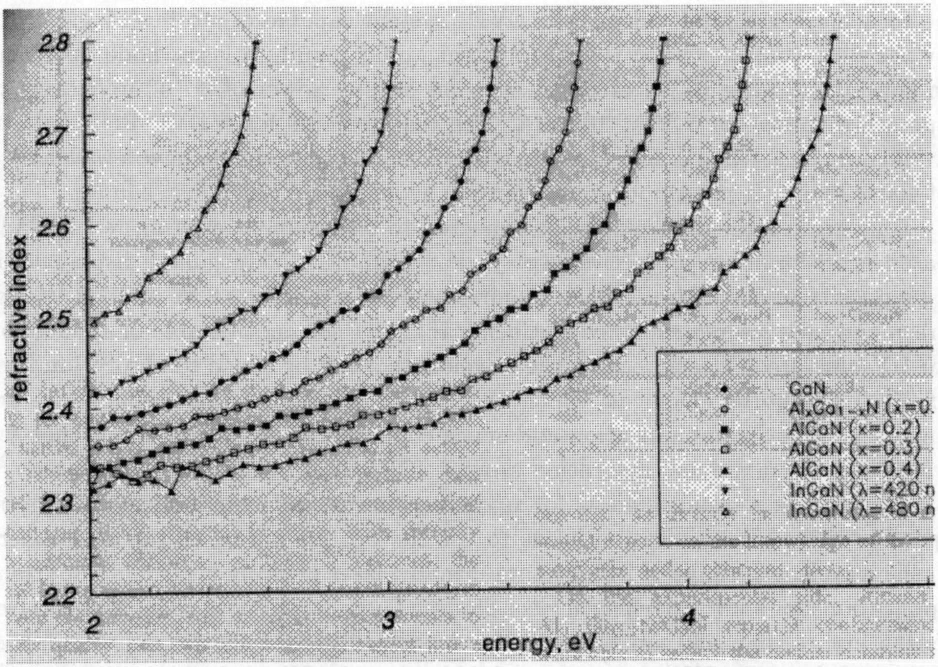

Fig. 2. Refractive indices of GaN, AlGaN and InGaN.[22]

transverse mode, a remarkable simple expression

$$\Gamma_T = \frac{D^2}{2 + D^2} \tag{2}$$

can be used with an accuracy of 1.5%. Here D is the normalized waveguide thickness

$$D = k_0(\mu_2^2 - \mu_1^2)^{1/2} d\,. \tag{3}$$

Here k_0 is the vacuum wave number $k_0 = 2\pi/\lambda$. μ_1 and μ_2 are the refractive indices of the cladding layers and the active layers respectively, d is the active layer thickness.

For SQW, because the active layer thickness is so small that $D \ll 1$. Using Eqs. (2) and (3), the confinement factor can be simplified as

$$\Gamma_T = \frac{2\pi^2(\mu_a^2 - \mu_c^2)d^2}{\lambda_0^2} \tag{4}$$

where μ_a and μ_c are the refractive indices of the active layer and cladding layer.

The confinement factor for MQW can be calculated in a similar procedure as the DH. The procedure is quite tedious because of the large number of layers involved. However, when only consider the fundamental mode, the confinement factor of a MQW structure is similar to that of a three-layer slab waveguide as shown in Fig. 3 except that, in a MQW structure, the total thickness, average refractive index are used.

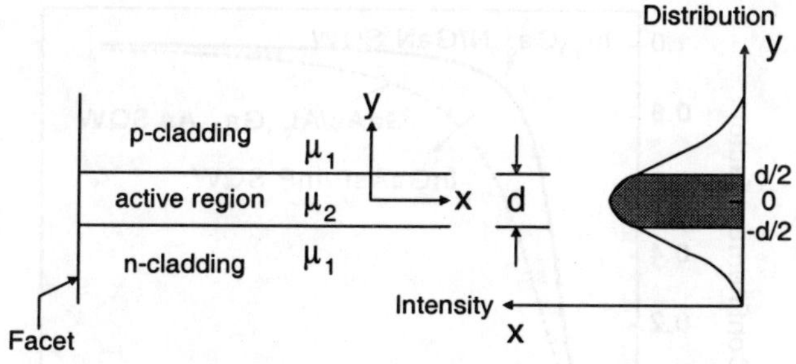

Fig. 3. Three-layer slab waveguide model of a DH semiconductor LED or LD. $\mu_2 > \mu_1$ is needed for waveguide in the active region.

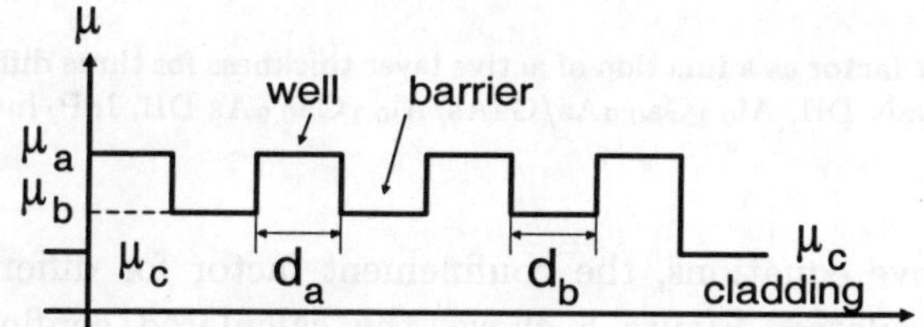

Fig. 4. Illustration of the refractive index variation in a MQW structure.

As shown in Fig. 4, suppose N_a and N_b are the number of active (well) and barrier layers in the MQW structure, d_a and d_b (and μ_a and μ_b) are the thickness (and refractive indices) of the active (well) and barrier layers respectively. μ_c is the refractive index of the cladding layer. $\bar{\mu}$ is the average refractive index of the MQW active region. Then the total thickness is

$$\bar{d} = N_a d_a + N_b d_b \,. \tag{5}$$

The average refractive index is

$$\bar{\mu} = \frac{N_a d_a \mu_a + N_b d_b \mu_b}{\bar{d}} \,. \tag{6}$$

The average total effective waveguide thickness is

$$\bar{D} = k_0(\bar{\mu}^2 - \mu_c^2)^{1/2}\bar{d} \,. \tag{7}$$

Three-layer equivalent confinement factor is

$$\gamma = \frac{\bar{D}^2}{2 + \bar{D}^2} \,. \tag{8}$$

Then the confinement factor of the fundamental transverse mode in the MQW structure is

$$\Gamma_T = \gamma \frac{N_a d_a}{\bar{d}} \,. \tag{9}$$

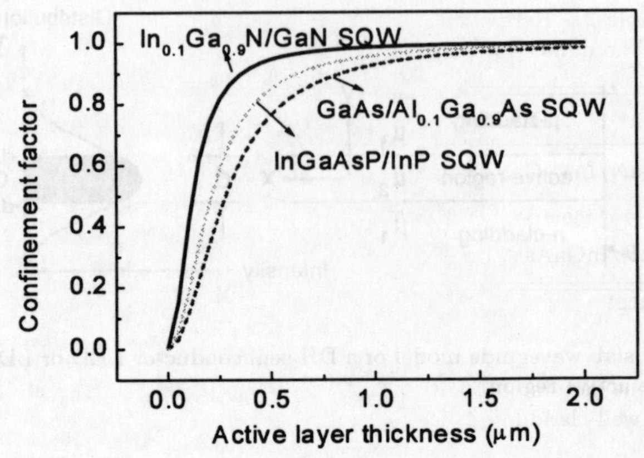

Fig. 5. Confinement factor as a function of active layer thickness for three different DH structures: GaN/$In_{0.1}Ga_{0.9}N$/GaN DH, $Al_{0.1}Ga_{0.9}As$/GaAs/$Al_{0.1}Ga_{0.9}As$ DH, InP/InGaAsP/InP DH.

Using the above equations, the confinement factor for different device structures can be calculated. Figure 5 shows the calculated confinement factor for GaN/$In_{0.1}Ga_{0.9}N$/GaN DH (solid line). For comparison, the confinement factors of two other DHs: $Al_{0.1}Ga_{0.9}As$/GaAs/$Al_{0.1}Ga_{0.9}As$ DH, InP/InGaAsP/InP DH (emitting at 1.3 μm) are also shown. The refractive indices used in the calculation are listed in Table 2.

Using Table 2, the confinement factors for SQW structures of the above three structures can also be calculated as shown in Fig. 6. From the calculation, it can be seen that, InGaN/GaN DH or SQW can have better optical confinement than conventional AlGaAs/GaAs, InGaAsP/InP DH or SQW with comparable composition variation.

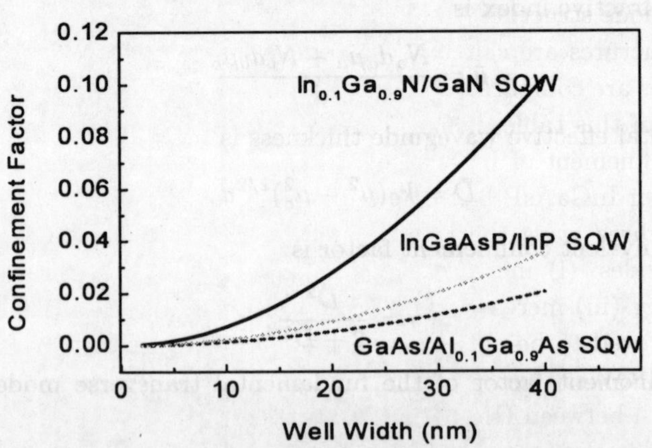

Fig. 6. Confinement factor as a function of well thickness for three different SQW structures: GaN/$In_{0.1}Ga_{0.9}N$/GaN SQW, $Al_{0.1}Ga_{0.9}As$/GaAs/$Al_{0.1}Ga_{0.9}As$ DH, InP/InGaAsP/InP SQW.

Table 2. Refractive indices of the three DH structures used in the numerical calculation.

DH Structure	μ_1	μ_2	$\lambda_0(\mu m)$
$GaN/In_{0.1}Ga_{0.9}N/GaN$	2.57	2.68	0.4
$Al_{0.1}Ga_{0.9}As/GaAs/Al_{0.1}Ga_{0.9}As$	3.52	3.59	0.87
InP/InGaAsP/InP	3.22	3.51	1.3

Table 3. Confinement factors of different MQW laser diode structures. Here Δa_{ab} and Δa_{ac} are the well-barrier and well-cladding lattice mismatch respectively.

MQW structure	Cladding	N_a	μ_a	μ_b	μ_c	Δa_{ab} (%)	Δa_{ac} (%)	λ_0 (μm)	Γ_T (%)
$In_{0.1}Ga_{0.9}N$ (33 Å)	GaN	6	2.68	2.57	2.57	−1.1	−1.1	0.4	2.6
/GaN (66 Å)*	$Al_{0.1}Ga_{0.9}N$	6	2.68	2.57	2.51	−1.1	−1.4	0.4	5.7
	GaN	10	2.68	2.57	2.57	−1.1	−1.1	0.4	6.4
$In_{0.2}Ga_{0.8}N$ (25 Å)	GaN	4	2.73	2.57	2.54	−1.7	−2.2	0.415	2.6
$/In_{0.05}Ga_{0.95}N$ (50 Å)**	$Al_{0.1}Ga_{0.9}N$	4	2.73	2.57	2.48	−1.7	−2.5	0.415	3.9
GaAs (120 Å)	$Al_{0.42}Ga_{0.58}As$	6	3.59	3.385	3.308	0.05	0.05	0.87	20.1
$/Al_{0.3}Ga_{0.7}As$ (120 Å)+									

As can be also seen from the above calculation, the confinement factor of SQW is very small because of the active layer (well) is very thin. In order to improve optical confinement and reduce the threshold current density, MQW structure is usually used for laser diode structure. In the following, confinement factors of four different laser diode structures are calculated in Table 3. Again, InGaN-based MQW laser diode structures are compared with AlGaAs-based and InGaAsP-based laser diode structures. From this table, it can be seen that, in real MQW-based laser diodes, the optical confinement of InGaN-based laser diodes are several times lower than AlGaAs-based or InGaAsP-based laser diodes.

There are three ways that could enhance the optical confinement of InGaN MQW laser diodes: (i) increase the number of wells, (ii) increase the well and barrier width, or (iii) increase the Al composition of the cladding layer. However, all these are proved to be extremely challenging because of two characteristics of InGaN-based MQWs: thermal instability of InGaN at growth temperature and lattice mismatch between III-Nitride alloys.

As discussed in the previous part, it is extremely challenging to grow large number of InGaN-based quantum wells while maintaining uniformity between top and bottom wells. The first generation of InGaN-based laser diodes had more than 20

InGaN-based quantum wells. These laser diodes can only operate at low temperature in pulsed mode.

Their linewidth are as wide as 2 nm. Multiple longitudinal modes were observed with strange mode spacing. All these could be related to the nonuniformity in the MQWs. The uniformity of thick well or barrier layer thickness could be also affected by the thermal instability of InGaN.

The large lattice mismatch between III-Nitride alloys prevent the growth of large well or barrier width or cladding layers with high Al composition. As shown in Table 3, the lattice mismatch between well and barrier, well and cladding layers in InGaN-based MQWs laser diodes is much larger than those in AlGaAs-based or InGaAsP-based laser diodes. To maintain the crystal quality of different layers, the well, barrier and cladding layers have to be thin to prevent defect formation.

An important consequence of the poor optical confinement in InGaN-based laser diodes is the leakage of the optical modes into the passive layers such as bottom GaN layer. This can introduce large cavity loss and increase the threshold current density.[22] Extra multi-mode far-field pattern (FFP) can be also generated because of the mode leakage.[23,24] To increase the cladding layer thickness without generating large amount of defects, modulation-doped strained-layer superlattice (MD-SLS) has been used as the cladding layer.

Band offset (carrier confinement)

Band offset between heterojunctions is an important parameter for laser diode design. The band offsets between GaN/InN, GaN/AlN, and AlN/InN have been calculated and experimentally measured by different groups, as shown in Table 4. There are large discrepencies between values reported by different groups.

Table 4. Band offsets in III-Nitride binaries.

Heterojunction	ΔE_c (eV)	ΔE_v (eV)	Reference
GaN/InN		1.05 ± 0.25	26*
	1.09	0.51	27
		0.3	28
		0.48	29
AlN/InN		1.81 ± 0.20	26*
	3.14	1.09	27
		1.25	29
GaN/AlN		0.70 ± 0.24	26*
	1.78	0.85	27
		0.7	28
		0.81	30
		1.36 ± 0.07	29

* From experiments. All others are from theoretical calculations.

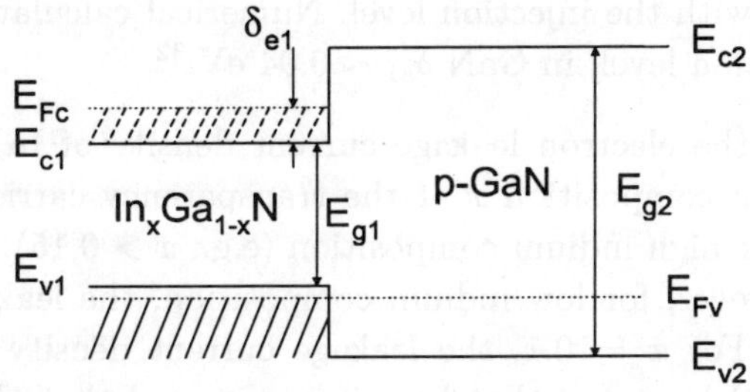

Fig. 7. $In_xGa_{1-x}N/p$-GaN heterojunction model used to estimate the leakage current.

The carrier confinement in a heterojunction can be characterized by the leakage current over the heterojunction barrier. The leakage current is due to the loss of carriers by diffusion. To estimate the leakage current in InGaN-based laser diodes, a simplified heterojunction model shown in Fig. 7 is used. The diffusion of electrons from the InGaN layer to p-GaN layer is given by

$$J_d = \frac{eD_{e2}}{L_{e2}} N_2 \,. \tag{10}$$

Here D_{e2}, L_{e2} are the electron diffusion coefficient and diffusion length in the p-type region (minority carrier diffusion length). e is the electron charge. N_2 is the electron density at the edge of the high bandgap p-type side, including the injected carriers. It is proved that when $E_{c2} - E_{Fc} > 4kT$ and under high injection condition, the above equation can be expressed as[25]

$$J_d = \frac{eD_{e2}}{L_{e2}} \left(\frac{N_{v2}}{P_2}\right) N_{c2} \exp\left[\frac{-(\Delta E_g - \delta_{e1})}{kT}\right] . \tag{11}$$

Here N_{c2} and N_{v2} are the effective density of states in the conduction band and valence band of the high bandgap p-type layer. P_2 is the hole concentration of the high bandgap p-type region. $\Delta E_g = E_{g2} - E_{g1}$ is the bandgap difference of the two layers. $\delta_{e1} = E_{c1} - E_{Fc}$ is relative quasi-Fermi level in the low bandgap layer. It represents the level of carrier injection in this layer. High carrier injection raises the quasi-Fermi level and increases δ_{e1}, which in turn reduce the carrier confinement barrier and increase the leakage current.

To calculate the leakage current in $In_xGa_{1-x}N/p$-GaN heterojunction, the following numerical parameters are used:

- The bowing parameter for InGaN is $b = 1.0$ eV, which gives

$$\begin{aligned} E_g(x) &= xE_{g,InN} + (1-x)E_{g,GaN} - bx(1-x) \\ E_{g,InN} &= 1.9 \text{ eV and } E_{g,GaN} = 3.4 \text{ eV} . \end{aligned} \tag{12}$$

- The minority carrier diffusion length $L_{e1} = 0.1$ μm,[31] diffusion coefficient $D_{e1} = 8.8$ cm^2/s (corresponding to mobility $\mu_{e1} \sim 330$ cm^2/V.s). P-type concentration $P2 = 5 \times 10^{17}$ cm^{-3}. At 300 K, the effective density of states are $N_{c1} = 3.5 \times 10^{18}$ cm^{-3}, $N_{v1} = 1.8 \times 10^{19}$ cm^{-3}.

- For δ_{e1}, it varies with the injection level. Numerical calculation showed that, at transparent injection level, in GaN $\delta_{e1} \sim 0.04$ eV.[32]

Figure 8 shows the electron leakage current density of $In_xGa_{1-x}N/p$-GaN as a function of indium composition x at the transparency carrier injection level. It can be seen that, for high indium composition (e.g., $x > 0.15$), the leakage current density is low. However, for low indium composition, the leakage current density is extremely large. For $x = 0.1$, the leakage current density can be as large as 8 kA/cm^2. It should be noted that here it is assumed that the current injection level is low (at the transparency condition level). As injection level increases, the confinement barrier lowers and the leakage will be substantially increased. Shown in Fig. 9 is the electron leakage current as a function of relative quasi-Fermi level δ_{e1} (injection level) for two indium compositions: $x = 0.1$ and $x = 0.2$. For $\delta_{e1} =$

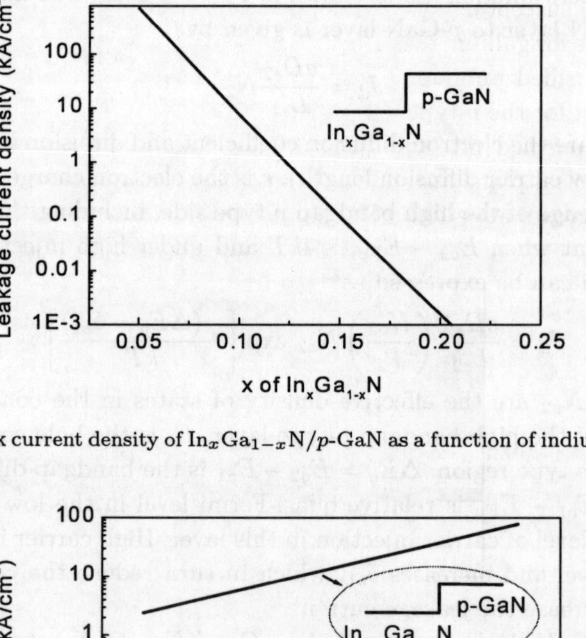

Fig. 8. Electron leak current density of $In_xGa_{1-x}N/p$-GaN as a function of indium composition x.

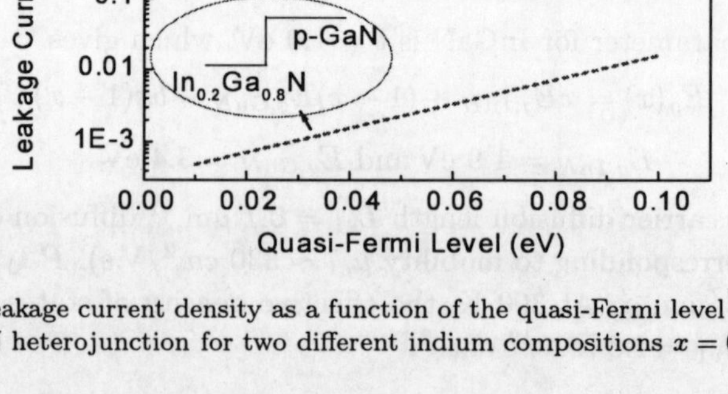

Fig. 9. Electron leakage current density as a function of the quasi-Fermi level (injection level) in $In_xGa_{1-x}N/p$-GaN heterojunction for two different indium compositions $x = 0.1$ and $x = 0.2$.

0.08 eV, the leakage current density for $x = 0.1$ is as large as 38 kA/cm^2, while for $x = 0.2$, the leakage current is $\sim$ 7 A/cm^2, which is very low and will not cause significant increase of lasing threshold current density.

The large current leakage in InGaN-based heterojunctions is mainly due to poor p-type doping, which lead to low minority diffusion length in the p-type region and low p-type carrier concentration. InGaN active layer with large indium composition is expected to reduce the current leakage. However, this is challenged by the large lattice mismatch, as in the case of optical confinement.

2. GaInN/GaN MQW Device Growth, Characterization and Fabrication

The device structures studied here were grown in a horizontal low pressure metal-organic chemical vapor deposition (MOCVD) reactor. The inductively heated SiC-coated graphite susceptor was rotated at about 40 rpm to improve uniformity. The source materials for gallium and nitrogen included high purity trimethyl- or triethyl-gallium and purified ammonia, respectively. Bis-cyclopentadienylmagnesium and silane were used for the p-type and n-type doping sources, respectively. Pd diffused hydrogen and purified nitrogen were used as the carrier gases. The growth was performed on basal plane sapphire substrates. A 3 μm-thick n-type Si-doped GaN layer was first grown at $\sim$ 1000°C. Then, the temperature was lowered to $\sim$ 750°C

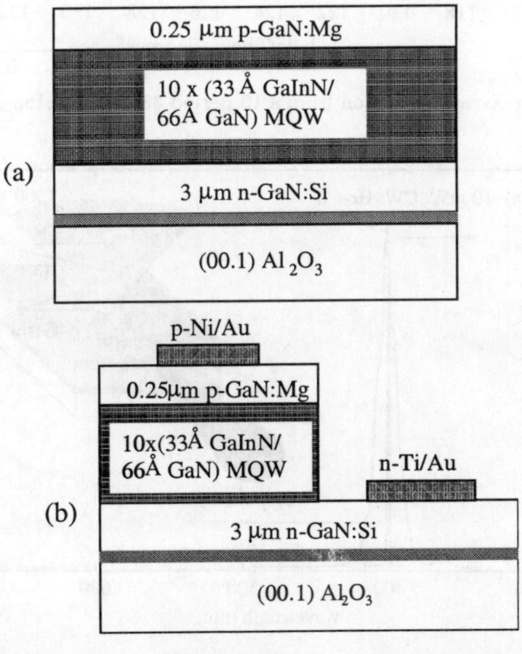

Fig. 10. (a) Schematic diagram of a 10 period 33 Å $Ga_{0.89}In_{0.11}N$/66 Å GaN MQW laser structure, and (b) Cross-section of a processed broad-area MQW laser with metal contacts.

to grow the GaInN multi-quantum wells or laser diode active layers. The growth temperature was then increased to ~ 1000°C to grow a 2500 Å-thick *p*-type GaN:Mg layer. A schematic diagram of the device structure is shown in Fig. 10(a).

The multi-quantum well structures consisted of a 10 or 15 periods of 33 Å $Ga_{0.89}In_{0.11}N$/66 Å GaN. Figure 11 shows an open detector X-ray diffraction spectrum from this structure. The −1 satellite peak on the left allowed to estimate the superlattice period to be 97 Å, confirming the thickness predicted from the growth rates. In order to be able to assess the optical properties of the multi-quantum wells in the device structures, the top *p*-type GaN:Mg layer was partially removed by dry etching to allow penetration of the laser beams. Figure 12(a) shows the RT

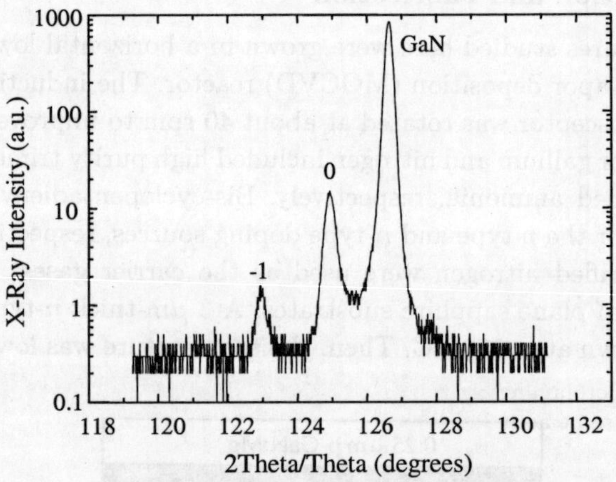

Fig. 11. Open detector X-ray diffraction from a 10 period 33 Å $Ga_{0.89}In_{0.11}N$/66 Å GaN MQW.

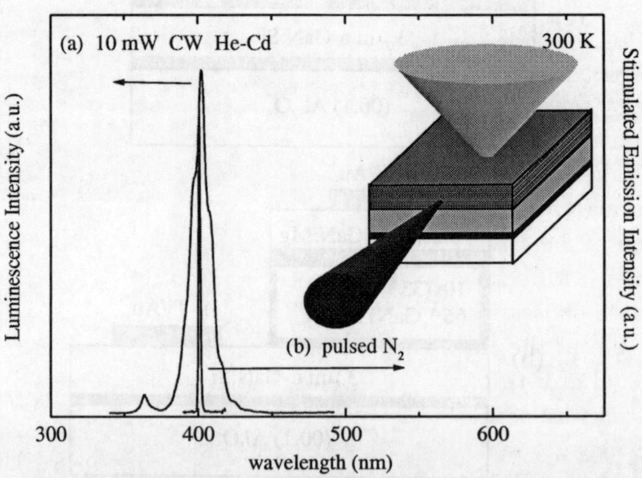

Fig. 12. Room temperature (a) photoluminescence and (b) stimulated emission from a 10 period 33 Å $Ga_{0.89}In_{0.11}N$/66 Å GaN MQW.

photoluminescence of this structure which exhibits a small peak at 365 nm corresponding to the cap GaN and an intense peak at 402.3 nm corresponding to the MQW. The MQW luminescence peak is located at a longer wavelength than for a 33 Å single quantum well. This may be because the MQW structure (1000 Å thick) is less strained. The MQW luminescence peak is much more intense than for the single well case, indicating good electrical confinement.

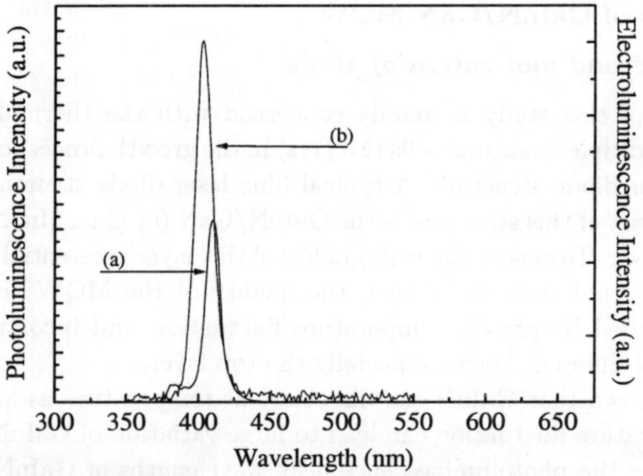

Fig. 13. (a) 79 K photoluminescence from a GaInN/GaN MQW and (b) 79 K electroluminescence from a LD with the same active region.

Figure 13(a) shows the 79 K photoluminescence of the MQW structure. The spectrum exhibited at 410 nm. The full width at half maximum of the peak at 410 nm was ~ 11 nm.

A 1 mm wide bar was cut from the MQW structure wafer and the long sides were mechanically polished. Room temperature optical pumping was then carried out using a pulsed nitrogen laser (337 nm) with a pulse width of 600 ps and a repetition rate of 6 Hz. Neutral density filters were used to attenuate the optical power. The edge emitting stimulated emission was collected from the polished edges through a monochromator coupled with a photomultiplier tube which has a resolution of ±0.3 nm. Stimulated emission was observed for pumping densities higher than a threshold estimated at ~ 100 kW/cm^2. Figure 12(b) shows a typical stimulated emission spectrum. The peak position was 401 nm and the peak width was ~ 1 nm at the threshold.

After epitaxial growth of the device structures, rapid thermal annealing under nitrogen ambient was performed to achieve low resistivity *p*-type GaN:Mg.[14] In order to make the *n*-type contact, the surface of the sample was etched until the underlying *n*-type GaN layer was exposed using ECR-RF dry etching as shown in Fig. 10(b). 100 μm and 10 μm-wide stripes were thus fabricated. Ti/Au and Ni/Au metal contacts were deposited on the *n*-type and *p*-type GaN, respectively, using an electron-beam evaporator and were defined by a conventional lift-off process. Mirror facets were mechanically polished for various cavity lengths from 650 μm to 1800 μm. The roughness of the facet surface was in the range of 50 nm. No antireflection or high-reflection coatings were applied on the mirror facets. No passivation was used.

3. Effects of Temperature and Top Cap Layer on the Optical Properties of GaInN/GaN MQW

3.1. *Objective and motivation of study*

The objective of this study is mainly concerned with the thermal instability of GaInN/GaN multiple quantum wells (MQWs) in the growth process of a III-Nitride-based blue laser diode structure. A typical blue laser diode structure is shown in Fig. 14. The heart of this structure is the GaInN/GaN (or $Ga_{1-x}In_xN/Ga_{1-y}In_yN$) MQW active layer. To ensure the high quality of this layer is essential. However, due to the poor thermal stability of InN, the quality of the MQW active layer can be easily degraded by growth temperature fluctuation and incompatible growth temperatures of different layers, especially the top layer.

It is well known that GaInN can dissociate at temperature as low as 500°C.[33] Growth temperature fluctuation can lead to large variation of GaInN composition. Figure 15 shows the photoluminescence peak wavelengths of GaInN/GaN MQWs grown at different reactor heater output power (corresponding to different growth temperatures). It can be seen that a temperature fluctuation of 10°C can shift the peak wavelength more than 22 nm. Therefore the peak wavelength shift is 2.2 nm/°C, or peak energy shift 9 meV/°C. This shift reflects the large compositional variation inside the GaInN/GaN MQW active layers if growth temperature is not controlled tightly.

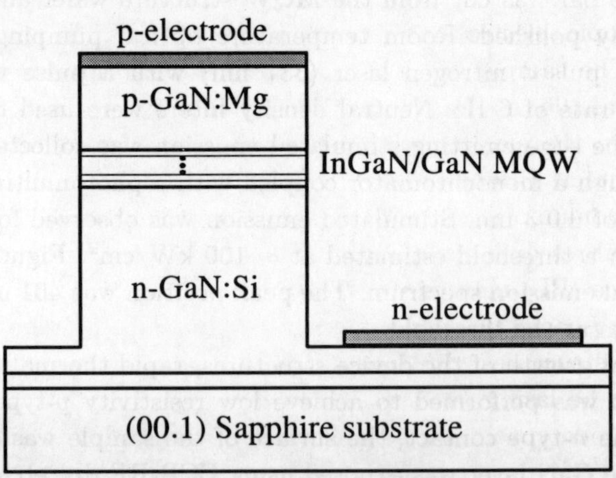

Fig. 14. A typical GaInN/GaN MQW based blue laser diode structure. Not drawn here are some other layers such as *p*-AlGaN, *n*-AlGaN layers.

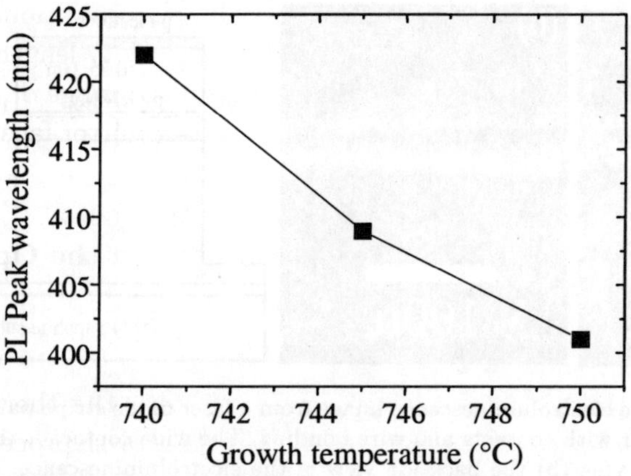

Fig. 15. Room temperature photoluminescence peak wavelength of GaInN/GaN MQWs grown at different growth temperatures.

In addition to growth temperature fluctuation, the incompatible growth temperatures for different layers can also lead to the compositional variation and defect formation in the MQW active layer. This can be summarized as follows:

- The *p*-type layer is usually grown at a higher temperature than that for GaInN MQW. During the top layer growth, apparently the MQW is subject to high temperature additional thermal annealing.
- The *p*-type has to be obtained by annealing the top GaN:Mg layer, which again puts the whole laser diode structure under thermal annealing.
- Even undoped GaInN layers usually have a high *n*-type carrier concentration. During the doping of the top layer, Mg could diffuse into GaInN layers and introduce compensation.

The microscopic electroluminescence image in Fig. 16 clearly shows the nonuniformity. The sample has a typical laser diode structure with a 10-period GaInN(33 Å)/GaN(66 Å) MQW sandwiched between *n*-type and *p*-type GaN (Fig. 16(c)). Here the electroluminescence image is taken from the back of the *p*-type stripe contact. Bright spots, bright edge and shaded areas can be clearly seen, indicating nonuniformity of the emitting layer.

In fact, several research groups have studied the compositional variation using photoluminescence (PL) mapping, cathodoluminescence (CL) mapping, transmission electron microscope (TEM), and X-ray.[34–36] It has been found that there exists quantum-dot like structures with dimensions of 2–5 nm inside the MQW, which has been supposed to be lasing regions. Various research groups have observed strong mode-hopping, mode spacing not corresponding to the cavity length, and broad emission spectra.[37–42]

It should be noted that TEM itself sometimes couldn't tell the compositional variations. Abrupt interface could be still observed with a moderate compositional

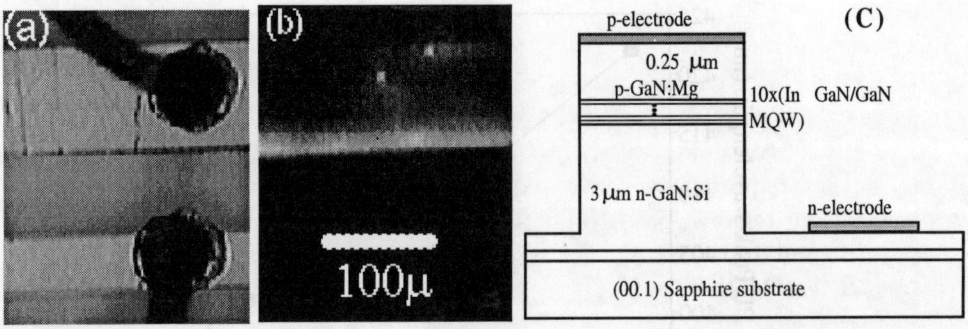

Fig. 16. Microscopic electroluminescence image from a laser diode stripe pair. (a) The top view of part of the stripe pair with contacts and wire bonding. The wide contact is the p-type contact and narrow one is for n-type; (b) the backside view of the electroluminescence. Nonuniform emission can be seen from the contrast, especially the bright spots and p-type edge. (c) The schematic of the stripe pair.

variation.[34] Nishio *et al.*,[43] showed abrupt GaInN/GaN interfaces in TEM in both bare GaInN/GaN MQW structure (without top layer) and GaInN/GaN MQW based laser diode structure (with top p-type layer). They concluded that the top layer growth would not degrade the MQW quality based on the observation of abrupt interfaces in TEM. However, with similar TEM images, Chichibu *et al.*, showed In-compositional variations with the help of atomic force microscopy (AFM) and CL.[34]

In addition to compositional variations, Wu *et al.*, and Chen *et al.*, have studied the so called "V" defects (pits) formation in GaInN MQWs using AFM, TEM.[44,45] Dislocations may induce these pits. These pits introduce a long wavelength shoulder in addition to the main peak in GaInN/GaN MQW PL spectrum. They could also be one of the important failure mechanisms in blue laser diode as will be discussed later in this section.

From the above discussion, it can be seen that indium compositional variation and defect formation in the GaInN/GaN or $Ga_{1-x}In_xN/Ga_{1-y}In_yN$ MQW active layer are of critical importance. It could be good for laser diode because of the formation of quantum-dot like structure could lower the lasing threshold. Therefore, the study of the effects of the top layer thermal treatment on active layer is important to improve the quality of GaInN MQW active layer.

3.2. *Experiment*

The effects of the top layer on the GaInN/GaN MQW grown by low pressure MOCVD as described earlier were studied using optical pumping and photoluminescence. The following experiments have been carried out:

- Thermal annealing of MQW structures. These MQW structures are the same as a typical laser diode structure except that they have a 500 Å undoped GaN on the top of the MQW instead of a thicker *p*-type layer. Polished bars with feedback mirrors and a cavity length around 1mm were fabricated just as laser diode bars. Then the bars were annealed under N_2 in a rapid thermal annealing (RTA) furnace at 1000°C for different duration of time. After each annealing, photoluminescence was measured to study the degradation of the sample. No mesas were required in optical pumping since a well focused line-shape beam provides the regions like gain-guided laser stripes (Fig. 17). MQW structures with different top layer thickness (corresponding to different top layer growth time). These MQW samples were polished into bars with mirrors and studied by optical pumping. The absorption by the top layer was calibrated using the absorption coefficient of GaN.[31]

- Various laser diode structures. The top *p*-type layers of these layer diode structures were etched away using reactive etching. Polished bars with mirrors were fabricated from these samples and then studied by optical pumping. Their lasing characteristics were compared with the samples studied above.
- Photoluminescence study of the above structures at room temperature and 77 K (liquid nitrogen temperature).

3.3. *Results and discussion*

Figure 18 shows the results of optically pumped stimulated emission of a MQW structure (Fig. 17(a)) after annealing at 1000°C for 0.5 minutes up to 16 minutes. Shown here is the emission intensity as a function of optical pumping density (will be called *P–I* curve hereafter). As a reference, the *P–I* curve of the bar before annealing is shown too. The crossing between the dashed line and the horizontal line is the threshold of each *P–I* curve. The slope efficiency is the slope of the dashed line. Figure 19 shows typical emission spectra at different optical pumping levels. It can be seen that the stimulated emission is around 402 nm and it is from the MQW active layer.

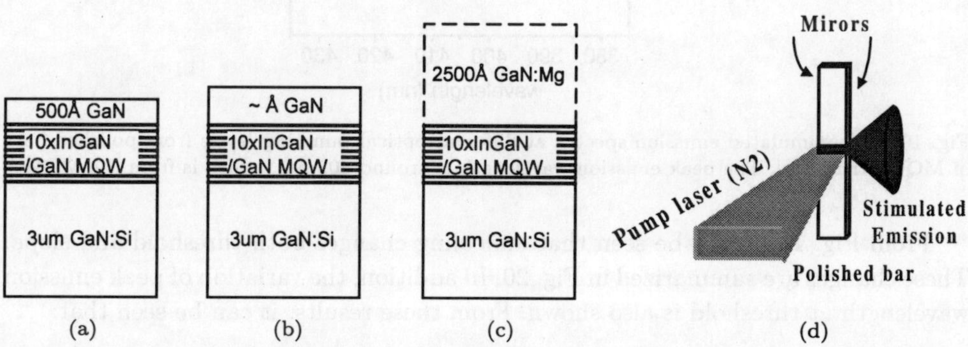

Fig. 17. Sample structures used in (a) annealing study; (b) different top layer thickness study; (c) laser diodes structures with etched top. (d) All the above samples were cut and polished into bars with mirrors and studied using optical pumping.

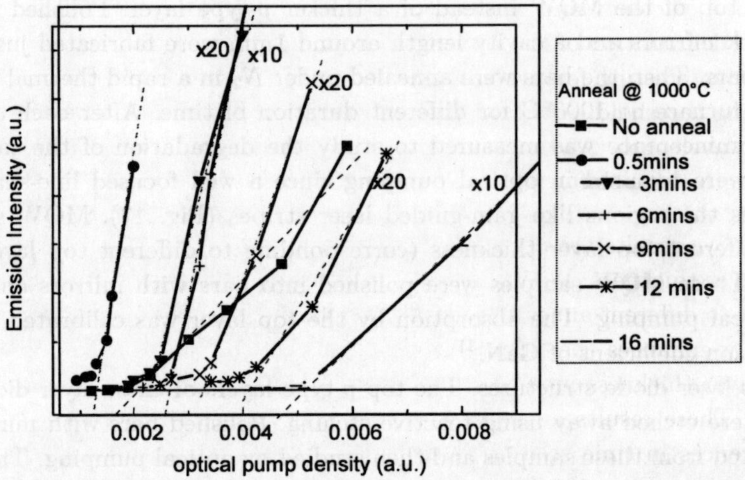

Fig. 18. Emission intensity as a function of optical pumping density (*P–I* curves) for a MQW structure annealed at 1000°C for different period of time. The dashed line and line are for the guide of eyes. Annealing changes the threshold and slope of the *P–I* curves (also see Fig. 20).

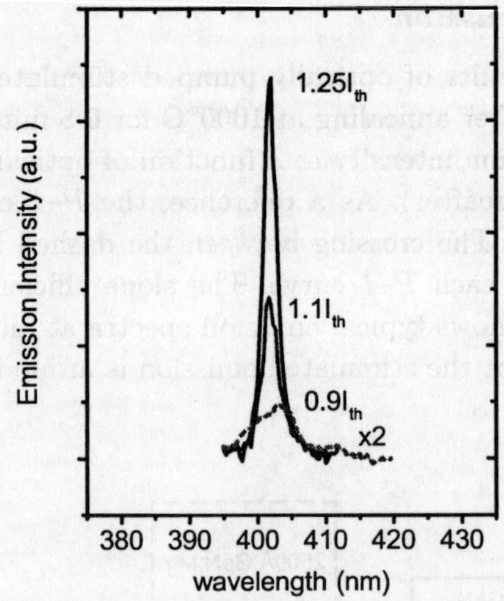

Fig. 19. The simulated emission spectra at different optical pumping levels from polished bars of MQW structures. The peak emission wavelength is around 402 nm, which is from the MQW.

From Fig. 18, it can be seen that annealing changes both threshold and slope. These changes are summarized in Fig. 20. In addition, the variation of peak emission wavelength at threshold is also shown. From these results, it can be seen that:

- Annealing is not necessarily degrading the lasing characteristics of the MQW. In fact the first half minutes annealing reduces the lasing threshold and improves the slope efficiency. From longer than half minutes to 6 minutes annealing, the

high slope efficiency and low threshold keep almost constant. Further annealing leads to degradation of the slope efficiency and increase of threshold.
- Another interesting feature is the variation of the emission peak wavelength at the lasing threshold. After the first half minutes, the emission peak is shifted to longer wavelength. As the annealing time increases, it keeps shifting to shorter wavelength (blue-shift).

The above results are confirmed by the study of MQW structures with different top layer thickness. With the same growth rate, different top layer thickness is equivalent to different annealing time. As can be seen from Fig. 19, a thin top layer can greatly enhances the slope efficiency and reduces the threshold. However, as the top layer thickness increases, the slope efficiency drops sharply and threshold increases. These results are consistent with the observations in annealing study. The differences in time scale, the extent of slope enhancement and threshold reduction between Figs. 20 and 21 can be attributed to the annealing characteristics difference between the reactor and the rapid thermal annealing (RTA) furnace. RTA has a much fast heating and cooling time than the reactor.

The annealing effects discussed above can be related to indium segregation in GaInN-based MQWs. As discussed in the beginning of this part, InN is thermally unstable at temperature higher than 500°C. When the MQW structures are annealed either by RTA or during top layer growth in the reactor, InN will dissociate and evaporate. In a MQW structure, GaInN is confined by GaN or GaInN with lower indium composition so the evaporation will go nowhere but moving in the confined plane perpendicular to the growth plane. During the annealing and cooling processes, indium atoms could choose to go to their favorable sites. These favorable

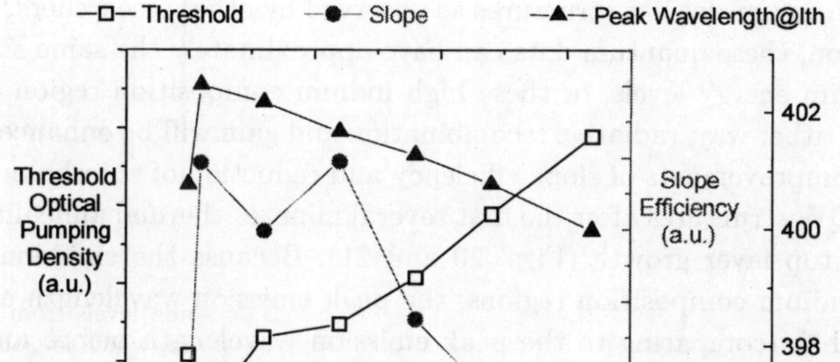

Fig. 20. Threshold optical pumping density, slope efficiency and stimulated emission peak wavelength at threshold, as a function of annealing time of the polished MQW bar (also see Fig. 18).

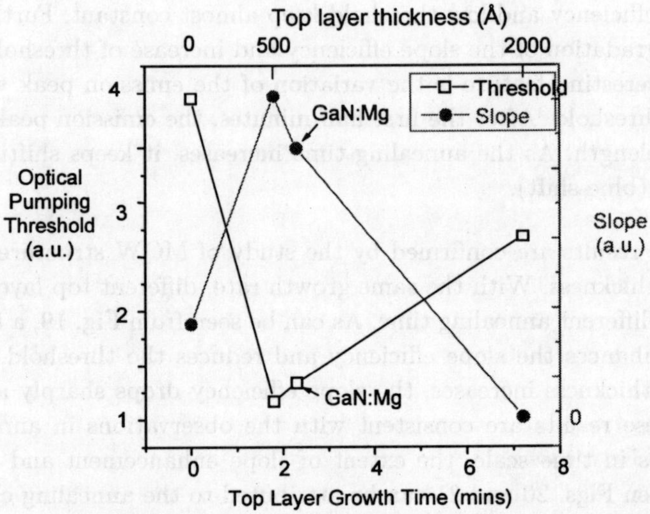

Fig. 21. Threshold optical pumping density and slope efficiency as a function of top layer growth time (top layer thickness) of MQW structures (see Fig. 17(b)). Here the data labeled GaN:Mg are from a MQW structure with only 500 Å GaN:Mg on top but it was annealed at 1000°C by rapid thermal annealing to activate *p*-type.

sites could be determined by the high density of defects, or by the intrinsic nature of GaInN ternary alloys since the compositional modulation due to phase separation would be favored in this system.[46,47]

The compositional variation in the active layer due to indium segregation can be beneficial or harmful for lasing, dependent on the annealing conditions. Under proper thermal annealing, these variations can lead to high indium composition regions or quantum-dot like structures as observed by other researchers.[34–40] In the best condition, these quantum dots can have approximately the same size and thus close quantum energy levels, or these high indium composition region can confine excitons. In either way, radiative recombination and gain will be enhanced. This can explain the improvements of slope efficiency and reduction of threshold of optically pumped MQW structures after the first several minutes thermal annealing or short-time (thin) top layer growth (Figs. 20 and 21). Because the emission originated from high indium composition regions, the peak emission wavelength at threshold shows red shift (comparing to the peak emission wavelength before annealing) in the first several minutes of annealing (Fig. 20). As the annealing time increases, the size of these self-formed quantum dots can be varied and radiative efficiency can be reduced by dislocation propagation (remember that the MQW is based on a layer with defect density $> 108\ \text{cm}^{-2}$). This certainly leads to degradation of the lasing characteristics of the MQW structures — the increase of lasing threshold and decrease of slope efficiency (Figs. 20 and 21). Further annealing reduces the role of dots, making In distribution more homogeneous. This can explain the blue shift of the emission peak wavelength after the first half minute of annealing.

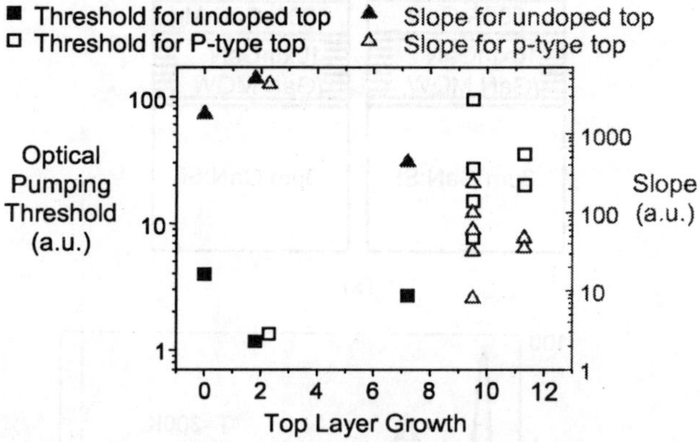

Fig. 22. Statistics of lasing characteristics of MQW structures (undoped top) and MQW laser diode structures (*p*-type top). The top *p*-type layers of laser diode structures have been etched away using reactive ion etching. The top layer thickness is converted to the growth time.

The above results demonstrate that it is very important to plan the growth of the top layer in order to optimize the GaInN-based MQW active layer for laser diodes. Figure 22 is a statistical study of optically pumped lasing characteristics (threshold and slope) as a function of top layer growth time for a variety GaInN-MQW based samples. These samples have the same MQW active layer[48] but different top layers as shown in Fig. 17. The laser diode structures have thick *p*-type top layers (usually 0.25 μm). As can be seen from this figure, laser diode structures have thresholds 1–3 orders of magnitude higher and slope efficiency 1–2 orders of magnitude lower than those of the best annealed MQW structures.

The threshold and slope efficiency of laser diode structures are much worse than those expected from annealing study and top layer thickness study of MQW structures with undoped top layer. Therefore there must be some other factors other than annealing that are degrading the laser diode performances. Next, the PL study will show that this is due to the *p*-type doping of the top layer.

From previous experiences, the effects of Mg-doping can be best studied using low temperature photoluminescence since Mg-related emission from unsuccessfully Mg-doped samples is easily quenched at room temperature. So room temperature and 77 K PL have been carried out to study the effect of Mg-doping of the top layer on MQW active layer.

Figure 23 is a comparison of room temperature (Fig. 23(b)) and 77 K (Fig. 23(c)) photoluminescence of two MQW structures. These two structures are the same except for the top layer doping: one is undoped and the other one is Mg-doped and annealed at 1000°C for 30 secs by RTA (see Fig. 23(a)). In the next, the sample with top layer undoped will be called TUD. The sample with Mg-doped top layer will be called TDMg. In these spectra, the short wavelength peak is GaN bandgap emission. The peak in the middle is the emission from GaInN/GaN MQW. The broad long wavelength peak is the yellow luminescence from the bottom GaN:Si layer.

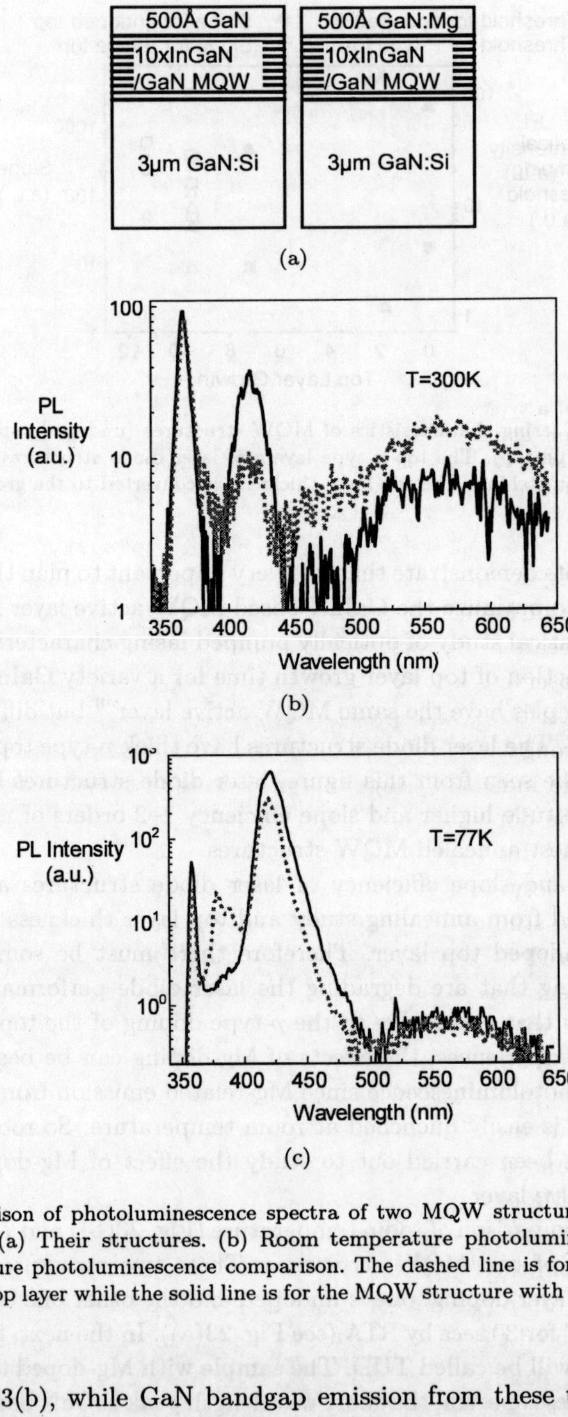

Fig. 23. Comparison of photoluminescence spectra of two MQW structures with thin GaN or GaN:Mg on top. (a) Their structures. (b) Room temperature photoluminescence comparison. (c) Low temperature photoluminescence comparison. The dashed line is for the MQW structure with GaN:Mg as top layer while the solid line is for the MQW structure with undoped GaN on top.

From Fig. 23(b), while GaN bandgap emission from these two structures are the same, MQW related emission in TDMg is more than 4 times lower than that of TUD. This indicates that somehow the MQW related emission in TDMg has been

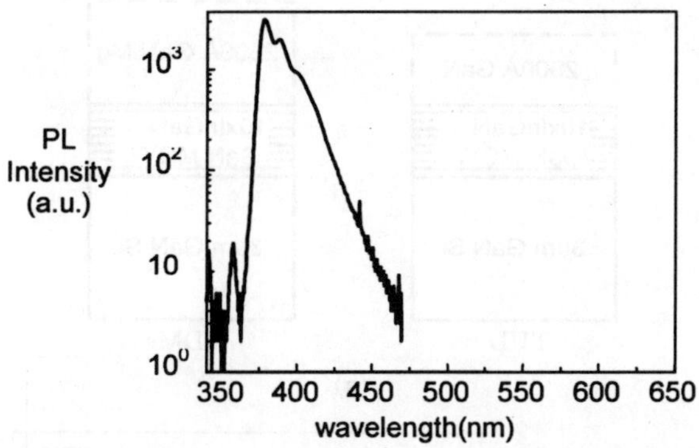

Fig. 24. 77 K PL of a typical lightly Mg-doped sample. These samples typically show a weak GaN bandgap emission, and a strong emission peaked at 378 nm with phonon replicas. This peak has the same feature with the extra peak observed from TDMg MQW structures (see Fig. 23(c)). It is a D–A recombination.[51]

degraded by the Mg-doping. The effect of Mg-doping can also be seen from the 77 K PL spectra shown in Fig. 23(c). Here, in addition to the MQW related emission, an extra peak appears at 378 nm with satellite peaks. This feature is very similar to what can be observed in low temperature PL of lightly Mg-doped samples (see Fig. 24). It was attributed to donor-to-acceptor (D–A) emission.[49] Those samples usually showed strong emission around 378 nm with phonon assisted satellite peaks. The similarity here indicates either this peak is from the top layer itself, because it is too thin to be doped well, or from a thin layer between GaN:Mg layer and GaInN/GaN MQW due to diffusion of Mg into the well region.

To distinguish between the above two possibilities, room temperature and 77 K photoluminescence studies have been carried out to characterize a series of laser diode structures with etched top layer (0.25 μm *p*-type GaN). Figure 25(c) shows the typical room temperature and 77 K photoluminescence spectra. For comparison, a MQW structure with thicker (0.2 μm undoped GaN) undoped top layer is also studied. The room temperature and 77 K PL spectra from this sample with etched top are also shown in Fig. 25(b). Again, a strong peak at 378 nm is observed in the laser diode structure but not in the MQW structure with undoped top layer. In addition, it is enhanced in the laser diode structures comparing to the sample with only 500 Å GaN:Mg on top (see Fig. 23(c) dashed line). This peak is consistently observed in all laser diode structures with etched top, which demonstrates that the emission is not from the top layer, it is from a layer somewhere near the interface of MQW and the top *p*-type layer.

It should be noted that a careful look at the 77 K spectra of MQW structures with undoped top layers also reveals a very low intensity shoulder around 378 nm (see solid lines of Figs. 23(c) and 25(b)). In fact, in Fig. 25(b), this shoulder is enhanced with thicker top layer, which is similar to the case of Mg-doped top layer. In summary, from these observations, it can be seen that:

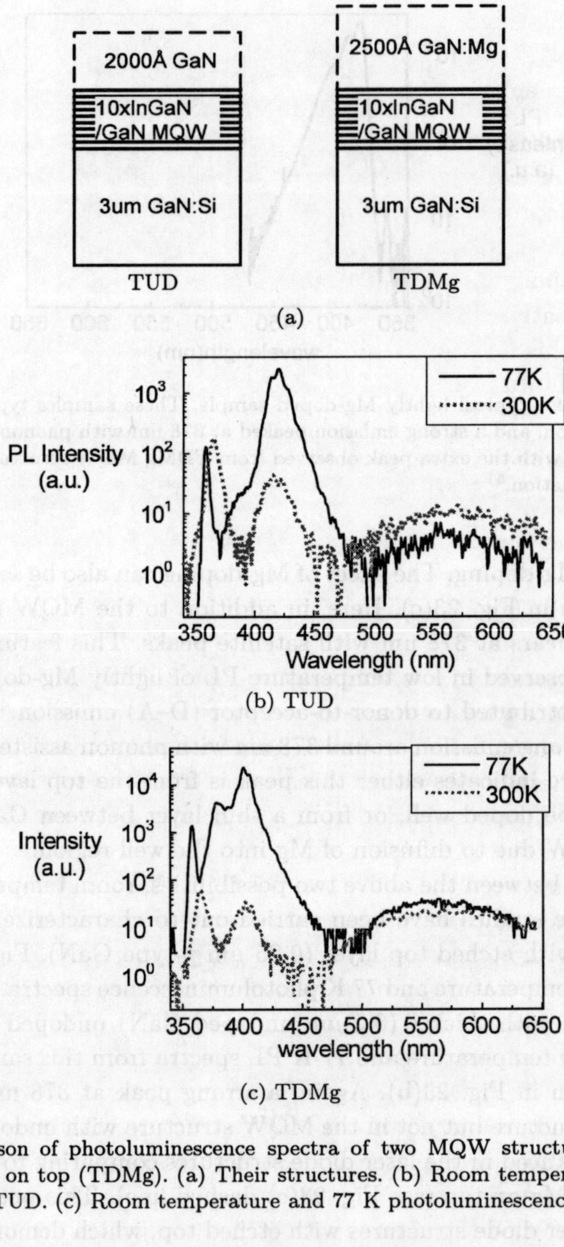

Fig. 25. Comparison of photoluminescence spectra of two MQW structures with thick GaN (TUD) or GaN:Mg on top (TDMg). (a) Their structures. (b) Room temperature and 77 K photoluminescence of TUD. (c) Room temperature and 77 K photoluminescence of TDMg.

– The 378 nm peak is a donor-to-acceptor (D–A) recombination. It is strong at low temperature and quenched at room temperature.
– This peak exists in all MQW structures and laser diode structures. It is extremely low in MQW structures with undoped top and greatly enhanced in MQW structures with Mg-doped top layer.

- It is not from the top layer itself. Instead, it is from the MQW or at least the top of the MQW.
- Thicker top layer enhances this peak.

The observation of a peak around 378 nm and its characteristics strongly suggest that there is a thin defective layer on the top of the MQW. It could be formed because of the compressive stress caused by the top GaN layer. As the top layer gets thicker, this layer becomes more defective because of the relaxation of stress. During the Mg-doping of the top layer, Mg can diffuse into this layer. Since Mg is an acceptor, it greatly enhances the D–A recombination around 378 nm at 77 K. The existence of this layer could be a strong factor that degrades the laser diode performances as shown in Fig. 22.

3.4. *Conclusions*

In a GaInN-based MQW blue laser diode structure, the growth of top layer has three effects on the GaInN-based MQW active layer: annealing, defect formation, and Mg diffusion.

Annealing can be beneficial or harmful to the active layer depending on annealing conditions. A proper annealing could lead to the formation of quantum dots like structure and localization of excitons at high indium composition region, both of which can enhance gain and radiative recombination. Over-annealing could lead to random variation of indium composition and energy bandgap variation in the active layer and therefore degrade the laser diode performance.

Due to lattice mismatch, the top *p*-type layer growth could also result in the formation of a thin defective, compensated layer around the active and top layer interface. Mg-doping enhances the compensation and D–A recombination in this layer. The existence of this thin layer could be a very important factor for the degradation of the lasing performance of laser diode structures.

4. Study of the Effects of Si-Doping

4.1. *Objective and motivation of study*

The aim of this study is to investigate if (a) Si-doping can improve the performance of GaInN-based MQW laser diodes, (b) what mechanism is responsible if there is an improvement, and (c) what are the causes if there is no improvement. The successful commercialization of GaN-based (including GaInN) LEDs made use of the heavy co-doping of Si and Zn in the recombination zone.[17,18,50]

Figure 26 is a good example illustrating the limitation of impurity-related emission. Here two samples were studied using photoluminescence and optical pumping at room temperature. These two samples are the same except the doping. Sample (1) is nominally undoped GaInN while sample (2) is co-doped with Zn and Si. PL was done using a CW He-Cd laser with an output power of 4 mW. Figure 26(a) shows the PL of these two samples at the same laser intensity. It can be seen that

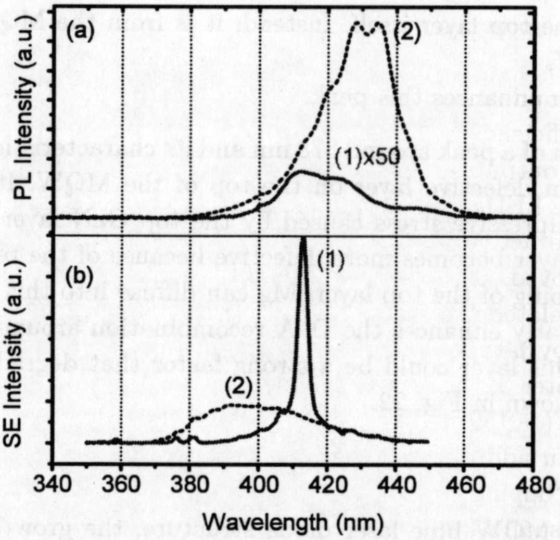

Fig. 26. Comparison of room temperature photoluminescence (PL) and optically pumped stimulated emission (SE) of two GaInN samples: (1) undoped GaInN; (2) Zn and Si co-doped GaInN:Zn,Si. Shown here are (a) PL (b) SE of these two samples. In fact, in (b) the emission given by sample (2) (GaInN:Zn, Si) is not SE.

PL from the doped sample is almost 300 times higher than that of the undoped one. In addition, the peak emission wavelength of the co-doped sample is more than 15 nm longer than that of the undoped one (Stokes shift), indicating the emission from the co-doped sample is impurity-related. Figure 26(b) is the emission spectra of these two samples under optical pumping. The optical pumping was done with a pulsed N_2 laser capable of delivering a peak power output of 2.0 MW. Here the spectra were obtained by attenuating the laser output 30 times (with some focusing).

Opposite to the PL results, the undoped sample showed a strong, sharp stimulated emission while the co-doped showed only a weak, broad spontaneous emission. The co-doped sample did not show any stimulated emission even to the damaging optical pumping density. It is also important to notice the emission peak wavelength difference between PL and optical pumping. In undoped sample, SE occurs at the short wavelength side of the PL (also the peak of the PL) with no Stokes shift, indicating the emission is from band-to-band. In the case of undoped sample, the optically pumped emission peak wavelength is much shorter than the PL peak, indicating the PL is from impurity-related emission. However, under strong optical pumping, emission is from band to band after the filling of impurity levels. It appears that the bandgap of the co-doped sample is larger than that of the undoped one. Heavy co-doping might be responsible for the low indium composition and thus the larger bandgap in this sample.

In fact, Si-doping substantially improves properties of III-Nitrides, including electrical, optical, and structural properties. Basically, using TEM, X-ray, photoluminescence, and optical pumping, other researchers have found the effects of Si-doping as follows:

Near band-to-band recombination is not only required for laser diodes, but also in high quantum efficiency LEDs. Nakamura *et al.*[18,51] demonstrated that the quantum efficiency and output power of visible LEDs, especially short wavelength LEDs such as violet, blue LEDs, can be enhanced using undoped GaInN single quantum well (SQW) as the active layer. The emission linewidth or color purity of these LEDs is also improved.

The impurity-related saturation at high injection in GaInN-based light-emitting devices is caused by p-type dopants (Zn, Mg) instead of n-type dopants (Si, Ge). As studied in chapter 2, the activation energies for p-type dopants such as Zn and Mg are very large (> 150 meV),[57] which means that their energy levels are deep in the band gap. In addition to that, the incorporation of these dopants are very hard therefore the doping concentration is usually low ($\sim 10^{18}$ cm^{-3}). As a result, these impurities usually form an energy band deep in the forbidden gap without merging with the valence band to form bandtails. The density of states (DOS) in these impurity bands is too low to provide enough gain for lasing. At high current injection, recombination increases the laser threshold.

N-type dopants such as Si and Ge have low ionization energies[49] and can be easily incorporated into III-Nitrides. Si has ionization energy around 22 meV, and is fully ionized at room temperature. The impurity band is so close to the conduction band that it can easily merge with conduction band to form bandtail. Therefore, Si-doping can provide more free carriers, and this is very important for wide bandgap III-Nitrides because of their low intrinsic carrier concentration.

- Dislocation reduction or suppression. Doping level up to a concentration of 3×10^{18} cm^{-3} can decrease the dislocation density and change the dislocation arrangement toward a more random distribution.[53] This is especially important for improving hetero-junction interface quality for layer thickness above critical thickness.[54]
- Stress relaxation. It was found that compressive stress in doped sample is lower than that in undoped samples.[53]
- Band-to-band recombination enhancement. In the doping level range from 5×10^{16} to 7×10^{18} cm^{-3}, band-to-band emission can be enhanced without Stokes shift.[55,56]
- Tunneling-assisted carrier injection. Tunneling through the impurity bandtails could lower the turn-on voltage and enhance carrier injection.[57,58]
- Laser diode performance improvements. Nakamura *et al.*,[38] reported a room temperature blue laser diode with doped GaInN well and doped GaInN barrier. Comparing to undoped laser diodes, the improvements are: operating voltage is reduced from 20 V to 5.5 V, threshold current density is reduced from 10 to 3.6 kA/cm^2, operation mode is changed from room temperature pulsed mode to room temperature continuous mode with 27 hours lifetime. Optically pumped

GaInN MQW lasers also showed threshold reduction after moderate Si-doping (2×10^{18} cm^{-3}).[59]

It should be noted that there exists side effects for Si-doping too. In low injection region, such as EL and PL, the increase of free carriers density by doping are beneficial for saturation of non-radiative recombination centers and thus enhancement of radiative recombination. However, in the active layer of a laser diode, high doping could introduce large free carrier absorption and increase the threshold.[59] Spectral broadening and shift have been observed in GaN:Si and these have been attributed to potential fluctuation.[55,60–61] Even in demonstrated GaInN MQW blue laser diodes with doped well and barrier, it was pointed out that the improvements of performance could be due to enhancement of exciton localization because of well-formed potential fluctuation.[38] Generally, random potential fluctuation leads to bandgap variation and thus gain broadening,[62] which is not desired in the active layer of a laser diode. Finally, the relaxation of stress or strain should also be treated cautiously. It could be closely related to growth process. If dislocation density is low (for example within the critical thickness), the addition of Si atoms could relax the strain and prevent further dislocation formation. However, if defect density is already high in the layer, the addition of Si atoms and their propagation along dislocation could enhance defect density. This has been observed by Lee *et al.*[60] They found that in the doping range from 4×10^{17} cm^{-3} to 1.6×10^{19} cm^{-3}, Si-doping relaxed stress but also introduced more defects. As a result, band edge emission to yellow luminescence (YL) emission ratio decreases as doping increases.

From the above discussion, it can be seen that the effects of Si-doping on III-Nitrides, especially GaInN-based MQWs, can be good or bad, depending on the trade-off between different factors. This trade-off has not been well understood. A controversial situation is the optimum doping scheme for GaInN quantum well based LEDs and blue laser diodes. The best LEDs have been based on undoped single quantum well while better performance is achieved in MQW-based blue laser diodes with Si-doped wells and barriers. This is unexpected from common knowledge of LEDs and laser diodes in other material systems. Therefore a good understanding of the effects of Si-doping on GaInN based MQWs is needed and this is the motivation of this study.

The effect of Si-doping on the GaInN/GaN MQW grown by low pressure MOCVD as described earlier have been studied. The following characterization experiments have been conducted.

1. Photoluminescence and optical pumping of MQW structures with different Si-doping profile: undoped GaInN/GaN MQW, barrier doped GaInN/GaN:Si MQW, and both well and barrier doped GaInN:Si/GaN:Si MQW. The doping concentration is the lowest controllable doping level in this lab and is estimated in the range of 10^{18} cm^{-3} in GaInN:Si and 10^{17} cm^{-3} in GaN:Si. Their structures are shown in Figs. 27(a)–(c). Optical pumping was carried out on polished bars with Fabry–Perot cavity mirrors (Fig. 27(d)).

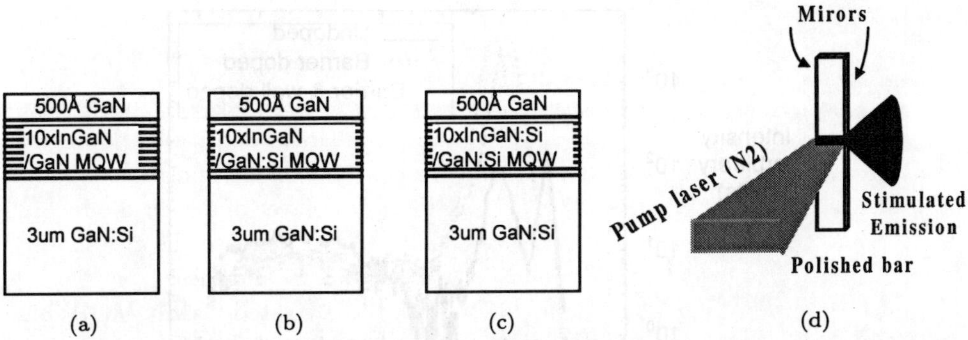

Fig. 27. MQW structures with different doping profile. (a) Undoped GaInN/GaN MQW. (b) Barrier doped GaInN/GaN:Si MQW. (c) Well and barrier doped GaInN:Si/GaN:Si MQW. (d) Schematic of optical pumping. These samples all have 500 Å undoped GaN on top.

2. PL and optical pumping of laser diodes structures whose active layers have different doping profile: undoped GaInN/GaN MQW and barrier-doped GaInN/GaN:Si MQW (at this moment, the barrier-well-doped MQW laser diode structure was not grown in the same batch thus will not be compared with the other two laser diode structures. On the other hand, it will be seen in the next section that barrier-doped and barrier-well-doped structures behave similarly). The doping concentration is the same as that of MQW structures mentioned in experiment 1. These two structures are the same as the corresponding structures in experiment 1 (Figs. 27(a) and (b)) except the top layer. The top layers of these laser diode structures are all the same: 0.25 μm *p*-type GaN (GaN:Mg). In PL and optical pumping study, their top layers are etched away to avoid top layer absorption.
3. Electroluminescence and *I–V* curve study of laser diode structures. LEDs were fabricated from existing laser stripes (with contacts and mesas). The mesa width is 100 μm and stripe length is around 1 mm. *I–V* curves were obtained at different temperatures by loading samples into a cryostat which can go as low as 77 K.

In the next part, "GaInN/GaN MQW structures" will designate GaInN/GaN MQW with thin top cap layers, while "GaInN/GaN MQW laser diode structures" will designate GaInN/GaN MQW laser diodes with originally thick top cap layer (0.25 μm) which were etched to allow penetration of the probing laser beams.

4.2. *Results and discussion*

GaInN/GaN MQW structures

Figure 28 is the PL spectra of the three GaInN/GaN MQW with different doping schemes. The short wavelength peak is the GaN band edge emission, which is nearly the same for all three samples. The mid-wavelength peak is from the MQW and

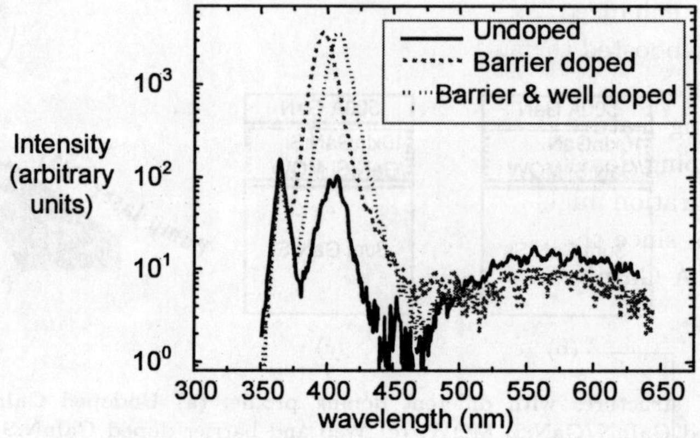

Fig. 28. 300 K PL of GaInN/GaN MQW structures with different Si-doping schemes.

the long wavelength peak is yellow luminescence from bottom GaN:Si layer. Clearly shown is the enhancement of MQW related emission after Si-doping of barrier or both barrier and well. It appears that doping the barrier introduced the most significant improvement. While doping both the barrier and the well did not improve the emission intensity any further.

The PL peak wavelength was also shifted. The confusing part is the blue shift in the barrier-doped MQW and the red shift in the barrier-well-doped sample. Red shift was expected because of either potential fluctuation or stress relaxation.[53,60] This can explain the red shift in the barrier-well-doped sample.

A possible reason for the blue shift in barrier-doped sample is the improvement of the GaInN/GaN:Si interface by Si-doping, resulting in better quantum confinement and higher bound-state energy in the well. This has been demonstrated in Si-doped AlGaN/GaN:Si MQW.[54] However, growth temperature variations from run to run could be the ultimate reason for these emission wavelength variations.

Optical pumping results are shown in Figs. 29 and 30. Figure 29 shows the *P–I* curves of these samples. Figure 29(a) is the *P–I* curve of these samples up to very high optical pumping densities. Saturation was observed from all three samples. TE/TM polarization ratio up to 50 has been observed from all three samples. Figure 29(b) is the detail *P–I* curves around the threshold. The stimulated emission spectra at around 1.1 × Threshold is shown in Fig. 30. The corresponding PL spectra of these samples are also shown here for comparison. The lasing characteristics of three samples are summarized in Table 5.

From these results, it can be seen that:

- The SE results are consistent with the PL results. Doping indeed can improve the spontaneous emission as well as the stimulated emission in GaInN/GaN MQW structures. As summarized in Table 5, Si-doping reduced the threshold. The barrier-well-doped sample had the lowest threshold, less than half of that of the undoped one. The most significant change was the slope efficiency. Doping dramatically enhanced the slope efficiency more than five times. Doping also substantially boosted the emission intensity by more than 8 times as can be seen from Fig. 29(a).

- An interesting feature is the saturation of these three samples. The saturation optical pumping density is almost the same for all three samples. This indicates that the saturation might be due to the split-over of excess carriers after the wells are filled up, since the barrier height and well width are the same for all three samples (33 Å GaInN/66 Å GaN).

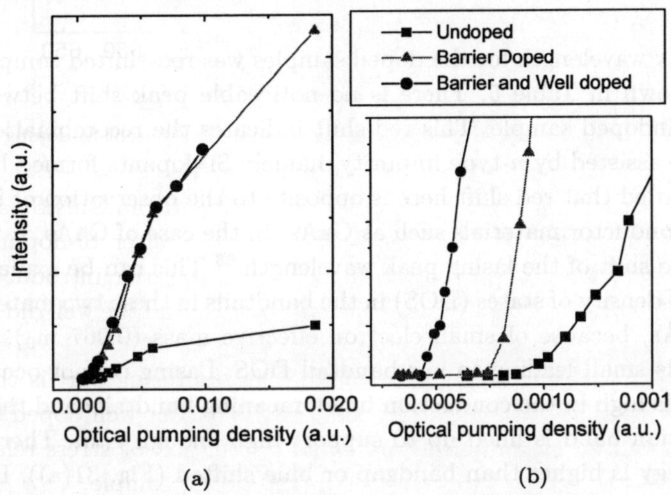

Fig. 29. Optically pumped *P–I* curve of three GaInN/GaN MQW structures with different doping schemes. (a) Whole *P–I* curve with saturation at very high optical pumping density; (b) detail view of the *P–I* curve near threshold.

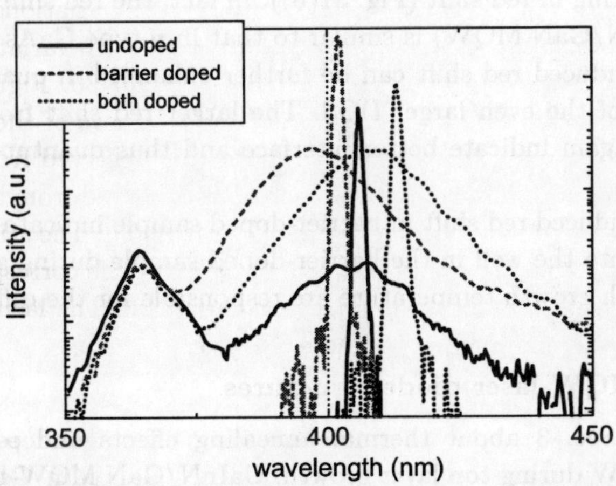

Fig. 30. Optically-pumped stimulated emission spectra at 1.1 × Threshold of three GaInN/GaN MQW structures with different doping scheme. Their corresponding PL spectra are also shown for comparison.

Table 5. Threshold and slope efficiency of three GaInN MQW structures with different doping schemes. For comparison, the SE peak emission wavelength at 1.1 × threshold (λ_{SE}), PL peak wavelength (λ_{PL}), and their difference are also listed.

Sample	Threshold (a.u.)	Slope (a.u.)	SE Emission peak λ_{SE}@$1.1I_{th}$ (nm)	PL peak λ_{PL} (nm)	red shift (meV)
Undoped	10^{-3}	325	405.9	~ 405.9	~ 0
Barrier doped	7.8×10^{-4}	1728	402.4	~ 396.9	~ 43
Barrier and well doped	4.4×10^{-4}	1552	413.2	~ 408.1	~ 38

- The SE peak wavelength for the doped samples was red-shifted comparing to PL peak, as shown in Table 5. There is no noticeable peak shift between SE and PL in the undoped sample. This red shift indicates the recombination in doped samples are assisted by *n*-type impurity, namely Si dopants formed bandtails. It should be noted that red shift here is opposite to the observation of blue shift in other semiconductor materials such as GaAs. In the case of GaAs, *n*-type doping leads to blue shift of the lasing peak wavelength.[63] This can be explained by the difference of density of states (DOS) in the bandtails in these two materials. In the case of GaAs, because of small electron effective mass (0.067 m_0), conduction band DOS is small leading to low bandtail DOS. Lasing cannot occur until the Fermi-level is high in the conduction band, meaning bandtails and the bottom of the conduction band is filled up to support high enough gain. Therefore lasing photon energy is higher than bandgap or blue shifted (Fig. 31(a)). For GaN, or GaInN, because of the large electron effective mass (0.27 m_0 for GaN, 4 times that of GaAs), conduction band DOS is high leading to high bandtail DOS. Thus conduction bandtails in GaN can support lasing with photon energy lower than bandgap, resulting in red shift (Fig. 31(b)). In fact, the red shift in *n*-type doped GaN (or GaInN/GaN MQW) is similar to that in *p*-type GaAs.[63]
- The bandtail-induced red shift can be further enhanced in quantum well structures because of the even larger DOS. The larger red shift from barrier-doped sample could again indicate better interface and thus quantum confinement in this sample.
- The bandtail-induced red shift in barrier-doped sample indicates that Si dopants have diffused into the well in the barrier-doped sample during growth. The thin barrier and high growth temperature are responsible for the diffusion.

GaInN/GaN MQW laser diode structures

As discussed in Sec. 3 about thermal annealing effects and *p*-type doping on GaInN/GaN MQW during top layer growth, GaInN/GaN MQW-based laser diode could respond differently to Si-doping comparing to GaInN/GaN MQW structures with only thin top layer discussed above. As will be shown below, this in fact is the case.

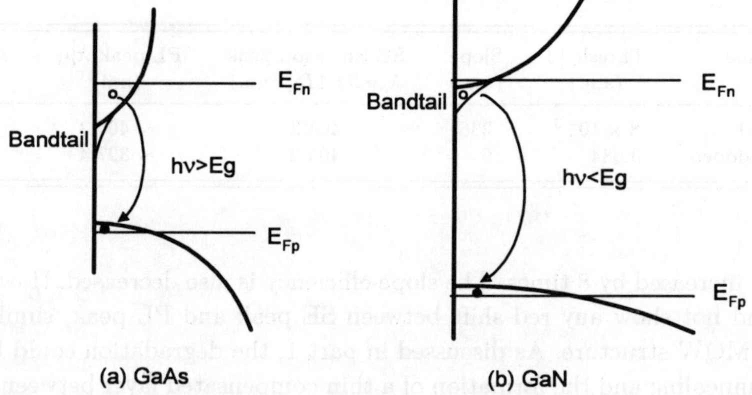

Fig. 31. Comparison of bandtail-assisted lasing in n-type GaAs and n-type GaN. (a) GaAs, DOS in bandtail is too low to support lasing. Lasing occurs between conduction and valence band so lasing photon energy is higher than bandgap (or blue shift). (b) GaN. DOS in bandtail is large enough to support lasing. Lasing energy is lower than bandgap (or red shift).

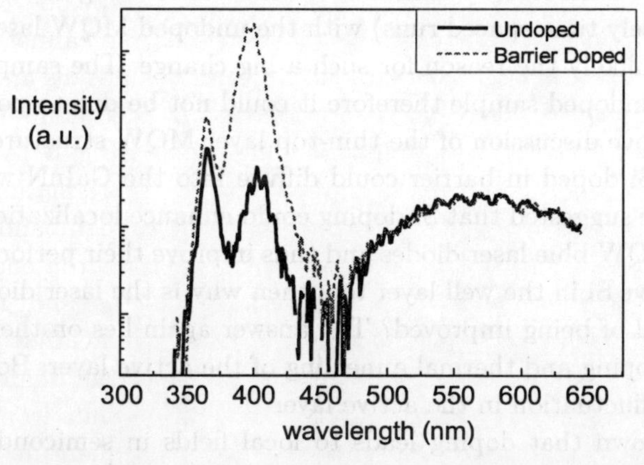

Fig. 32. Room temperature PL of undoped (solid line) and barrier doped GaInN/GaN MQW laser diode structure (dashed line). The top layers of both of the samples have been etched away.

The room temperature photoluminescence spectra of two GaInN/GaN MQW laser diode structures are shown in Fig. 32. One laser diode structure has undoped GaInN/GaN MQW as active layer while the other has barrier-doped GaInN/GaN:Si as active layer. It can be seen again that Si-doping enhances MQW emission almost two orders of magnitude, consistent with the PL results of GaInN/GaN MQW structures discussed above. However, the optical pumping results are quite different.

Table 6 is the summary of SE characteristics of these two structures. Comparing to the corresponding MQW structures (see Table 5), both of the samples showed degraded performances. For the undoped sample, the threshold optical pumping

Table 6. Threshold and slope efficiency of two GaInN MQW laser diode structures with different doping schemes. For comparison, the SE peak emission wavelength at 1.1 × threshold (λ_{SE}), PL peak wavelength (λ_{PL}), and their difference are also listed.

Sample	Threshold (a.u.)	Slope (a.u.)	SE Emission peak λ_{SE}@$1.1I_{th}$ (nm)	PL peak λ_{PL} (nm)	red shift (meV)
Undoped	8×10^{-3}	236	404.3	~ 404.3	~ 0
Barrier doped	0.084	6	403.2	~ 397.4	~ 45

density is increased by 8 times. The slope efficiency is also decreased. However, this sample did not show any red shift between SE peak and PL peak, similar to the undoped MQW structure. As discussed in part 1, the degradation could be due to thermal annealing and the formation of a thin compensated layer between GaN:Mg and the top of MQW active layer.

The surprising results are from the barrier-doped laser diode structure, which showed much worse degradation. Comparing to barrier-doped MQW structures with thin top layer, threshold is increased by more than two orders of magnitude. Slope efficiency is decreased by almost 300 times. Since this growth was done in the same batch (closely time-spaced runs) with the undoped MQW laser diode, growth variations are unlikely the reason for such a big change. The sample was polished along with the undoped sample therefore it could not be due to polishing.

From the above discussion of the thin-top layer MQW structures, it is strongly suggested that Si doped in barrier could diffuse into the GaInN well layer. Nakamura *et al.*, have suggested that Si-doping could enhance localization of excitons in GaInN-based MQW blue laser diodes and thus improve their performance.[68] In our case here, we have Si in the well layer too; then why is the laser diode performance degraded instead of being improved? The answer again lies on the combination of two effects: Si-doping and thermal annealing of the active layer. Both of them lead to energy band fluctuation in the active layer.

It is well known that doping leads to local fields in semiconductor materials and thus the perturbation of the band edges as shown in Fig. 33(a). The local fields result in separation of electron-hole pairs and therefore reduce the recombination efficiency.

Compositional fluctuation of GaInN due to thermal annealing has a different effect on the energy band from that of doping. In this case, the local energy bandgap varies with the variation of the GaInN composition, as shown in Fig. 33(b). Electrons and holes can be trapped in the low indium composition region. If the feature size of these fluctuations are small enough, they behave like quantum dots and can enhance the recombination efficiency.

In the MQW structures studied above, both of doping and compositional fluctuation are present in the MQW active layer. Whether or not the recombination efficiency can be enhanced or deteriorated depends on the relative feature size of the

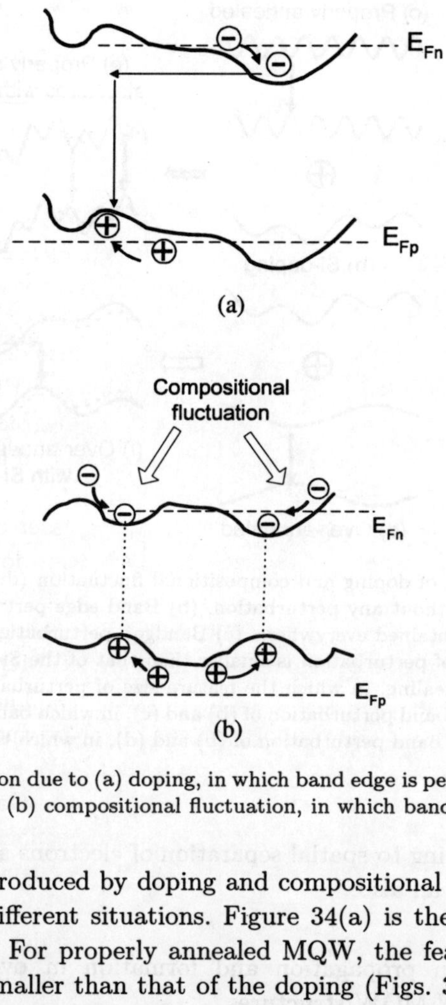

Fig. 33. Band perturbation due to (a) doping, in which band edge is perturbed and local bandgap is maintained everywhere; (b) compositional fluctuation, in which band gap is perturbed locally.

band perturbation introduced by doping and compositional fluctuation. Figure 34 is an illustration of different situations. Figure 34(a) is the ideal band structure without perturbation. For properly annealed MQW, the feature size of compositional fluctuation is smaller than that of the doping (Figs. 34(b) and (c)). In this case, the combination of doping and compositional fluctuation is still dominated by the smaller scale compositional fluctuation as shown in Fig. 34(e). Electron-hole pairs will be still confined in the "dot"-like structures and recombination efficiency is enhanced. For over-annealed MQW, the compositional fluctuation feature size is larger than that of doping, as shown in Figs. 34(b) and (d). In this case, the band perturbation is dominated by doping effect or band edge perturbation. In other words, the "well" like structure due to annealing is smashed by doping. Therefore electron-hole pairs are separated by local fields and recombination efficiency is reduced.

The above pictures can be used to explain the controversial doping effects on GaInN MQW structures and GaInN MQW laser diodes. The MQW structures with thin top layer are not over-annealed and their top is not *p*-type doped. Therefore Si-doping can enhance their performance. While laser diode structures are over-annealed and thus Si-doping degraded their performances.

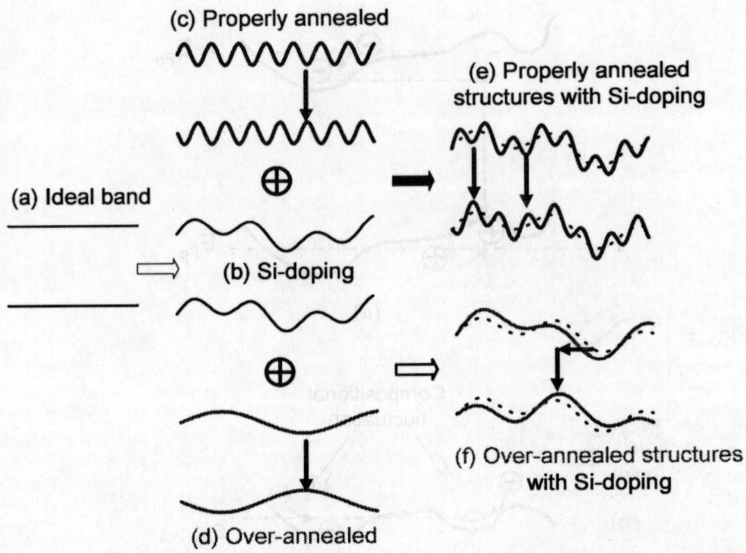

Fig. 34. Combined effect of doping and compositional fluctuation (due to annealing) in GaInN MQWs. (a) Ideal band without any perturbation. (b) Band edge perturbation due to doping, in which the bandgap is maintained everywhere. (c) Bandgap perturbation due to proper annealing in which the feature size of perturbation is smaller than that of the Si-doping. (d) Bandgap perturbation due to over-annealing, in which the feature size of perturbation is larger than that of doping. (e) The combined band perturbation of (b) and (c), in which bandgap perturbation is dominating. (f) The combined band perturbation of (b) and (d), in which the band edge perturbation is dominating.

In addition to leading to spatial separation of electrons and holes, doping in an over-annealed MQW can also:

- Enhance dislocation propagation and formation in over-annealed thermally relaxed GaInN/GaN MQW structures.
- Enhance compensation in the possible thin compensated layer around the MQW and *p*-type layer interface. This compensated layer and *n*-type layer act as barriers for holes thus reduce hole injection efficiency (except for tunneling-assisted injection if properly designed). The reduction of hole injection efficiency is detrimental for III-Nitride-base laser diodes since *p*-type doping is already poor.

In the following part, using electroluminescence and *I–V* curves, it will be demonstrated there are indeed more deep-levels or potential fluctuations in the Si-doped laser diode structure. Lasing is the ultimate way to determine the quality of a laser diode structure. However, because of the high injection, a lot of phenomena other than stimulated emission can be covered but can affect laser diode performances. To study these phenomena, low injection techniques such as photoluminescence, electroluminescence, *I–V* curve, can yield more information.

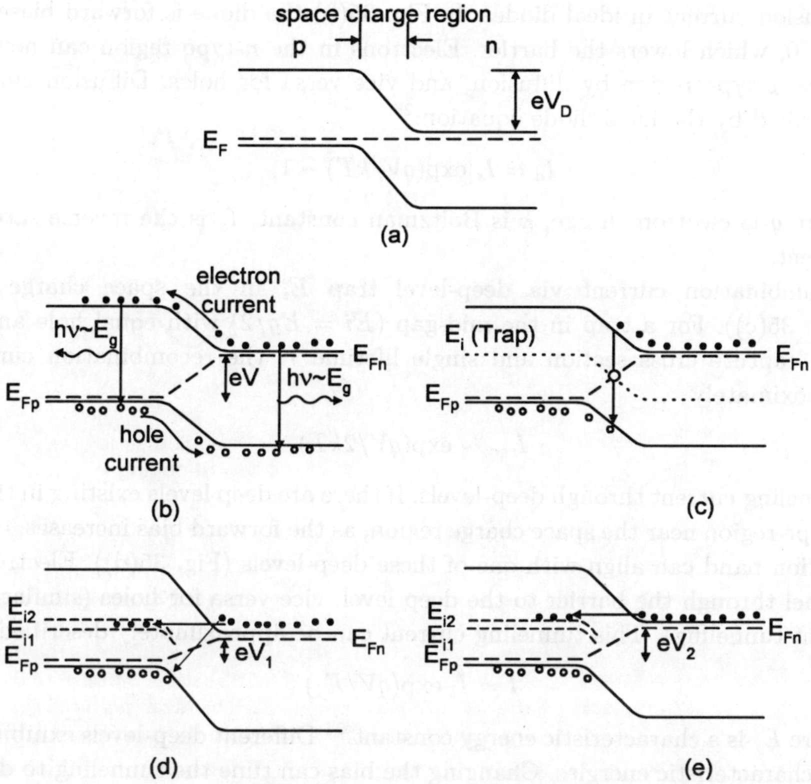

Fig. 35. Various current generation mechanisms in a *p–n* junction under forward bias. (a) Unbiased, no current flow (dark current neglected). (b) Forward biased with $V > 0$. Diffusion current (thermal emission over the barrier); (c) Forward biased with $V > 0$. Generation-recombination current through space charge region generation and recombination via trap with activation energy E_i. (d) Forward biased with $V_1 > 0$. *N*-type conduction band edge is aligned with deep-level E_{i1} in *p*-type region. Electrons in *n*-type conduction band tunnel to this deep level (similar to resonant tunneling), generating tunneling current through deep-level E_{i1}. (e) Forward biased with $V_2 > 0$. *N*-type conduction band edge is aligned with deep-level E_{i2} in *p*-type region. Electrons in *n*-type conduction band tunnel to this deep level (similar to resonant tunneling), generating tunneling current through deep-level E_{i2}.

Electroluminescence and *I–V* curves of GaInN/GaN MQW laser diode structures

Electroluminescence and *I–V* curves can very well characterize the quality of a non-lasing LED. Before going to experimental results, we will first review some theory. Let's start with a simple *p–n* junction diode, whose *I–V* curve yields detail current generation mechanisms. Figure 35 shows various current generation mechanisms in a *p–n* junction under forward bias.

- Figure 35(a) shows the diode without bias so no current flows (dark current neglected).

- Diffusion current in ideal diodes. In Fig. 35(b) the diode is forward biased with $V > 0$, which lowers the barrier. Electrons in the n-type region can now move to the p-type region by diffusion, and vice versa for holes. Diffusion current is described by the ideal diode equation:[25]

$$I_d = I_s[\exp(qV/kT) - 1] \tag{13}$$

where q is electron charge, k is Boltzman constant, I_s is the reverse saturation current.

- Recombination current via deep-level trap E_i in the space charge region (Fig. 35(c)). For a trap in the mid-gap ($Ei = Eg/2$) with equal hole and electron capture cross section and single lifetime τ, the recombination current is approximately

$$I_{\rm rec} \sim \exp(qV/2kT) \tag{14}$$

- Tunneling current through deep-levels. If there are deep levels existing in the p- or n-type region near the space charge region, as the forward bias increases, the conduction band can align with one of these deep-levels (Fig. 35(d)). Electrons can tunnel through the barrier to the deep level, vice versa for holes (similar to resonant tunneling). This tunneling current can be approximately described by:[64]

$$I = I_1 \exp(qV/E_j) \tag{15}$$

where E_j is a characteristic energy constant.[64] Different deep-levels exhibit different characteristic energies. Changing the bias can tune the tunneling to different deep levels and as shown in Figs. 35(c) and (d).

At large biases, diode series resistance can reduce the current. For example, with series resistance Rs, the I–V relation for diffusion current Eq. (13) will be reduced to

$$I_d = I_s\{\exp[q(V - IR_s)/kT] - 1\}\,. \tag{16}$$

From the above equations, it can be seen that, except for the case with series resistance, they all have a common feature: current I is always exponentially dependent on voltage. Therefore they will be lines in semi-log plot with different slope, as shown in Fig. 36. Here I–V curves are corresponding to different current generation mechanisms shown in Fig. 35. Also shown are the thermal diffusion current with series resistance, which leads to turn-over at relatively high current. The experimentally measured I–V curve from a GaInN/GaN MQW-based laser diode is also shown for comparison. From Fig. 36, it can be seen as follows:

- The most efficient current generation mechanism is by thermal diffusion. While the least efficient is by tunneling. The larger the characteristic energy, the lower current it can generate. Apparently, a combination of these mechanisms will reduce the current to some value below ideal diode current because of the competition between thermal diffusion and other mechanisms.

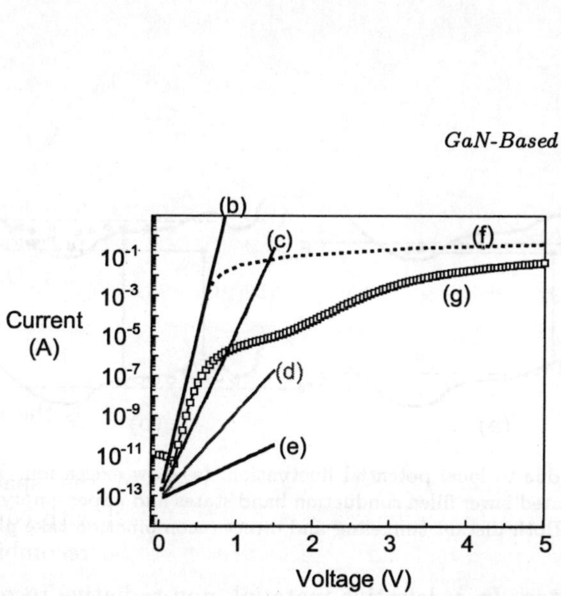

Fig. 36. *I–V* curves with different current generation mechanisms. The labels are corresponding to drawings in Fig. 35. (b) Thermal diffusion current; (c) recombination current through mid-gap trap in space charge region; (d) tunneling current through deep level with $E_{i1} = 100$ meV; (e) tunneling current through deep-level with $E_{i2} = 240$ meV. (f) Thermal diffusion current with series resistance $R_s = 8\ \Omega$. (g) Measured *I–V* curve of a GaInN/GaN MQW based MQW laser diode.

- With the exception of tunneling process, the other two mechanisms are temperature dependent. Therefore temperature-dependent *I–V* curves should be useful to determine tunneling process.
- *I–V* curve slope in semi-log plot basically tells the mechanism of current generation. An example is curve (f). Below 0.6 V the current generation is a combination of thermal diffusion and recombination via trap. Or recombination with traps other than those in the mid-gap with single lifetime. From 0.6 V to 1.6 V, the slope is too low to be described by either thermal diffusion or recombination via trap. It is described by tunneling to a deep-level. The same thing for *I–V* curve from 1.6 V to 3 V, which has a different slope and different characteristic energy, meaning tunneling to another deep level. For $V > 3$ V, *I–V* is dominated by series resistance.
- An important conclusion for us here is that the number of slopes in the tunneling range in the *I–V* curve corresponds to the number of deep-levels in the band (Figs. 35(d) and (e)). Of course, instead of deep levels, tunneling could be due to local potential variations, which could spatially separate upper and lower states (Fig. 37).[65] From Fig. 37, it can be seen that nonuniform potential fluctuation can also lead to different and multiple tunneling processes at different excitation levels. The number of different slopes indicates the extent of this nonuniformity.

The *I–V* curve of an LED is directly related to its power output–current characteristic (*P–I* curve), whose linearity characterizes the quality of LED. In the case of injection into a region with a radiative lifetime independent of the injection carrier density, the light output is a linear function of the injection current since each injected carrier is assumed to lead, via recombination with a hole, to the

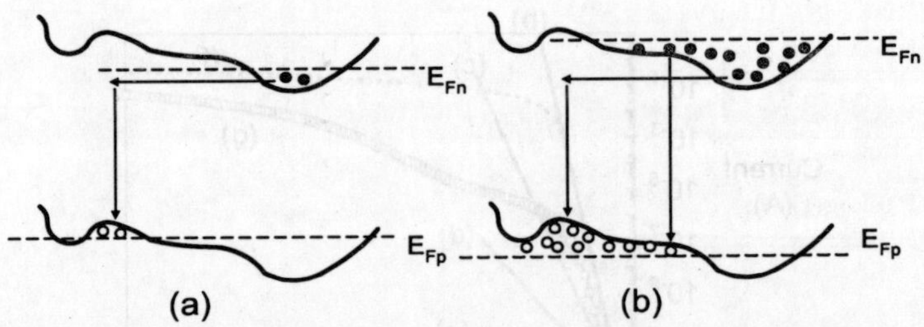

Fig. 37. Tunneling due to local potential fluctuation. (a) Low excitation. Tunneling occurs between spatially separated lower filled conduction band states and upper empty valence band states. (b) High excitation. Both distant tunneling and direct recombination take place.

emission of a photon. In a defective material, non-radiative recombination occurs in the space-charge region due to defects, the light output power P_{out} will remain proportional to the carriers that can flow over the barrier, and thus varies with the applied voltage V as:[25]

$$P_{\text{out}} \propto \exp(qV/kT)\,. \tag{17}$$

From Eqs. (13) and (14) in generation-recombination or diffusion region of the diode,

$$I \propto \exp(qV/nkT) \tag{18}$$

where n is the ideality factor and it is between 1 and 2. Therefore in this case

$$P_{\text{out}} \propto I^n\,. \tag{19}$$

Only in the ideal case with $n = 1$ will P–I curve be linear. Otherwise it is a super-linear and the deviation from the linearity tells the quality of the diodes. In this case the interpretation of the super-linear P–I curve should be the super-linear (dramatic) reduction (quenching) of radiative recombination as current decreases (or voltage drops), instead of the super-linear enhancement of output power by gain or amplification in the case of super-radiance or lasing above threshold.

The P–I curve could be even more nonlinear if the current injection is through tunneling by deep-levels or local potentials. In this case, I–V relation follows

$$I = I_1 \exp(qV/E_j)\,. \tag{15}$$

Similar to I–V curve, here again the number of slopes reflects the number of deep-levels or different sizes of potential fluctuations.

Table 7 summarizes the slope of I–V curve and P–I curve for three different carrier generation mechanisms. From slope s in I–V curve ($= \Delta \log I/\Delta V$) in the diffusion, generation and recombination regions, ideality factor can be calculated as:

$$n = q/(skT)\,. \tag{20}$$

Or for tunneling process, characteristic energy in I–V curve can be calculated as:

$$E_j(eV) = q/s\,. \tag{21}$$

Table 7. Slopes of I–V curve and P–I curve for three typical current or generation or photon emission mechanisms.

Curve	Thermal Diffusion	Generation and Recombination	Tunneling
I–V (Log(I)–V)	1	2	> 2
P–I (Log(P)–Log(I))	1	2	> 2

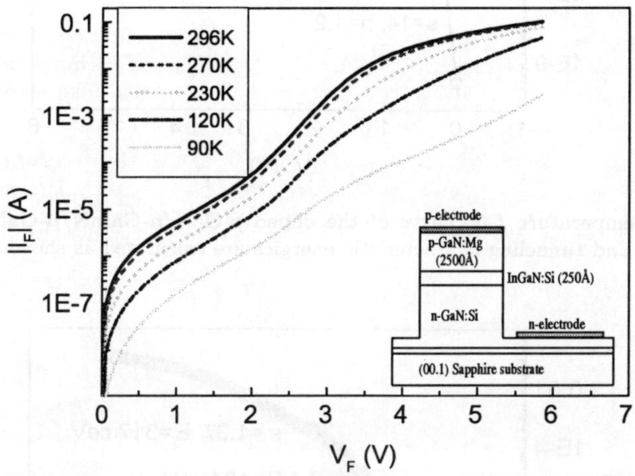

Fig. 38. I–V curve of a p-GaN/n-GaInN/n-GaN heterostructure at different temperatures. Temperature independent slope suggests tunneling, rather than thermal diffusion is the conduction mechanism.

Now let us study the I–V curves and P–I curves of GaInN/GaN heterostructure or MQW LEDs. Figure 38 is the temperature-dependent I–V curve of a p-GaN/n-GaInN/n-GaN heterostructure LED. Its structure is shown in the inset of the figure. The active layer is doped with Si. A distinctive feature of these I–V curves is the constant slope at different voltage region up to around 3.2 V. This feature suggest that the carrier injection is through tunneling rather than thermal diffusion. For higher voltage, series resistance mainly limits the current injection.

Figure 39 is the detailed room temperature I–V curve of the doped heterostructure. Slope s in each voltage region is calculated. The low injection region can be described by thermal diffusion with an ideality factor of 1.2. Moving to higher current regions, the slope cannot be described by thermal diffusion but by tunneling. The different slopes at different voltage range correspond to different tunneling characteristic energies E_j, which are calculated according to Ref. 20 and shown in Fig. 39.

GaInN/GaN MQW LEDs showed similar I–V curves. Figure 40 shows the I–V curves of the barrier-doped LED. Figure 41 is the I–V curve of the undoped MQW LED. In the low injection region, they both showed carrier injections due to thermal diffusion with ideality factors of 1.1 for undoped LED and 1.5 for doped

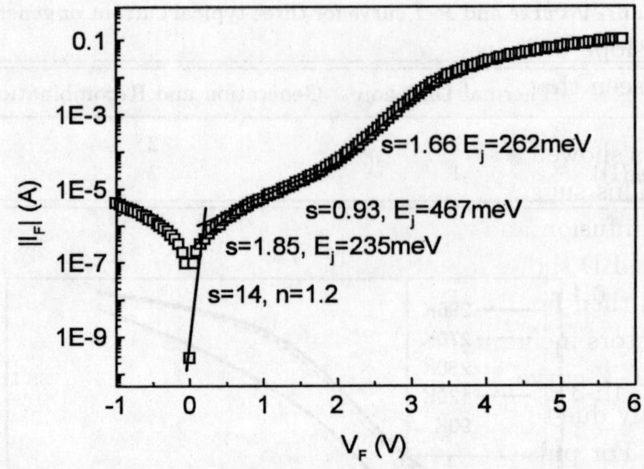

Fig. 39. Room temperature *I–V* curve of the doped *p*-GaN/*n*-GaInN/*n*-GaN heterostructure LED. The slope *s* and tunneling characteristic energies are calculated as shown.

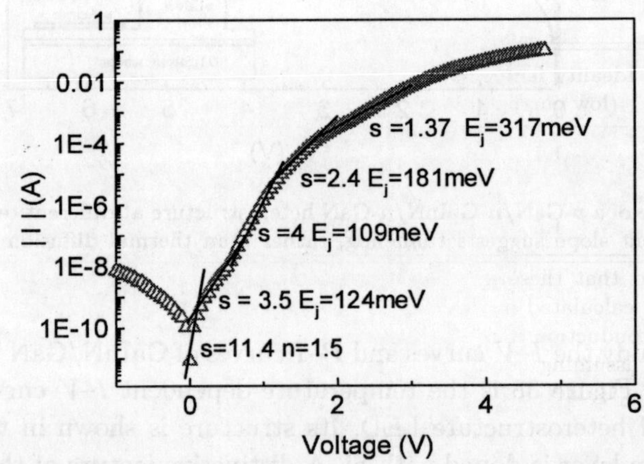

Fig. 40. *I–V* curve of barrier doped MQW LED. Shown also are the slope *s*, ideality factor *n* and characteristic energies E_j in different voltage regions.

LED. However, as voltage increases, tunneling becomes the dominating mechanism of current injection. Undoped LED showed a single tunneling characteristic energy while barrier-doped LED showed multiple tunneling characteristic energy at different current injection levels. In addition, the barrier-doped LED showed much lower current at the same bias voltage up to around 2 volts. Table 8 summarizes *I–V* characteristics of these two LEDs.

For both LEDs, series resistance limits current injection at relatively high voltages. In this case, the LEDs were characterized by *P–I* curve. Figure 42 shows room temperature *P–I* curves of the two LED at different injection region. Figure 42(a) shows CW *P–I* curves of the two LEDs in low injection region up to 70 mA. To avoid heating, pulsed injection was used. The pulse width is 1 μs with a repetition rate of 200 Hz. Figure 42(b) shows pulsed *P–I* curves in high injection region. s is the slope of in the log scale in the corresponding section. From this figure, it can be seen that:

- Undoped LEDs showed a single slope $s = 1.6$, which corresponding to an ideality factor of 1.6. This suggests that the radiative recombination is affected through both thermal diffusion and deep-level defects.
- Barrier-doped LED showed multiple slopes in different current range from 1.6 to 4.3 to 1.9, and then finally to 0.8, indicating radiative recombination is affected by multiple factors including thermal diffusion, deep level defects and tunneling. Comparing to *I–V* curve, the *P–I* curve of this LED showed less tunneling related process, possibly due to the saturation of defects or local potential fluctuation at high injection. For pulse current > 500 mA, $s = 0.8 < 1$, indicating saturation. This saturation could be related to defects.

Table 8. Summary of *I–V* characteristics of the two GaInN/GaN MQW LEDs with different doping schemes.

Sample	Ideality factor n (low current)	Tunneling characteristic energy E_j (meV)*	General reference energy (meV)(theory)**
Undoped	1.1	255	$\Delta Ec \sim 227, \Delta Ev \sim 106$
			$E_{je} \sim 238, E_{jh} \sim 138$
Barrier doped	1.5	109, 124, 181, 317	$\Delta Ec \sim 191, \Delta Ev \sim 89$

*It should be noted that these calculated tunneling characteristic energies are only for rough references, since E_j calculated in this way can only applied to tunneling to deep-levels.
** ΔEc, ΔEv are conduction band and valence band offsets. They are estimated from the band offset of InN/GaN assuming conduction and valence band offsets in GaInN/GaN are reduced proportionally to the GaInN/InN band gap energy ratio.

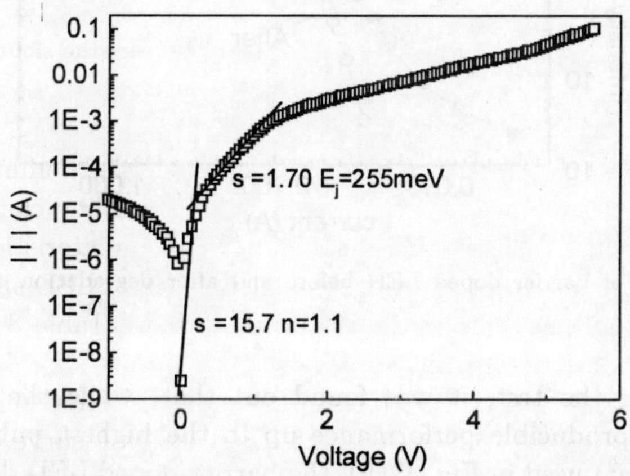

Fig. 41. *I–V* curve of barrier undoped MQW LED. Shown also are the slope s, ideality factor n and characteristic energies E_j in different voltage regions.

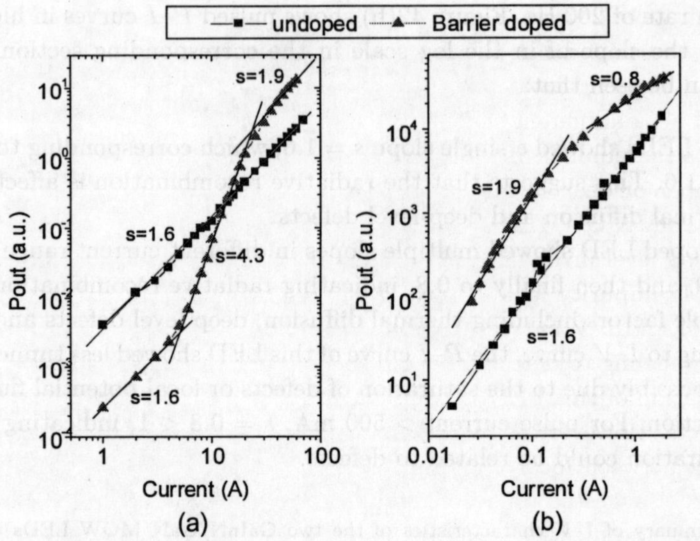

Fig. 42. Room temperature electroluminescence (EL) output power as a function of injection current (P–I curve) in the two GaInN/GaN MQW laser diode structures (no mirrors) at different injection region: one has undoped MQW and the other has Si-doped barrier in the MQW. (a) CW P–I curve in low injection region; (b) pulsed P–I curve in high injection region. s is the slope of in the log scale in the corresponding section.

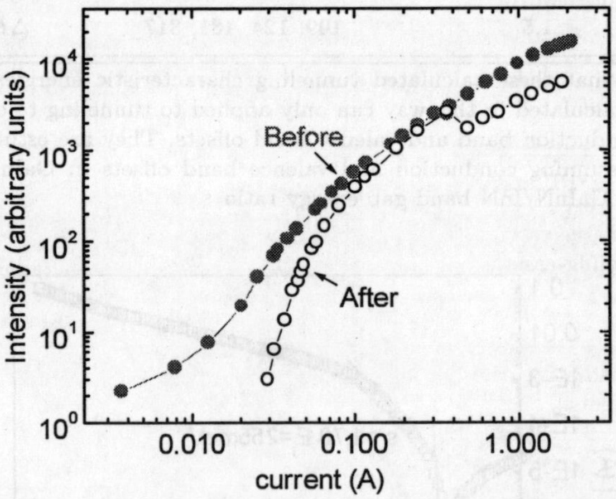

Fig. 43. P–I curve of barrier doped LED before and after degradation of under high pulse current density.

In fact, during the test, it was found out that, while the undoped MQW LED sustained reproducible performance up to the highest pulsed current density ($\sim$ 1.8 kA/cm^2) used in Fig. 42(b), the barrier-doped LED degraded easily at high pulsed current density. Figure 43 shows the P–I curve of the barrier doped LED before and after degradation by high current density ($<$ 1.8 kA/cm^2). It can be seen that, after degradation, not only the output power decreased, but also the slope at low current densities increased, strongly indicating the defect density is increased by high current density.

The number of different slopes (tunneling characteristic energies or ideality factors) reflects the number of tunneling mechanism, or the variation of barrier sizes, or the number of deep levels. From the above discussion, it can be seen that in undoped GaInN MQW LED, independent of current density, there appears to be only one deep level involved in the recombination process (which could be simply related to Mg impurity level). While in doped GaInN MQW, there exist multiple slopes in P–I and I–V curves. Right now it is not clear if these multiples slopes are related to tunneling to defect-related deep levels or through local potential barrier. The degradation of this LED at high current density strongly suggested high density of defects. However, this does not exclude the possibility of the co-existence of tunneling between local potential barriers. These barriers could be of different sizes, which is suggested by different slopes.

The above arguments can also explain the controversy on PL, EL and SE from GaInN/GaN MQW laser diode structures. It should have been noted that both PL, EL are much stronger in the barrier-doped MQW laser diode structure than the undoped one. Opposite to this, SE in the undoped MQW laser diode structure is much better than barrier-doped one. This is not the case for GaInN/GaN MQW structures with thin top layers studied earlier in this part. There PL, EL, and SE are all enhanced by doping.

Table 9. Comparison of helpful and harmful factors for PL, EL and lasing in GaInN/GaN MQW-based structures.

	PL and EL	SE or lasing
Beneficial	– Initial free carriers by doping; – Tunneling-assisted recombination.	– Initial free carriers by doping (limited); – Localized excitons; – Quantum dot like structures (near uniform potential fluctuations).
Harmful	– Non radiative deep levels or defects.	– Any defects (radiative or non-radiative). – Nonuniformity.

The beneficial and harmful factors for luminescence and stimulated emission are summarized in Table 9.

4.3. *Conclusions*

The study of the effects of Si-doping on GaInN/GaN MQW grown by low pressure MOCVD showed that doping enhanced the optical properties, including PL, EL, and SE of GaInN/GaN MQW structures with thin top layers due to four mechanisms: (1) the improvement of interface quality, (2) the injection of initial free carriers due to bandtail formation, (3) the promotion of self-organized uniform local potential fluctuations, (4) tunneling-assisted carrier injection between local potentials. The last two mechanisms could be the most important and unique mechanisms for GaInN-based MQW lasers.

Si-doping in GaInN/GaN MQW laser diode structures improves PL, EL but dramatically degrades SE characteristics. It has been demonstrated earlier that these laser diode structures were over-annealed because of thicker top layer growth and Mg-doping doping of the top layer. Over-annealing led to nonuniform distribution of potential fluctuations and defect formation in the MQW and Si-doping make this even worse. The nonuniform distribution of local potentials or high density defects has been revealed by the multiple tunneling characteristic energies in low current injection region, and multi-slope super-linear *P–I* curve in high injection region in doped MQW laser diode structures. Tunneling processes may help improving PL, EL but not SE, which requires much higher current injection. High-density defects or nonuniformity degrade lasing performances.

5. Low Temperature Characteristics of GaInN/GaN MQW Lasers

The light emission properties of the GaInN/GaN MQW under current injection were first tested at low temperatures, i.e. 79 K, without mirror facet. Figure 13(b) shows the resulting electroluminescence spectrum, exhibiting a peak at ~ 403 nm.

Laser diodes of 100 μm width with cavity lengths varying from 700 to 1800 μm were prepared and bonded with indium to copper heatsinks. Light–current characteristics were recorded in pulse operation (pulse width 2–6 μs, repetition rate 200 Hz) using a silicon detector. From the current versus voltage (*I–V*) curve, the series resistance of the laser (L $\sim$ 1800 μm) was estimated to be 13 Ω at 300 K (turn-on voltage $\sim$ 3.6 V) and 14 Ω (turn on voltage $\sim$ 6 V) at 79 K, as shown in Fig. 44. When decreasing the sample temperature from 300 to 79 K, the resistivity of the *p*-type GaN:Mg is expected to increase drastically, by at least two orders of magnitude, as a result of the relatively high activation energy of Mg levels in GaN.[15,66] The fact that the series resistance does not increase significantly in these diodes may be directly due to the small thickness of the GaN:Mg layer, to the absence of AlGaN cladding layers or to higher *p*-type doping of GaN:Mg which results in lower device resistances.[15]

Figure 44 illustrates the light output power versus injection current of an uncoated 1800 μm-long laser at 79 K. Stimulated emission was observed at currents of 2.5 A, which corresponds to a threshold current density of 1.4 kA/cm^2. Because the laser stripes used in this study are 100 μm wide, the lasing action may occur in filaments and the threshold current density may be locally higher than the value measured above. The voltage of this laser at threshold was 25 V. The peak wavelength of the measured lasers was 405 nm at 3.4 A as shown in Fig. 44. At currents lower than 4 A, no degradation was observed for these diodes under pulse operation.

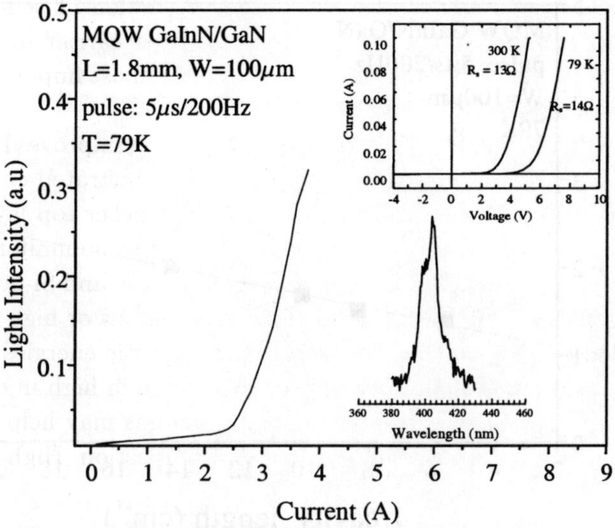

Fig. 44. Output power versus injection current for a MQW GaInN/GaN 405 nm uncoated diode laser measured at 79 K (pulse operation: 5 μs–200 Hz) with cavity length of 1800 μm and 100 μm -wide aperture. Inset: Current–voltage characteristics of the laser diode with series resistance of 13 Ω at 300 K and 14 Ω at 79 K. Spectrum of the laser diode slightly above threshold.

Table 10. Emission wavelengths from GaInN/GaN MQW and lasers using different measurement techniques at low temperature (79 K).

Measurement technique	Emission wavelength
Photoluminescence	410 ± 2 nm
Electroluminescence	403 ± 2 nm
Current injection stimulated emission	405 ± 2 nm

Table 10 summarizes the emission wavelengths from the same laser wafer measured at low temperature (79 K) by photoluminescence, electroluminescence and current injection stimulated emission. A good agreement between the electroluminescence and current injection stimulated emission is observed. The threshold current density for several lasers of different cavity length was measured and the results are summarized in Fig. 45.

6. Room Temperature Characteristics of GaInN/GaN Based MQW Lasers

The light emission properties of the GaInN/GaN MQW under current injection at room temperature are shown in Fig. 46. The spectra correspond to increasing injected currents under continuous wave operation. The light was collected from the back of the sample. An intense electroluminescence peak at $\sim$ 416 nm could be observed.

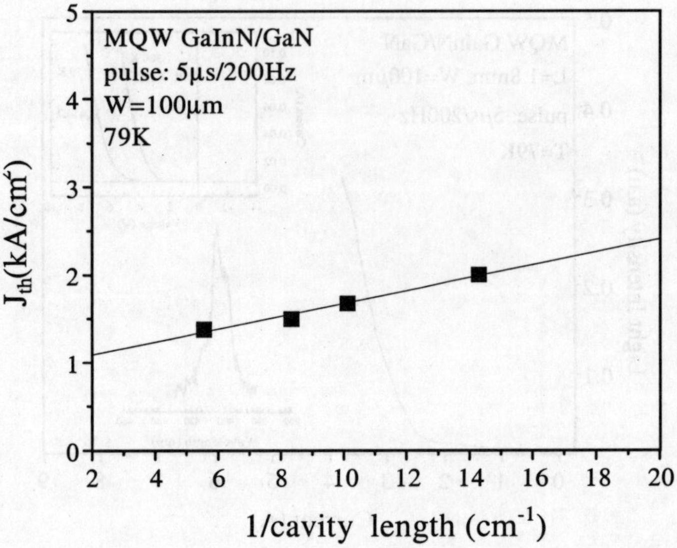

Fig. 45. Threshold current density versus inverse cavity length for 100 μm wide GaInN/GaN MQW lasers at 79 K.

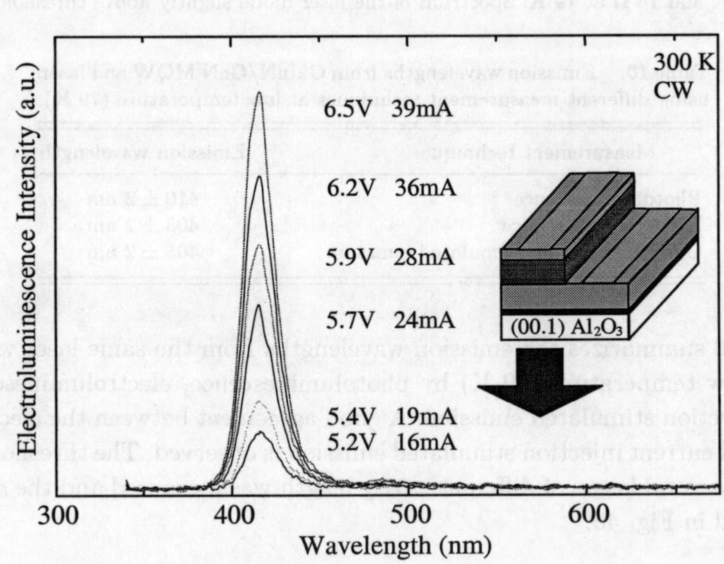

Fig. 46. Room temperature continuous wave electroluminescence from a 10 period 33 Å $Ga_{0.89}In_{0.11}N$/66 Å GaN MQW.

Laser diodes of 100 μm width were first prepared and bonded with indium to copper heatsinks. The *P–I*, and *I–V* curves of a typical laser diode with 100 μm width and 1200 μm cavity length under pulse operation (5 μs/200 Hz) are shown in Fig. 47(a), respectively. The *P–I* curve of the laser diode at CW operation is shown in Fig. 47(b). The I–V curve (dotted line) in a forward direction shows the series resistance as low as 6 Ω and turn-on voltage of 6.2 V. As shown in Fig. 47(a), the laser showed a sudden increase in the slope of optical output power versus injection current curve by one or two orders of magnitude at around $I \sim$ 300 mA, indicating the onset of a laser operation. The laser in pulse-operation showed saturation of output power around 2 mW of output power (Fig. 47(a)).

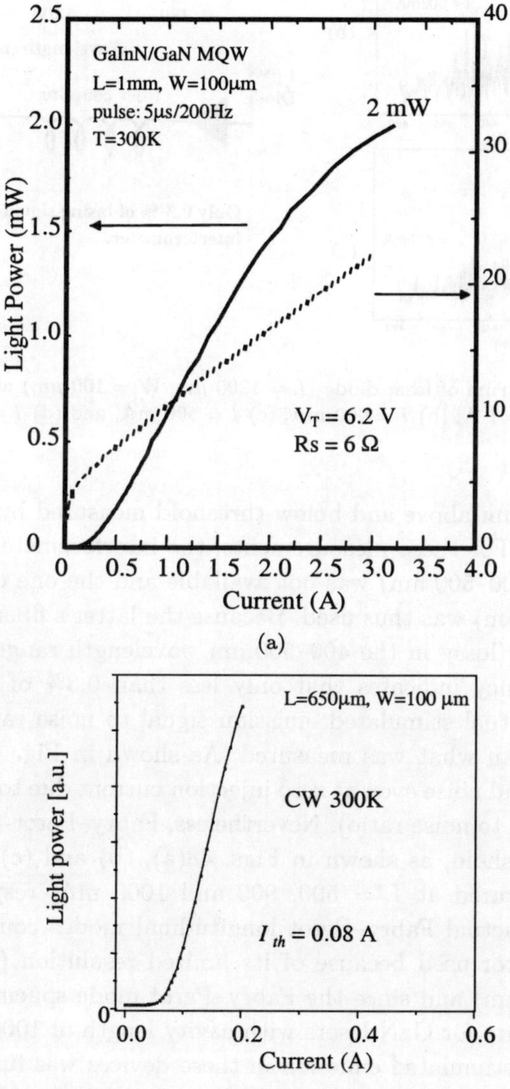

Fig. 47. (a) Light–current (P–I) and voltage–current (I–V) characteristics of GaInN/GaN MQW laser under pulse operation at 300 K. (b) Light–current (P–I) curve for GaInN/GaN MQW laser under CW operation at 300 K ($L = 650$ μm, $W = 100$ μm).

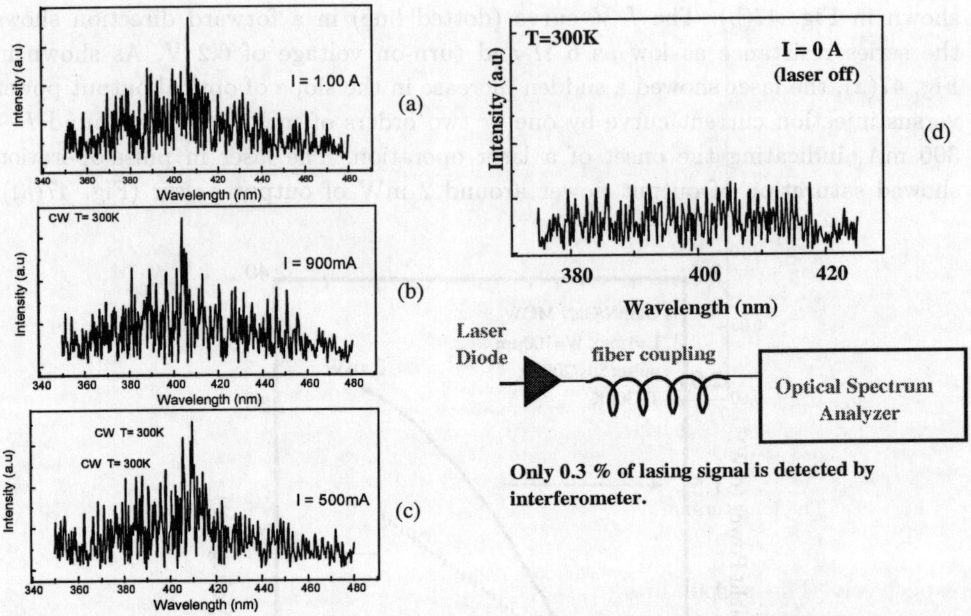

Fig. 48. Emission spectrum of laser diode ($L = 1200$ μm, $W = 100$ μm) at 300 K, measured by interferometer at (a) $I = 1$ A, (b) $I = 900$ mA, (c) $I = 500$ mA, and (d) $I = 0$.

The emission spectrum above and below threshold measured by an interferometer is shown in Fig. 48. For these measurements, the interferometer designed for this wavelength range (400–500 nm) was not available and the one designed for longer wavelengths (> 0.8 μm) was thus used. Because the latter's fiber and other optical components are very lossy in the 400–500 nm wavelength range, an estimation of the collection efficiency indicates that only less than 0.3% of light will actually be analyzed. The actual stimulated emission signal to noise ratio should thus be 300 times higher than what was measured. As shown in Fig. 48(d), there was a significant background noise even at zero injection current due to this low collection efficiency (low signal to noise ratio). Nevertheless, Fabry–Perot-like peaks could be observed above threshold, as shown in Figs. 48(a), (b) and (c) which correspond to the spectra measured at I = 500, 900 and 1000 mA, respectively. In these measurements, the actual Fabry–Perot longitudinal modes could not be resolved in the monochromator used because of its limited resolution (~ 0.1 nm) at this wavelength (~ 410 nm) and since the Fabry–Perot mode spacing was expected to be only about 0.03 nm for GaN lasers with cavity length of 1000 μm (Fig. 49).

The presence of stimulated emission in these devices was further confirmed by measurements of far-field patterns and polarization properties of the emitted beam, as these properties have very distinctive features between stimulated and spontaneous emission beams. Figure 50 shows far-field patterns with beam divergence of 13° and 20° for parallel and perpendicular directions to the epilayer plane,

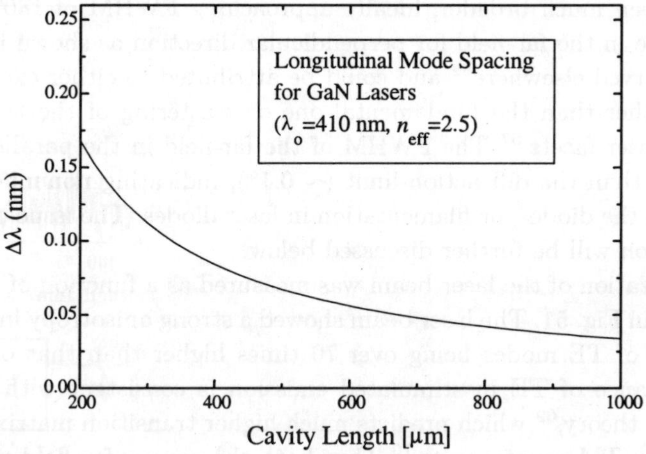

Fig. 49. The longitudinal mode spacing $\Delta\lambda$ of GaInN lasers ($\lambda_o = 410$ nm, $n_{\text{eff}} = 2.5$).

respectively. The perpendicular direction far-field pattern was calculated using a slab waveguide model,[67] assuming the difference in refractive index between GaN and $In_{0.1}Ga_{0.9}N$ to be 0.025. The calculation showed that the far-field full width at half maximum (FWHM) was about 15°, which is not far from the measured one (20°) considering the inaccuracy of refractive indices for each layer. This comparison strongly indicates that the emission is dominated by a few fundamental optical modes amplified by stimulated emission. If spontaneous emission were the dominant emission (i.e. all optical modes are equally excited), the far-field pattern

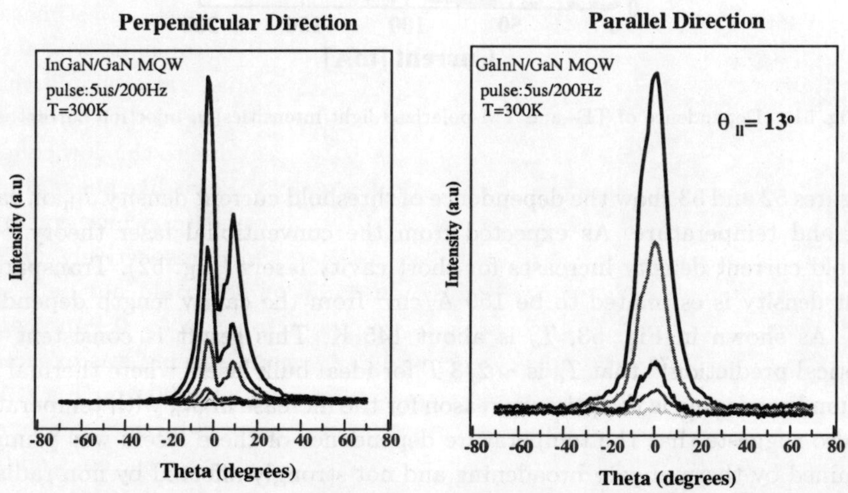

Fig. 50. (a) Far-field in the perpendicular direction of a GaInN/GaN MQW laser diode at 300 K. (b) Far-field in the parallel direction of a GaInN/GaN MQW laser diode at 300 K. (c) Theoretical perpendicular far-field pattern of GaN lasers (FWHM: 26°).

would have been much broader, ideally approaching FWHM of 180°. The multi-lobed structure in the far-field for perpendicular direction as shown in Fig. 50 has also been observed elsewhere[68] and could be attributed to either excitation of optical modes other than the fundamental one or scattering of the laser beam due to imperfect laser facets.[67] The FWHM of the far-field in the parallel direction is much broader than the diffraction-limit ($\sim 0.1°$), indicating nonuniform injection of current into the diodes, or filamentation in laser diodes. The issue of nonuniform current injection will be further discussed below.

The polarization of the laser beam was measured as a function of injection current as shown in Fig. 51. The laser beam showed a strong anisotropy in polarization, with intensity of TE modes being over 70 times higher than that of TM modes. Such a dominance of TE in stimulated emission is consistent with conventional QW laser gain theory,[69] which predicts much higher transition matrix and gain for TE modes than TM counterparts.[70] Thus both the narrow far-field and the polarization anisotropy is consistent with lasing operation of the lasers above threshold.

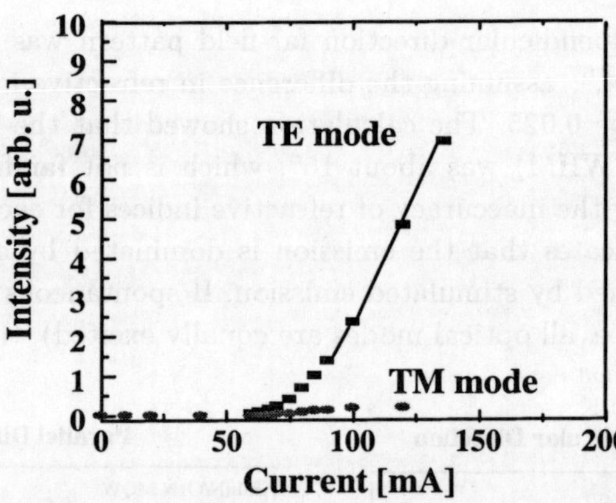

Fig. 51. Dependence of TE- and TM-polarized light intensities on injection current.

Figures 52 and 53 show the dependence of threshold current density J_{th} on cavity length and temperature. As expected from the conventional laser theory,[71] the threshold current density increases for short cavity lasers (Fig. 52). Transparency current density is estimated to be 150 A/cm^2 from the cavity length dependence of J_{th}. As shown in Fig. 53, T_o is about 145 K. This result is consistent with theoretical prediction[72] that T_o is $\sim 2/3\ T$ for ideal bulk lasers where thermal gain spectrum broadening is the primary reason for the increase in J_{th} with temperature. This also suggests that the temperature dependence of these lasers was primarily determined by thermal gain broadening and not strongly affected by non-radiative recombination or leakage current with increase in temperature.

If it is assumed that the current is uniformly injected into the diodes, the threshold current (~ 300 mA) of our lasers with stripe area of 100 μm $\times$ 1200 μm

(Fig. 47(b)) corresponds to threshold current density of 250 A/cm^2. This is about one order of magnitude lower than previous both experimental[68] and theoretical results.[73,74] Theoretically predicted values for threshold current density and threshold carrier density are around $1 \sim 2$ kA/cm^2 and $1 \sim 2 \times 10^{19}$ cm^{-3}, respectively,[73,74] when using free-electron recombination and Kane's band model. Considering the high effective mass of conduction band electrons (~ 0.2 free electron mass),[73,74] this estimated *apparent* threshold current density is unlikely to be realistic. All of the measurements of optical power and current were performed in high speed oscilloscope (400 MHz bandwidth) and no pulsation was observed.

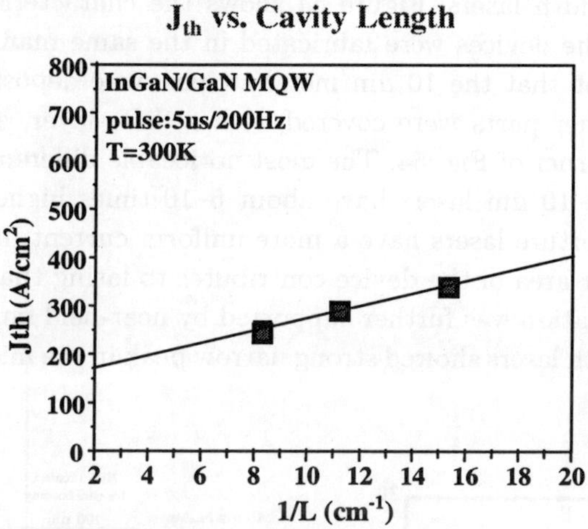

Fig. 52. Threshold current density versus inverse cavity length of GaInN/GaN laser diodes ($W =$ 100 μm aperture) at 300 K.

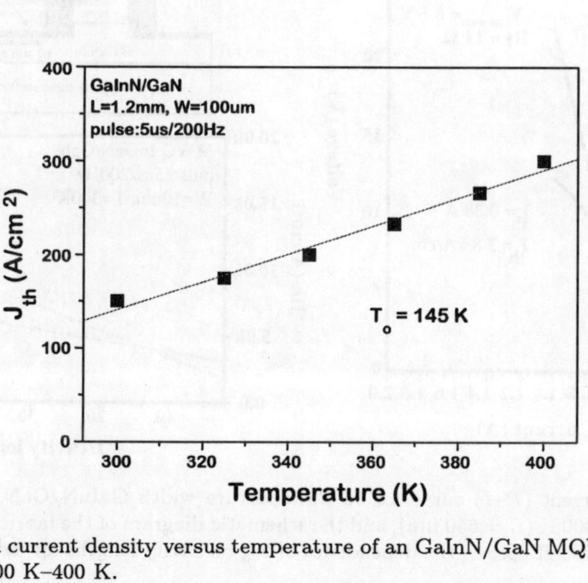

Fig. 53. Threshold current density versus temperature of an GaInN/GaN MQW laser diode with $T_o \sim 145$ K from 300 K–400 K.

This excludes the possibility that the low threshold current may result from the self-pulsation, such as observed in other material systems.[75] It is believed that this low threshold current density may result from the fact that injection current could be nonuniformly distributed, and thus only a fraction of the diode contributes to the lasing operation. In this case, the threshold current density can be grossly overestimated. The nonuniform current distribution could be observed from near-field measurements performed for both LED and laser structures fabricated from the same GaInN/GaN MQW lasers. In these measurements, the near-field patterns of lasers with different aperture widths of 10 and 100 μm were compared. The 10 μm width lasers are expected to have less nonuniformity in current distribution compared to 100 μm width lasers. Figure 54 shows the characteristics of the 10 μm aperture devices. The devices were fabricated in the same manner as the 100 μm width devices except that the 10 μm metal stripes were deposited on the p-GaN cap layers while other parts were covered with a SiO_2 layer, as indicated in the upper right-hand corner of Fig. 54. The most noticeable difference between the P–I curves is that the 10 μm lasers have about 5–10 times higher J_{th}. This means that the narrow-aperture lasers have a more uniform current injection and consequently that a wider area of the device contributes to lasing than in 100 μm width devices. This observation was further supported by near-field imaging. As shown in Fig. 55, 100 μm width lasers showed strong narrow peak in the mirror, indicating the

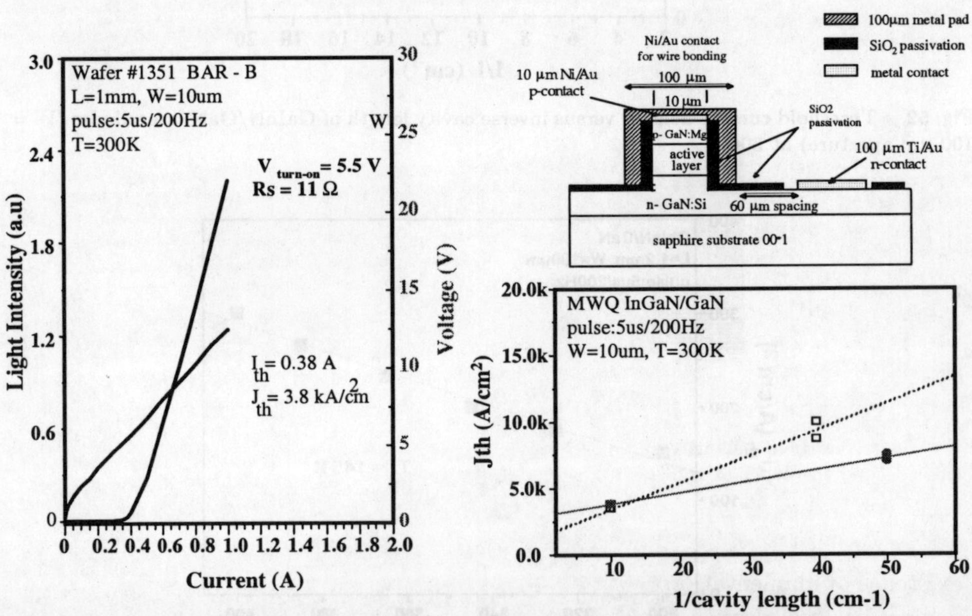

Fig. 54. Light–current (P–I) curve for 10 μm aperture width GaInN/GaN MQW laser under pulse operation at 300 K ($L = 650$ μm), and the schematic diagram of the fabricated laser structure (right corner). In the left corner, the dependence of J_{th} on cavity lengths for lasers from two wafers were shown.

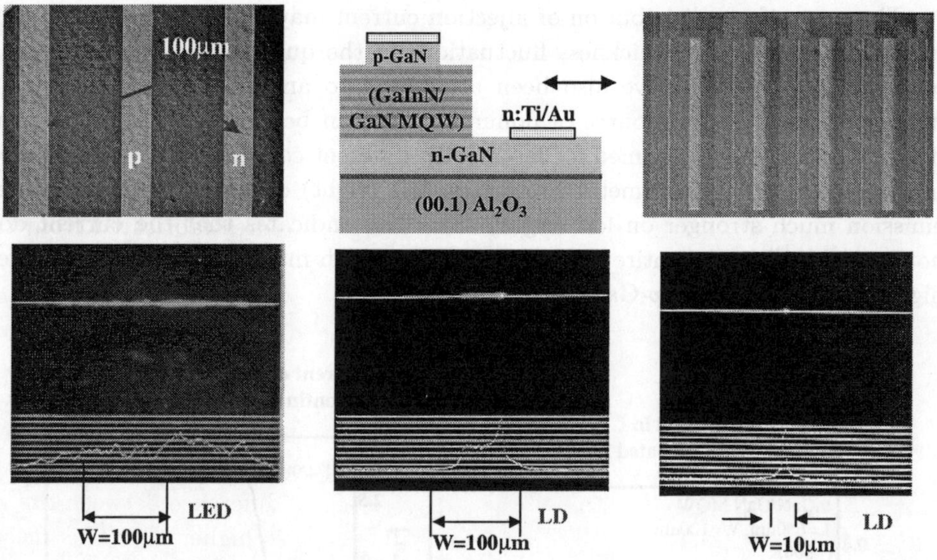

Fig. 55. Near-field patterns of LED, 100 μm aperture laser, and 10 μm aperture laser fabricated from a single wafer.

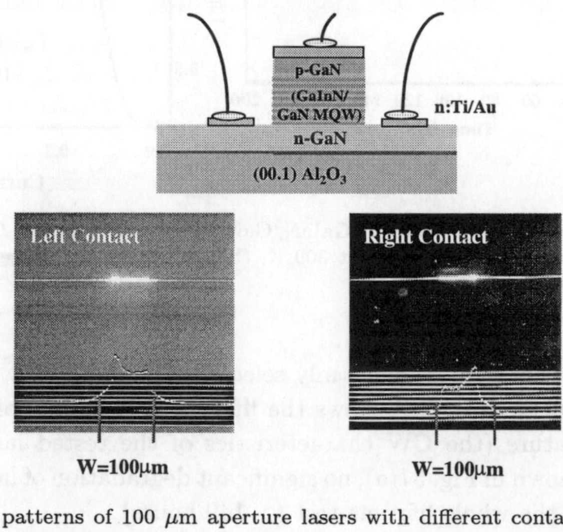

Fig. 56. Near-field patterns of 100 μm aperture lasers with different contact positions (one at left side and the other at right side).

existence of filamentation resulting from laser instability. By contrast, 10 μm width devices showed a much more uniform near-field emission. For comparison, a LED showed a rather structure-less near-field spectrum without filamentation. Note that filamentation is one of the important characteristics of wide-aperture lasers, and is a good indication of the presence of stimulated emission.

The nonuniform distribution of injection current may result from, for example, indium composition or thickness fluctuations in the quantum wells, formation of quantum dots. These have also been attributed to anomalous emission spectra in other reports.[68] One source of nonuniformity can be explained from the near-field measurements performed for lasers with different contact positions. As shown in Fig. 56, lasers with n-metal contact on left (right) corner yielded a near-field emission much stronger on left (right) side. This indicates that the current does not spread well in the entire area of the lasers, which may be due to the relatively higher resistivity of the p-GaN.

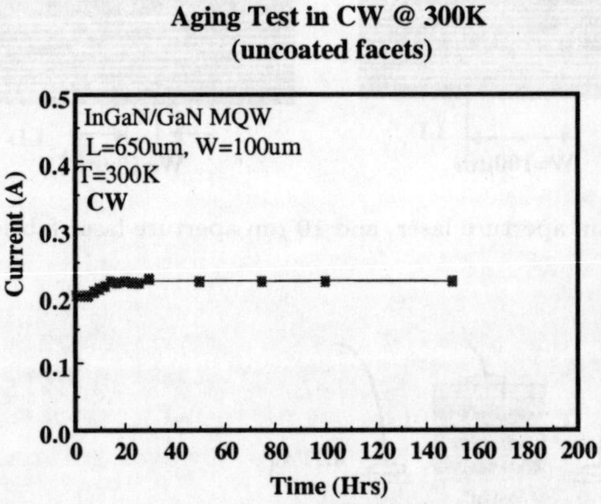

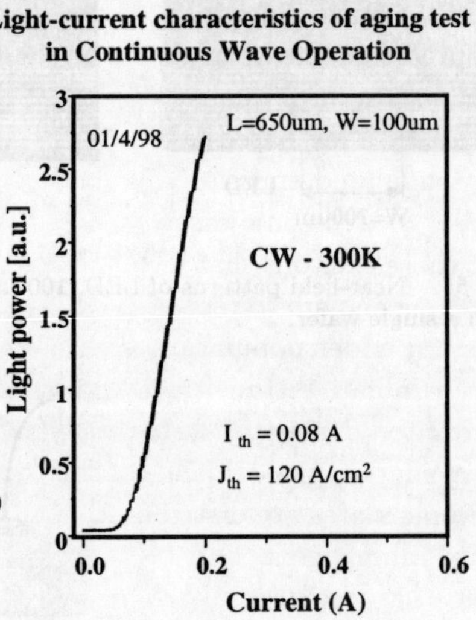

Fig. 57. (a) Lifetime testing result of GaInN/GaN MQW laser diode (L = 1200 μm, W = 100 μm) in continuous wave operation at 300 K. (b) P–I curve of a tested laser at CW room temperature operation.

A number of devices were randomly selected, and subjected to lifetime testing in CW operation. Figure 57(a) shows the lifetime test results of one of the lasers at room temperature (the CW characteristics of the tested lasers are shown in Fig. 57(b)). As shown in Fig. 57(a), no significant degradation of laser characteristics was observed for the whole lifetime test (> 140 hours).

7. Exciton Localization in GaInN/GaN Quantum Wells

The recent progress in III-Nitride-based optoelectronic applications such as light-emitting and laser diodes,[24,68] mates the discussion of the nature of radiative transitions in III-Nitrides-based quantum wells (QW) exciting. Below we will discuss the relation of the optical properties of the III-Nitride structures to the compositional disorder in ternary alloy-based QWs.

It is known that compositional inhomogeneities cause the broadening of the photoluminescence (PL) spectra that has been widely discussed with respect to GaAs-based alloys[76–78] and QWs.[79] In the GaAs-based structures the inhomogeneous linewidth due to compositional fluctuations and interface roughness appears to be small, not larger than several meV and it does not change significantly electron energy distribution near the band edge. In this case, the inhomogeneous broadening is not a main source of the exciton PL linewidth which is mainly attributed to homogeneous broadening due to the exciton-optical phonon interaction.

Wide bandgap III-Nitrides reveal quite different behavior. The activation energy of the LO-phonons is about 90 meV, and exciton-phonon interaction gives small contribution to linewidth up to room temperatures. To the moment it is widely accepted that the optical properties of the InGaN/GaN structures strongly depend on the structural quality of the layers, namely, on compositional inhomogeneities in ternary alloy-based QW.[62,80–82] The nature of the radiative states, optical gain and stimulated emission in InGaN-based quantum wells are assumed to be closely related to electronic disorder due to In-content fluctuations. It has been supposed that those excitons that localized in fluctuations induced potential minima have lifetime long enough to recombine radiatively and even to provide gain and stimulation emission under population inversion condition. In order to play such an important role in radiative transitions the alloy disorder should have significant influence on the exciton density of states and hence on the PL linewidth. The experiments cited above support the evidence of the wide energy domain (≈ 100 meV) where localized excitonic states are distributed near the excitonic band edge. However, estimations of the inhomogeneous PL linewidth due to the alloy disorder in InGaN give the value of about 10 meV that is too far from experimental values and does not support the idea of the importance of compositional fluctuations in the distribution of the near bandgap exciton states responsible for spontaneous emission and lasing.

The effect of the compositional fluctuations on excitons of different dimensions has been considered considered in Ref. 83. The fluctuations induced excitonic PL linewidths for two limiting cases of two-(2D) and three (3D)-dimensional excitons in QW have been calculated. We have obtained coincidence with the estimation made in Ref. 78 for 3D exciton PL linewidth as well as much larger broadening of the PL line and the density of states due to 2D exciton localization. This removes the contradiction mentioned above and shows that in thin QW we may expect strong exciton localization and significant influence of the compositional inhomogeneities on near band edge energy states distribution.

The tail of the density of states in the energy gap region is the measure of the fluctuation influence on excitonic spectrum. Below we calculate the exciton density of states inhomogeneously broadened by composition fluctuations and present the broadening as a function of InGaN material parameters.

Let us consider the impact of the crystalline disorder on exciton PL line broadening. The interaction of the composition fluctuations with electrons and holes can be described by parameters[84] $\alpha_{c,v} = \frac{dE_{c,v}}{dx}$, where $E_{c,v}$ is the conduction (valence) band edge. Quantum localization in short-range white-noise-type potential has been considered in Refs. 85 and 86 by optimal fluctuation method. Here we assume that exciton feels smooth classical fluctuations that causes the edge of the excitonic band $V(\vec{R})$ to be a random variable ($\vec{R}$ is the position of the exciton center of mass). Since we assume the bandgap fluctuations to be classical for excitons, the random field $V(\vec{R})$ is Gaussian with distribution function[87]

$$F(V) = \frac{1}{W\sqrt{2\pi}} \exp\left(-\frac{V^2}{2W^2}\right), \tag{22}$$

where

$$W^2 = \frac{\alpha^2 x(1-x)}{N} \int d^3 r \varphi^2(\vec{r}), \alpha - \frac{dE_g}{dx},$$

where N is the density of cation sites, $\varphi(\vec{r})$ is the wave function of the relative motion of electron and hole.

We will calculate the PL linewidth related to W for two limiting cases of 2D and 3D A-excitons, that corresponds to optical transitions between heavy-hole and conduction bands. Wave function of the 2D exciton in hydrogenic approximation for relative motion has the form

$$\phi(\vec{r}) = \sqrt{\frac{2}{\pi}} \frac{1}{a_0} \exp\left(-\frac{r}{a_0}\right), a_B = 2a_0, \tag{23}$$

where a_B is the Bohr radius.

In 2D case from Eq. (12) we obtain

$$W_{2\mathrm{D}}^2 = \frac{\alpha^2 x(1-x)}{N_s} \frac{1}{2\pi a_0^2} \tag{24}$$

where N_s is the sheet density of the cation sites.

The density of excitonic states is given as:

$$D_i^{2\mathrm{D}}(\varepsilon) = \left\langle \sum_{\vec{k}} \delta\left(\varepsilon - \frac{\eta^2 k^2}{2M} - V\right) \right\rangle = \frac{M}{4\pi\eta^2} \left[1 + erf\left(\frac{\varepsilon}{\gamma_{2\mathrm{D}}}\right)\right],$$

$$\gamma_{2\mathrm{D}} = W_{2\mathrm{D}}\sqrt{2} \tag{25}$$

where $\langle \ldots \rangle$ means average with distribution function Eq. (11), $\varepsilon = E - E_{ex}$ is the energy of exciton center-of-mass motion, and function $1 + erf(z)$ describes the tail of 2D density of states for $\varepsilon < 0$. The energy domain where localized excitonic states are distributed is $\approx \gamma$. In this region there are localized and extended exciton states divided by mobility edge. Equation (14) is the density of states of excitons localized by classical fluctuations.

For 3D excitons the calculation similar to that done in Eq. (14) gives

$$D_i^{3D}(\varepsilon) = \frac{\sqrt{2}M^{3/2}\sqrt{\gamma_{3D}}}{\pi^2\eta^3}G\left(\frac{\varepsilon}{\gamma_{3D}}\right);$$

$$W_{3D}^2 = \frac{\alpha^2 x(1-x)}{N}\frac{1}{8\pi a_B^3}; \tag{26}$$

$$G(z) = \frac{1}{\sqrt{\pi}}\int_{-\infty}^{z}\exp(-y^2)(z-y)^{1/2}dy.$$

The function $G(\varepsilon/\gamma_{3D})$ describes the tail of the 3D density of states for $\varepsilon < 0$[88] and for large positive ε tends to $\sqrt{\varepsilon/\gamma_{3D}}$, which corresponds to the high-energy 3D free-like states in the excitonic band.

Using the natural valence band offset between InN and GaN $\Delta E_v = 0.26$ eV[89] and $\alpha = -2.147$ eV ($T = 300$ K, $x = 0.2$) we obtain $\alpha_c = -1.887$ eV, $\alpha_v = 0.26$ eV. The inequality $|\alpha_c| > \alpha_v$ means that exciton interacts with fluctuations mostly via electronic component.

In order to calculate the fluctuations induced broadening of the excitonic PL line we assume parameters as follow. Effective masses: $m_c = 0.2$ (GaN); $m_c = 0.11$ (InN); $m_h = 0.8$ (GaN); $m_h = 1.6$ (InN).[90] Lattice constants: $a = 3.19$ A (GaN); $a = 3.53$ A (InN); $c = 5.185$ A (GaN); $c = 5.7$ A (InN). Dielectric constant $\varepsilon = 9.38$. Energy gaps: $E_g(x) = (1-x)E_g\,(\text{GaN}) + xE_g(\text{InN}) + x(x-1)$,[91] E_g(GaN)= 3.44 eV;[92] E_g(InN)= 1.89 eV,[93] $T = 300$ K, $d = 30$ A. For $x = 0.2$ the calculated exciton Rydberg $Ry = 23.8$ meV, Bohr radius $a_B = 32$ A, binding energy of 2D 1s exciton $E_{ex} = 95.2$ meV, radius of 2D exciton $a_0 = 16$ A.

With above parameters the calculated exciton PL linewidths due to the compositional fluctuations for $x = 0.2$: $\Delta = 2W\sqrt{2\ln 2} \approx 11$ meV (3D exciton) and 107 meV (2D exciton). For $x = 0.1$ $\Delta \approx 9$ meV (3D) and 89 meV (2D). The estimations of 3D exciton linewidth are close to that done in Ref. 62 (10 meV). The obtained results for linewidth in InGaN are presented in Fig. 58 as functions of In mole fraction. The exciton binding energies are shown in Fig. 59 in order to follow the relation between them and linewidths in Fig. 58.

We have also made similar calculations for AlGaN alloy using data for effective masses from Ref. 94. Results are shown in Figs. 60 and 61. The exciton binding energy is determined mostly by electron effective mass. Different composition dependences of the electron effective mass in $In_x\ Ga_{1-x}$ N and $Al_x\ Ga_{1-x}$ N result in different behavior of the exciton binding energies with composition (Figs. 59 and 61).

Here we have considered limiting cases and shown that the broadening of 2D exciton line is much larger than that of 3D exciton. This reflects more strong coupling of 2D exciton to composition fluctuations due to (a) its small radius as compared to 3D exciton, and (b) small size in growth direction. The character of the exciton in the real QW structure (binding energy, exciton radius) depends on confining potentials for electrons and holes and is in between three- and two-dimensional ones, thus actual fluctuations induced linewidth lies between limiting values shown in Fig. 58. In thin and deep enough QW's, where the optical transitions occur between well resolved first conduction and valence subbands, the character of the exciton is expected to be close to the two-dimensional. Large effective masses as compared to

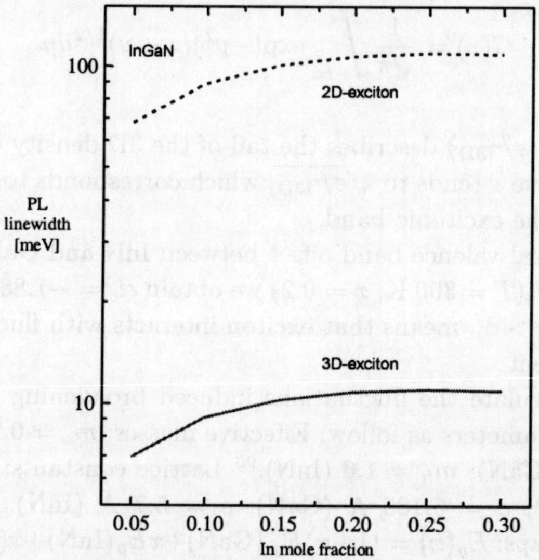

Fig. 58. Inhomogeneous broadening of the photoluminescence line in $In_xGa_{1-x}N$ quantum well.

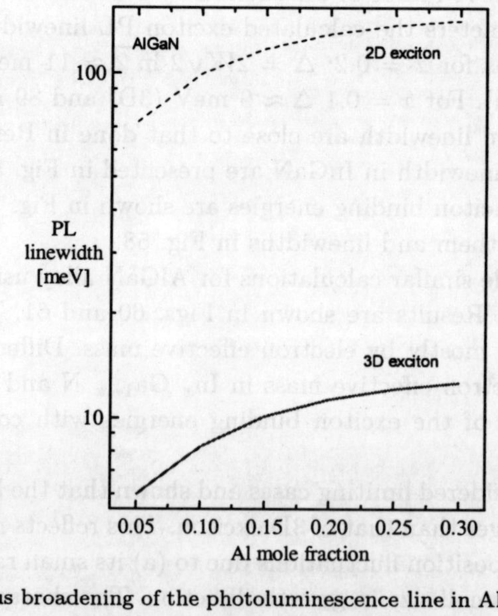

Fig. 59. Inhomogeneous broadening of the photoluminescence line in $Al_xGa_{1-x}N$ quantum well.

GaAs-based materials also decrease the penetration of the electron and hole wave functions into barrier and makes the character of the exciton in III-Nitride QW closer to the 2D one. In this case the near band edge exciton states are greatly influenced by compositional fluctuations.

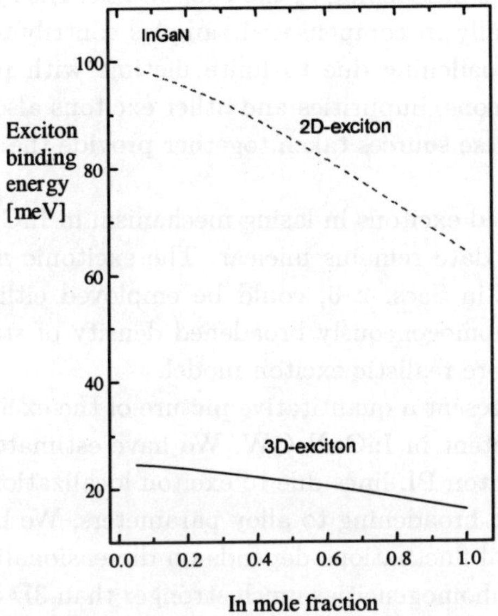

Fig. 60. A-exciton binding energies in $In_xGa_{1-x}N$ quantum well.

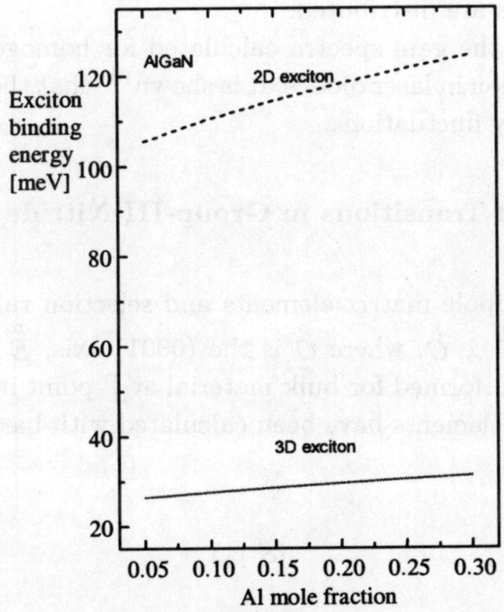

Fig. 61. A-exciton binding energies in $Al_xGa_{1-x}N$ quantum well.

The experimental PL linewidths, obtained in InGaN QW's, are given as (in meV units) 20,[95] 45,[96] 50–200,[97] 75,[98] 98,[99] 100, 90–150.[100] Most of these data are in the range obtained in this paper (see Fig. 58). For detailed comparison to a particular experiment, we need to take into account that the linewidth calculated here is the contribution from the alloy disorder. Compositional fluctuations is not the only reason for inhomogeneous broadening of the exciton line. Also the electronic disorder due to doping especially in compensated samples contributes the linewidth.[55,101] The homogeneous broadening due to finite lifetime with respect to collisions of the exciton with phonons, impurities and other excitons also increases the observable linewidth. All these sources taken together provide the observable exciton PL linewidth in InGaN.

The role of localized excitons in lasing mechanism in InGaN MQW lasers could be important and to date remains unclear. The excitonic model of gain,[102] used in materials mention in Secs. 2–6, could be employed either in this problem in combination with inhomogeneously broadened density of states Eq. (14) or other one, calculated for more realistic exciton model.

In summary, we present a quantitative picture of the exciton broadening due to fluctuations of In content in InGaN QW. We have estimated the inhomogeneous broadening of the exciton PL lines due to exciton localization on composition fluctuation and relate the broadening to alloy parameters. We have shown that sensitivity to compositional fluctuations depends on dimensionality of the exciton, and 2D exciton feels the inhomogeneities much stronger than 3D one. As it follows from presented calculations, the contribution to the linewidth for quasi-2D exciton is in the agreement with the experimental data. We have also found the inhomogeneously broadened density of exciton states which determines the energy domain where radiative states are distributed.

Below we discuss the gain spectra calculated for homogeneous and inhomogeneous alloy active region in laser diodes. It is shown[103] that the TM/TE polarization ratio depends on alloy fluctuations.

8. TE/TM Optical Transitions in Group-III Nitride Quantum Wells

Calculations of the dipole matrix elements and selection rules for different polarizations ($\overset{\rho}{E}||\overset{\rho}{C}$ and $\overset{\rho}{E} \perp \overset{\rho}{C}$, where $\overset{\rho}{C}$ is the (0001) axis, $\overset{\rho}{E}$ is the electric field in the wave) has been performed for bulk material at Γ-point in Brillouin zone.[104,105] These optical matrix elements have been calculated with basis wave functions near the band edges;

Conduction band:

$$|S \uparrow\downarrow\rangle$$

Valence band:

$$\begin{aligned}
&\Gamma_9(HH) \qquad |1\rangle = -\frac{1}{\sqrt{2}}|(X+iY)\uparrow\downarrow\rangle\,,\\
&\Gamma_7^1(LH) \qquad |2\rangle = \frac{1}{\sqrt{2}}|(X+iY)\uparrow\downarrow\rangle\,,\\
&\Gamma_7^2(CH) \qquad |3\rangle = |Z\uparrow\downarrow\rangle
\end{aligned} \tag{27}$$

where S, X, Y, Z are zero-momentum an s-orbital and three p-orbitals, respectively. Optical matrix elements for transitions between edges of conduction and HH bands are

$$|M_{C-HH}|^2 = \begin{cases} \dfrac{P_x^2}{2}, E \perp c \\ 0, E \| c \end{cases} \tag{28}$$

where $P_x = \langle S|p_x|X\rangle$.

For quantum well structure grown in c-direction, Eq. (17) means that transition C-HH does not contain TM component. It is correct only for in-plane momentum $k_t = 0$ (transition between top of HH valence subband and bottom of conduction subband). Apart from subband edges the selection rules for TE and TM transitions differentiate from those in Γ point and they have to be calculated in order to obtain the observable TE and TM emission intensties. In order to calculate spectral gain one has to take into account k_t-dependence of the matrix elements:

$$|M|^2_{C-HH} = \begin{cases} \dfrac{P_2^2}{8}\langle\phi(z)|g_{11}(z)\rangle^2(1+\cos^2\theta_k), & TE \\ \dfrac{P_1^2}{4}\langle\phi(z)|g_{11}(z)\rangle^2\sin^2\theta_k, & TM \end{cases}$$

where $\cos^2\theta_k = \frac{k_{z1}^2}{k_{z1}^2+k_t^2}$, where k_{z1} is the size-quantized wavenumber in growth direction. Spectral dependence of the material gain is given as:

$$g(\eta\omega) = \left[1 - \exp\left(\frac{\eta\omega - \Delta F}{k_b T}\right)\right] \times \frac{q^2\gamma}{2\pi n_r c\varepsilon_0 m_0^2 d\eta\omega} \times \int_0^\infty |M_{C-HH}|^2 k_t dk_t \frac{f_{1c}(1-f_{1v})}{(E_{1c}-E_{1v}-\eta\omega)^2)+\gamma^2}\,, \tag{29}$$

where $E_{1c,v}$ and $f_{1c,v}$ are energy bands and electron filling factors for first conduction and heavy hole energy levels, ΔF is the quasi-Fermi levels separation, d is the width of the QW, n_r is the refractive index of the well material, γ is the homogeneous broadening due to electron scattering on impurities and phonons. Results for TE and TM material gain at different carrier densities are shown in Figs. 62 and 63, respectively ($d = 33$ Å, $\gamma = 30$ meV). In result, both polarizations are present in C-HH gain spectrum. Peak TE gain is 8 times larger than peak TM gain for $n = 1.5\times10^{19}$ cm^{-3}, and 7 times for $n = 2\times10^{19}$ cm^{-3}.

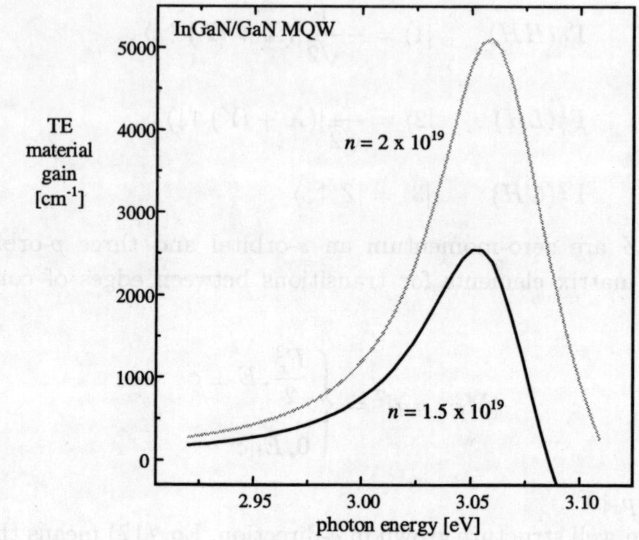

Fig. 62. TE spectral gain of GaN at $n = 2 \times 10^{19}$ cm^{-3} and 1.5×10^{19} cm^{-3}.

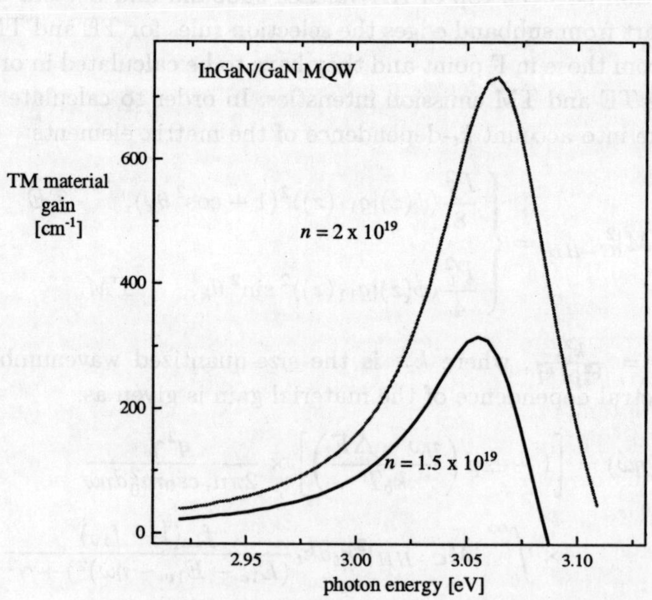

Fig. 63. TM spectral gain of GaN at $n = 2 \times 10^{19}$ cm^{-3} and 1.5×10^{19} cm^{-3}.

We have also calculated TE/TM peak gain ratio in inhomogeneous quantum wells. The measure of inhomogeneity is the broadening of the density of states W. Results are shown in Fig. 64 for different carrier concentrations. Lasing starts from TE-modes and stimulated emission reveals high TE/TM ratio due to amplification of the TE component.

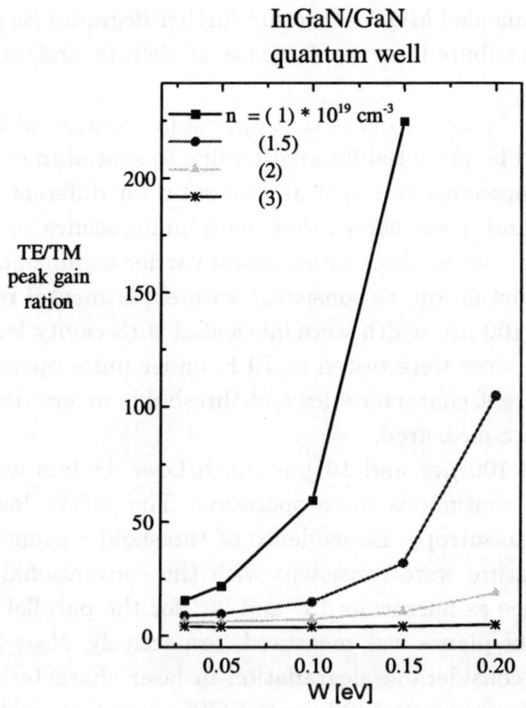

Fig. 64. Peak gain ratio in inhomogeneous quantum wells.

9. Conclusions

The growth of GaInN/GaN multi-quantum well structures for laser diodes was conducted on (00.1) sapphire substrates by low pressure metalorganic chemical vapor deposition. The structures were structurally and optically characterized through X-ray diffraction, photoluminescence and optical pumping.

The effects of temperature on the optical properties of GaInN/GaN MQW have been investigated. The In concentration in GaInN was shown to be strongly sensitive to the growth temperature as expected due to the low melting point of InN. The photoluminescence spectra and the optically pumped stimulated emission threshold and slope efficiency from the MQW structures were investigated as a function of thermal annealing time and top cap layer properties (doping, thickness).

The effects of Si doping of the GaInN/GaN MQW have been investigated. Several doping schemes were studied: undoped structures, doped barriers (i.e. GaN), doped barriers and wells (i.e. GaN and GaInN).

It was shown that proper annealing or Si-doping of GaInN/GaN MQW structures could reduce the lasing threshold and improve the slope efficiency. Over-annealing of these MQW can lead to thermal degradation of the active layer. *P*-type doping on the top of the MQW using a cap GaN:Mg could lead to the formation of compensation layer which degrades laser diode performances.

Si-doping in over-annealed MQW structure further degrades its quality. The degradation has been attributed to the increase of defects and/or nonuniform local potential formation.

Our numerical calculation shows that optical propoerties of InGaN laser diodes are stronlgy affected by potential fluctuation due to spatial In composition inhomogeneity. Photoluminescence and optical gain ratio for different beam polaizations are calculated and and it was shown that photoluminescence spectral width can be as high as 100 meV due to three-dimensional carrier confinement as in quantum-dot-like potential fluctuation, as consistent with experimental results.

Laser stripes of 100 μm width were fabricated with cavity lengths from 650 μm to 1800 μm. The devices were tested at 79 K under pulse operation. Output light versus injection current characteristics and threshold current density dependencies on cavity length were measured.

Laser devices of 100 μm and 10 μm width laser devices were tested at room temperature under continuous wave operation. The MQW lasers showed strong beam polarization anisotropy. Dependence of threshold current density on cavity-length and temperature were consistent with the conventional laser theory. Far-field beam divergence as narrow as 13° and 20° for the parallel and perpendicular directions to epilayer planes was measured, respectively. Near-field patterns were also measured. No considerable degradation in laser characteristics was observed during lifetime testing for over 140 hours of CW room temperature operation.

References

1. B. Daudin, J. L. Rouvire, and M. Arlery, "Polarity determination of GaN films by ion channeling and convergent beam electron diffraction", *Appl. Phys. Lett.* **69** (1996) 2480–2482.
2. A. D. Bykhovski, V. V. Kaminski, M. S. Shur, Q. C. Chen, and M. A. Khan, "Pyroelectricity in gallium nitride thin films", *Appl. Phys. Lett.* **69** (1996) 3254–3256.
3. J. Miragliotta, D. K. Wickenden, T. J. Kistenmacher, and W. A. Bryden, "Linear optical and nonlinear-optical properties of GaN thin films", *J. Opt. Soc. Am.* **B10** (1993) 1447–14456.
4. H. Y. Zhang, X. H. He, Y. H. Shih, M. Schurman, Z. C. Feng, and R. A. Stall, "Study of nonlinear optical effects in GaN:Mg epitaxial film", *Appl. Phys. Lett.* **69** (1996) 2953–55.
5. S. Nakamura, M. Senoh, S. Nagahama, N. Iwasa, T. Yamada, T. Matsushita, H. Kiyoku, Y. Sugimoto, T. Kozaki, H. Umemoto, M. Sano, and K. Chocho, "InGaN/GaN/AlGaN-based laser diodes with modulation-doped strained-layer superlattices grown on an epitaxially laterally overgrown GaN substrate", *Appl. Phys. Lett.* **72** (1998) 211–213.
6. O. Madelung (ed.), *Semiconductors: Group IV Elements and III–V Compounds*, Springer-Verlag, Berlin, 1991.
7. O. Madelung, *Semiconductors — Basic Data*, 2nd ed., Springer-Verlag, Berlin, 1996.
8. M. Leszczynski, T. Suski, P. Perlin, H. Teisseyre, I. Grzegory, M. Bockowski, J. Jun, S. Porowski, and J. Major, "Lattice constants, thermal expansion, and compressibility of gallium nitride", *J. Phys.* **D28** (1995) A149–A153.

9. H. P. Maruska and J. J. Tietjen, "The preparation and properties of vapor-deposited single-crystalline GaN", *Appl. Phys. Lett.* **15** (1969) 327–329.
10. H. M. Manasevit, F. M. Erdmann, and W. I. Simpson, "The use of metalorganics in the preparation of semiconductor materials. IV. The nitrides of aluminum and gallium", *J. Electrochem. Soc.* **118** (1971) 1864–68 .
11. H. Amano, N. Sawaki, I. Akasaki, and Y. Toyoda, "Metalorganic vapor phase epitaxial growth of high quality GaN film using an AlN buffer", *Appl. Phys. Lett.* **48** (1986) 353–355.
12. S. Nakamura, "GaN growth using GaN buffer layer", *Jpn. J. Appl. Phys.* **30** (1991) L1705–L1707.
13. H. Amano, M. Kito, K. Hiramatsu, and I. Akasaki, "*P*-type conduction in Mg-doped GaN treated with low-energy electron beam irradiation (LEEBI)", *Jpn. J. Appl. Phys.* **28** (1989) L2112–L2114.
14. S. Nakamura, T. Mukai, M. Senoh, and N. Iwasa, "Thermal annealing effects on *p*-type Mg-doped GaN films", *Jpn. J. Appl. Phys.* **31** (1992) L139–L142.
15. P. Kung, A. Saxler, D. Walker, X. Zhang, R. Lavado, K. S. Kim, and M. Razeghi, "AlGaN-based materials and heterostructures", *Mat. Res. Soc. Symp. Proc.* **449** (1997) 79–84.
16. S. Nakamura and T. Mukai, "High-quality InGaN films growm on GaN films", *Jpn. J. Appl. Phys.* **31** (1992) L1457–L1459.
17. S. Nakamura, T. Mukai, and M. Senoh, "Candela-class high-brightness InGaN/AlGaN double-heterostructure blue-emitting diodes", *Appl. Phys. Lett.* **64** (1994) 1687–1689.
18. S. Nakamura, M. Senoh, N. Iwasa, and S. Nagahama, "High-brightness InGaN blue, green and yellow light-emitting diodes with quantum well structures", *Jpn. J. Appl. Phys.* **34** (1995) L797–L799.
19. S. Nakamura, M. Senoh, S. Nagahama, N. Iwasa, T. Yamada, T. Matsushita, H. Kiyoku, and Y. Sugimoto, "InGaN based muliquantum well structure laser diodes", *Jpn. J. Appl. Phys.* **35** (1996) L74–L76.
20. L. H. Galligan, "Range gated uderwater viewing", *SPIE Proc.* **980** (1988) 36–40.
21. S.N. Mohammad, A. A. Salvador, and H. Morkoc, "Emerging gallium nitride based devices", *Proc. IEEE* **83** (1995) 1306–1355.
22. V. E. Bougrov and A. S. Zubrilov, "Optical confinement and threshold currents in III–V nitride heterostructures: Simulation", *J. Appl. Phys.* **81** (1997) 2952–2956.
23. D. Hofstetter, D. P. Bour, R. L. Thornton, and N. M. Johnson, "Excitation of a higher order transverse mode in an optically pumped $In_{0.15}Ga_{0.85}N/In_{0.05}Ga_{0.95}N$ multiquantum well laser structure", *Appl. Phys. Lett.* **70** (1998) 1650–1652.
24. S. Nakamura, M. Senoh, S. Nagahama, N. Iwasa, T. Yamada, T. Matsushita, H. Kiyoku, Y. Sugimoto, T. Kozaki, H. Umemoto, M. Sano, and K. Chocho, "Continuous-wave operation of InGaN/GaN/AlGaN-based laser diodes grown on GaN substrates", *Appl. Phys. Lett.* **72** (1998) 2014–2016.
25. H. Kressel and J. K. Butler, *Semiconductor Lasers and Heterojunction LEDs*, Academic Press, 1977.
26. G. Martin, A. Botchkarev, A. Rockett, and H. Morkoc, "Valence-band discontinuities of wurtzite GaN, AlN, and InN heterojunctions measured by X-ray photoemission spectroscopy", *Appl. Phys. Lett.* **68** (1996) 2541–2543.
27. E. A. Albanesi, W. R. L. Lambrecht, and B. Segall, "Band offsets between group III nitrides", *Proc. Mat. Res. Conf.* **339** (1994) 137–145.
28. C. G. Van de Walle and J. Neugebauer, "Small valence-band offsets at GaN/InGaN heterojunctions", *Appl. Phys. Lett.* **70** (1997) 2577–2579.

29. J. R. Waldrop and R. W. Grant, "Measurement of AlN/GaN (0001) heterojunction band offsets by X-ray photoemission spectroscopy", *Appl. Phys. Lett.* **68** (1996) 2879–2881.
30. S.-H. Wei and A. Zunger, "Valence band splittings and band offsets of AlN, GaN, and InN", *Appl. Phys. Lett.* **69** (1996) 2719–2721.
31. X. Zhang, P. Kung, D. Walker, J. Piotrowski, A. Rogalski, A. Saxler, and M. Razeghi, "Photovoltaic effects in GaN structures with *p–n* junction", *Appl. Phys. Lett.* **67** (1995) 2038–2040.
32. X. Zhang, "Optical properties of GaN/InGaN/AlGaN structures grown by metal organic chemical vapor deposition", PhD Dissertation, Northwestern University, 1999.
33. Q. Guo, O. Kato, and A. Yoshida, "Thermal stability of indium nitride single crystal films", *J. Appl. Phys.* **73** (1993) 7969–7971.
34. S. Chichibu, K. Wada, and S. Nakamura, "Spatially resolved cathodoluminescence spectra of InGaN quantum wells", *Appl. Phys. Lett.* **71** (1997) 2346–2349.
35. Y. Narukawa, Y. Kawakami, M. Funato, S. Fujita, S. Fujita, and S. Nakamura, "Role of self-formed InGaN quantum dots for exciton localization in the purple laser diode emitting at 420 nm", *Appl. Phys. Lett.* **70** (1997) 981–983.
36. K. Doman, A. Kuramata, and T. Tanahashi, "Lasing mechanism of InGaN/GaN/AlGaN multiquantum well laser diode", *Appl. Phys. Lett.* **72** (1998) 1359–1361.
37. S. Nakamura, M. Senoh, S. Nagahama, N. Iwasa, T. Yamada, T. Matsushita, Y. Sugimoto, and H. Kiyoku, "Longitudinal mode spectra and ultrashort pulse generation of InGaN multiquantum well structure laser diodes", *Appl. Phys. Lett.* **70** (1996) 616–618.
38. S. Nakamura, M. Senoh, S. Nagahama, N. Iwasa, T. Yamada, T. Matsushita, H. Kiyoku, and Y. Sugimoto, "Room-temperature continuous-wave operation of InGaN multi-quantum-well structure laser diodes with a lifetime of 27 hours", *Appl. Phys. Lett.* **70** (1996) 1417–1419.
39. S. Nakamura, M. Senoh, S. Nagahama, N. Iwasa, T. Yamada, T. Matsushita, Y. Sugimoto, and H. Kiyoku, "Subband emissions of InGaN multi-quantum-well laser diodes under room-temperature continuous wave operation", *Appl. Phys. Lett.* **70** (1997) 2753–2755.
40. S. Nakamura, "Characteristics of room temperature-CW operated InGaN multi-quantum-well-structure laser diodes", *Internet J. Nitride Semiconductor Res.* **2** (1997) 5.
41. P. Kung, A. Saxler, D. Walker, A. Rybaltowski, X. Zhang, and M. Razeghi, "GaInN/GaN multi-quantum well laser diodes grown by low-pressure metalorganic chemical vapor deposition", *Internet J. Nitride Semiconductor Res.* **3** (1997) 1.
42. M. P. Mack, A. Abare, M. Aizcorbe, Peter Kozodoy, S. Keller, U. K. Mishra, L. Coldren, and Steven DenBaars, "Characteristics of indium-gallium-nitride multiple-quantum-well blue laser diodes grown by MOCVD", *Internet J. Nitride Semiconductor Res.* **2** (1997) 41.
43. J. Nishio, L. Sugiura, H. Fujimoto, Y. Kokubun, and K. Itaya, "Characterization of InGaN multiquantum well structures for blue semiconductor laser diodes", *Appl. Phys. Lett.* **70** (1997) 3431–3433.
44. X. H. Wu, R. Elsass, A. Abare, M. Mack, S. Keller, P. M. Petroff, S. P. DenBaars, and J. S. Speck, "Structural origin of V-defects and correlation with localized excitonic centers in InGaN/GaN multiple quantum wells", *Appl. Phys. Lett.* **72** (1998) 692–694.

45. Y. Chen, T. Takeuchi, H. Amano, I. Akasaki, N. Yamada, Y. Kaneko, and S. Y. Wang, "Pit formation in GaInN quantum wells", *Appl. Phys. Lett.* **72** (1998) 710–712.
46. I. H. Ho and G. B. Stringfellow, "Solid phase immiscibility in GaInN", *Appl. Phys. Lett.* **69** (1996) 2701–2703.
47. A. Koukitsu, N. Takahashi, T. Taki, and H. Seki, "Thermodynamic analysis of InGaN alloy composition grown by MVPE", *Jpn. J. Appl. Phys.* **35** (1996) L673–L675.
48. Some MQW based laser diode structures have Si-doped barriers or Si-doped wells or both. As will be discussed in the next part, doping in fact can reduce the lasing threshold.
49. X. Zhang, P. Kung, A. Saxler, D. Walker, T. Wang, and M. Razeghi, "Photoluminescence study of GaN", *Acta Physica Polonica* **88** (1995) 601–606.
50. S. Nakamura, T. Mukai, and M. Senoh, "High-power GaN *p–n* junction blue-emitting diodes", *Jpn. J. Appl. Phys.* **30** (1991) L1998–L2001.
51. S. Nakamura, M. Senoh, N. Iwasa, and S. Nagahama, "High-power InGaN single-quantum-well-structure blue and violet light-emitting diodes", *Appl. Phys. Lett.* **67** (1995) 1868–1870.
52. P. Kung, X. Zhang, E. Bigan, M. Razeghi, and A. Saxler, "Low pressure metalorganic chemical vapor deposition of high AlN and GaN thin films on sapphire and silicon substrates", *SPIE Proc.* **2397** (1995) 311–320.
53. S. Ruvimov, Z. L. Weber, T. Suski, J. W. Ager III, J. Washburn, J. Krueger, C. Kisielowski, E. R. Weber, H. Amano, and I. Akasaki, "Effect of Si doping on the dislocation structure of GaN grown on the A-face of sapphire", *Appl. Phys. Lett.* **69** (1996) 990–992.
54. K. C. Zeng, J. Y. Lin, H. X. Jiang, A. Salvador, G. Popovici, H. Tang, W. Kim, and H. Morkoc, "Effects of well thickness and Si doping on the optical properties of GaN/AlGaN multiple quantum wells", *Appl. Phys. Lett.* **71** (1997) 1368–1370.
55. E. F. Schubert, I. D. Goepfert, W. Grieshaber, and J. M. Redwing, "Optical properties of Si-doped GaN", *Appl. Phys. Lett.* **71** (1997) 921–923.
56. Y. H. Cho, J. J. Song, S. Keller, M. S. Minsky, E. Hu, U. K. Mishra, and S. P. DenBaars, "Influence of Si doping on characteristics of InGaN/GaN multiple quantum wells", *Appl. Phys. Lett.* **73** (1998) 1128–1130.
57. H. C. Casey, Jr., J. Muth, S. Krishnankutty, and J. M. Zavada, "Dominance of tunneling current and band filling in InGaN/AlGaN double heterostructure blue light-emitting diodes", *Appl. Phys. Lett.* **68** (1996) 2867–2869.
58. J. T. Torvik, C. H. Qiu, M. Leksono, and J. I. Pankove, "Optical characterization of GaN/SiC *n–p* heterojunctions and *p*-SiC", *Appl. Phys. Lett.* **72** (1998) 945–947.
59. S. Bidnyk, T. J. Schmidt, Y. H. Cho, G. H. Gainer, J. J. Song, S. Keller, U. K. Mishra, and S. P. DenBaars, "High-temperature stimulated emission in optically pumped InGaN/GaN multiquantum wells", *Appl. Phys. Lett.* **72** (1998) 1623–25.
60. I. H. Lee, I. H. Choi, C. R. Lee, and S. K. Noh, "Evolution of stress relaxation and yellow luminescence in GaN/sapphire by Si incorporation", *Appl. Phys. Lett.* **71** (1997) 1359–1361.
61. E. Oh, H. Park, and Y. Park, "Influence of potential fluctuation on optical and electrical properties in GaN", *Appl. Phys. Lett.* **72** (1998) 1848–1850.
62. Y.-K. Song, M. Kuball, A. V. Nurmikko, G. E. Bulman, K. Doverspike, S. T. Sheppard, T. W. Weeks, M. Leonard, H. S. Kong, H. Dieringer, and J. Edmond, "Gain characteristics of InGaN/GaN quantum well diode lasers", *Appl. Phys. Lett.* **72** (1998) 1418–1420.

63. N. Holonyak, Jr. and M. H. Lee, "Photopumped III–V Semiconductor Lasers", in *Semiconductors and Semimetals* **14**, Academic Press, New York, 1979, pp. 1–64.
64. T. N. Morgan, "Recombination by tunneling in electroluminescent diodes", *Phys. Rev.* **148** (1966) 890–903.
65. J. I. Pankove, *Optical Processes in Semiconductors*, Dover Publications, New York, 1971.
66. H. Nakayama, P. Hacke, M. R. H. Khan, T. Detchprohm, K. Hiramatsu, and N. Sawaki, "Electrical transport properties of *p*-GaN", *Jpn. J. Appl. Phys.* **35** (1996) L282–L284.
67. H. Yi, A. Rybaltowski, J. Diaz, D. Wu, B. Lane, Y. Xiao, and M. Razeghi, "Stability of far fields in double heterostructure and multiple quantum well InAsSb/InPAsSb/InAs midinfrared lasers", *Appl. Phys. Lett.* **70** (1997) 3236–3238.
68. S. Nakamura and G. Fasol, *The Blue Laser Diode*, Springer, Berlin, 1997.
69. D. Ahn, S. L. Chuang, and Y. C. Chang, "Valence-band mixing effects on the gain and the refractive index changes of quantum well lasers", *J. Appl. Phys.* **64** (1988) 4056–4064.
70. Polarization anisotropy would be observed even for spontaneous emission because of higher transition matrix in QW structure. However, the difference should be less than a factor of 3.
71. H. Casey, *Heterostructure Lasers*, Academic Press, New York, 1978.
72. M. Razeghi, H. Yi, and V. Litvinov (unpublished).
73. M. Suzuki and T. Uenoyama, "Biaxial strain effect on wurtzite GaN/AlGaN quantum well lasers", *Jpn. J. Appl. Phys.* **35** (1996) 1420–1423.
74. P. Rees, C. Cooper, P. Smowton, P. Blood, and J. Hegarty, "Calculated threshold currents of nitrides and phospides based quantum well lasers", *IEEE Photon. Technol. Lett.* **8** (1996) 197–199.
75. K. D. Choquette, H. Hou, K. Lear, H. Chui, K. Geib, A. Mar, and B. Hammons, "Self-pulsed oxide-confined vertical cavity lasers with ultralow operating current", *Electron. Lett.* **32** (1996) 459–460.
76. O. Goede, L. John, and D. H. Hennig, "Composition disorder-induced broadening for free excitons in II–VI semiconducting mixed crystals", *Phys. Stat. Sol.* **B89** (1978) K183–K186.
77. E. F. Schubert and W. T. Tsang, "Photoluminescence lineshape of excitons in alloy semiconductors", *Phys. Rev.* **B34** (1986) 2991–94.
78. J. Singh and K. K. Bajaj, "Quantum mechanical theory of linewidths of localized radiative transition in semiconductor alloys", *Appl. Phys. Lett.* **48** (1986) 1077–79.
79. J. Christen and D. Bimberg, "Lineshapes of intersubband and excitonic recombination in quantum wells: Influence of final-state interaction, statistical broadening, and momentum conservation", *Phys. Rev.* **B42** (1990) 7213–7219.
80. Y. Narukawa, Y. Kawakami, S. Fujita, Sg. Fujita, and S. Nakamura, *Phys. Rev.* **B55** (1997) R1938; T. Deguchi, T. Azunata, T. Sota, S. Chichibu, M. Arita, H. Nakanishi, and S. Kawamura, *Semicond. Sci. Technol.* **13** (1998) 97.
81. M. Kuball, E. S. Jeon, Y. K. Song, A. V. Nurmikko, P. Kozodoy, A. Abare, S. Keller, L. A. Coldren, U. K. Mishra, S. P. DenBaars, and D. A. Steigerwald, "Gain spectroscopy on InGaN/GaN quantum well diodes", *Appl. Phys. Lett.* **70** (1997) 2580–2582; Y. Cho, G. H. Gainer, A. J. Fischer, J. J. Song, S. Keller, U. K. Mishra, and S. P. DenBaars, "S-shaped" temperature-dependent emission shift and carrier dynamics in InGaN/GaN multiple quantum wells", *Appl. Phys. Lett.* **73** (1998) 1370–1372; T. J. Schmidt, Y. Cho, G. H. Gainer, J. J. Song, S. Keller, U. K. Mishra,

and S. P. DenBaars, "Pump-probe spectroscopy of band tail states in metalorganic chemical vapor deposition-grown InGaN", *Appl. Phys. Lett.* **73** (1998) 1892–1894.

82. E. S. Jeon, V. Kozlov, Y. K. Song, A. Vertikov, M. Kuball, A. V. Nurmikko, H. Liu, C. Chen, R. S. Kern, C. P. Kuo, and M. G. Craford, "Recombination dynamics in InGaN quantum wells", *Appl. Phys. Lett.* **69** (1996) 4194–96.
83. V. Litvnov and M. Razeghi, "Exciton localization in group–III nitride quantum wells", *Phys. Rev. B* **59** (1999) 9783–9786.
84. A. L. Efros and M. E. Raikh, "Effect of composition disorder on the electronic properties of semiconducting mixed crystals", in *Optical Properties of Mixed Crystals*, eds. R. J. Elliot and I. P. Ipatova, North-Holland, Amsterdam, 1988.
85. Al. Efros, C. Wetzel, and J. M. Worlock, "Effect of a random adiabatic potential on optical properties of two-dimensional excitons", *Phys. Rev.* **B52** (1995) 8384–8390.
86. D. J. Thouless and M. E. Elzain, "The two-dimensional white noise problem and localisation in an inversion layer", *J. Phys.* **C11** (1978) 3425–3438.
87. E. V. Tsyper, "Optical properties of arrays of quantum dots with internal disorder", *Phys. Rev.* **B54** (1996) 1959–1966.
88. B. I. Shklovskii and A. L. Efros, *Electronic Properties of Doped Semiconductors*, Springer, Berlin, 1984.
89. S. H. Wei and A. Zunger, "Calculated natural band offsets of all IIVI and IIIV semiconductors: Chemical trends and the role of cation *d*-orbitals", *Appl. Phys. Lett.* **72** (1998) 2011–2013.
90. S. Strite and H. Morkoc, "GaN, AlN, and InN: A review", *J. Vac. Sci. Technol.* **B10** (1992) 1237.
91. S. Nakamura, T. Mukai, M. Senoh, S. Nagayama, and N. Iwasa, "InGaN/InGaN superlattices grown on GaN films", *J. Appl. Phys.* **74** (1993) 3911–3915.
92. B. Monemar "Fundamental gap of GaN from photoluminescence excitation spectra", *Phys. Rev.* **10** (1974) 676–681.
93. T. L. Tansley and C. P. Foley, "Optical bnad gap of indium nitride", *J. Appl. Phys.* **59** (1986) 3241–43.
94. M. Suzuki, T. Uenoyama, and A. Yanase, "First-principles calculations of effective-mass parameters of AlN and GaN", *Phys. Rev.* **B52** (1995) 8132–8139.
95. H. Yu, H. Htoon, A. deLozanne, C. K. Shin, P. A. Gudowski, R. D. Dupuis, K. Zeng, R. Mair, J. Y. Lin, and H. X. Jiang, *J. Vac. Sci. Technol.* **B16** (1998) 2215.
96. Y. Yamada, T. Taguchi, F. Sasaki, S. Kobayashi, and T. Tani, "Effect of Si-doping on luminescence properties of InGaN epitaxial layers", *J. Cryst. Growth* **189/190** (1998) 611.
97. P. G. Middleton, K. P. O'Donnel, T. Breitkopf, H. Kalt, Van der Stricht, I. Moerman, and P. Deemster, *Mat. Sci. Eng.* **B50** (1997) 285.
98. S. Chichibu, T. Sota, K. Wada, and S. Nakamura, *J. Vac. Sci. Technol.* **B16** (1998) 2204; S. Chichibu, T. Azunata, T. Sota, and S. Nakamura, "Luminescences from localized states in InGaN epilayers", *Appl. Phys. Lett.* **70** (1997) 2822–2824.
99. K. C. Zeng, M. Smith, J. Y. Lin, and H. X. Jiang, "Collective effects of interface roughness and alloy disorder in $In_xGa_{1-x}N$/GaN multiple quantum wells", *Appl. Phys. Lett.* **73** (1998) 1724–1726.
100. Y. Narukawa, S. Saijo, Y. Kawakami, S. Fujita, S. Fujita, and S. Nakamura, "Time-resolved electroluminescence spectra of InGaN single quantum well LED's", *J. Cryst. Growth* **189/190** (1998) 593–596.
101. E. Iliopoulos, D. Doppalapudi, H. M. Ng, and T. D. Moustakas, "Broadening of near-band-gap photoluminescence in *n*-GaN films", *Appl. Phys. Lett.* **73** (1998) 375–377.

102. J. Ding, M. Hagerrot, T. Ishihara, H. Jeon, and A. V. Nurmikko, "(Zn,Cd)/ZnSe quantum well lasers: Excitonic gain in an inhomogeneously broadened quasi-wo-dimensional system", *Phys. Rev.* **B47** (1993) 10528–10542.
103. V. I. Litvinov and M. Razeghi, "Polarization dependent gain spectra in inhomogeneous InGaN quantum wells", to be submitted (1999).
104. G. L. Bir and G. E. Pikus, *Symmetry and Strain Effects in Semiconductors*, Wiley, New York, 1974.
105. S. Chuang and C. S. Chang, "k.p method for strained wurtzite semiconductors", *Phys. Rev.* **B54** (1996) 2491–2504.

International Journal of High Speed Electronics and Systems, Vol. 9, No. 4 (1998) 1081–1107

QUANTUM DOT SEMICONDUCTOR LASERS

PALLAB BHATTACHARYA
Department of Electrical Engineering and Computer Science,
University of Michigan, 1301 Beal Avenue,
Ann Arbor, Michigan 48109-2122, USA

The formation of coherently strained islands during the growth of strained heterostructures has been exploited to form an array of quantum dots. The shape and size of the islands vary with growth parameters, but exhibit the electronic properties of zero-dimensional systems. One or multiple, vertically coupled, layers of such quantum dots can form the gain region of a separately confined heterostructure (SCH) laser. The properties of such InGaAs/GaAs self-organized quantum dot lasers are described here. The lasers exhibit temperature independent operation up to 100 K and beyond. Typical threshold currents of 200 μm long room temperature lasers vary from 6 to 20 mA. The small-signal modulation bandwidths of ridge waveguide lasers are 5–7.5 GHz at 300 K and increased to > 20 GHz at 80 K. We believe that electron-hole scattering intrinsically limits the high-speed performance of these devices, in spite of differential gains as high as $\sim 7 \times 10^{-14}$ cm^2 at room temperature. Wavelength switching is demonstrated in these devices and preliminary results on long-wavelengths intersubband quantum dot light emitters are also presented.

1. Introduction

Some of the desirable attributes of semiconductor lasers are low threshold currents, high output power and efficiency, temperature-independent operation, large modulation bandwidth, and negligible chirp. Many of these characteristics are determined by the density of states and electronic properties of the gain medium. The use of gain media of reduced dimensionality, in particular quantum wells, have greatly enhanced laser performance, particularly due to the non-zero density of states at the band edge and the resulting large gain and differential gain.[1,2] Three dimensional confinement of carriers in an ideal quantum dot gives rise to atomic like discrete states with a delta function density of states. Therefore, a semiconductor laser with a quantum dot active region[3] promises some of the desirable characteristics mentioned above. There has therefore been worldwide interest in the development of quantum dot lasers, whose performance would eventually surpass that of other semiconductor lasers.

Conventionally, the most straightforward technique to realize quantum dots is the controlled etching of epitaxially grown quantum wells. However, even the best of dry etching techniques produce sufficient damage to preclude sufficient light emission.[4] Recently, *self-organized* quantum dots[5–7] have been proven to be the structures which best approach the desired properties, and almost all groups actively pursuing quantum dot device development are utilizing strain-induced self-organized growth. Much progress has been made in the area of growth[8–11] — where the focus has been size control, and optical characterization[12–20] — where the goal has been the application to quantum dot lasers and detectors. The use of defect-free strain-induced self-organized quantum dots provides several advantages. Due to the pyramidal shape of these dots and the complicated strain tensor with a strong hydrostatic component within them, as will be described in Sec. 2, large modulation of the interband photon energy can be produced. For example, InAs/GaAs quantum dot lasers emit at ~ 1 μm, a wavelength much smaller than that corresponding to the bandgap. Recently, 1.3 μm emission has also been achieved.[21] Since the first demonstration of room temperature self-organized quantum dot lasers,[22–25] much progress has been made in improving the device characteristics. Thus, in spite of the broad spontaneous emission spectrum resulting from the size non-uniformity in the self-organization process, threshold current densities as low as 62 A/cm^2 at 300 K[26] and 20 A/cm^2 at 77 K[27] and T_0 as high as 385 K up to 330 K[26] in InGaAs/GaAs quantum dot lasers been reported. Low threshold currents are being realized in surface-emitting quantum dot lasers also.[28] The recent progress in the DC characteristics of these devices has been summarized by Bimberg *et al.*[26]

In what follows, the growth and luminescent properties of self-organized In(Ga)As/Al(Ga)As quantum dots will be first described. This is followed by the characteristics of high-speed lasers and their relationship to the unusual carrier dynamics in these devices. Finally, recent results on 1.3 and 1.9 μm lasers and quantum dot vertical-cavity surface-emitting lasers (VCSELs) will be briefly described.

2. Self-Organized Growth of Quantum Dots

It has been shown that highly lattice-mismatched Ga(In)As epitaxially grows on GaAs in the so-called Stranski-Krastanow[29] growth mode, where self-organized islands are formed after a few monolayers of layer-by-layer growth. From RHEED measurements during molecular beam epitaxial (MBE) growth of InGaAs on GaAs and from energy minimization considerations in a unit cell of the growing layer, we had shown for the first time that for a misfit $f > 1.8\%$, the island mode of growth is preferred.[30] For typical growth parameters used in MBE or metal-organic vapor phase epitaxy (MOVPE), an array of pyramidal islands of lateral size 10–40 nm and heights 5–8 nm are formed. Elastic relaxation on the facet edges, renormalization of the surface energy of the facets, and interaction between neighboring islands via the substrate are the driving forces for self-organized growth. There are considerable strain fields within the pyramidal dots, in the substrate underneath, and in the overlayer, if the latter is grown. *In-situ* atomic force microscope (AFM) studies[10] during growth of InAs on GaAs have given valuable insight into the evolution of the size distribution between dots, as growth proceeds, and a tendency for eventual size equalization. Careful studies of growth in the InAs-GaAs system[6] have also

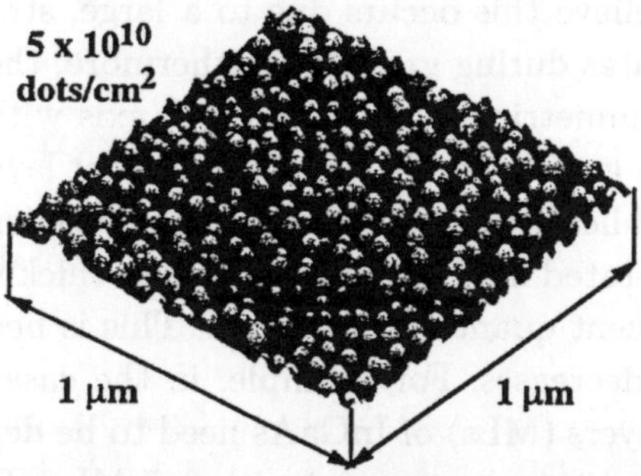

(a)

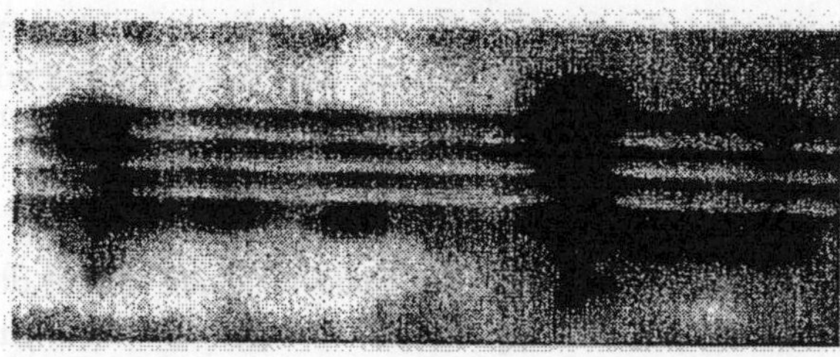

(b)

Fig. 1. (a) AFM image of a single layer of exposed self-organized In(Ga)As quantum dots. Dot density estimated from this image is 5×10^{10} cm^{-2} and (b) cross-sectional TEM image of a stack of four layers of InAs quantum dots separated by nominally 5-nm GaAs spacers. Alignment of the dots in the growth direction is evident.

shown that there exists a relatively narrow range of deposition parameters, where the islands are small ($\sim$ 10 nm), have very similar size and shape and form dense arrays. Interaction of the islands via the substrate also makes their lateral ordering favorable.[31] By virtue size and shape, the self-organized islands best approach the desired properties of zero-dimensional quantum dots. It is important to realize, for laser and other device applications, that the pyramidal quantum dots are defect-free. An AFM image of an array of $In_{0.4}Ga_{0.7}As/GaAs$ dots grown by MBE at 540°C at a rate of 0.25 monolayer/sec. is shown in Fig. 1(a). From this image, the dot density is estimated to be 5×10^{10} cm^{-2}. The pyramids have a base diagonal of 20 nm and height of 7 nm.

If a layer of InGaAs dots is covered with a thin layer of GaAs and another InGaAs growth cycle is initiated, the dots in the second sheet are formed exactly on top of the dots in the first layer and this trend continues, resulting in a 3-D array of vertically aligned and electronically coupled dots.[32–33] Multiple layer quantum dots (MLQD) are very useful for laser applications. The cross-sectional transmission electron microscope (TEM) image of a 4-layer stack of InAs dots grown by MBE at 500°C is shown in Fig. 1(b). The strain driven alignment of the dots is evident. We have also observed from recent cross-sectional scanning tunneling microscopy studies[34] that the separation between the dot stack decreases with increasing number of dot layers. We believe this occurs due to a large, strain-modulated increase in the In-Ga diffusion rates during growth. Furthermore, the dots become progressively larger and less symmetrical about the (001) axis with increasing number of dot layers. The optimum growth conditions for multi-dot layers have been described in detail in previous publications by us[19,20] and some salient features, relevant to laser operation, are reiterated here. Usually, a smaller thickness of InGaAs needs to be deposited for subsequent quantum dot layers. This is because the wetting layer thickness progressively decreases. For example, in the case of $In_{0.4}Ga_{0.6}As/GaAs$ quantum dots, 7 monolayers (MLs) of InGaAs need to be deposited for the first dot layer and subsequent dot layers are formed with 3–5 MLs of InGaAs, depending on the GaAs barrier layer thickness.

Self-organized quantum dots can also be grown by other techniques of deposition during epitaxy. For example, slightly layer dots are obtained when a monolayer superlattice, or sub-monolayer superlattice of InAs/GaAs is grown.[21,35,36] The number of periods needs to be optimized, and can vary between 6–15, depending on the growth conditions. A quantum dot is formed by interdiffusion and intermixing of In and Ga. More importantly, because of the increase in the dot size, the photoluminescence emission at room temperature is observed at 1.3 μm, an important wavelength for lightwave networks.

3. Photoluminescence and Bandstructure

Heterostructures grown for PL measurements typically consist of a single layer of $In_xGa_{1-x}As$ ($0.4 \leq \times \leq 1$) with 0.1 μm GaAs on each side. A superlattice buffer is grown in all the samples, to smoothen the growth front, before growing the actual structures. Substrate temperatures used are $\sim 620°C$ and 520°C for GaAs and InGaAs, respectively, and the V/III flux ratio is maintained at ~ 5. The number of monolayers for the transition of the RHEED spectra from streaky (2D) to spotty (3D) pattern, where island growth commences, varies from 1.7 monolayers (InAs) to 11 monolayers ($In_{0.35}Ga_{0.65}As$). Typically 1.5–2 monolayers are grown to form the dot.

Photoluminescence spectra are measured at different temperatures using the standard technique. A 632.8 nm HeNe laser with a variable attenuator is used as an excitation source. A 0.75 m scanning spectrometer, liquid-nitrogen-cooled photomultiplier, and lock-in amplification are employed to scan the emission spectra with a resolution of better than 0.1 nm. Photoluminescence spectra from quantum dots are expected to be broader than those from quantum wells due to the statistical nature of size fluctuation of a large number of quantum dots sampled. The spectral shape is found to be approximately Gaussian for low excitation. At higher excitation levels, contributions from excited-state transitions give rise to a second peak due to the saturation behavior of the ground-state transition.[37] Figure 2 shows the PL spectra obtained at two different excitation levels. The energy separation between the two peaks of the $In_{0.35}Ga_{0.65}As$ sample is 62 meV. Moreover, there

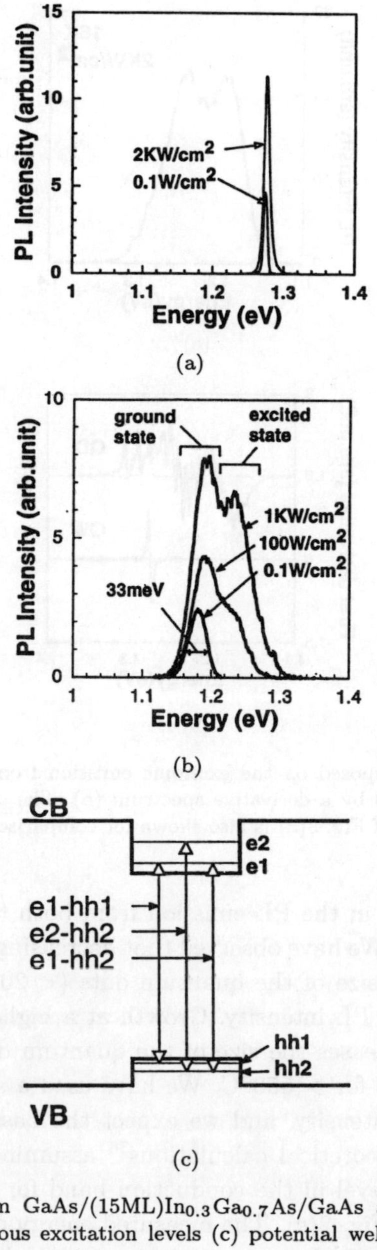

Fig. 2. PL spectra (a) from GaAs/(15ML)$In_{0.3}Ga_{0.7}As$/GaAs and (b) from GaAs/(15ML) $In_{0.35}Ga_{0.65}As$/GaAs at various excitation levels (c) potential well and different transitions in self-organized quantum dots.

are fine excitonic features superimposed on the emission peaks form QDs, which represent groups of dots with identical dimensions. This is clearly resolved by a derivative spectrum in Fig. 3. The overall linewidth of the ground-state emission of the QDs is 33 meV. This indicates the high degree of uniformity in the QD size. The

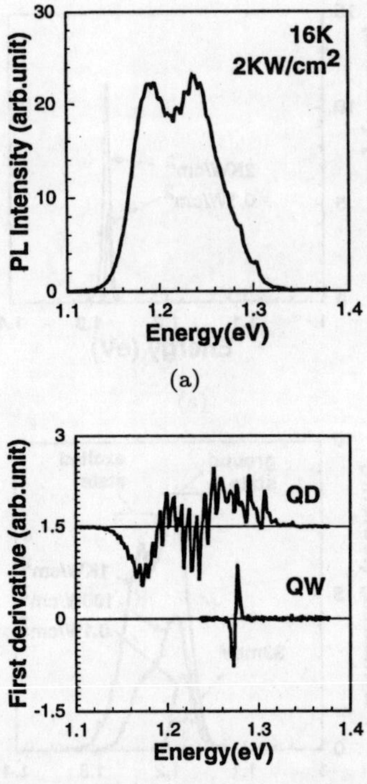

Fig. 3. Sharp peaks superimposed on the excitonic emission from quantum dots observed in PL spectrum (a) and resolved by a derivative spectrum (b). The derivative spectrum from the quantum-well emission peak of Fig. 3(a) is also shown for comparison.

fine structure is observed in the PL emission from both the ground-state and the excited-state transitions. We have observed that decreasing the growth temperature ($< 500°C$) decreases the size of the quantum dots (< 20 nm base width), at the same time decreasing the PL intensity. Growth at a higher temperature improves the PL intensity and increases the size of the quantum dots, ultimately resulting in a large size fluctuation for $> 550°C$. We have used a substrate temperature of 520°C for the best PL intensity, and we expect the base width of the quantum dots to be 20–30 nm. Theoretical calculations[38] assuming a biaxial strain tensor predict a second energy level in the conduction band for these dimensions. These transitions are shown in Fig. 2(c). The measured energy difference of 62 meV, between ground-state and excited-state transitions, is found to agree with the energy difference between e1-hh1 and e2-hh2 in QDs with a pyramidal base width of 21 nm and a height of 10 nm.

Figure 4(a) shows low-temperature photoluminescence (PL) spectra from $In_{0.4}Ga_{0.6}As$ quantum dots, the wetting layers and the GaAs barriers in a 6-layer QD sample. Figure 4(b) shows an interesting feature in the low temperature PL emission linewidth (FWHM) from $In_{0.4}Ga_{0.6}As$ single and multiple dot layer

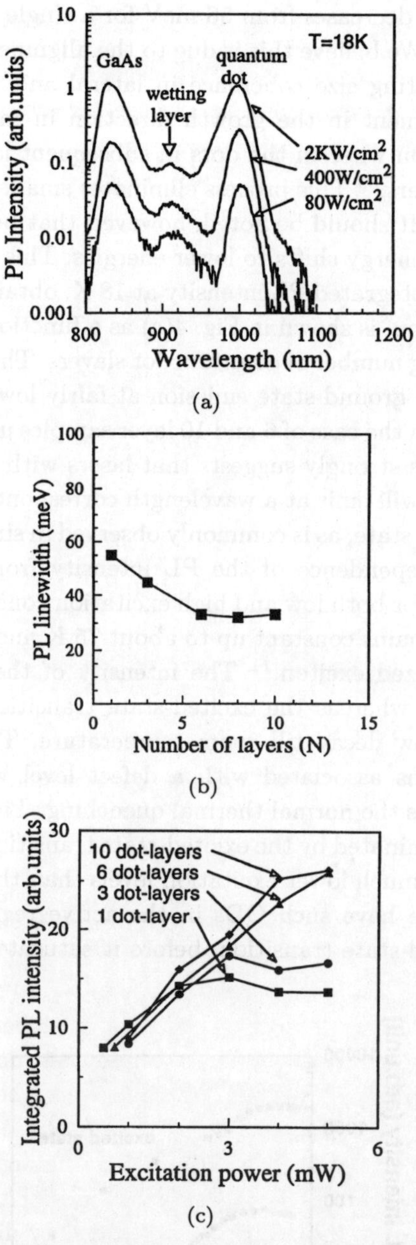

Fig. 4. 18 K photoluminescence (PL) spectra of $In_{0.4}Ga_{0.6}As$/GaAs quantum dots: (a) PL spectra showing emission from ground state, excited state and wetting layer of quantum dots as well as GaAs barrier and (b) ground state emission linewidth as a function of number of dot-layers. There is a significant improvement in the PL linewidth due to the improved size coherence in multilayer quantum dots, and (c) integrated PL intensity of ground state emission as a function of excitation intensity for various number of dot layers. In the case of single- and 3-layer dots saturation of ground state emission is observed while for 6- and 10-layer dots no saturation is observed up to an excitation intensity of 2 KW/cm^2.

samples. The linewidth decreases from 56 meV for a single dot layer to 33 meV for a six dot layer sample. We believe this is due to the alignment of dots in the growth direction and the resulting size coherence in lateral and vertical directions. The strain driven self-alignment in the growth direction in multilayer quantum dots leads to a filtering action wherein the dots in subsequent layers align to the larger dots in the previous layers.[39] This process eliminates smaller dots and improves the overall size uniformity. It should be noted, however, that because of this size filtering, the peak emission energy shifts to lower energies. This shift is 65 meV for a 10 dot layer sample. The integrated PL intensity at 18 K, obtained by a Gaussian fit to the ground state emission, is shown in Fig. 4(c) as a function of excitation intensity for samples with varying number of quantum dot slayers. The single dot layer sample shows saturation of the ground state emission at fairly low intensities, whereas no saturation is observed in the case of 6 and 10 layer samples up to an excitation intensity of 2 KW/cm^2. This strongly suggests that lasers with multiple layer quantum dots in the gain region will emit at a wavelength corresponding to the ground state and not from an excited state, as is commonly observed in single dot layer lasers.[22,23]

The temperature dependence of the PL intensity from $In_{0.35}Ga_{0.65}As/GaAs$ QDs is shown in Fig. 5 for both low and high excitation conditions. The wavelength-integrated intensity remains constant up to about 75 K and then decays due to the ionization of the localized exciton.[11] The intensity of the ground-state emission drops off dramatically, whereas the excited-state transition first decays at a fast rate, followed by a slow decay till room temperature. The fast decay indicates non-radiative transitions associated with a defect level while the slow decay of the excited state reflects the normal thermal quenching. The room-temperature PL spectrum (Fig. 6) is dominated by the excited-state transition since the ground-state transition saturates at much lower excitation levels than that at low temperatures. This implies that if we have such QDs in the active region of a laser, the gain provided by the ground-state transition, before it saturates, may not be sufficient

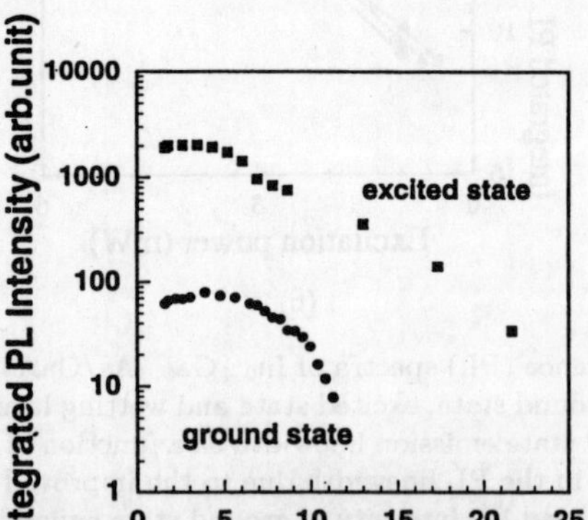

Fig. 5. Temperature variation of the wavelength-integrated intensity for the ground-state and excited-state transition.

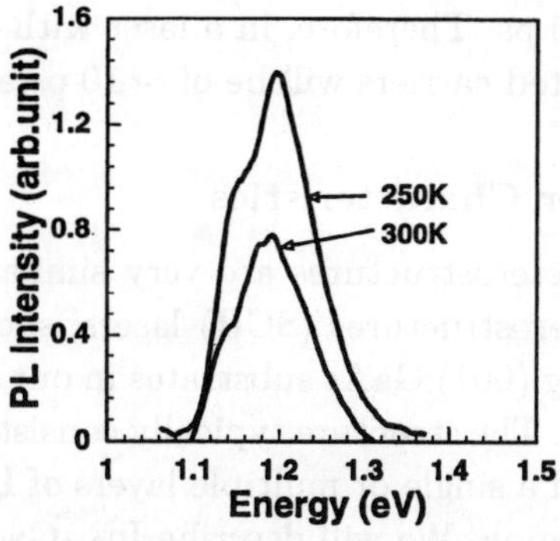

Fig. 6. PL spectra at room temperature (300 K) and at 250 K from $In_{0.35}Ga_{0.65}As/GaAs$ QD structure.

for the onset of lasing. We have made broad-area lasers from an SCH laser structure with $In_{0.4}Ga_{0.6}As$ QDs in the active region, which had a threshold current density of 650 A/cm^2 for 1 mm cavity length. The emission spectra from these lasers showed[12] that the ground-state transition saturates with increasing injection current and lasing indeed takes place from the excited state transition.

4. Carrier Dynamics in Quantum Dots

In order to study the carrier dynamics in these quantum dots, we have made low-temperature time-resolved PL measurements.[20] A Ti-sapphire laser was used as the excitation source and the spectrally dispersed time-resolved image was obtained with a 0.5 m monochromator, Hamamatsu streak camera, and a high-speed CCD camera. It is found that the decay time constants for the ground state and excited state are $\sim$ 700 ps and $\sim$ 200 ps, respectively. We believe that the slow decay of the ground state may be responsible for the saturation behavior observed in the time-integrated PL measurements. It is interesting to note that the excited state has an order of magnitude shorter decay time constant than the ground state. This predicts a better high-speed performance of the lasers emitting from excited states compared to that from ground states.

We have also performed low-temperature (10 K) femtosecond differential transmission spectroscopy (DTS) with a 4 $In_{0.35}Ga_{0.65}As/GaAs$ dot-layer sample using the pump-probe technique. The sample was essentially a laser heterostructure except that it was undoped. The GaAs barrier regions were excited with 800 nm pulses, and the population of the ground and excited state were measured using probe pulses of 925 nm and 980 nm, respectively. Other details of the measurement system and interpretation of results reported elsewhere;[40] only the salient features are reiterated here. From analysis of the data with rate equations, the relaxation times of electrons form the GaAs barrier to the excited and ground state are determined to be 2.8 and 8 ps respectively. Because the electrons and holes contribute equally to the DT spectrum, the effect of both are included. The hole relaxation time is estimated to be 0.5 ps. Therefore, in a laser with such an active region, the total capture time of injected carriers will be of $\sim$ 10 ps at cryogenic temperatures.

5. Edge-Emitting Laser Characteristics

The quantum dot laser heterostructures are very similar to quantum well lasers. Separate confinement heterostructures (SCH) lasers, shown in Fig. 7, were grown by MBE on semi-insulating (001) GaAs substrates in our studies. These devices can also be grown by MOCVD. The structure typically consists of an $Al_{0.3}Ga_{0.7}As$ outer clad, GaAs inner clad, and a single or multiple layers of $In_xGa_{1-x}As$ self-organized quantum dots as active region. We will describe $In_{0.4}Ga_{0.6}As$ QD lasers here. The nominal thickness of the $In_{0.4}Ga_{0.6}As$ layer is 8 ML. Most of the lasers studied by us have 4 dot layers. With a dot density of 5×10^{10} cm^{-2} determined by AFM, we have estimated the areal fill factor to be 0.1. The corresponding confinement factor for the SCH laser is estimated to be 10^{-3} per dot layer. Broad area lasers with dimensions of 50 μm by 1 mm are fabricated for material characterization. Room temperature lasing was observed at $\sim$ 1 μm with threshold current density of 650 A/cm^2 and 280 A/cm^2 for single- and 4-layer dot lasers, respectively. Three μm-wide ridge waveguide lasers of variable length were made by standard photolithographic, wet and dry etching, and metallization techniques. Typical light-current characteristics, measured under pulsed current conditions at room temperature, of single- and 4-layer dot lasers are shown in Fig. 8. The slope efficiencies are, respectively, 0.29 and 0.4 and the threshold currents are, respectively, 14.4 and 6.9 mA.

It is worthwhile to note that the single-layer dot laser emission is at a smaller wavelength (0.98 μm) compared to that of the 4-layer dot laser (1.033 μm). The lasing of single dot layer lasers form an excited state and that of multi-dot layer devices from a ground state transition have been observed before.[22,23,26] As in the PL data of Fig. 4(c), this behavior is related to the saturation of the ground states in single dot layer devices. The energy difference between the emission peaks of these

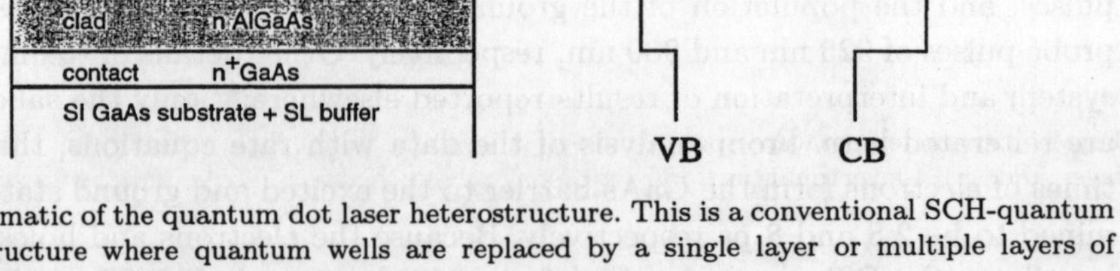

Fig. 7. Schematic of the quantum dot laser heterostructure. This is a conventional SCH-quantum well laser structure where quantum wells are replaced by a single layer or multiple layers of quantum dots.

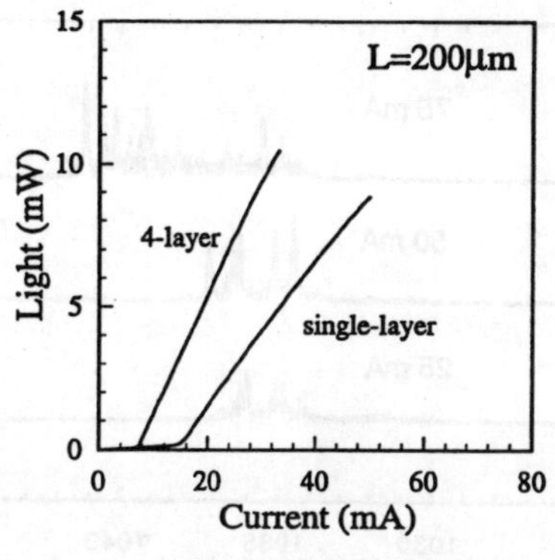

Fig. 8. Light-current characteristics of 200-μm cavity single and 4-layer quantum dot ridge waveguide lasers. The threshold currents are 14.4 and 6.9 mA, respectively.

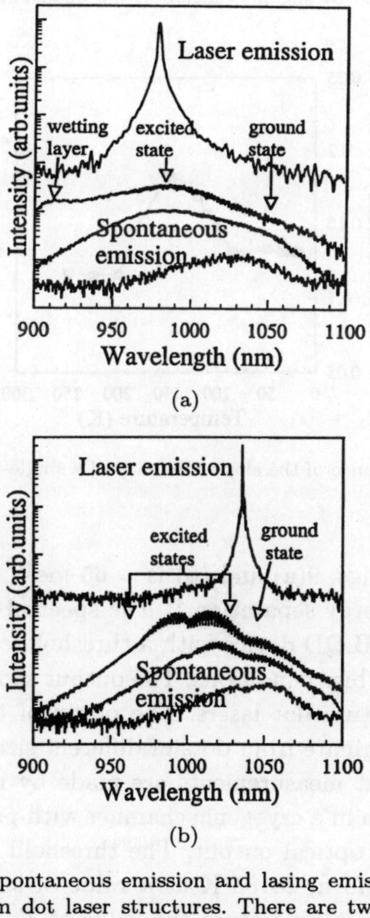

Fig. 9. Room temperature spontaneous emission and lasing emission spectra from: (a) single-layer and (b) 4-layer quantum dot laser structures. There are two broad peaks in the case of single-layer quantum dots, corresponding to the ground state and excited state emission, and the lasing occurs from the excited state. In four-layer quantum dots, at least three peaks are clearly observed and lasing occurs from ground state.

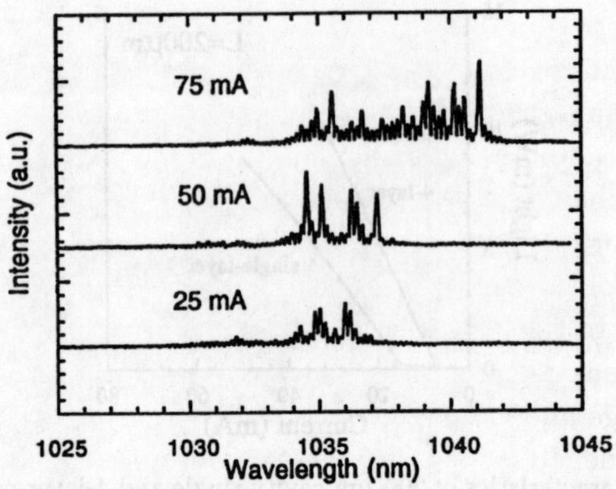

Fig. 10. Laser emission spectra at three different currents above threshold from 4-layer quantum dot ridge laser. A large number of longitudinal modes are observed due to the inhomogeneously broadened gain spectrum caused by size fluctuation of the quantum dots.

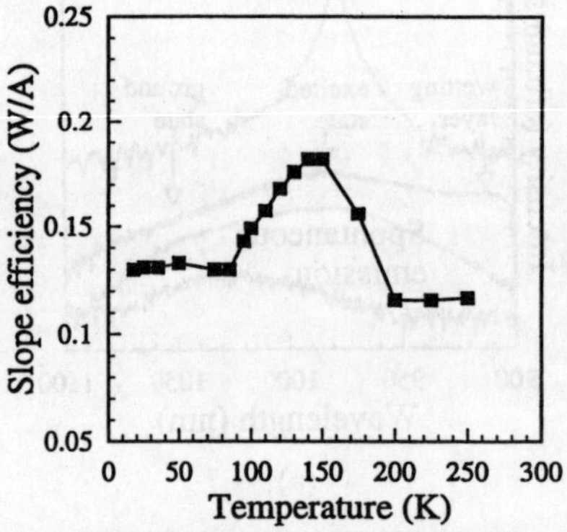

Fig. 11. Temperature dependence of the slope efficiency of a single-layer quantum dot broad area laser.

two different samples of Figs. 9(a) and (b) is $\sim$ 65 meV, close to the theoretically calculated value of the energy separation.[41] The spectral lasing characteristics are shown in Fig. 10, for an MLQD device with a threshold current of 20 mA at room temperature under pulsed bias conditions. The output shows multiple peaks, which are characteristic of quantum dot lasers. The origin of these peaks are not fully understood, but could originate from dots of different size.

Temperature-dependent measurements are made by mounting the lasers on a heat sink and placing them in a cryogenic chamber with provision for DC and high-frequency biasing and for optical output. The threshold current density is nearly temperature independent form 20–85 K at a value of 50 A/cm^2, with a value of $T_0 > 670$ K. Around room temperature the value of $T_0 = 80$ K in these devices. The temperature-independent behavior at low temperatures confirms atomic-like discrete states in the dots. The temperature-independent behavior can be extended to higher temperature ranges by using AlGaAs barrier layers. The slope efficiency of the single-dot layer device as a function of temperature is plotted in Fig. 11. The slope efficiency remains constant up to 85 K, beyond which it abruptly increases. In fact, the temperature for the change in slope coincides exactly with that at which the $J_{\rm th}$-vs.-T curve changes slope. At low temperatures the carriers are frozen in the dots in a non-equilibrium distribution. The increase in slope efficiency is attributed to the equilibration of the carrier distribution which results in an increasing number of dots contributing to the lasing press. This behavior saturates at 150 K, when the equilibration process is complete. Beyond this the slope efficiency decreases steadily due to carrier leakage and change in energy distribution of the carriers.

The optical frequency response was measured as a function of temperature using a photoreceiver, amplifier and network analyzer. The temperature dependent modulation characteristics of quantum dot lasers are markedly different from those of quantum well lasers. Figures 12(a), (b) and (c) show the measured modulation response of a quantum dot laser at various temperatures 90 K, 155 K, and 245 K, respectively. The maximum modulation bandwidth increases from 5–6 GHz at room temperature to 7 GHz at 245 K and 20 GHz at 90 K. This factor of four increase in the ultimate modulation bandwidth is accompanied by a dramatic change in shape of the curves, from a highly capture-time limited response at room temperature to a normal semiconductor laser response at cryogenic temperatures. We believe this is due to the temperature dependence of the carrier capture mechanism and the consequent temperature dependence of gain compression. The differential gain is extracted from the relationship between resonance frequency and net injection current, $I - I_{\rm th}$, and the internal conversion efficiency and optical confinement.[42] Values for internal conversion efficiency and optical confinement factor calculated

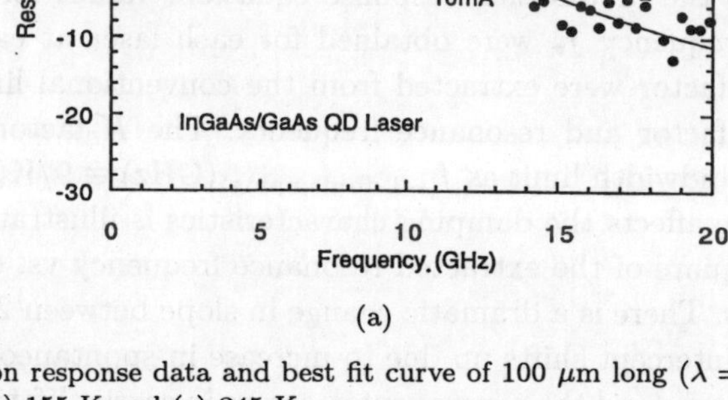

(a)

Fig. 12. Modulation response data and best fit curve of 100 μm long ($\lambda = 1$ μm) quantum dot laser at (a) 90 K, (b) 155 K and (c) 245 K.

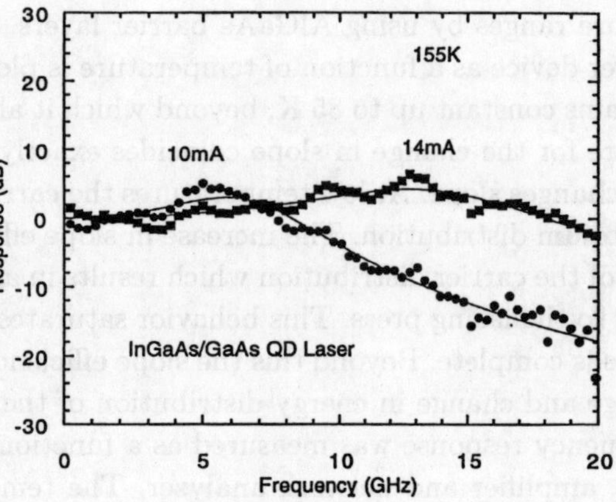

(b)

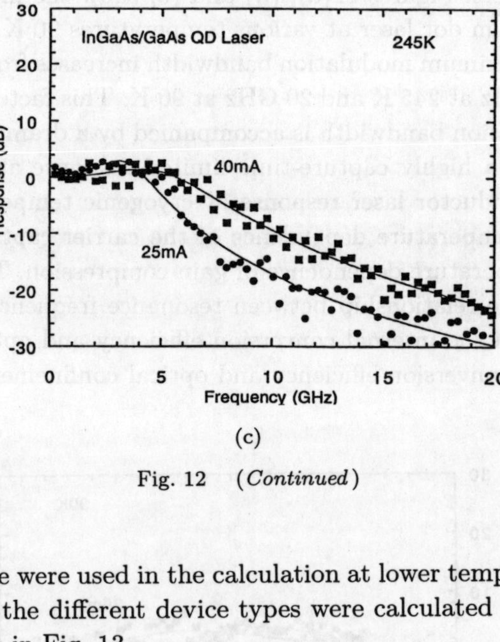

(c)

Fig. 12 (*Continued*)

at room temperature were used in the calculation at lower temperatures. Values for differential gain for the different device types were calculated at various temperatures and are shown in Fig. 13.

From the fit to the modulation response equation, values for damping factor γ and resonance frequency f_r were obtained for each laser at each bias current. Values for the K-factor were extracted from the conventional linear dependence between damping factor and resonance frequency. The K-factor determines the damping-limited bandwidth limit as $f_{(-3\ \text{db-damping})}(\text{GHz}) \equiv 9/\text{K(ns)}$. An example of how temperature affects the damping characteristics is illustrated in Fig. 14(a), which shows the square of the extracted resonance frequency vs. damping rate for quantum dot lasers. There is a dramatic change in slope between 245 K and 155 K. In addition, the y-intercept shifts up due to increase in spontaneous emission with decreasing temperature and the consequent decrease in carrier lifetime. Figure 14(b)

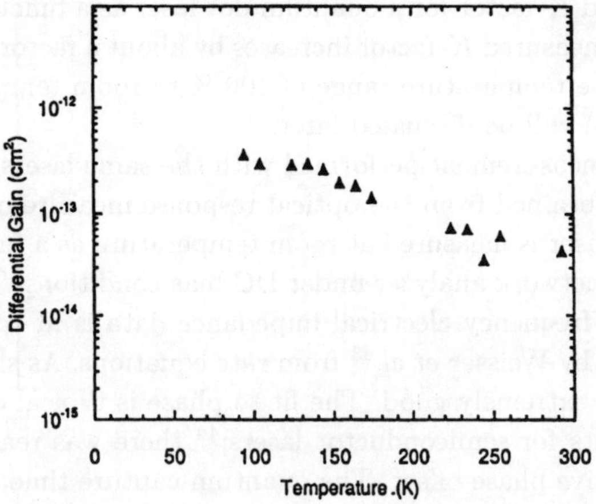

Fig. 13. Measured differential gain as a function of temperature of a 200 μm long ($\lambda = 1$ μm) quantum dot laser.

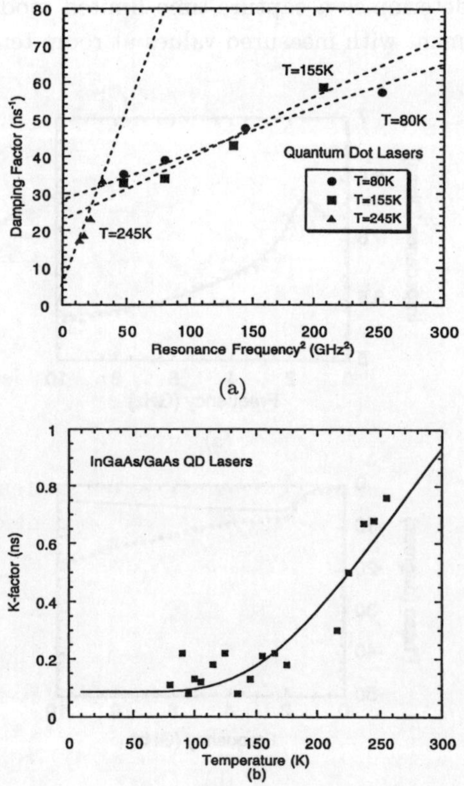

(b)

Fig. 14. (a) Measured damping factor vs. resonance frequency squared as a function of temperature for 200 μm long quantum dot laser; and (b) measured K-factor as a function of temperature for InGaAs/GaAs quantum dot lasers.

shows the measured K-factor for a quantum dot laser as a function of temperature. It is seen that the measured K-factor increases by about a factor of five for quantum dot devices over the temperature range of 100 K to room temperature. The curve shown in Fig. 14(b) will be discussed later.

An additional measurement performed with the same lasers, helps us to understand the results obtained from the optical response measurements. The electrical impedance of the laser is measured at room temperature as a function of frequency with the help of a network analyzer under DC bias conditions. The measured magnitude of the high-frequency electrical impedance data is fit to the expression for impedance derived by Weisser *et al.*[43] from rate equations. As shown in Fig. 15, the fit to magnitude is extremely good. The fit to phase is typical of what was seen in other impedance fits for semiconductor lasers;[44] there was reasonable agreement, with a linear negative phase offset. The quantum capture time of injected carriers, determined in this way, varies from 30–100 ps. These values are, of course, dependent on injection current, and typical values at high bias currents are quoted here. The capture time decreases with increase of injection current. The value of 30 ps for a MLQD device determines a capture-time limited modulation bandwidth of 5 GHz, in good agreement with measured values at room temperature.

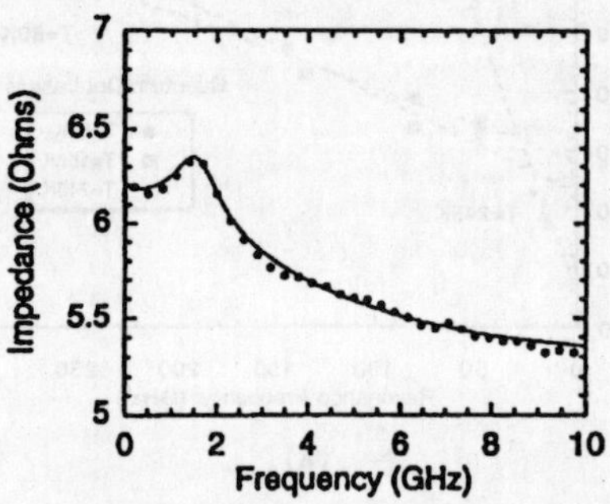

(a)

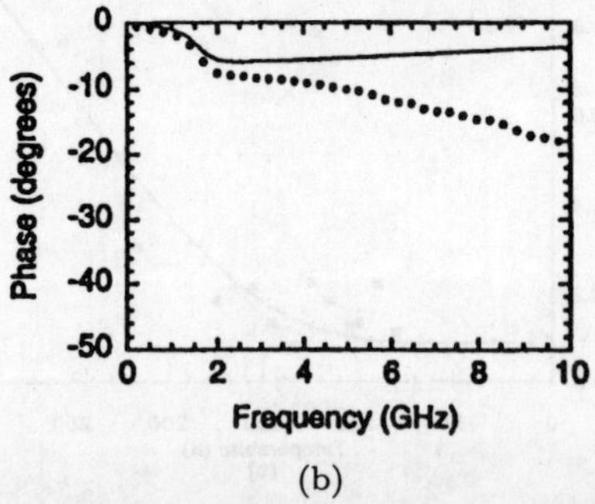

(b)

Fig. 15. (a) Magnitude and (b) phase of electrical impedance of a single-layer quantum dot ridge laser with threshold current of 20 mA at a bias of 38 mA measured at room temperature. The best fit to the expressions derived from rate equations is also shown. The carrier capture times determined in this way is $\sim$ 40 ps for single-layer devices and $\sim$ 30 ps for multilayer devices.

Some very interesting phenomena are apparent in comparing the temperature dependent dynamics in quantum dot devices. From the pump-probe measurements and from analysis of the small signal electrical impedance, the magnitude of the capture time at room temperature is about 30–100 ps, reducing to about 10 ps at cryogenic temperature. From the magnitude and temperature dependence of the capture time, the capture mechanism is inferred to be electron-hole scattering.[45] The graph of K-factor vs. temperature for quantum dot lasers, Fig. 14(b), quantitatively confirms that the mechanism for carrier capture is electron-hole scattering. From the model proposed in Ref. 46, the K-factor scales with the capture time.

The mechanism of electron-hole scattering is illustrated in Fig. 16. An excited electron transfer energy to a ground state hole, which then rapidly relaxes through the closely spaced hole states. The first step in this mechanism is the promotion of the hole to an excited energy level. Hence, it is reasonable to assume that the scattering time should depend inversely on the number of hole vacancies at that energy level. Assuming that the holes are thermally distributed and that the quasi-Fermi level for holes is the ground hole state, the measured K-factor as a function of temperature can be fit to an expression based on Fermi statistics,

$$\tau_c \propto \frac{1}{1+\exp(\delta E/kT)} \tag{1}$$

$$K(T) = a\tau_c(T) + b$$

for the number of vacancies at the energy separation δE between the excited and ground electron states, with a, b, and δE as fit parameters. The parameter δE is the energy separation between the excited and ground electron state. The extracted separation is 60 meV, and the line is the best fit curve. This should be compared to the value of 56 meV calculated using an 8-band k.p model[41] for $In_{0.4}Ga_{0.6}As/GaAs$

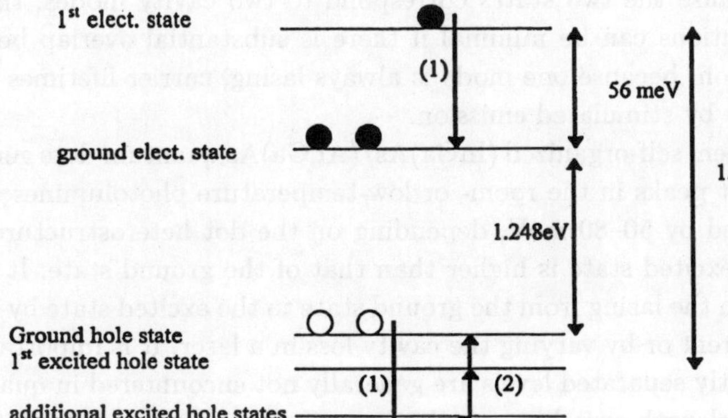

Fig. 16. The mechanism for electron-hole scattering, illustrated with the calculated energy levels of an $In_{0.4}Ga_{0.6}As$ quantum dot. An excited electron interacts with a ground state hole (1) which then rapidly relaxes through the closely spaced hole states (2).

dots. This agreement between calculated and experimental value is excellent corroboration for both the calculated energy level spacing and the determined, dominant mechanism for carrier capture being electron-hole scattering. The actual increase in K-factor (from 0.2 ns to 1 ns) is about the right order as the increase in capture time from 10 ps to 30 ps, and the rest of the difference could be explained by the moderate thermal dependence of the carrier population in the gain region.

We note the implications of Eq. (1) for device applications purposes. For example, it suggests that operating at a higher temperature will increase the capture time. Such an increase would be undesirable for a direct modulated interband semiconductor laser but might be ideal for an intersubband photodetector or laser. In fact, such an intersubband laser would more easily achieve inversion, and hence lasing, at *high* temperatures. It also suggests that bandstructure engineering which places the electron levels closer in energy may reduce the capture time at room temperature to the point where performances of quantum dot lasers at room temperature may approach these measured cryogenic performances.

6. Wavelength Switching and Bistability

There have been extensive studies of optical bistability in semiconductor lasers with intracavity saturable absorbers.[47,48] Such devices have been successfully used as memory elements in optical communication systems using time-division multiplexing. Because the bistability in these devices corresponds to the laser on and off states, however, the ultimate switching speed is limited by the spontaneous carrier lifetime in the active region, which is of the order of a few nanoseconds. One approach to improving the switching speed of bistable semiconductor lasers is to make use of two-mode bistability,[49] where the two states can correspond to two modes of the laser having different wavelengths, polarization, or directional orientation. Because the two states correspond to two cavity modes, the changes in carrier populations can be minimal if there is substantial overlap between the modes. In addition, because one mode is always lasing, carrier lifetimes are shortened appreciably by stimulated emission.

As we have seen, self-organized (InGa)As/(Al,Ga)As quantum dots generally exhibit two distinct peaks in the room- or low-temperature photoluminescence (PL) spectra, separated by 50–80 meV, depending on the dot heterostructure. The degeneracy of the excited state is higher than that of the ground state. It should be possible to switch the lasing from the ground state to the excited state by increasing the injection current or by varying the cavity loss in a laser. It is important to note that such distinctly separated levels are generally not encountered in quantum well lasers. Such wavelength switching can be investigated in a coupled-cavity system, made with quantum dot active regions, in which two sections are separately biased. By changing the bias across the absorbing region, the overall cavity loss can be made high enough so that the ground state gain can be made zero or negative and the excited state gain remains positive.

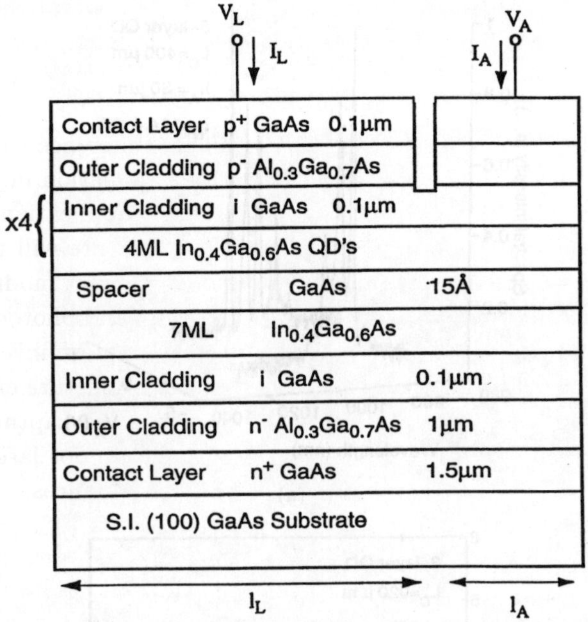

Fig. 17. Heterostructure for quantum dot laser and intracavity absorber grown by molecular beam epitaxy, where l_L and l_A are the laser and the absorber lengths, respectively I_L is the laser current, and V_A is the absorber applied voltage.

The laser and the intracavity saturable absorber (SA) device were fabricated with the SCH laser heterostructure in the ground-signal-ground (GSG) configuration by using standard photolithography and lift-off techniques and a combination of dry and wet etching. The device is shown schematically in Fig. 17. The gap between the gain and SA region is 10 μm in width and extends almost to the inner cladding layer just above the quantum dot active region. The laser and absorber single-mode waveguides are etched to almost the same depth. The deep etch in the gap ensures a resistance larger than 2 KΩ between the two sections. The width of the waveguide is 3 μm and the lengths L_C and L_A, as indicated in Fig. 17, vary from 400 to 1200 μm and 15 to 100 μm, respectively. The laser and SA end facets are not coated. The measured spectral switching characteristics are displayed in Fig. 18(a) for a coupled-cavity device with 5 quantum dot layers in the active region. The laser current is kept constant at 114.4 mA and when the SA voltage V_A is changed from -4 V to 2 V, the output switches from 0.992 μm to 1.008 μm. The shift corresponds to almost 20 meV.

It is observed that the energy difference of the lasing shift is $\sim$ 20 meV and not $\sim$ 50 meV. It is possible that the bound states of the dots are made up of contributions form the various dot sizes and switching occurs from the ground state of larger dots to those of smaller dots, and not from ground to excited states. Another possibility is that we see switching from the uppermost ground states to

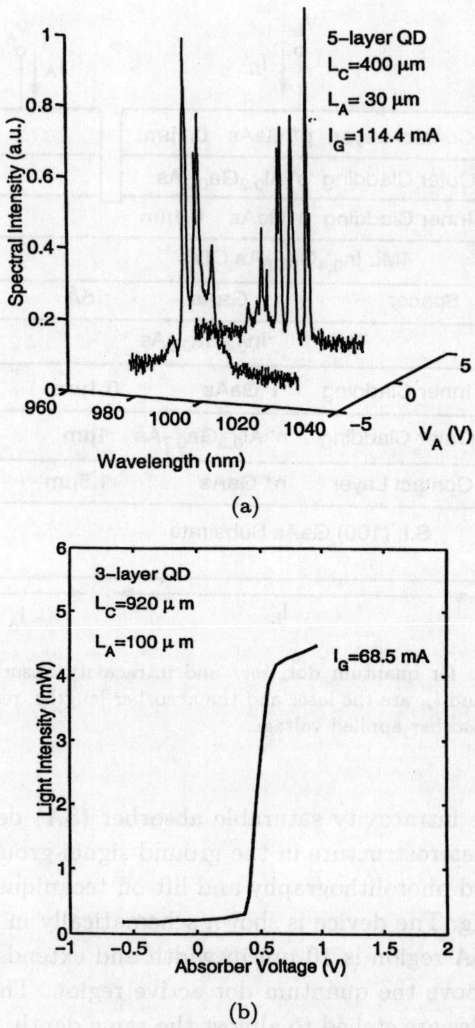

Fig. 18. (a) Lasing wavelengths at two different saturable absorber biasing voltages; (b) modulation of ground state laser emission with saturable absorber biasing voltage change.

the lowermost excited states of the dots, the spread in each subband being caused by the 15–25% size non-uniformity. It is also evident that the coupled-cavity device can be operated as an efficient light modulator, with emission from the ground state transitions. As the SA voltage decreases, the light output at $\sim 1\ \mu$m switches from high to low intensity for an increment of only 0.2 V in the SA voltage. This is illustrated in Fig. 18(b).

Measured light-current (L-I) characteristics of the coupled-cavity device, for different values of saturable absorber bias V_A, are shown in Fig. 19. The light exiting from the SA end is measured. There are two significant features in the data. First, the threshold current linearly increases from the lowest value of 18 mA (where

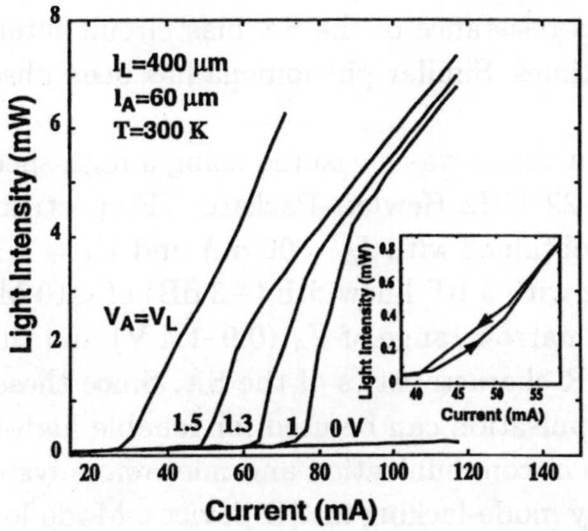

Fig. 19. Measured room-temperature light-current (L-I) characteristics of the device with l_L = 400 μm and l_A – 60 μm. The inset in the figure shows bistability in the L-I characteristics of a device having l_L = 400 μm, l_A – 50 μm and V_A = 1 V.

the gain and SA regions are shorted) as the SA bias is tuned to increase the cavity losses. Secondly, a distinct discontinuity is observed in the L-I characteristics at higher output powers. These two features are attributed, respectively, to non-saturable and saturable losses in the SA.[50] We have also observed bistability in the L-I characteristics, as shown in the expanded inset of Fig. 19, for specific lengths of the SA and voltage applied to it. The data shown in the inset of Fig. 19 are with a device having l_L = 400 μm, l_A = 50 μm and V_A = 1 V. The observed bistability can be understood in terms of the saturable absorber (photo) current-voltage characteristics, with excitation from the laser. A typical characteristic is shown in Fig. 20, which has the same form as the photocurrent-voltage characteristics of quantum well electroabsorption modulator, originating from the quantum confined Stark effect. Bistability in the L-I characteristics occurs when the load line corre-

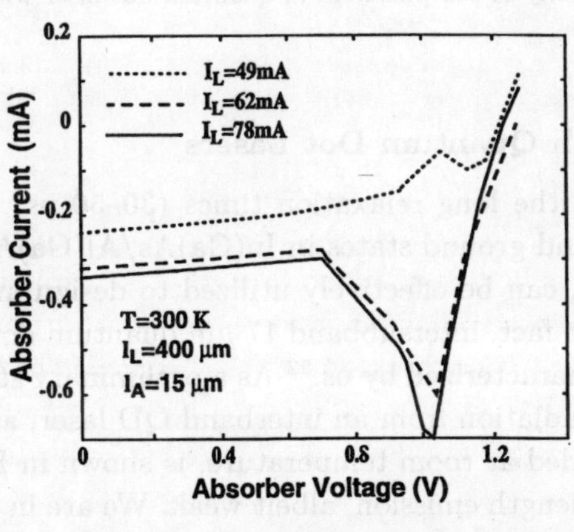

Fig. 20. Room-temperature voltage-current characteristics of the absorber for different values of laser bias current.

sponding to the series resistance of the SA bias circuit intersects the SA $I_A - V_A$ characteristics three times. Similar phenomena has been observed in quantum well devices.[51]

Self-pulsation phenomena was measured using a high-speed (40 GHz) Newfocus photodetector and a 22 GHz Hewlett-Packard RF spectrum analyzer. The trace shown in Fig. 21 was obtained with $I_L = 90$ mA and $V_A = -1$ V. The self-pulsation is centered at 1.6 GHz with a RF linewidth (−3 dB) of <10 MHz. Self-pulsation was obtained over a fairly narrow range of V_A (0.9–1.4 V) and this depends on the bias load line and the NDR characteristics of the SA. Since these can be controlled, to some extent, the self-pulsation can be used for reliable high-frequency sources. For potential applications in communication and microwave systems, it is of interest to explore high-frequency mode-locking in QD devices. Mode-locking can be achieved, without self-pulsation, at vary large intracavity photon density and low threshold with high reflectivity coatings and large absorber length.

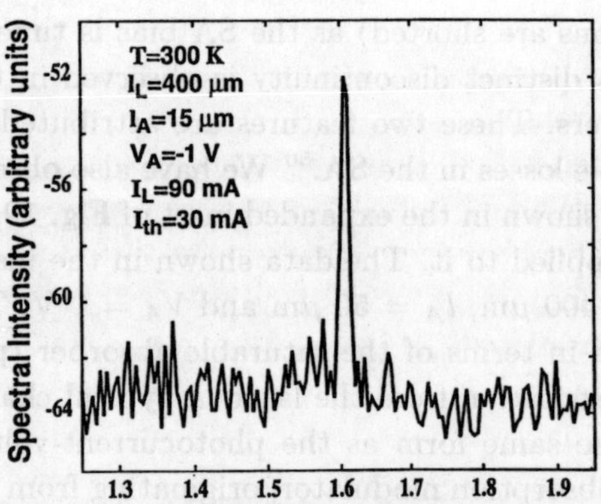

Fig. 21. RF spectral density of self-pulsation in quantum dot laser with intracavity absorber at room temperature.

7. Long Wavelength Quantum Dot Lasers

As mentioned earlier, the long relaxation times (30–50 ps) at room temperature between the excited and ground states in In(Ga)As/Al(Ga)As quantum dots, separated by 60–80 meV, can be effectively utilized to design and realize far-infrared intersubband lasers. In fact, intersubband 17 μm quantum dot photodetectors have been fabricated and characterized by us.[52] As a preliminary step, we have measured the long-wavelength radiation from an interband QD laser, as described in this review. The data, recorded at room temperature, is shown in Fig. 22. There is clear evidence of long-wavelength emission, albeit weak. We are in the process of designing an appropriate cavity for making lasers, utilizing the favorable intersubband carrier dynamics in quantum dots.

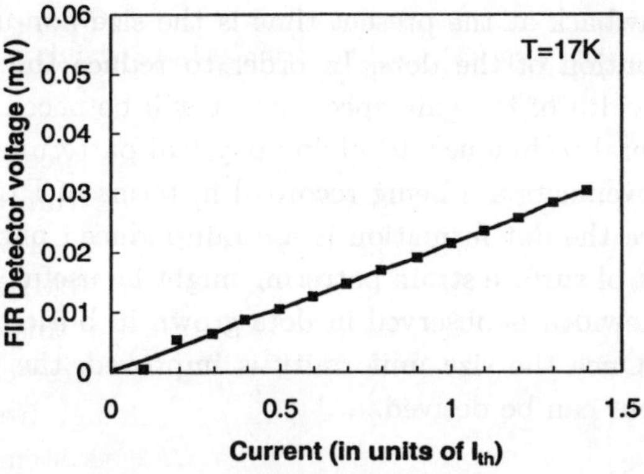

Fig. 22. Long-wavelength (~ 13 μm) spontaneous emission from electron intersubband transitions in $In_{0.4}Ga_{0.6}As/GaAs$ quantum dots at room temperature.

8. Future Prospects

An advantage in using quantum dots as the active (gain) medium of lasers is the tunability. Using conventional strained In(Ga,Al)As growth on GaAs, AlGaAs, or InP emission ranging from 600 nm[53] to ~ 2 μm[54] have been demonstrated. More importantly, as described in Sec. 2, by growing the quantum dots with monolayer or submonolayer InAs/GaAs deposition, peak emission at 1.3 μm is achieved at room temperature. Optical fibers have zero dispersion at this wavelength and hence lasers emitting at 1.3 μm are important for fiber optical communication systems. The first 1.3 μm quantum dot lasers have been reported.[21]

This review has mainly described growth and properties of edge-emitting quantum dot lasers. Quantum dot active regions have also been incorporated in VCSELs[28,54] and these devices, at 0.98 μm and 1.3 μm will also be important for lightwave network applications.

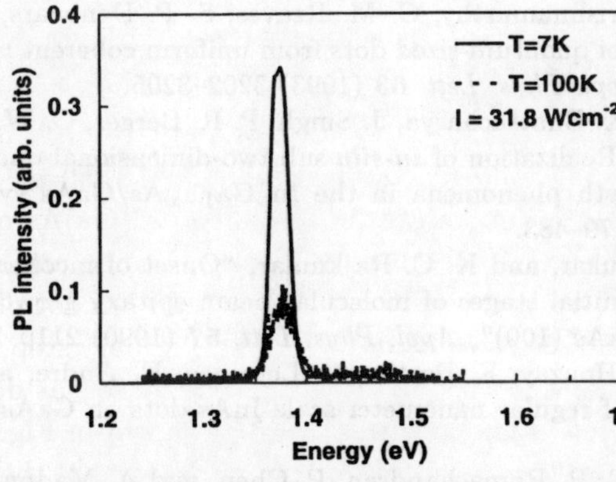

Fig. 23. Photoluminescence spectra of $In_{0.4}Ga_{0.6}As/GaAs$ quantum dots grown on InGaAlAs stressor dots. The measured linewidth (FWHM) is 21 meV.

The biggest drawback at the present time is the size non-uniformity and random spatial distribution of the dots. In order to reduce the photoluminescence linewidth and the width of the gain spectrum, it will be necessary to improve the size coherence. Several techniques, involving physical patterning are being investigated. While improvements are being recorded in terms of PL linewidth, the dot density is low. Since the dot formation is a strain-induced phenomenon, it would seem that some sort of surface strain pattering might be useful. In fact, substantial reduction of PL linewidth is observed in dots grown in buried "stressor" dots, as shown in Fig. 23. Once the size uniformity is improved, the full benefits of 3-d quantum confinement can be derived.

Acknowledgments

The author thanks K. Kamath, J. Phillips, D. Klotzkin, S. Krishna, O. Qasaimeh, W.-D. Zhou and D.-H. Zhu for their contributions. The work is being supported by the Army Research Office and the National Science Foundation.

References

1. Y. Arakawa and H. Sakaki, "Multidimensional quantum well lasers and temperature dependence of its threshold current", *Appl. Phys. Lett.* **40** (1982) 939–941.
2. Y. Arakawa and A. Yariv, "Theory of gain, modulation response, and spectral linewidth in AlGaAs quantum well lasers", *IEEE J. Quantum Electron.* **21** (1985) 1666–1674.
3. M. Asada, Y. Miyamoto, and Y. Suematsu, "Gain and the threshold of three-dimensional quantum-box lasers", *IEEE J. Quantum Electron.* **22** (1986) 1915–1921.
4. H. Benisty, C. M. Sotomayor-Torres, and C. Weisbuch, "Intrinsic mechanism for the poor luminescence properties of quantum-box systems", *Phys. Rev.* **B44** (1991) 10945–10948.
5. L. Goldstein, F. Glas, J. Y. Marzin, M. N. Charasse, and G. Leroux, "Growth by molecular beam epitaxy and characterization of InAs/GaAs strained-layer superlattices", *Appl. Phys. Lett.* **47** (1985) 1099–1101.
6. D. Leonard, M. Krishnamurthy, C. M. Reaves, S. P. Denbaars, and P. M. Petroff, "Direct formation of quantum-sized dots from uniform coherent islands of InGaAs on GaAs surfaces", *Appl. Phys. Lett.* **63** (1993) 3202–3205.
7. J. Pamulapati, P. K. Bhattacharya, J. Singh, P. R. Berger, C. W. Snyder, B. G. Orr, and R. L. Tober, "Realization of *in-situ* sub two-dimensional quantum structures by strained layer growth phenomena in the $In_xGa_{1-x}As/GaAs$ system", *J. Electron. Mater.* **25** (1996) 479–483.
8. S. Guha, A. Madhukar, and K. C. Rajkumar, "Onset of incoherency and defect introduction in the initial stages of molecular beam epitaxy growth of highly strained $In_xGa_{1-x}As$ on GaAs (100)", *Appl. Phys. Lett.* **57** (1990) 2110–2112.
9. J. M. Moison, F. Houzay, F. Barthe, L. Leprince, E. Andre, and O. Vatel, "Self-organized growth of regular nanometer-scale InAs dots on GaAs", *Appl. Phys. Lett.* **64** (1994) 196–198.
10. N. P. Kobayashi, T. R. Ramachandran, P. Chen, and A. Madhukar, "*In-situ* atomic force microscope studies of the InAs three-dimensional islands on GaAs (100)", *Appl. Phys. Lett.* **68** (1996) 3299–3301.

11. Q. Xie, P. Chen, A. Kalburge, T. R. Ramachandran, A. Nayfonov, A. Konkar, and A. Madhukar, "Realization of optically active strained InAs island quantum boxes on GaAs (100) via molecular beam epitaxy and the role of island induced strain fields", *J. Cryst. Growth* **150** (1995) 357–363.
12. J.-Y. Marzin, J.-M. Gerard, A. Izrael, and D. Barrier, "Photoluminescence of single InAs quantum dots obtained by self-organized growth on GaAs", *Phys. Rev. Lett.* **73** (1994) 716–719.
13. S. Fafard, D. Leonard, J. L. Merz, and P. M. Petroff, "Selective excitation of photoluminescence and the energy levels of ultrasmall InGaAs/GaAs quantum dots", *Appl. Phys. Lett.* **65** (1994) 1388–1390.
14. G. Wang, S. Fafard, D. Leonard, J. L. Merz, and P. M. Petroff, "Time-resolved optical characterization of InGaAs/GaAs quantum dots", *Appl. Phys. Lett.* **64** (1994) 2815–2817.
15. M. Grundmann, "InAs/GaAs quantum dots radiative recombination from zero-dimensional states", *Phys. Stat. Sol.* **188** (1995) 249–258.
16. D. I. Lubyshev, P. P. Gonzales-Borrero, E. Marega, E. Petitprez, N. Lascala, and P. Basmaji, "Exciton localization and temperature stability in self-organized InAs quantum dots", *Appl. Phys. Lett.* **68** (1996) 205–207.
17. R. Heitz, M. Grundmann, N. N. Ledenstov, L. Eckey, M. Veit, D. Bimberg, V. M. Ustinov, A. Egorov, A. E. Zhukov, P. S. Kop'ev, and Z. I. Alferov, "Multiphonon relaxation processes in self-organized InAs/GaAs quantum dots", *Appl. Phys. Lett.* **68** (1996) 361–363.
18. M. Grundmann, N. N. Ledenstov, O. Stier, D. Bimberg, V. M. Ustinov, P. S. Kop'ev, and Z. I. Alferov, "Excited states in self-organized InAs/GaAs quantum dots: Theory and experiment", *Appl. Phys. Lett.* **68** (1996) 979–981.
19. K. Kamath, P. Bhattacharya, and J. Phillips, "Room temperature luminescence from self-organized $In_xGa_{1-x}As/GaAs$ ($0.35 < x < 0.45$) quantum dots with high size uniformity", *J. Cryst. Growth* **175/176** (1997) 720–724.
20. K. Kamath, N. Chervela, K. K. Linder, T. Sosnowski, H.-T. Jiang, T. Norris, J. Singh, and P. Bhattacharya, "Photoluminescence and time-resolved photoluminescence characteristics of $In_xGa_{1-x}As/GaAs$ self-organized single- and multiple-layer quantum dot laser structures", *Appl. Phys. Lett.* **71** (1997) 927–929.
21. D. L. Huffaker, G. Park, Z. Zou, O. B. Shchekin, and D. G. Deppe, "1.3 μm room-temperature GaAs-based quantum-dot laser", *Appl. Phys. Lett.* **73** (1998) 2564–2566.
22. K. Kamath, P. Bhattacharya, T. Sosnowski, T. Norris, and J. Phillips, "Room temperature operation of $In_{0.4}Ga_{0.6}As/GaAs$ self-organized quantum dot lasers", *Electron. Lett.* **32** (1996) 1374–1375.
23. R. Mirin, A. Gossard, and J. Bowers, "Room temperature lasing from InGaAs quantum dots", *Electron. Lett.* **32** (1996) 1732–1734.
24. D. Bimberg, N. N. Ledenstov, O. Stier, D. Bimberg, V. M. Ustinov, P. S. Kop'ev, and Z. I. Alferov, "InAs-GaAs quantum pyramid lasers: *In-situ* growth, radiative lifetimes and polarization properties", *Jpn. J. Appl. Phys.* **35** (1996) 1311–1319.
25. H. Shoji, Y. Nakata, K. Mukai, Y. Sugiyama, M. Sugawara, N. Yokoyama, and H. Ishikawa, "Temperature dependent lasing characteristics of multi-stacked quantum dot lasers", *Appl. Phys. Lett.* **71** (1997) 193–195.
26. D. Bimberg, N. Kirstaedter, N. N. Ledenstov, Z. I. Alferov, P. S. Kop'ev, and V. M. Ustinov, "InGaAs-GaAs quantum dot lasers", *IEEE J. Select. Topics Quantum Electron.* **3** (1997) 196–205.
27. D. Bimberg, N. N. Ledenstov, M. Grundmann, and F. Heinrichsdorff, "Edge and surface-emitting quantum dot lasers", *IEDM Tech. Dig.* (1997) 381–384.

28. D. L. Huffaker, O. Baklenov, L. A. Graham, B. G. Streetman, and D. G. Deppe, "Quantum dot vertical-cavity surface emitting laser with a dielectric aperture", *Appl. Phys. Lett.* **70** (1997) 2356–2358.
29. I. N. Stranski and L. V. Krastanow, *Akad. Wiss. Lett. Mainz Math. Natur.* **146** (1939) 797.
30. P. R. Berger, K. Chang, P. Bhattacharya, J. Singh, and K. K. Bajaj, "Role of strain and growth conditions on the growth front profile of $In_xGa_{1-x}As$ on GaAs during the pseudomorphic growth regime", *Appl. Phys. Lett.* **53** (1988) 684–686.
31. V. A. Shchukin and N. N. Ledentsov *et al.* "Spontaneous ordering of arrays of coherent strained islands", *Phys. Rev. Lett.* **75** (1995) 2968–2971.
32. Q. Xie, A. Madhukar, P. Chen, and N. P. Kobayashi, "Vertically self-organized InAs quantum box islands on GaAs(001)", *Phys. Rev. Lett.* **75** (1995) 2542–2545.
33. G. S. Solomon, J. A. Trezza, A. F. Marshall, and J. S. Harris, Jr., "Vertically aligned and electronically coupled growth induced InAs islands in GaAs", *Phys. Rev. Lett.* **76** (1996) 952–955.
34. B. Lita, R. S. Goldman, J. D. Phillips, and P. Bhattacharya, "Nanometer-scale studies of vertical organization and evolution of stacked self-assembled InAs/GaAs quantum dots", '*Appl. Phys. Lett.* **74** (1999) 2824–2826.
35. K. Mukai, O. Nobuyuki, S. Mitsuru, and S. Yamazaki, "Self-formed $In_{0.5}Ga_{0.5}As$ quantum dots on GaAs substrates emitting at 1.3 μm", *Jpn. J. Appl. Phys.* **33** (1994) L1710–L1712.
36. R. Mirin, J. Ibbetson, K. Nishi, A. Gossard, and J. Bowers, "1.3 μm photoluminescence from InGaAs quantum dots on GaAs", *Appl. Phys. Lett.* **67** (1995) 3795–3797.
37. M. Grundmann, N. N. Ledentsov, O. Stier, D. Bimberg, V. M. Ustinov, P. S. Kop'ev, and Z. I. Alferov, "Excited states in self-organized InAs/GaAs quantum dots: Theory and experiment", *Appl. Phys. Lett.* **68** (1996) 979–981.
38. J. Singh and J. Hongtao, private communication.
39. J. Tersoff, C. Teichert, and M. G. Lagally, "Self-organization in growth of quantum dot superlattices", *Phys. Rev. Lett.* **76** (1996) 1675–1678.
40. T. S. Sosnowski, T. B. Norris, H. Jiang, J. Singh, K. Kamath, and P. Bhattacharya, "Rapid carrier relaxation in $In_{0.4}Ga_{0.6}As$/GaAs quantum dots characterized by differential transmission spectroscopy", *Phys. Rev.* **B57** (1998) R9423–R9426.
41. H.-T. Jiang and J. Singh, "Strain distribution and electronic spectra of InAs/GaAs self- assembled dots: An eight-band study", *Phys. Rev.* **B56** (1997) 4696–4701.
42. R. Nagarajan, M. Ishikawa, T. Fukushima, R. Geels, and J. Bowers, "High speed quantum well lasers and carrier transport effects", *IEEE J. Quantum Electron.* **28** (1992) 1990–2008.
43. S. Weisser, I. Esquivias, P. J. Tasker, J. D. Ralston, B. Romero, and J. Rosenzweig, "Impedence characteristics of quantum well lasers", *IEEE Photon. Technol. Lett.* **6** (1994) 1421–1423.
44. D. Klotzkin, K. Kamath, and P. Bhattacharya, "Quantum capture times at room temperature in high-speed $In_{0.4}Ga_{0.6}As$-GaAs self-organized quantum dot lasers", *IEEE Photon. Technol. Lett.* **9** (1997) 1301–1303.
45. I. Vurgaftman, Y. Lam, and J. Singh, "Carrier thermalization in sub-three-dimensional electronic systems: fundamental limits on modulation bandwidth in semiconductor lasers", *Phys. Rev.* **B50** (1994) 309–326.
46. W. Rideout, W. Sharfin, E. Koteles, M. Vassell, and B. Elman, "Well-barrier hole burning in quantum well lasers", *IEEE Photon. Technol. Lett.* **3** (1991) 784–786.
47. H. Kawaguchi, "Bistable operation of semiconductor lasers by optical injection", *Electron. Lett.* **17** (1981) 741–742.

48. S. Tarucha and H. Okamoto, "Voltage-controlled optical bistability associated with two-dimensional exciton in GaAs-AlGaAs multiple quantum well lasers", *Appl. Phys. Lett.* **49** (1986) 543–545.
49. W. T. Tsang, N. A. Olsson, and R. A. Logan, "Optoelectronic logic operations by cleaved-coupled-cavity semiconductor lasers", *IEEE J. Quantum Electron.* **19** (1983) 1621–1625.
50. J. O'Gorman, A. F. J. Levi, R. N. Nottenburg, T. Tanbun-Ek, and R. A. Logan, "Dynamic and static response of multielectrode lasers", *Appl. Phys. Lett.* **57** (1990) 968–973.
51. S. Lim, J. Hudgings, G. Li, W. Yuen, K. Lau, and C. Chang-Hasnain, "Self-pulsating and bistable VCSEL with controllable intracavity quantum-well saturable absorber", *Electron. Lett.* **33** (1997) 1708–1710.
52. J. Phillips, P. Bhattacharya, S. W. Kennerly, D. W. Beekman, and M. Dutta, "Self-assembled InAs-GaAs quantum-dot intersubband detectors", *IEEE J. Quantum Electron.* **35** (1999) 936–943.
53. R. Leon, S. Fafard, D. Leonard, J. L. Merz, and P. M. Petroff, "Visible luminescence from semiconductor quantum dots in large ensembles", *Appl. Phys. Lett.* **67** (1995) 521–523.
54. H. Saito, K. Nishi, I. Ogura, S. Sugou, and Y. Sugimoto, "Room-temperature lasing operation of a quantum-dot vertical-cavity surface-emitting laser", *Appl. Phys. Lett.* **69** (1996) 3140–3145.

International Journal of High Speed Electronics and Systems, Vol. 9, No. 4 (1998) 1109–1138

DEVICE CHARACTERISTICS OF LOW-THRESHOLD QUANTUM-DOT LASERS

ALEXEY E. ZHUKOV, VICTOR M. USTINOV, and ZHORES I. ALFEROV
A. F. Ioffe Physico-Technical Institute, Russian Academy of Sciences,
26 Politekhnicheskaya, St. Petersburg 194021, Russia

This paper addresses the issues associated with physics and technology of diode lasers based on self-organized quantum dots (QDs). Theoretically predicted advantages of a QD array as the active region of a semiconductor laser and basic principles of QD formation using self-organization phenomena are discussed. Special attention is paid to relationship between structural and electronic properties of QDs and laser characteristics. Recent achievements in controlling these parameters including the effects of vertical stacking, changing the matrix bandgap and the surface density of QDs are reviewed. The threshold and power characteristics of the state-of-the-art QD lasers are demonstrated.

1. Introduction

One of the main characteristics of an injection laser is the dependence of the optical gain on the current density. At a certain level of the total loss determined by the cavity length, mirror coating, and the internal loss, the threshold current density can be easily estimated using this dependence. This characteristic is strongly connected with the band structure of the active medium. The invention of a quantum-well laser[1] showed that the band structure can be effectively changed by artificial quantum-sized region resulting in the improvement in the device performance. When the freedom of electron motion is restricted by reducing the dimensionality of the active region from the bulk to the quantum well (QW), quantum wire, and, quantum dot (QD), the density of states (DOS) is strongly modified. In the ultimate case of a QD ensemble it represents a set of delta-function peaks centered at the atomic-like energy levels. This DOS redistribution leads to the accumulation of more and more states near the ground level. This means that in the lasing regime the most part of carriers will occupy the "working state" at the expense of the higher energy "parasitic states" of the continuum.

A quantum dot laser was proposed for the first time by Arakawa and Sakaki in 1982.[2] They indicated that the threshold current density, $J_{\rm th}$, of such a laser should be independent of temperature. This property is directly associated with the absence of the higher energy states that can be thermally populated. Arakawa and Sakaki originally proposed to create QD-like states by applying the magnetic filed to the QW laser. Up to date the most promising results have been achieved using spontaneous formation of three-dimensional islands at early stages of strained layer epitaxy. Since the pioneering work of Goldstein *et al.*,[3] arrays of self-organized QDs have been synthesized in a number of systems including InGaAs/GaAs,[4] InGaAs/AlGaAs,[5] InAlAs/AlGaAs,[6,7] and InP/InGaP[8,9] on GaAs substrates as well as InAs/InGaAs,[10] InAs/InAlAs[10,11] and InAs/InP[12] on InP, GaInP/GaP,[13] InAs/GaP,[14] and InP/GaP[14] on GaP.

The first stage of investigation mostly directed on the accumulation of the primary knowledge on structural and optical properties of self-organized quantum dots seems to be nearly finished now. The main direction of further research will lie in device applications of QDs and, first of all, lasers. Low-threshold injection lasers based on self-organized QDs have been already created and preliminary studied in many of the above material systems.[15–22]

This paper is organized as follows. In Sec. 2, we briefly review the main advantages theoretically predicted for the QD array as the active region of a semiconductor laser. In Sec. 3, the basic principles of QD formation using self-organization phenomena are discussed. In Sec. 4, we describe the relationship between the optical gain and the current density for a real QD laser taking into account the inhomogeneous broadening and the presence of the higher energy states. Section 5 summarizes recent results on injection lasers based on self-organized QDs emphasizing the methods to improve the threshold characteristics. Power characteristics of QD diode lasers are described in Sec. 6. Section 7 addresses the issues related to the long-wavelength QD lasers on a GaAs substrate. Section 8 is devoted to QD lasers on InP substrates.

2. Theoretical Advantages of Quantum Dots for Laser Applications

DOS of the ideal QD array consists of the ground-state delta-function-like level sufficiently separated in energy from the excited states. The relationship between the peak modal gain, g_{QD}, and the radiative recombination current, J_{QD}, for the ideal QD array can be written as[23]

$$g_{QD} = g^{sat}\frac{J_{QD} - J_{tr}}{J_{tr}}, \tag{1}$$

where J_{tr} is the transparency current density corresponding to the zero gain and g^{sat} is the saturated (maximal available) optical gain. J_{tr} can be easily estimated as:

$$J_{tr}^{QD} = \frac{eN_e}{2\tau}, \tag{2}$$

where e is the electron charge, τ is the radiative recombination lifetime, N_e is the number of the states per unit surface. For the ground level N_e is equal to the doubled surface density of the array, N_{QD}, taking into account the spin degeneracy. Each QD is filled by one exciton if the current density equals to J_{tr}^{QD}. The gain increases linearly with increasing the current density until $2J_{tr}^{QD}$, that corresponds to the maximal filling of the QD state by two excitons, and then saturates at the level of g^{sat}.

It should be noted that the equation for the transparency current density is free of any temperature-related terms. Assuming the sufficient energy separation between the ground and possible excited level(s), there are no additional states that can be thermally populated. Therefore temperature does not affect a QD laser. The characteristic temperature, T_0, describing temperature dependence of the threshold current density ($1/T_0 = d \ln j_{\text{th}}/dT$) is infinite. This is unlike the QW case where the transparency current density principally depends on temperature due to the thermal spreading of the Fermi function.

The carrier density in the active region of the ideal QD laser cannot exceed a certain value. It is determined by N_{QD} and can be less than the temperature-dependent population of QW states. Since N_{QD} of the order of 1–10 $\times 10^{10}$ cm^{-2} and τ of about 1–2 ns[24] are typical for self-organized QDs (Sec. 3), J_{tr} as low as 5 A/cm^2 can be expected. On the other hand, simple estimation for the QW laser shows that J_{tr}^{QW} is approximately 50 A/cm^2, in a good agreement with the best experimental results for InGaAs-based QW lasers[25] and about one order of magnitude higher than the J_{tr}^{QD}.

The maximal achievable gain, g^{sat}, is proportional to the amplitude of the (reduced) density of states, $\rho^{\max}$. For an ideal QD ensemble $\rho^{\max}$ is given by the homogeneous broadening, Δ, (Fig. 1(a)) and can be written as:

$$g^{sat} \propto \rho_{QD}^{\max} \approx \frac{N_e}{\Delta}\,. \tag{3}$$

Because the value of Δ can be at least 0.1 meV or less, and taking into account the typical value of the surface density of self-organized QDs, $\rho_{QD}^{\max}$ is expected to

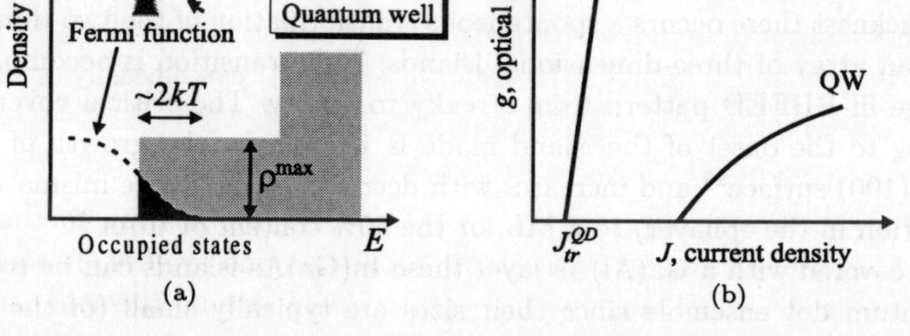

Fig. 1. Population of states (a) and gain versus current characteristics (b) for the laser active region based on QW and ideal QD array.

be higher than 1×10^{12} cm^{-2}meV^{-1}. At the same time the estimation for a QW case ($\rho_{QW}^{\max} = 2m/\tau\hbar^2$, m is reduced effective mass) gives the value of 5×10^{10} cm^{-2}meV^{-1}, about one order of magnitude lower than the corresponding figure for the QDs. The gain versus current characteristics for QD and QW lasers are schematically presented in Fig. 1(b). At any level of the total loss (for both long and short diodes) the ideal QD laser offers much lower (and temperature independent) threshold current density compared with its bulk and quantum well counterparts owing to a possibility to achieve low transparency current density, high differential and high saturated gain.

3. Self-Organization of Quantum Islands in Stranski–Krastanow Growth Mode

The most important issue of the QD laser technology is how to create an array of zero-dimensional objects in the active region. This array should provide a possibility of direct current injection, be free of non-radiative centers (including point defects and dislocations), have reasonable surface density and narrow size distribution. Each quantum dot of this array should be small enough to exhibit the effects of zero dimensionality. In addition, it is very important to increase as much as possible the energy gap between the ground-state level of QDs and the excited states and the matrix.

A number of different technological attempts have been proposed to satisfy these requirements. These approaches can be basically divided into two main parts. First is lithographic methods[26] whereas the second is based on using the self-organization phenomena.[3,27,28] The main advantages of the latter approach are its relative simplicity, principal possibility to avoid contamination and damage associated with the post-growth treatment, and the fabrication of device structure in one growth run.

The driving force for self-organization process is the lattice mismatch between the epitaxial layer and the substrate (buffer layer). For instance, when highly strained InAs layer is deposited on top of Ga(Al)As the difference in lattice constants is about 7%. Standard MBE equipment permits of a possibility to control the surface morphology by high energy electron diffraction (RHEED). At the first stage of growth InAs forms a pseudomorphic film. However, upon the deposition of some critical thickness there occurs a spontaneous transformation of the two-dimensional film into an array of three-dimensional islands. This transition is accompanied by the change in RHEED pattern from streaky to spotty. The critical coverage corresponding to the onset of the island mode is 1.7 ML for the growth of InAs on the GaAs(100) surface[29] and increases with decreasing the lattice mismatch (InAs mole fraction in the epilayer) to 4 ML for the 50% content of InAs.[30]

Being covered with a Ga(Al)As layer these In(Ga)As islands can be considered as a quantum dot ensemble since their sizes are typically small (of the order of 10 nm in the base length and 4–6 nm in the height) to demonstrate the effects of three-dimensional quantization. According to transmission electron microscopy

(TEM) studies[31] In(Ga)As QDs in Ga(Al)As matrix are pyramidally shaped with the typical surface density of about 5×10^{10} cm^{-2}. In-plane distribution of the islands is characterized by the short-range ordering in the form of primitive square lattice with [001] basic vectors. The islands lie on a residual two-dimensional film of In(Ga)As few monolayer thick — the wetting layer (WL). The theory of self-organization at the initial stages of strained-layer epitaxy was given by Schukin *et al.*[32] They pointed out that the island formation is governed by the internal energetic parameters of the system including surface energies, strain energy, etc., rather than the kinetically controlled factors. The size, shape and ordering can be understood in terms of partial (elastic) strain relaxation at top of the island. In addition, interaction of strained islands via the substrate should be taken into account.

Goldstein *et al.* observed these InAs islands in a GaAs matrix for the first time in 1985[3] when they studied a possibility to grow short-period (2 ML) InAs/GaAs superlattices. However, it took several years to find that these clusters can be coherently strained, i.e., free of dislocations[33] and, as a result, demonstrate strong luminescence.[4] It was found that the key point in the creation of defect-free InAs islands is the precise control of the effective thickness of In(Ga)As deposited, $Q_{\mathrm{In(Ga)As}}$.[34] The results obtained show that optically active islands can be formed within a very limited range of the coverage (from 1.7 to about 4 ML for InAs). Within this interval the QD luminescence covers the wavelengths from 1.05 to 1.24 μm (Fig. 2) reflecting the increase in QD size.[35] Note, that InGaAs/GaAs strained QW usually emit not longer than 1.1 μm because of the inherent limitation of the In composition and the QW width due to the plastic strain relaxation. PL spectra of the samples with different Q_{InAs} taken at room temperature are shown

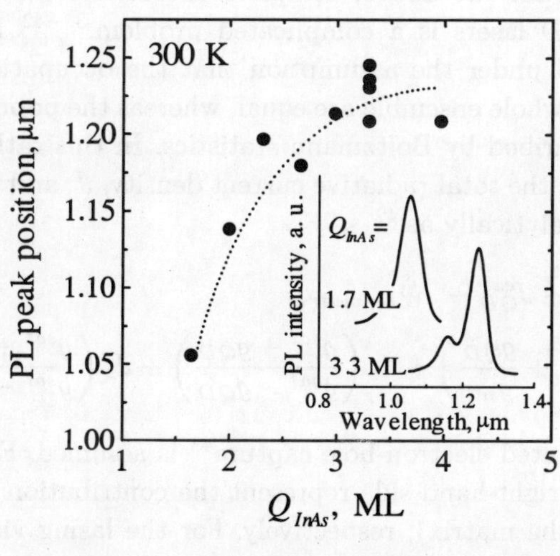

Fig. 2. PL peak position and spectra (inset) of InAs QD structures in GaAs matrix at different effective thickness of InAs, Q_{InAS}.

in the inset of Fig. 2. In addition to the intensive long-wavelength line originated from the ground state QD luminescence, the shorter wavelength line appears in samples with higher Q_{InAs} indicating that the island volume becomes large enough to localize the excited level. Typical linewidth of these PL lines is 40–60 meV reflecting the size fluctuations in the QD ensemble. PL intensity drops rapidly with increasing the effective thickness of InAs, Q_{InAs}, beyond 4–5 MLs as a result of the plastic strain relaxation in islands with the formation of misfit dislocations.

4. Optical Gain and Threshold Characteristics of Injection Lasers Based on Self-Organized Quantum Dots

The QD array is characterized by a certain surface density. Therefore, only a finite number of carriers can be localized in QDs and, consequently, contribute to lasing. This should lead to low values of the transparency current density. On the other hand, problem of the gain saturation can occur when the total cavity loss reaches g^{sat}, i.e., maximal gain achievable for the given QD array. The energy level of the QD ensemble formed by self-organization process is additionally inhomogeneously broadened due to size fluctuation of the islands. This proportionally decreases the value of g^{sat} according to Eq. (3). Thus, the gain saturation problem can be much more serious for real lasers based on self-organized QDs. In addition to the ground-state (GS) level, one or more excited-state (ES) levels can be localized in self-organized QDs.[36–38] These states can be thermally populated at temperatures of interest giving additional contribution to the threshold current density and its temperature dependence. Also, these excited levels have the higher degeneracy and, consequently, the higher saturated gain. Thus the transition from ground-state to excited-state lasing can be observed with increasing the loss.

Taking into account the above, complete theoretical description of the gain characteristics of QD lasers is a complicated problem.[39–42] However, it can be essentially simplified under the assumption that the occupation probabilities for each QD within the whole ensemble are equal, whereas the population of the higher energy states is described by Boltzmann statistics. In this rather typical case the relationship between the total radiative current density, J, and the peak gain, g_{QD}, can be expressed analytically as:[42]

$$\begin{aligned} J &= J_{QD} + J_{QD}^{ex} + J_{WL+matrix} \\ &= J_0\left(1 + \frac{g_{QD}}{g_{sat}}\right) + J_1\left(\frac{g^{sat} + g_{QD}}{g^{sat} - g_{QD}}\right) + J_2\left(\frac{g^{sat} + g_{QD}}{g^{sat} - g_{QD}}\right)^2 . \end{aligned} \tag{4}$$

In Eq. (4) the correlated electron-hole capture[43] is assumed. For the ground state lasing, the terms in right-hand side represent the contribution of GS, ES, and the wetting layer (and the matrix), respectively. For the lasing via the excited state, the second term should be excluded, whereas the first term acquire a sense of the collective contribution of both the GS and ES.

Under the condition of the low gain ($g_{QD} \ll g^{sat}$), Eq. (4) is simplified to:

$$g_{QD} = g^{sat} \frac{J - J_{tr}}{J_{tr}} \gamma \,, \tag{5}$$

$$J_{tr} = J_0 + J_1 + J_2 \,, \tag{6}$$

$$\gamma = \frac{J_{tr}}{J_{tr} + J_1 + 3J_2} \,. \tag{7}$$

Here J_{tr} is the transparency current density. Comparison with the linear gain behavior for an ideal QD array (Eq. (1)), shows that the existence of the upper states increases the transparency current density. The γ-factor evaluates the relative contribution of the higher energy states in the linear gain regime. Note that g^{sat} is inversely scaled with inhomogeneous broadening of QD.

Figure 3 shows the dependence of the peak modal gain and the lasing wavelength on the threshold current density measured experimentally[44] in lasers of different cavity length based on (In,Ga)As QDs in an AlGaAs/GaAs matrix (see Sec. 5.3). Two distinct regions are clearly seen in the g-J curve due to the switching from the ground state to the excited state lasing. In each region the gain saturates at certain level. It should be noted that the onset of the second region corresponds to the abrupt blue shift of the lasing wavelength. The size fluctuations mask to a degree this transition because the lasing wavelength and the gain can be also changed within the same inhomogeneously broadened level of QD ensemble.

Using Eq. (4) and properly chosen parameters J_0, J_1, J_2, g^{sat} one can fit with a good accuracy the gain versus current density characteristic for both GS and ES lasing, as shown in Fig. 3. It is interesting to note that the ratio of the saturated gain levels for the ES and GS lasing is very close to four. Assuming the same

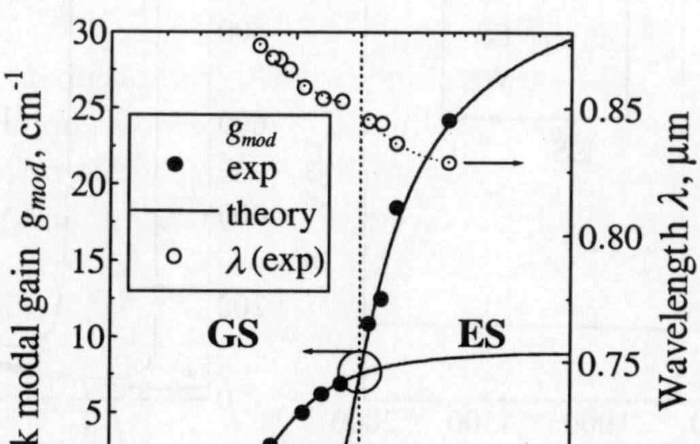

Fig. 3. Peak modal gain (experiment and theory, Eq. (4)) and lasing wavelength versus threshold current density for the lasers based on InGaAs/AlGaAs QDs. Fitting parameters are: $g^{sat} = 4.5$ (17) cm^{-1}; $J_0 = 19.8$ (80), $J_1 = 6.3$ (–), $J_2 = 0.2$ (11) A/cm^2 for GS (ES) lasing.

oscillator strength for both sates, this implies the four-fold degeneracy (in addition to the spin degeneracy) of the ES in a self-assembled QD in a good agreement with the simple hydrogen-like atomic model of electronic structure of a QD. Note that the splitting of the ES caused by non-Coulomb potential and non-spherical shape of QD is negligibly small when comparing with the inhomogeneous broadening due to QD size fluctuations.

Figure 4(a) shows the current components of the threshold current density associated with the radiative recombination in the different states (GS, ES, and the wetting layer). It is seen that the low threshold current is mainly determined by the GS recombination. This contribution saturates as $J_{\rm th}$ increases and the ES recombination becomes dominant. Simultaneously, the WL-related component is also enhanced. This component increases rapidly and dominates at the higher thresholds because the contribution of the ES recombination also saturates. This saturation level is higher as compared with the GS recombination due to the higher degeneracy. It should be pointed out that the lasing can still proceed via the QD ground state even if the threshold current is mainly governed by the recombination in the higher energy states. The similar situation can take place for the ES lasing.

N-fold change in the surface density of QDs results in the corresponding N-fold change in J_0, J_1, g^{sat}, whereas the effect of the wetting layer and matrix (J_2) is not scaled. Figure 4(b) shows the calculated dependence of the threshold current density on N_{QD} for different threshold gain. It is seen that this dependence is qualitatively the same as that for a QW laser. There exists some optimal QD density, N_{opt}, corresponding to the minimal $J_{\rm th}$ at the given threshold gain $G_{\rm th}$. If

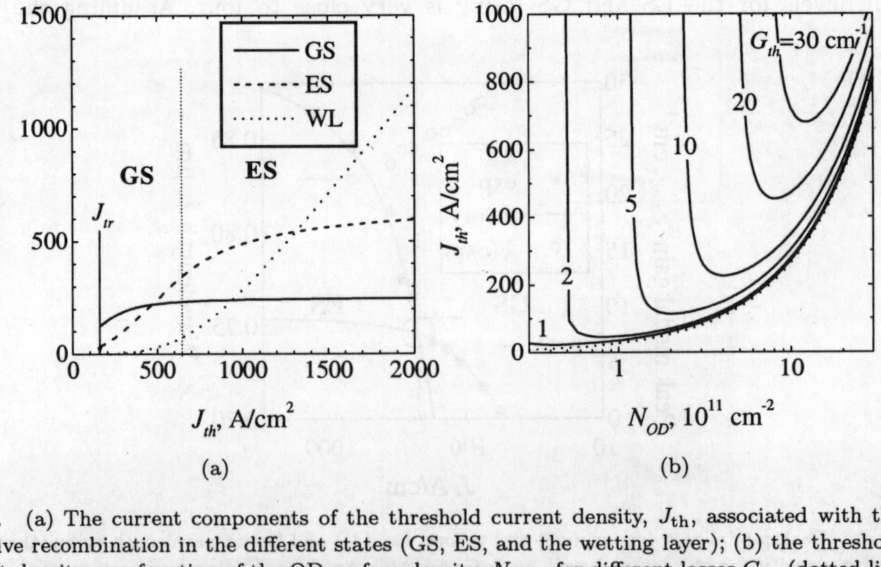

Fig. 4. (a) The current components of the threshold current density, $J_{\rm th}$, associated with the radiative recombination in the different states (GS, ES, and the wetting layer); (b) the threshold current density as a function of the QD surface density, N_{QD}, for different losses $G_{\rm th}$ (dotted line corresponds to the transparency current density).

the necessary gain is too high for the given QD surface density the lasing through the GS of QDs will be never achieved at any current density but the lasing may be possible via the excited state. For the most of device applications the threshold gain is relatively high because the shorter cavities are required for achieving the high differential efficiency. Therefore, denser arrays of QDs would be desired for practical applications.

5. Threshold Characteristics of QD Lasers

Laser oscillation from self-organized QDs was realized for the first time in 1994 by the team from Ioffe Institute, St. Petersburg, Russia in cooperation with Technical University of Berlin, Germany first for the optically pumped structure[45] and then for current-injected diode laser.[15,30] Schematic cross-section of the structure is presented in Fig. 5. One or several planes of QDs are formed in the middle of (Al)GaAs matrix layer (waveguide) of conventional GRIN SCH AlGaAs/GaAs laser grown by solid-source MBE on n-type GaAs(100) substrate. InAs or $In_{0.5}Ga_{0.5}As$ are usually used as the QD material. Conventional photolithography and chemical etching

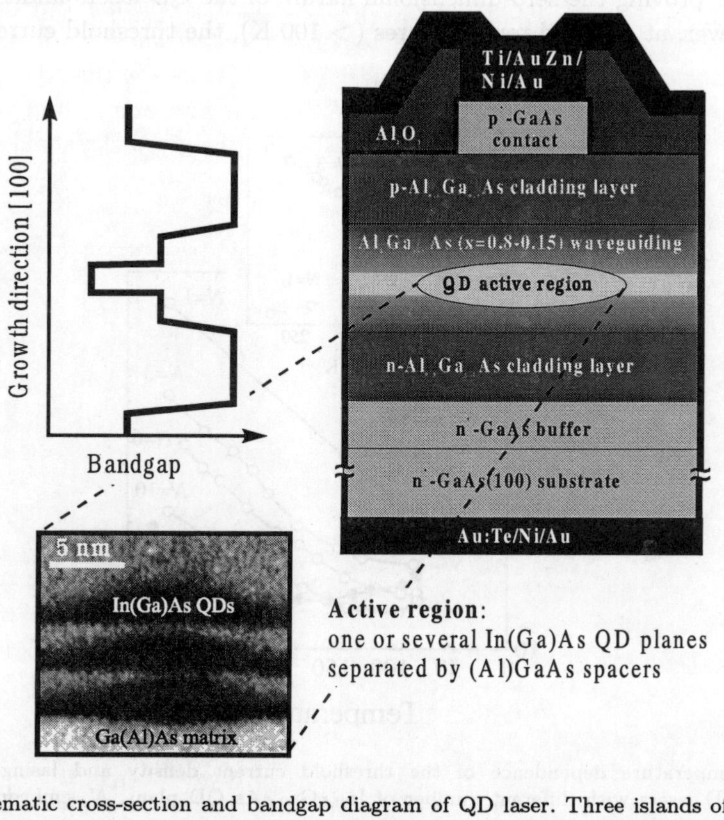

Fig. 5. Schematic cross-section and bandgap diagram of QD laser. Three islands of InAs with thin GaAs spacers located along one axis are clearly observed in TEM image (see inset).

technique were used to fabricate broad-area (100- or 200-μm-wide) devices. Laser chips of various length ($L = 0.5$–2 mm) with uncoated facets are tested under pulsed excitation (0.3–1 μs pulse duration, 5 kHz repetition frequency) in the wide temperature range. To minimize the mirror loss, lasers with four cleaved facets are also studied. The structure of this kind is nearly equivalent to the infinite cavity length because ring-type cavity modes are formed. To eliminate possible effect of filamentation, the threshold current density, J_{th}, was averaged over several devices with different dimensions (typically, 300×300 μm^2). No correlation between J_{th} and the diode size is observed. The power–current characteristics were measured in stripe devices mounted p-side down on a Cu heatsink by In solder under continuous-wave (CW) operation at room temperature.

Figure 6 shows the threshold current density and the lasing wavelength (inset) as functions of temperature for the lasers (four-cleaved facets) containing single plane of QD ($N = 1$). Relatively low and practically temperature independent threshold current density of 80 A/cm^2 was observed in the 30–100 K temperature range. In this temperature interval, the lasing wavelength follows the temperature dependence of the QD luminescence line. The characteristic temperature, T_0, is approximately 400 K that is much higher than the value calculated theoretically for a QW laser proving the zero-dimensional nature of the QD lasers under investigation. However, at elevated temperatures (> 100 K), the threshold current density

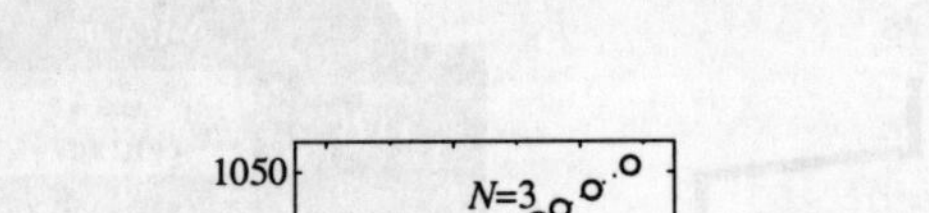

Fig. 6. Temperature dependence of the threshold current density and lasing wavelength (inset) of QD lasers with different number of $In_{0.5}Ga_{0.5}As$ QD plans, N, embedded in GaAs or $Al_{0.15}Ga_{0.85}As$ matrix.

demonstrates much stronger temperature dependence. J_{th} exceeds 1 kA/cm^2 at room temperature making it impossible to achieve CW operation. At the same time, the lasing wavelength becomes increasingly shorter and at room temperature nearly reaches the position of the wetting layer luminescence band.

Capacitance spectroscopy[46] and calculation[36] of electronic levels in QDs reveal that the reason of this behavior is the carrier evaporation out of the QDs to the wetting layer and the matrix. The energy separation between the ground levels of QD and the wetting layer in the range of several tens meV for the typical QD size that is well comparable to the thermal energy k_BT at room temperature.

The contribution of the higher energy states to the threshold current density is governed by the term $\exp((F - \Delta E)/k_BT)$, where $F(\Delta E)$ is the energy separation between QD lasing level and the Fermi level (the higher states, respectively). To prevent carrier evaporation, one should make the QD level deeper with respect to the band edges of the matrix and the wetting layer. In the laser diode, the Fermi level goes up to increase the population of the active region and, consequently, to achieve the necessary optical gain matching the total cavity loss. It is also important to keep the Fermi level as low as possible. This problem is closely connected with the insufficient gain in a QD laser and, as it will be shown below, can be partly overcome by increasing the surface density of the QD array. The further optimization of device characteristics of QD lasers went along the direction of creation denser and deeper localized arrays of QDs.

5.1. *Effect of vertical stacking*

To increase the density of the working states of a QD array, it was proposed to repeat QD planes for several times separating them by spacers of the matrix material.[47,48] The spacer thickness, d_{sp}, cannot be chosen arbitrarily high. To arrange all the QD planes near the maximum of the optical mode in the SCH region, d_{sp} has to be about 10 nm or less. It was found that during the successive deposition of several In(Ga)As QD sheets with relatively thin (Al)GaAs spacers, the islands of a next sheet are formed just above the dots of the preceding one. The cross section TEM image of a three-times stacked QD structure ($N = 3$) is given in the inset of Fig. 5. The reason for such vertical coupling is the preferable migration of the In(Ga) adatoms to (from) the QDs of the first plane caused by the strain fields.[48,49] Since the strain energy progressively accumulates during the island stacking there is a risk of plastic relaxation. Thus, the effective thickness of InAs deposited should be taken into account when choosing the spacer thickness.

Increasing the number of QD planes, N, in the active region remarkably decreases the room temperature J_{th} from 950 A/cm^2 ($N = 1$) to 98 A/cm^2 ($N = 10$)[50] as indicated in Fig. 6. It is worth mentioning that low-temperature J_{th} also decreases with N reaching the value as low as 18 A/cm^2 (120 K, $N = 10$). This behavior is directly connected with the increase in the number of QD states and, in the turn, the gain in the ground state of QD array with stacking. The temperature dependence of the lasing wavelength in the multi-stacked QD laser is essentially different from that of the single-QD laser (see Fig. 6, inset): the lasing wavelength follows the temperature dependence of the GS line in the whole temperature range. Thus, lasing proceeds via the ground state of the quantum dots up to room temperature. However, increasing N does not significantly change the range of high temperature stability of J_{th} as the localization energy is increased only slightly. The increase in the gain in the stacked QD ensemble is most indicative when comparing the dependence of the threshold current density on the cavity length (Fig. 7). Steep superlinear increase in J_{th} caused by gain saturation is observed with decreasing the cavity length. However, the higher the number of QD planes, the less pronounced is the effect of gain saturation.[51]

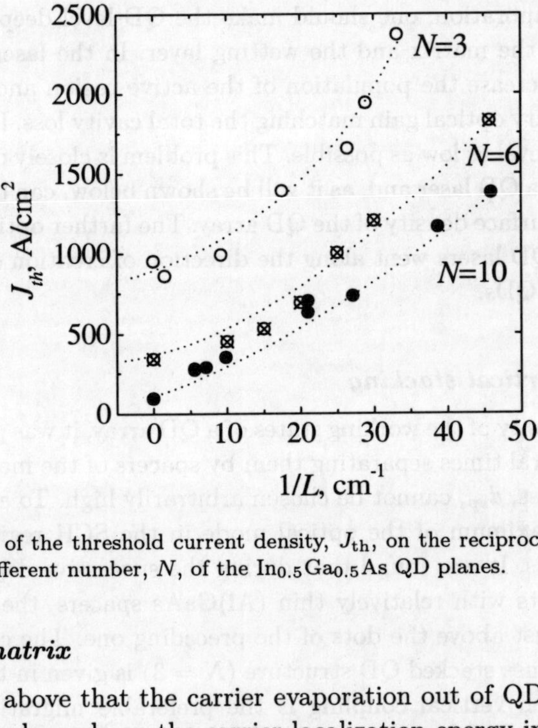

Fig. 7. Dependence of the threshold current density, J_{th}, on the reciprocal cavity length, $1/L$, for lasers with the different number, N, of the $In_{0.5}Ga_{0.5}As$ QD planes.

5.2. *Effect of matrix*

It was mentioned above that the carrier evaporation out of QDs at elevated temperatures strongly depends on the carrier localization energy in QDs, ΔE. It can be effectively increased by using AlGaAs instead of GaAs as a matrix material.[20] The formation of QDs in the $Al_XGa_{1-X}As$ matrix is very similar to the GaAs case in the critical thickness for the island growth mode and structural properties of a QD array. However, electronic properties are strongly modified. PL study revealed that the increase in the band gap energy of the matrix by increasing x affects slightly the PL peak position of QD line (by 120 meV for $x = 0.3$), whereas the blue shift of the WL line is much stronger (300 meV). This effect is very similar to the well-known behavior of energy levels in a quantum well with increasing barrier height. The lower localization energy and, consequently, larger penetration of carrier wave functions from the wetting layer into the barriers as compared to that of the quantum dot leads to stronger perturbation of energy levels of the wetting layer. In the turn, this should lead to lower thermal population of the upper states at elevated temperatures thereby reducing the threshold current density (see Fig. 6). The threshold current density for the structure based on three QD planes in $Al_{0.15}Ga_{0.85}As$ matrix was as low as 63 (300 K) and 18 A/cm^2 (150 K). It should be mentioned that this value of J_{th} at room temperature is approximately six times less than the corresponding value obtained for the case of GaAs matrix ($N = 3$).

The threshold current density decreases with temperature within the 77–150 K range (Fig. 6), i.e., these low-threshold QD lasers have a negative characteristic temperature. This unusual behavior is associated with the lack of an equilibrium dot population at low temperatures,[19] since the thermal evaporation of carriers out of QDs is strongly suppressed at low temperatures due to its phonon-induced nature. There exists some boundary temperature, which separates the regions of the equilibrium and non-equilibrium populations of QDs.[39] Under the latter condition the same gain is achieved at the higher current density, because the population of QD states is controlled by the capture probability rather than the Fermi distribution. Therefore J_{th} decreases when temperature passes through the boundary temperature. This boundary temperature increases with increasing the localization energy. Thus, the region of the negative characteristic temperature is more pronounced in the case of an AlGaAs matrix than in that of GaAs ($N = 10$, Fig. 6).

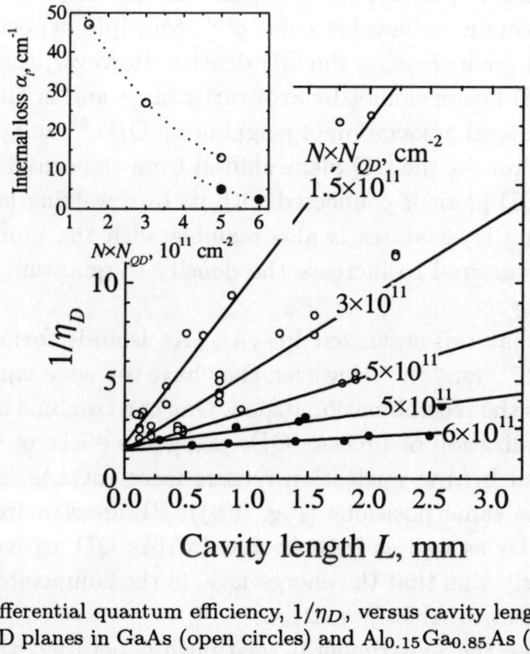

Fig. 8. Reciprocal differential quantum efficiency, $1/\eta_D$, versus cavity length, L, for lasers based on N-times stacked QD planes in GaAs (open circles) and $Al_{0.15}Ga_{0.85}As$ (solid circles) matrices. Internal loss, α_i, as a function of the total surface density of QDs in the active region, $N \times N_{QD}$, is shown in the inset.

Lasers based on multiply stacked QDs offers much better efficiency characteristics.[51] Figure 8 shows the dependence of the differential quantum efficiency, η_D, on the cavity length, L, for the lasers with the different total surface density of QDs in the active region (product of the number of QD planes, N, and the QD density in each plane, N_{QD}). The differential efficiency increases with $N \times N_{QD}$, the internal differential efficiency being practically unchanged in between 60–70%. The improvement in η_D is mainly due to the dramatic decrease in the internal loss, α_i, with increasing the total surface density of QDs in the active region (inset of Fig. 8). These results imply that self-assembled islands themselves do not contribute significantly to the scattering loss in the waveguide region and α_i is mainly associated with the free carrier absorption in the matrix and the wetting layer. When the QD density increases the same gain is achieved at the lower position of the Fermi level. Thus, the carrier concentration in the higher energy states is reduced. For the same reason the replacement of GaAs to AlGaAs as the matrix material (also in combination with denser array of QDs, see Sec. 5.3) leads to further reduction in the internal loss to 2.5 cm^{-1} as indicated in Fig. 8.

5.3. *Effect of quantum dot surface density*

The QD surface density, N_{QD}, is one of the key factors determining the threshold characteristics of QD lasers (see Sec. 4). The relatively low N_{QD} of 5×10^{10} cm^{-2} is characteristic of In(Ga)As QD array in an (Al)GaAs matrix. The higher values of N_{QD} would make it possible to overcome the influence of gain saturation by increasing the maximum achievable gain, g^{sat}. Multiple deposition of several QD planes can be used for increasing the QD density. However, the number of successively deposited QD layers cannot be arbitrarily large and is limited by the plastic strain relaxation, lateral association of neighboring QDs,[52] and decreasing the optical confinement factor for the QD plane shifted from the center of the SCH region. In addition, each QD plane is connected with its own wetting layer. Therefore, the effect of the wetting layer states is also scalable with the number of QD planes. Thus it is strongly desired to increase the density of quantum dots in each plane rather than stacking.

It was found that self-organized $In_{0.5}Al_{0.5}As$ islands form essentially denser array ($N_{QD} \sim 2 \times 10^{11}$ cm^{-2}).[53] However, they have too wide bandgap (Fig. 9(a)) to be solely applied as the recombination region. One can combine high N_{QD} of InAlAs QDs and deep localization of InGaAs QDs using the effect of vertical coupling.[54] These pre-deposited InAlAs nucleation centers force InGaAs of the next plane to form islands at the same positions (Fig. 9(b)). PL spectra from such composite InGaAs/InAlAs QDs as well as InGaAs and InAlAs QD arrays are also shown in Fig. 9(c). It is clearly seen that the energy level in the composite QDs is sufficiently localized with respect to AlGaAs matrix.

Figure 10 shows the experimental relationship between the modal gain and the threshold current density for lasers based on conventional InGaAs QDs and

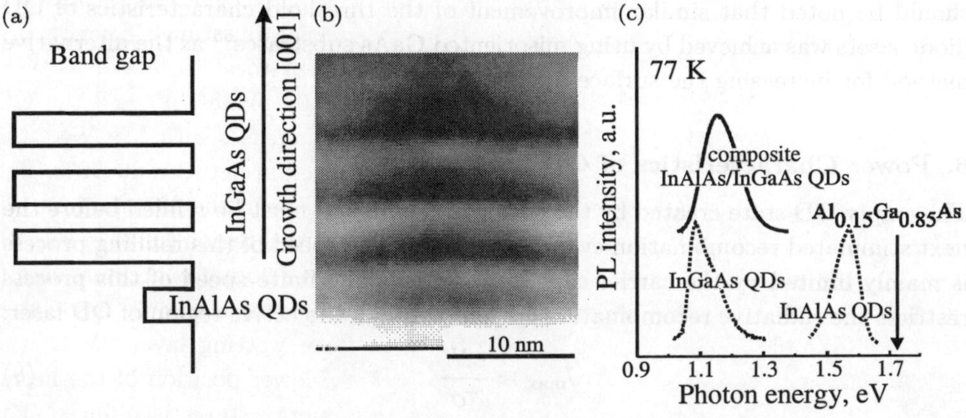

Fig. 9. Schematic bandgap diagram (a), cross-section TEM image (b) and PL spectrum (c) of composite $In_{0.5}Ga_{0.5}As/In_{0.5}Al_{0.5}As$ QDs in $Al_{0.15}Ga_{0.85}As$ matrix. Pre-deposition of InAlAs QD array results in a weak blue-shift indicating decrease in the size of InGaAs islands (redistribution of the same amount of the material deposited to denser array leads to smaller size of the islands).

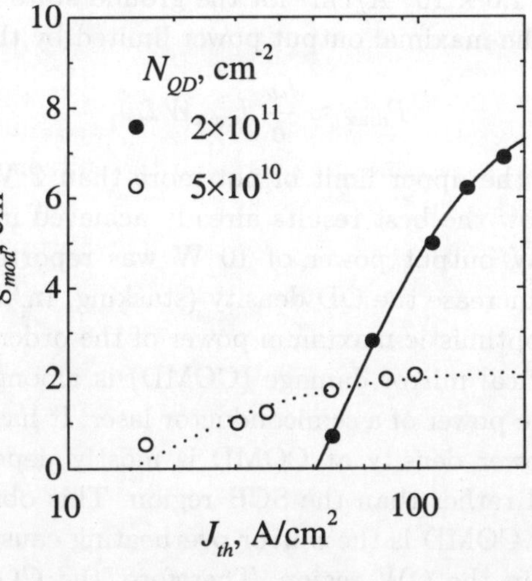

Fig. 10. Relationship between the threshold current density, J_{th}, and the peak modal gain, g_{mod} (both are normalized per number of QD planes) for QD lasers with different surface density of QDs, N_{QD}.

composite InGaAs/InAlAs QDs in $Al_{0.15}Ga_{0.85}As$ matrix ($N_{QD} = 5 \times 10^{10}$ and 2×10^{11} cm^{-2} per QD plane, respectively).[44] When comparing these characteristics, one can see that, under low loss condition (longer stripes), the diluted array offers lower threshold current due to lower transparency current density. On the other hand, in the most applicable regime of high loss (shorter stripes) the denser array of QDs is more preferable for laser application owing to higher saturated gain. It should be noted that similar improvement of the threshold characteristics of QD diode lasers was achieved by using misoriented GaAs substrates[55] as the alternative method for increasing the surface density of QDs.

6. Power Characteristics of QD Lasers

The empty QD state created by the stimulated emission must be refilled before the next stimulated recombination event takes place. The speed of this refilling process is mainly limited by the carrier capture time, τ_C. The finite speed of this process restricts the radiative recombination current through the active region of QD laser:

$$J_{\max} \approx \frac{eN_e}{\tau_C}. \tag{8}$$

This process is similar to spectral hole burning[56] and seems to be most important for limiting the maximal output power of QD laser due to the finite number of the active states themselves, N_e, mentioned above. Taking into account typical parameters of self-organized QDs (the surface density is about 5×10^{10} cm^{-2}, the time of relaxation to the ground state is approximately 20 ps[57]) the value of $J_{\max}$ can be estimated as 1.6×10^3 A/cm^2 for the ground state lasing. Assuming 100% wallplug efficiency, the maximal output power limited by the refilling process is:

$$P_{\max} \approx \frac{h\nu}{e} J_{\max} WL. \tag{9}$$

This equation gives the upper limit of not more than 2 W for 100 μm $\times$ 1 mm stripe. It is well below the best results already achieved in 0.98 μm InGaAs QW lasers where the CW output power of 10 W was reported.[58,59] However, using various methods to increase the QD density (stacking, InAlAs QD pre-deposition) one can expect the optimistic maximum power of the order of 20 W or higher.

Catastrophic optical mirror damage (COMD) is among the dominant mechanisms for limiting the power of a semiconductor laser. It has been shown recently[60] that the internal power density at COMD is mostly dependent on the material of the quantum well rather than the SCH region. This observation confirms that the driving force for COMD is the mirror overheating caused by nonradiative surface recombination in the QW region. Therefore, the COMD level expected for InGaAs-based QD laser should be at least not less than the corresponding value for InGaAs QW lasers. Moreover, the QD array can strongly prevent the transport of nonequilibrium carriers to facets thereby suppressing the mirror heating.

At the beginning of investigation of self-organized QDs the possibility to achieve a high output power in QD laser diodes has not been studied or discussed. Decreasing the threshold current density of self-organized QD lasers by vertical stacking allowed them to oscillate in CW mode at room temperature. In 1996 the output of 300 mW was reported for lasers based on 10-times stacked array of InGaAs QDs in GaAs matrix.[50] Improving the design of the QD active region discussed above increases this value. Total CW output power, P_{out}, as a function of the drive current

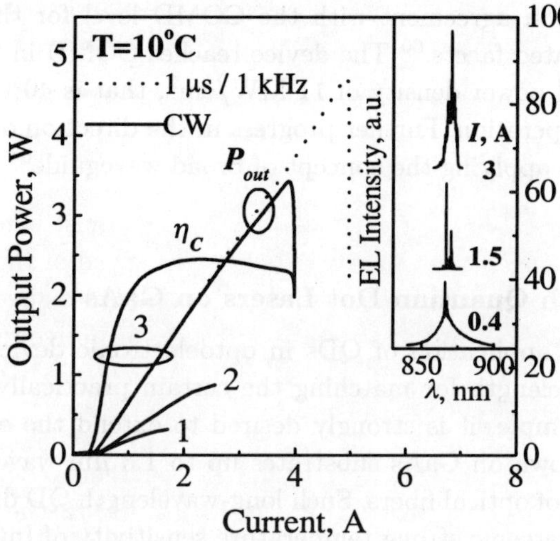

Fig. 11. Output power, P_{out}, and conversion efficiency, η_C, versus drive current for lasers with different QD active region: GaAs matrix, $N_{QD} = 5 \times 10^{10}$ cm^{-2} (curve 1); $Al_{0.15}Ga_{0.85}As$ matrix $N_{QD} = 5 \times 10^{10}$ (curve 2) or 2×10^{11} cm^{-2} (curve 3). Lasing spectra in the inset corresponds to (3).

is presented in Fig. 11 for broad-area (100 μm $\times$ 1 mm) stripe lasers with different types of the QD region. The heat sink temperature was 10°C. Thermal rollover caused by low wallplug efficiency restricts the power level at 0.3 W for the case of a GaAs matrix whereas it was 1 W for AlGaAs.[61] Maximum heatsink temperature for CW operation was 71°C in the latter case.

The output characteristics of QD lasers was improved drastically using the denser array of QDs formed by pre-deposition of InAlAs islands. It is seen in Fig. 11 that this concept leads to significant increase in the maximum output power up to 3.3–3.5 W.[62,63] In pulsed operation the slope efficiency is nearly constant and equals to 1.05 W/A ($\eta_D = 73\%$) up to the sudden failure due to COMD. CW power-current dependence becomes slightly sublinear beyond 2.5 A due to the heating of the active region, that is accompanied by a slight red-shift of the lasing line. The lasing spectra taken at different drive currents are shown in the inset. In the whole range of excitation levels lasing proceeds through the QD states. The maximum output power recorded in CW and pulsed regimes are 3.5 and 4.8 W at $I = 4$ and 5 A, respectively. The maximum conversion efficiency, η_C, of 45% occurs at 2 W.

This result indicates that QDs can be effectively used as the active region of injection lasers for high-power applications. Increasing the QD density leads to increase the maximum current flowing through the QD active region (see Eq. (8)). Moreover, this decreases the population of the wetting layer and matrix states at the same pumping level that, in the turn, reduces the parasitic current and structure overheating caused by the recombination in the higher energy states. It should be noted that the calculated internal density of the optical power at COMD per one facet (8 MW/cm^2) is in agreement with the COMD level for the InGaAs-based QW lasers with uncoated facets.[60] The device reaches COMD in pulsed operation at the internal optical power density of 11 MW/cm^2, that is 40% higher than the COMD level in CW operation. Further progress in the direction of high-power QD lasers can be achieved applying the concept of broad waveguide[64] and proper facet coating.

7. Long-Wavelength Quantum Dot Lasers on GaAs

The key point in the application of QDs in optoelectronic devices is the control of their emission wavelength for matching the certain practically desirable wavelength range. For example, it is strongly desired to extend the emission range of the QD structures grown on GaAs substrates up to 1.3 μm wavelength to fit the transparency window of optical fibers. Such long-wavelength QD diode lasers would make it possible to overcome strong temperature sensitivity of InGaAsP-based devices caused by insufficient electron confinement in the active region[65] and give more flexibility in design of vertically emitting structures. Simultaneously, they would enable to realize lasing in sub-100 A/cm^2 regime owing to very deep localization energy ($\geq$ 470 meV with respect to GaAs matrix). Huffacker *et al.* have already reported on the operation of 1.3 μm QD laser.[66,67] The authors used successive deposition of In and Ga atoms and As_4 molecules to extend the dislocation-free growth of InGaAs QDs. However, gain saturation of the ground-state QD transition associated with the low surface density of such QD array ($\sim 1.3 \times 10^{10}$ cm^{-2}) forced authors to use long stripe lasers with high-reflectivity facet coatings which are principally characterized by low external differential efficiency.

An alternative approach for achieving near-1.3-μm electroluminescence and lasing from QD-structures grown on GaAs substrates utilizes the sensitivity of the energy of optical transition in a quantum dot to the bandgap of the surrounding matrix (see Sec. 5.2). The longer wavelength border of the QD emission can be extended by placing a QD array into an external InGaAs strained quantum well.[68,69] The most important advantage of this approach is that the density of such QD array is as high as 4×10^{10} cm^{-2}. Moreover, several QD planes separated by GaAs spacer layers can be stacked in the active region (Fig. 12) further increasing the surface density and preventing gain saturation. The wavelength of the QD luminescence can be precisely controlled by varying the parameters of the external QW (well width and InAs mole fraction) reaching 1.3 μm under properly chosen conditions.[70]

Being used as the active region of diode laser these QD-in-QW arrays allow us to realize near-1.3 μm electroluminescence and lasing on GaAs substrates. Figure 13(a) shows near-threshold EL spectra for the structure based on three-times stacked QD-in-QW arrays. The ladder of well-resolved maxima corresponding to the ground (m_0 at 1.26 μm) and at least two excited (m_1 at 1.16 μm, m_2 at 1.09 μm) state transitions of the QD array can be observed in sub-threshold regime. Depending on L lasing proceeds through the ground state (m_0) or excited state (m_1, m_2) level.

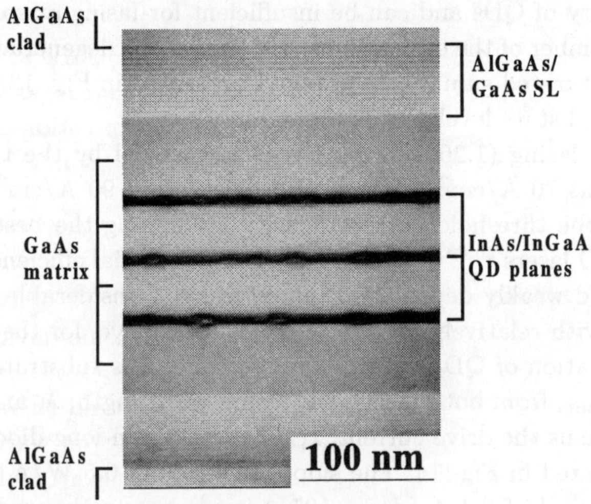

Fig. 12. Cross-section TEM image of the waveguide region of QD laser grown on GaAs substrate containing three InAs/InGaAs QD planes in GaAs matrix confined by short-period AlGaAs/GaAs superlattices (SLs).

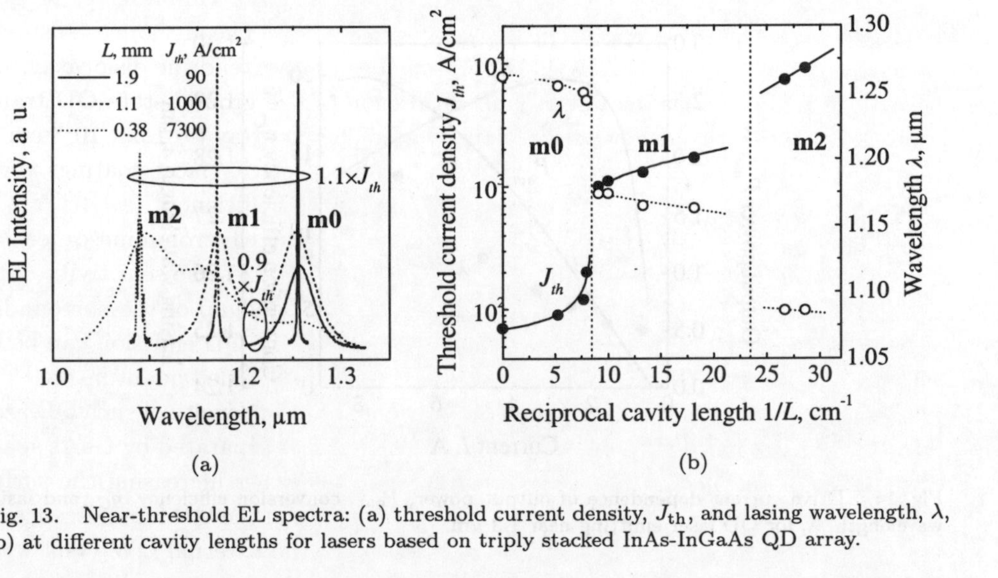

Fig. 13. Near-threshold EL spectra: (a) threshold current density, J_{th}, and lasing wavelength, λ, (b) at different cavity lengths for lasers based on triply stacked InAs-InGaAs QD array.

The higher the mirror loss, the shorter the wavelength of maximum involved into lasing. The threshold current density, J_{th}, and the lasing wavelength as functions of the reciprocal cavity length, $1/L$ at room temperature are presented in Fig. 13(b). The transitions of lasing from m_0 to m_1 at $L \sim 1$ mm and then from m_1 to m_2 at ~ 400 μm observed as sudden jumps of the lasing wavelength are also revealed as discontinuity points in J_{th}. This behavior can be explained in terms of gain saturation. The maximal achievable optical gain is limited by the degeneracy and the surface density of QDs and can be insufficient for lasing at some loss. Usually, the higher the number of the excited level, the higher the degeneracy and, therefore, the higher the saturated gain. From the data presented in Fig. 13(b) the saturated gain at the ground state level was estimated at 12 cm^{-1}.

Ground state lasing (1.26–1.28 μm) is characterized by the threshold current densities as low as 70 A/cm^2 (four cleaved facets) and 90 A/cm^2 ($L = 1.9$ mm). These values of the threshold current density are among the best results ever reported for the QD lasers.[20,66,67,71] The external differential efficiency was measured to be 57–53% and weakly depends on temperature. Considerable reduction in J_{th} in combination with relatively high η_D enables to achieve, for the first time, near-1.3-μm CW operation of QD-based diode laser on GaAs substrate. The measured output power, P_{out}, from both facets, the lasing wavelength, λ, and the conversion efficiency, η_C, versus the drive current, I, for the 1.9-mm-long diode in CW regime (17°C) are presented in Fig. 14. The slope efficiency of 0.5 W/A ($\eta_D = 51\%$) and the threshold current of 0.4 A ($J_{th} = 105$ A/cm^2) were estimated. The maximum output power of 2.7 W is limited by thermal rollover. The lasing wavelength is red-shifted with increasing the drive current to 1.28 μm at the point of the maximum power. The conversion efficiency reaches a maximum of 20% at P_{out} of 1.2 W.

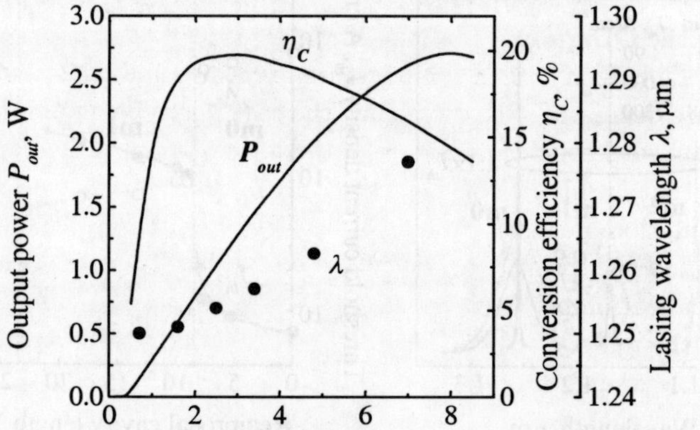

Fig. 14. Drive current dependence of output power, P_{out}, conversion efficiency, η_C, and lasing wavelength, λ, for QD laser emitting near 1.3 μm.

Temperature dependences of the threshold current density and the lasing wavelength are shown in Fig. 15 for devices characterized by different loss (stripe or four-cleaved-facets geometry). Lasing proceeds via the ground QD level in the whole studied temperature range up to 386 K in the sample with four cleaved facets ($J_{th} = 400$ A/cm^2, 1.3 μm) and to 333 K in long stripes ($J_{th} = 237$ A/cm^2, 1.28 μm). For both devices the threshold current density weakly depends on temperature at the lower temperatures and is probably due to the inhomogeneous

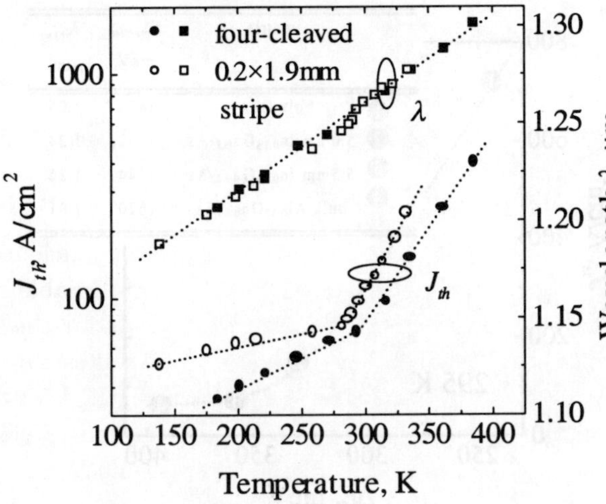

Fig. 15. Temperature dependence of the lasing wavelength (squares) and threshold current density (circles) for near-1.3-μm QD lasers (four-cleaved facets – solid marks, 200 μm × 1.9 mm stripe – hollow marks).

broadening of the QD ground state level. On the other hand, $J_{\rm th}$ rises steeply with temperature ($T_0 \approx 40$ K) when temperature exceeds some critical value. This critical temperature decreases with increasing the cavity loss being about 295 and 280 K for the samples with four cleaved facets and the 1.9-mm-long stripe, respectively. This suggests that, at higher temperatures, $J_{\rm th}$ is mainly governed by the thermally activated population of the higher energy states (the excited levels of QD array, the wetting layer and the matrix).

Deeper localization of the QD ground state level reduces this thermally excited population of the matrix. Figure 16 shows the dependence of the threshold current density (four-cleaved facets, 295 K) on the energy difference, ΔE, between the lasing energy and the matrix bandgap, E_{matrix}, for the lasers based on three-times stacked QD planes realized in different material systems. We use for comparison two InGaAs QD lasers with GaAs and $Al_{0.15}Ga_{0.85}As$ matrices[20,50] as well as two QD-in-QW lasers assuming the effective bandgap of strained $In_{0.15}Ga_{0.85}As/GaAs$ quantum well as the bandgap of the matrix material (see inset). It is seen that there is no correlation between $J_{\rm th}$ and the effective bandgap of the matrix or the lasing wavelength if one can try to use only these parameters. However, the threshold current density decreases monotonously with increasing ΔE.

The results obtained show applicability of the GaAs-based quantum dot lasers in optical fiber communications. Extremely low threshold current densities and acceptably high differential efficiency can be achieved owing to the QD nature of the active media, deep localization with respect to GaAs matrix, and prevention of gain saturation by multiple stacking of several QD planes.

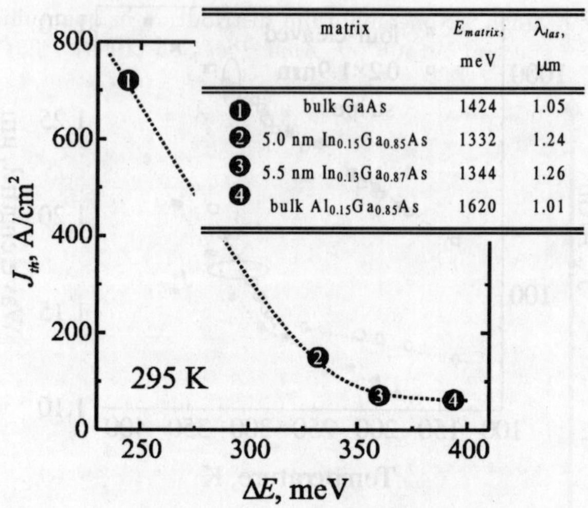

Fig. 16. Relationship between room temperature threshold current density, J_{th}, and carrier localization energy in the active region, ΔE, for four-cleaved QD lasers. Active regions are three-times stacked QDs formed in different matrix material. Inset shows matrix bandgap, E_{matrix}, and lasing wavelength, λ_{las}.

8. QD Lasers on InP Substrate

The bandgap energy of the surrounding matrix can be further decreased if growing lattice-matched InGaAs on InP substrates. In this section we present the results on InAs self-organized QDs grown in an $In_{0.53}Ga_{0.47}As/InP$ matrix.[72] The driving force for the island formation is the mismatch between lattice parameters of the substrate and the epitaxial layer. The 3.5% lattice mismatch between InAs and $In_{0.53}Ga_{0.47}As/InP$ is nearly the same as in the case of the $In_{0.5}Ga_{0.5}As$-on-GaAs deposition. It was found that the InAs film does transform into an array of three-dimensional islands upon the deposition of 3–3.3 MLs,[10] in a good agreement with the results for $In_{0.5}Ga_{0.5}As$ deposition on GaAs. However, the islands in this material system are characterized by lower height-to-base aspect ratio. The characteristic lateral size and height of the individual island revealed by TEM are about 50 and 4–5 nm, respectively, with the relatively small surface density of the order of 1×10^{10} cm^{-2}. Depending on the effective thickness of InAs deposited, Q_{InAs}, PL peak position from these quantum islands cover the 1.7–1.95-μm spectral range at 77 K. The full width at half maximum of PL spectrum is as low as 27 meV indicating relatively good uniformity of the QD size distribution. Integrated PL intensity is approximately constant until $Q_{\text{InAs}} = 9$ ML and then decreases dramatically due to strain relaxation.

To study the laser operation the structure was grown on n-type InP (100) substrate by solid source molecular beam epitaxy (MBE) (Fig 17).[21] The active region was inserted into the middle of the 0.6 μm InGaAs waveguide confined by a 1.5 μm InAlAs and the substrate as the top- and bottom-side cladding layers, respectively.

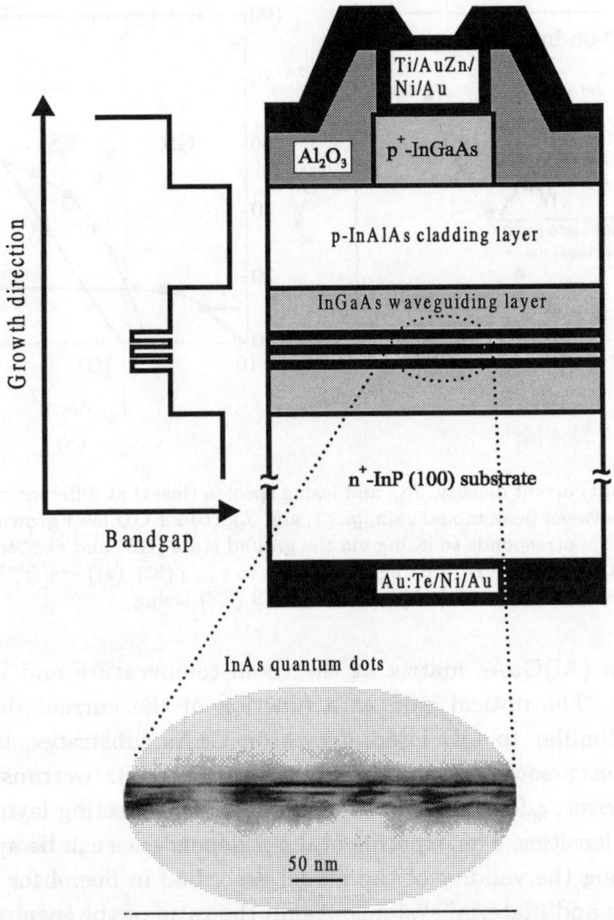

Fig. 17. Schematic cross-section and bandgap diagram of QD laser on InP substrate. TEM image of the active region is shown in the inset.

The active region consisted of the three QD planes (7 ML of InAs in each plane) separated by 5 nm-thick InGaAs spacers. Cross-section TEM image of the laser structure is shown in the inset of Fig. 17. Strained InAs islands in InGaAs matrix are clearly seen as the regions of the dark contrast.

Lasing line appears at 1.894 μm in the four-cleaved facet samples at 77 K at the threshold current density as low as 11.4 A/cm^2. To our knowledge, this is the longest wavelength ever reported for any kind of QD injection laser. Figure 18(a) shows the dependence of the threshold current density on the reciprocal cavity length. J_{th} weakly depends on the cavity length being lower than 100 A/cm^2 when L is longer than 0.25 mm. However, the steep increase in J_{th} accompanied by the blue shift of the lasing wavelength is observed when the cavity length is decreased below 240 μm. This behavior is very similar to that observed in lasers based on

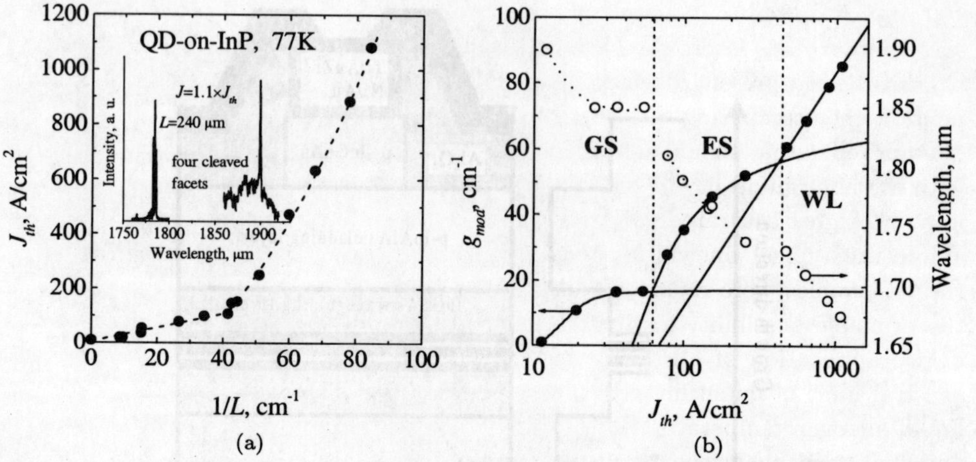

Fig. 18. Threshold current density, $J_{\rm th}$, and lasing spectra (inset) at different cavity length, L (a) and relationship between peak modal gain, g_{mod}, and $J_{\rm th}$ (b) for QD laser grown on InP substrate. Three regions in (b) corresponds to lasing via the ground state (GS) and excited state (ES) of QD array and the wetting layer (WL). Fitting parameters to g_{mod} (Eq. (4)) are: $g^{sat} = 5.7$ (23) cm^{-1}; $J_0 = 3.4$ (14), $J_1 = 2$ (–), $J_2 = 0$ (1.2) A/cm^2 for GS (ES) lasing.

InGaAs QDs in (Al)GaAs matrix at the room temperature and is attributed to gain saturation. The optical gain as a function of the current density is shown in Fig. 18(b). Similar to QD lasers grown on GaAs substrates, two pronounced saturating regions associated with the ground-to-excited states transition are clearly observed. Moreover, additional region of lasing via the wetting layer is observed at higher current densities. The experimental g-J dependence can be approximated by Eq. (4) indicating the validity of the model described in Sec. 4 for the wide range of temperatures and material systems. Again the ratio of the saturated gain values for ES and GS lasing is very close to four. This implies the four-fold degeneracy of the first excited state is the common feature of In(Ga)As self-organized QD.

9. Conclusions

Since the beginning of the 90s, very fast progress has been observed in a field of self-organized quantum dots.[73] Optically active quantum-sized islands were synthesized in a number of material systems and studied in detail. Injection lasers with the active region based on self-organized QD arrays demonstrate step-by-step improvement of their characteristics. We can summarize that the threshold current density of the lasers based on self-organized QDs is mainly governed by the surface density of QD array (due to the effect of gain saturation) and the energy separation between the "working" level and the matrix (due to the thermal population of the higher energy states). In particular, threshold (and power) characteristics were strongly improved when the concepts of vertical stacking and wide bandgap matrix were applied. Now, QD diode lasers with the threshold current densities less than

100 A/cm^2 and output power higher than 3 W were reported on. The operating wavelength of QD lasers covers a wide spectral range from ~ 0.8 to 1.9 μm.

However, at present, device performance of the lasers based on *real* QDs are still inferior when comparing with the theoretical predictions for the *ideal* QD case. The main problem that should be overcome is the inhomogeneous broadening associated with the non-uniformity in size of the islands. Improving this parameter will result in much higher saturated gain relieving QD lasers of the gain saturation problem. While the typical linewidth of the QD luminescence is as high as 40–60 meV, the best values were reported to be in the range of 20–30 meV[21,47,74] indicating the principle possibility for further improvement. Wide frontiers are expected for device application of QDs emitting in wavelength ranges which are beyond the spectral limit of quantum well. The way from first demonstration of QD laser at liquid nitrogen temperature to the current status took only several years. One can anticipate new achievements in the nearest future.

Acknowledgments

This work was supported in different parts by Russian Foundation for Basic Researches, Program of the Ministry of Science of Russia "Physics of Solid State Nanostructures", INTAS, Volkswagen foundation, BMBF, and NanoOp.

References

1. J. P. van der Ziel, R. Dingle, R. C. Miller, W. Wiegmann, and W. A. Nordland, Jr., "Laser oscillation from quantum states in very thin GaAs-$Al_{0.2}Ga_{0.8}As$ multilayer structures", *Appl. Phys. Lett.* **26** (1975) 463–465.
2. Y. Arakawa and H. Sakaki, "Multidimensional quantum well laser and temperature dependence of its threshold current", *Appl. Phys. Lett.* **40** (1982) 939–941.
3. L. Goldstein, F. Glas, J. Y. Marzin, M. N. Charasse, and G. Le Roux, "Growth by molecular beam epitaxy and characterization of InAs/GaAs strained-layer superlattices", *Appl. Phys. Lett.* **47** (1985) 1099–1101.
4. D. Leonard, M. Krishnamurthy, L. M. Reaves, S. P. Den Baars, and P. M. Petroff, "Direct formation of quantum-sized dots from uniform coherent islands of InGaAs on GaAs surfaces", *Appl. Phys. Lett.* **63** (1993) 3203–3205.
5. A. E. Zhukov, A. Yu. Egorov, A. R. Kovsh, V. M. Ustinov, N. N. Ledentsov, M. V. Maximov, A. F. Tsatsul'nikov, S. V. Zaitsev, N. Yu. Gordeev, P. S. Kop'ev, D. Bimberg, and Zh. I. Alferov, "Injection heterolaser based on an array of vertically aligned InGaAs quantum dots in an AlGaAs matrix", *Semiconductors* **31** (1997) 411–414.
6. R. Leon, S. Fafard, D. Leonard, J. L. Merz, and P. M. Petroff, "Visible luminescence from semiconductor quantum dots in large ensembles", *Appl. Phys. Lett.* **67** (1995) 521–523.
7. A. F. Tsatsul'nikov, A. Yu. Egorov, P. S. Kop'ev, A. R. Kovsh, N. N. Ledentsov, M. V. Maximov, A. A. Suvorova, V. M. Ustinov, B. V. Volovik, A. E. Zhukov, M. Grundmann, D. Bimberg, and Zh. I. Alferov, "Optical properties of InAlAs quantum dots in an AlGaAs matrix", *Appl. Surf. Sci.* **123/124** (1998) 381–384.
8. N. Carlsson, W. Seifert, A. Petersson, P. Castrillo, M. E. Pistol, and L. Samuelson,

"Study of the two-dimensional–three-dimensional growth mode transition in metal-organic vapor phase epitaxy of GaInP/InP quantum-sized structures", *Appl. Phys. Lett.* **65** (1994) 3093–3095.

9. M. K. Zundel, P. Specht, K. Eberl, N. Y. Jin-Phillipp, and F. Phillipp, "Structural and optical properties of vertically aligned InP quantum dots", *Appl. Phys. Lett.* **71** (1997) 2972–2974.
10. V. M. Ustinov, E. R. Weber, S. Ruvimov, Z. Liliental-Weber, A. E. Zhukov, A. Yu. Egorov, A. R. Kovsh, A. F. Tsatsul'nikov, and P. S. Kop'ev, "Effect of matrix on InAs self-organized quantum dots on InP substrate", *Appl. Phys. Lett.* **72** (1998) 362–364.
11. S. Fafard, Z. Wasilewski, J. Mc Caffrey, S. Raymond, and S. Charbonneau, "InAs self-asssembled quantum dots on InP by molecular beam epitaxy", *Appl. Phys. Lett.* **68** (1996) 991–993.
12. A. Ponchet, A. Le Corre, H. L'Haridon, B. Lambert, and S. Salaün, "Relationship between self-organization and size of InAs islands on InP(001) grown by gas-source molecular beam epitaxy", *Appl. Phys. Lett.* **67** (1995) 1850–1852.
13. J.-W. Lee, A. T. Schremer, D. Fekete, and J. M. Ballantyne, "GaInP/GaP partially ordered layer type-I strained quantum well", *Appl. Phys. Lett.* **69** (1997) 4236–4238.
14. B. Junno, T. Junno, M. S. Miller, and L. Samuelson, "A reflection high-energy electron diffraction and atomic force microscopy study of the chemical beam epitaxial growth of InAs and InP islands on (001) GaP", *Appl. Phys. Lett.* **72** (1998) 954–956.
15. N. Kirstaedter, N. N. Ledentsov, M. Grundmann, D. Bimberg, V. M. Ustinov, S. S. Ruvimov, M. V. Maximov, P. S. Kop'ev, Zh. I. Alferov, U. Richter, P. Werner, U. Gösele, and J. Heydenreich, "Low threshold, large T_0 injection laser emission from (InGa)As quntum dots", *Electron. Lett.* **30** (1994) 1416–1417.
16. H. Shoji, Y. Nakata, K. Mukai, Y. Sugiyama, M. Sugawara, N. Yokoyama, and H. Ishikawa, "Room temperature CW operation at the ground state of self-formed quantum dot lasers with multi-stacked dot layer", *Electron. Lett.* **32** (1996) 2023–2024.
17. K. Kamath, P. Bhattacharya, T. Sosnowski, T. Norris, and J. Phillips, "Room-temperature operation of $In_{0.4}Ga_{0.6}As$/GaAs self-organized quantum dot lasers", *Electron. Lett.* **32** (1996) 1374–1375.
18. R. Mirin, A. Cossard, and J. Bowers, "Room temperature lasing from InGaAs quantum dots", *Electron. Lett.* **32** (1996) 1732–1734.
19. A. E. Zhukov, V. M. Ustinov, A. Yu. Egorov, A. R. Kovsh, A. F. Tsatsul'nikov, N. N. Ledentsov, S. V. Zaitsev, N. Yu. Gordeev, P. S. Kop'ev, and Zh. I. Alferov, "Negative characteristic temperature of InGaAs quantum dot injection laser", *Jpn. J. Appl. Phys.* **B36** (1997) 4216–4218.
20. A. E. Zhukov, V. M. Ustinov, A. Yu. Egorov, A. R. Kovsh, A. F. Tsatsul'nikov, M. V. Maximov, N. N. Ledentsov, S. V. Zaitsev, N. Yu. Gordeev, V. I. Kopchatov, Y. M. Shernyakov, P. S. Kop'ev, D. Bimberg, and Zh. I. Alferov, "Injection lasers based on InGaAs quantum dots in an AlGaAs matrix", *J. Electron. Mat.* **27** (1998) 106–109.
21. V. M. Ustinov, A. E. Zhukov, A. Yu. Egorov, A. R. Kovsh, S. V. Zaitsev, N. Yu. Gordeev, V. I. Kopchatov, N. N. Ledentsov, A. F. Tsatsul'nikov, B. V. Volovik, P. S. Kop'ev, Z. I. Alferov, S. S. Ruvimov, Z. Liliental-Weber, and D. Bimberg, "Low threshold quantum dot injection laser emitting at 1.9 μm", *Electron. Lett.* **34** (1998) 670–672.
22. T. Riedl, E. Fehrenbacher, A. Hangleiter, M. K. Zundel, and K. Eberl, "Injection lasers with vertically aligned InP/GaInP quantum dots: Dependence of the threshold current on temperature and dot size", *Appl. Phys. Lett.* **73** (1998) 3730–3732.
23. D. Bimberg, N. Kirstaedter, N. N. Ledentsov, Zh. I. Alferov, P. S. Kop'ev, and V. M. Ustinov, "InGaAs-GaAs quantum-dot lasers", *IEEE J. Selected Topics Quantum*

Electron. **3** (1997) 196–205.

24. D. Bimberg, N. N. Ledentsov, N. Kirstaedter, O. Schmidt, M. Grundmann, V. M. Ustinov, A. Yu. Egorov, A. E. Zhukov, M. V. Maximov, P. S. Kop'ev, Zh. I. Alferov, S. S. Ruvimov, U. Gösele, and J. Heydenreich, "InAs-GaAs quantum dot lasers: *In situ* growth, radiative lifetimes and polarization properties", *Jpn. J. Appl. Phys.* **B35** (1996) 1311–1319.
25. N. Chand, E. E. Becker, J. P. Van der Ziel, S. N. G. Chu, and N. K. Dutta, "Excellent uniformity and very low (< 50 A/cm^2) threshold current density strained InGaAs quantum well diode lasers on GaAs substrate", *Appl. Phys. Lett.* **58** (1991) 1704–1706.
26. G. L. Snider, I.-H. Tan, and E. L. Hu, "Electron states in mesa-etched one-dimensional quantum-well wirtes", *J. Appl. Phys.* **68** (1990) 2849–2853.
27. P. M. Petroff and S. P. Den Baars, "MBE and MOCVD growth and properties of self-assembling quantum dot arrays in III-V semiconductor structures", *Superlat. Microstruct.* **15** (1994) 15–21.
28. J. M. Moison, F. Houzay, F. Barthe, L. Leprince, E. Andre, and O. Vatel, "Self-organized growth of regular nanometer-scale InAs dots on GaAs", *Appl. Phys. Lett.* **64** (1994) 196–198.
29. D. Leonard, K. Pond, and P. M. Petroff, "Critical layer thickness for self-assembled InAs islands on GaAs", *Phys. Rev.* **B50** (1995) 11687–11692.
30. A. Yu. Egorov, A. E. Zhukov, P. S. Kop'ev, N. N. Ledentsov, M. V. Maksimov, and V. M. Ustinov, "Effect of deposition conditions on the formation of (In,Ga)As quantum clusters in a GaAs matrix", *Semiconductors* **28** (1994) 809–811.
31. S. S. Ruvimov, P. Werner, K. Scheerschmidt, U. Gösele, J. Heydenreich, U. Richter, N. N. Ledentsov, M. Grundmann, D. Bimberg, V. M. Ustinov, A. Yu. Egorov, P. S. Kop'ev, and Zh. I. Alferov, "Structural characterization of (In,Ga)As quantum dots in a GaAs matrix", *Phys. Rev.* **B51** (1995) 14766–14769.
32. V. A. Shchukin, N. N. Ledentsov, P. S. Kop'ev and D. Bimberg, "Spontaneous ordering of arrays of coherent strained islands", *Phys. Rev. Lett.* **75** (1995) 2968–2971.
33. M. Lentzen, D. Gerthsen, A. Forster, and K. Urban, "Growth mode and strain relaxation during the initial stage of InGaAs growth on GaAs(100)", *Appl. Phys. Lett.* **60** (1992) 74–76.
34. A. Yu. Egorov, A. E. Zhukov, P. S. Kop'ev, N. N. Ledentsov, M. V. Maximov, V. M. Ustinov, A. F. Tsatsul'nikov, Zh. I. Alferov, D. L. Fedorov, and D. Bimberg, "Optical emission range of structures with strained InAs quantum dots in GaAs", *Semiconductors* **30** (1996) 707–710.
35. J. M. Moison, F. Houzay, F. Barthe, L. Leprince, E. Andre, and O. Vatel, "Self organized growth of regular nanometer-scale InAs dots on GaAs", *Appl. Phys. Lett.* **64** (1994) 196–198.
36. M. Grundman, O. Stier, and D. Bimberg, "InAs/GaAs quantum pyramids: Strain distribution, optical phonons and electronic structure", *Phys. Rev.* **B52** (1996) 11969–11978.
37. J.-Y. Marzin and G. Basstard, "Calculation of the energy levels in InAs/GaAs quantum dots", *Solid State Commun.* **92** (1995) 437–442.
38. L.-W. Wang, J. Kim, and A. Zunger, "Electronic structures of [110]-faceted self-assembled pyramidal InAs/GaAs quantum dots", *Phys. Rev.* **B59** (1999) 5678–5687.
39. L. V. Asryan and R. A. Suris, "Inhomogeneous line broadening and the threshold current density of a semiconductor quantum dot laser", *Semicond. Sci. Technol.* **11** (1996) 554–567.
40. M. Grundmann and D. Bimberg, "Theory of random population for quantum dots", *Phys. Rev.* **B55** (1997) 9740–9745.

41. H. Jiang and J. Singh, "Nonequilibrium distribution in quantum dots lasers and influence on laser spectral output", *J. Appl. Phys.* **85** (1999) 7438–7442.
42. A. R. Kovsh, A. E. Zhukov, A. Yu. Egorov, V. M. Ustinov, N. N. Ledentsov, M. V. Maximov, A. F. Tsatsul'nikov, and P. S. Kop'ev, "Gain in injection lasers based on self-organized quantum dots", *Semiconductors* **33** (1999) 184–191.
43. R. Heitz, M. Grundmann, N. N. Ledentsov, L. Eckey, M. Veit, D. Bimberg, V. M. Ustinov, A. Yu. Egorov, A. E. Zhukov, P. S. Kop'ev, and Zh. I. Alferov, "Exciton relaxation in self-organized InAs/GaAs quantum dots", *Surf. Sci.* **361/362** (1996) 770–773.
44. A. E. Zhukov, A. R. Kovsh, V. M. Ustinov, A. Yu. Egorov, N. N. Ledentsov, A. F. Tsatsul'nikov, M. V. Maximov, Yu. M. Shernyakov, V. I. Kopchatov, A. V. Lunev, P. S. Kop'ev, D. Bimberg, and Zh. I. Alferov, "Gain characteristics of quantum dot injection lasers", *Semicond. Sci. Technol.* **14** (1999) 118–123.
45. N. N. Ledentsov, V. M. Ustinov, A. Yu. Egorov, A. E. Zhukov, M. V. Maximov, I. G. Tabatadze, and P. S. Kop'ev, "Optical properties of heterostructures with InGaAs-GaAs quantum clusters", *Semiconductors* **28** (1994) 832–834.
46. P. N. Brounkov, S. G. Konnikov, V. M. Ustinov, A. E. Zhukov, A. Yu. Egorov, M. V. Maximov, N. N. Ledentsov, and P. S. Kop'ev, "Capacitance spectroscopy of electron energy levels in InAs quantum dots in a GaAs matrix", *Semiconductors* **30** (1996) 492–496.
47. A. Yu. Egorov, A. E. Zhukov, P. S. Kop'ev, N. N. Ledentsov, M. V. Maximov, V. M. Ustinov, A. F. Tsatsul'nikov, N. A. Bert, A. O. Kosogov, D. Bimberg, and Zh. I. Alferov, "Formation of vertically aligned arrays of strained InAs quantum dots in a GaAs(100) matrix", *Semiconductors* **30** (1996) 879–883.
48. Q. Xie, A. Madhukar, P. Chen, and N. Kobayashi, "Vertically self-organized InAs quantum box islands on GaAs (100)", *Phys. Rev. Lett.* **75** (1995) 2542–2545.
49. N. N. Ledentsov, V. A. Schukin, M. Grundmann, N. Kirstaedter, J. Böhrer, O. Schmidt, D. Bimberg, V. M. Ustinov, A. Yu. Egorov, A. E. Zhukov, P. S. Kop'ev, S. V. Zaitsev, Zh. I. Alferov, A. I. Borovkov, A. O. Kosogov, S. S. Ruvimov, P. Werner, U. Gösele, and J. Heydenreich, "Direct formation of vertically coupled quantum dots by Stranski–Krastanov growth", *Phys. Rev.* **B54** (1996) 8743–8750.
50. V. M. Ustinov, A. Yu. Egorov, A. R. Kovsh, A. E. Zhukov, M. V. Maksimov, A. F. Tsatsul'nikov, N. Yu. Gordeev, S. V. Zaitsev, Yu. M. Shernyakov, N. A. Bert, P. S. Kop'ev, Zh. I. Alferov, N. N. Ledentsov, J. Böhrer, D. Bimberg, A. O. Kosogov, P. Werner, and U. Gösele, "Low-threshold injection lasers based on vertically coupled quantum dots", *J. Cryst. Growth* **175/176** (1997) 689–695.
51. S. V. Zaitsev, N. Yu. Gordeev, Yu. M. Sherniakov, V. M. Ustinov, A. E. Zhukov, A. Yu. Egorov, M. V. Maximov, P. S. Kop'ev, Zh. I. Alferov, N. N. Ledentsov, N. Kirstaedter, and D. Bimberg, "Radiation characteristics of injection lasers based on vertically coupled quantum dots", *Superlat. Microstruct.* **21** (1997) 559–564.
52. A. F. Tsatsul'nikov, M. V. Belousov, N. A. Bert, A. Yu. Egorov, P. S. Kop'ev, A. R. Kovsh, I. L. Krestnikov, N. N. Ledentsov, M. V. Maximov, A. A. Suvorova, V. M. Ustinov, B. A. Volovik, A. E. Zhukov, M. Grundmann, D. Bimberg, and Zh. I. Alferov, "Lateral association of vertically-coupled quantum dots", *Microelectron. Eng.* **43/44** (1998) 37–43.
53. A. F. Tsatsul'nikov, A. Yu. Egorov, A. E. Zhukov, A. R. Kovsh, V. M. Ustinov, N. N. Ledentsov, M. V. Maksimov, A. V. Sakharov, A. A. Suvorova, P. S. Kop'ev, Zh. I. Alferov, and D. Bimberg, "Modulation of a quantum well potential by a quantum-dot array", *Semiconductors* **31** (1997) 88–91.
54. A. R. Kovsh, A. E. Zhukov, A. Yu. Egorov, V. M. Ustinov, Yu. M. Shernyakov, M.

V. Maximov, A. F. Tsatsul'nikov, B. V. Volovik, A. V. Lunev, N. N. Ledentsov, P. S. Kop'ev, Zh. I. Alferov, and D. Bimberg, "Effect of the quantum-dot surface density in the active region on injection-laser characteristics", *Semiconductors* **32** (1998) 997–1000.
55. V. P. Evtikhiev, I. V. Kudryashov, E. Yu. Kotel'nikov, V. E. Tokranov, A. N. Titkov, I. S. Tarasov, and Zh. I. Alferov, "Continuous wave at T = 293 K lasing of SCH lasers with a one sheet InAs quantum dots grown on the vicinal GaAs (001) surfaces misoriented to the [010] direction", *Semiconductors* **32** (1998) 1482–1486.
56. C.-Y. Tsai, C.-Y. Tsai, Y.-H. Lo, and R. M. Spencer, "Effects of spectral hole burning, carrier heating, and carrier transport on the small-signal modulation response of quantum well lasers", *Appl. Phys. Lett.* **67** (1995) 3084–3086.
57. D. Bimberg, N. N. Ledentsov, M. Grundmann, F. Heinrichsdorff, V. M. Ustinov, P. S. Kop'ev, M. V. Maximov, Zh. I. Alferov, and J. A. Lott, "Application of self-organized quantum dots to edge emitting and vertical cavity lasers", *Physica* **E3** (1998) 129–136.
58. S. O'Brien, H. Zhao, A. Schoenfelder, and R. J. Lang, "9.3 W CW (In)AlGaAs 100 μm wide lasers at 970 nm", *Electron. Lett.* **33** (1997) 1869–1870.
59. A. Al-Muhanna, L. J. Mawst, D. Botez, D. Z. Garbuzov, R. U. Martinelli, and J. C. Connolly, "High-power (> 10 W) continuous-wave operation from 100-mm-aperture 0.97-mm-emitting Al-free diode lasers", *Appl. Phys. Lett.* **73** (1998) 1182–1184.
60. L. J. Mawst, A. Bhattacharya, J. Lopez, D. Botez, D. Z. Garbuzov, L. De Marco, J. C. Connolly, M. Jansen, F. Fang, and R. F. Nabiev, "8 W continuous wave front-facet power from broad-waveguide Al-free 980 nm diode lasers", *Appl. Phys. Lett.* **69** (1996) 1532–1534.
61. Yu. M. Shernyakov, A. Yu. Egorov, A. E. Zhukov, S. V. Zaitsev, A. R. Kovsh, I. L. Krestnikov, A. V. Lunev, N. N. Ledentsov, M. V. Maximov, A. V. Sakharov, V. M. Ustinov, Zh. Zhao, P. S. Kop'ev, Zh. I. Alferov, and D. Bimberg, "Quantum dot CW heterojunction injection laser operating at room temperature with an output power of 1 W", *Tech. Phys. Lett.* **23** (1997) 149–150.
62. A. R. Kovsh, D. A. Livshits, A. E. Zhukov, A. Yu. Egorov, M. V. Maximov, V. M. Ustinov, I. S. Tarasov, N. N. Ledentsov, P. S. Kop'ev, Zh. I. Alferov, and D. Bimberg, "Injection heterolaser based on quantum dots with 3.3 W output power", *Tech. Phys. Lett.* **25** (1999) 41–46.
63. A. R. Kovsh, D. A. Livshits, A. E. Zhukov, A. Yu. Egorov, V. M. Ustinov, M. V. Maximov, Yu. G. Musikhin, N. N. Ledentsov, P. S. Kop'ev, Zh. I. Alferov, and D. Bimberg, "3.5 W CW operation of a quantum dot laser", *Electron. Lett.*, submitted.
64. M. R. Gokhale, J. C. Dries, P. V. Studenkov, S. R. Forrest, and D. Z. Garbuzov, "High-power high-efficiency 0.98-μm wavelength InGaAs-(In)GaAs(P)-InGaP broadened waveguide lasers grown by gas-source molecular beam epitaxy", *IEEE J. Quantum Electron.* **33** (1997) 2266–2276.
65. M. Kondow, K. Uomi, A. Niwa, T. Kitatani, S. Watahiki, and Y. Yazawa, "GaInNAs: A novel material for long-wavelength-range laser diodes with excellent high-temperature performance", *Jpn. J. Appl. Phys.* **B35** (1996) 1273–1275.
66. D. L. Huffaker, G. Park, Z. Zou, O. B. Shchekin, and D. G. Deppe, "1.3 μm room-temperature GaAs-based quantum-dot laser", *Appl. Phys. Lett.* **73** (1998) 2564–2566.
67. G. Park, D.L. Huffaker, Z. Zou, and D.G. Deppe, "Temperature dependence of lasing characteristics for long-wavelength (1.3-μm) GaAs-based quantum-dot lasers", *IEEE Photon. Tech. Lett.* **11** (1999) 301–303.
68. A. E. Zhukov, A. R. Kovsh, A. Yu. Egorov, N. A. Maleev, V. M. Ustinov, B. V. Volovik, M. V. Maximov, A. F. Tsatsul'nikov, N. N. Ledentsov,Yu. M. Shernyakov, A. V. Lunev, Yu. G. Musikhin, N. A. Bert, P. S. Kop'ev, Zh. I. Alferov, "Photo- and

electroluminescence in 1.3-μm range from quantum-dot structures grown on GaAs substrates", *Semiconductors* **33** (1999) 153–156.

69. V. M. Ustinov, N. A. Maleev, A. E. Zhukov, A. R. Kovsh, A. Yu. Egorov, A. V. Lunev, B. V. Volovik, I. L. Krestnikov, Yu. G. Musikhin, N. A. Bert, P. S. Kop'ev, Zh. I. Alferov, N. N. Ledentsov, and D. Bimberg, "InAs/InGaAs quantum dot structures on GaAs substrates emitting at 1.3 μm", *Appl. Phys. Lett.* **74** (1999) 2815–2817.
70. A. E. Zhukov, V. M. Ustinov, A. R. Kovsh, A. Yu. Egorov, N. A. Maleev, N. N. Ledentsov, A. F. Tsatsul'nikov, M. V. Maximov, Yu. G. Musikhin, N. A. Bert, P. S. Kop'ev, D. Bimberg, and Zh. I. Alferov, "Control of the emission wavelength of self-organized quantum dots: Main achievements and present status", *Semicond. Sci. Technol.* **14** (1999) 575–581.
71. F. Schafer, J. P. Reithmaier, and A. Forchel, "High-performance GaInAs/GaAs quantum-dot lasers based on a single active layer", *Appl. Phys. Lett.* **74** (1999) 2915–2917.
72. V. M. Ustinov, A. E. Zhukov, A. F. Tsatsul'nikov, A. Yu. Egorov, A. R. Kovsh, M. V. Maximov, A. A. Suvorova, N. A. Bert, and P. S. Kop'ev, "Arrays of strained InAs quantum dots in an (In,Ga)As matrix, grown on InP substrates by molecular-beam epitaxy", *Semiconductors* **31** (1997) 1080–1083.
73. D. Bimberg, M. Grundmann, and N. N. Ledentsov, *Quantum Dot Heterostructures*, Wiley, Chichester, 1999.
74. K. Nishi, H. Saito, S. Sugou, and J.-S. Lee, "A narrow photoluminescence linewidth of 21 meV at 1.35 μm from strain-reduced InAs quantum dots covered by $In_{0.2}Ga_{0.8}As$ grown on GaAs substrates", *Appl. Phys. Lett.* **74** (1999) 1111–1113.

International Journal of High Speed Electronics and Systems, Vol. 9, No. 4 (1998) 1139–1162

NITRIDE LASERS: OPTICAL GAIN AND DEVICE IMPLICATIONS

ARTO V. NURMIKKO and YOON-KYU SONG
Division of Engineering, Brown University,
Providence, RI 02912, USA

The blue nitride laser presents a fascinating case for the study of the microscopics of optical gain. For InGaN quantum wells, these are highlighted by intense competition between localized and extended electronic states in this unusual, heterogeneous active laser medium.

1. Introduction

As described in this volume, progress with GaN-based diode lasers has crossed the threshold of commercial availability at the Nichia Company in Japan.[1] In the US, there have been demonstrations of the continuous-wave operation[2,3] of the blue/violet laser. Other advances suggest a further broadening of the base of this emerging technology, for which potential applications abound. Thus, in spite of several technical, device-related challenges, optimism is warranted.

A striking feature of the present InGaN QW lasers is their high threshold current density, implying an unusually high electron-hole pair density for a semiconductor laser, typically in excess of 10^{19} cm^{-3}. The closest comparison that can be readily made (within the effective mass approximation to the electronic bandstructure) is with the blue-green II-VI QW diode lasers. These have achieved room temperature threshold current densities below 200 A/cm^2,[4] a value approximately one order of magnitude smaller than so far accomplished in the best InGaN QW lasers. Furthermore, and in notable contrast to the ZnCdSe active QW medium in the green II-VI lasers, the InGaN QW shows strong departures from a usual random alloy. A propensity for In-clustering has been at least qualitatively established via many microprobe studies, owing its origin the finite degree of thermodynamic "immiscibility"of the InN and GaN constituents in a normal solid solution. The clustering, or "compositional anomalies", in the InGaN system in turn profoundly effect the bandedge electronic states which form the "electronic power supply" for optical emission, both for light emitting diodes and diode lasers. In earlier literature, Nakamura and colleagues speculated that InGaN QWs be subject to spontaneous "quantum dot" formation,[5] with implications that diode laser action could derive a benefit in such a blue nitride quantum dot laser in a manner already clearly demonstrated in "conventional" III-VI self-assembled systems, most notably the

InAs/GaAs strained layer case (chapters by Bhattacharya and Zhukov *et al.* in this volume; see also Ref. 6).

We will show in this chapter how the compositional anomalies in InGaN QWs lead to a veritable competition of electronic excitations between localized and extended electronic states. The localized electronic states serve the case of an LED well from the viewpoint of high radiative efficiency. On the other hand, the situation for a diode laser in more ambiguous so that, unlike in a true quantum dot system, the potential energy landscape in an InGaN QW is usually far from being spatially ordered. This translates to a requirement for a high injection level for optical gain of sufficient magnitude to form in a real device. The overall picture is still not fully understood, given the added complications created by large piezoelectric and spontaneous dielectric polarization fields, which typify the wurtzite nitrides in general. On the other hand, within the past year or so, carefully controlled epitaxial growth of the binary GaN/AlGaN QW system has led to the clear demonstration of the synthesis of quantum dots in this strained nitride system, driven by the Stranski-Krastanov-type of mechanism. In many ways, the GaN QW lacks the idiosyncrasies of the InGaN media required for the blue/violet emitters. However, even rudimentary device progress has been very painstaking at these NUV wavelengths (~ 350 nm). As of now, there are no diode lasers or even LEDs (of any reasonable efficiency) that have been fabricated from the GaN QWs in the near ultraviolet. Curiously, the absence of indium creates a much higher degree of vulnerability to both point and extended defect states in GaN and AlGaN in terms of nonradiative recombination. While such defects are ubiquitous in all nitride compounds, the materials are still exceptionally resistant to high current densities.

This chapter is organized as follows. We first study the InGaN QW material system from the standpoint of optical gain in Sec. 2 and its correlation with the complex crystal microstructure in Sec. 3. A good deal of experimental evidence is now available, including gain spectroscopy of the diode laser devices. We also briefly examine the status of the quantum dots in the binary GaN QWs in Sec. 4. In Sec. 5, a comparison is made with the "other" prototypical wide bandgap diode laser, based on ZnSe compounds, where experimental evidence shown the striking enhancements to optical gain by excitonic effects. Such effects should, in principle, be found to be even more potent in properly designed nitride QW lasers. Finally, in Sec. 6, we consider optical gain in the nitride quantum wells in the context of current efforts to develop blue and violet vertical cavity surface emitting lasers.

2. Gain Spectroscopy of InGaN QW Diode Laser

In spite of significant rapid advances that have led to the demonstration of a long lived cw blue InGaN QW diode laser, major unanswered questions exist about the physics of optical gain and stimulated emission in these fascinating new sources. A key issue is the nature of those band edge electronic states that supply the requisite optical gain, given the demonstrably large departure of InGaN from a random alloy in terms of the In-concentration fluctuations ($x_{\rm In}$) on a microscopic (atomic) scale. Typically, the mean values of $x_{\rm In}$ in the laser devices are in the range of $x_{\rm In} \sim 0.1$–0.2; *ad hoc* arguments can be made for the probable partial cation segregation due to the differences in the bond energies and lattice constants for the InN and GaN binary endpoints. High-resolution electron microscopy studies have been evoked show the "clustering" that takes place in the InGaN QW and the term "quantum dot" has been applied to such clusters.[5] The question that arises when this type of nanoscale heterogeneous semiconductor forms the active laser material is the competition between localization and many electron correlations within the available electron-hole pair states, given the very high pair densities required in present devices to reach lasing threshold ($> 10^{19}$ cm^{-3}).

In this chapter, we provide the experimental results on the gain spectra of electrically active devices to obtain *spectroscopic* insight into the gain in InGaN/GaN QW *pn*-junction heterostructures. Recently, a number of reports have appeared on spectroscopic features of the InGaN QW at the bandedge, including the optical gain regime. Gain experiments have been performed by optical pumping especially by Hangleiter and co-workers, usually by the so-called stripe excitation method[7] which yields a limited spectroscopic view and may not be accurately correspond to the conditions in an electrically injected device. Nakamura and co-workers have used the Hakki-Paoli method to acquire gain spectra on actual diode laser devices[8] and an electrical injection-optical probe method has been applied to study the formation of optical gain in InGaN QW *pn*-junction heterostructures.[9]

Significant insight to the optical gain spectra of the InGaN blue laser active medium has been recently acquired, based on the analysis of the spontaneous emission spectra of the diode laser, in conjuction with its threshold characteristics. The approach makes use of the fundamental relationships between spontaneous emission, stimulated emission, and absorption. This method, initially introduced by Henry *et al.*,[10] to semiconductor lasers has been applied to the detailed study of different III-V QW semiconductor lasers,[11,12] as well as the II-VI green-blue diode lasers illustrated later in this chapter.[13] The technique has important advantages over the Hakki-Paoli technique, including the problem in the latter method to separate the nonradiative component of the injection current from the radiative one and the inability to reach into the absorbing regions of the spectrum at injection levels well below the lasing threshold. The application of the Hakki-Paoli method is further complicated for the InGaN diode lasers due to the frequent presence of extraneous longitudinal sub-cavity modes. For a gain medium possessing such idiosyncrasies as the InGaN QW, the "Henry" approach is advantageous, given the central question of the fundamental gain characteristics and their relationship with radiative processes.

Typical gain/absorption spectrum for an InGaN MQW diode laser is shown in Fig. 1, for ridge waveguide device with a indium concentration $x_{\rm In} = 0.15$ in the active region.[14] The values for the current range from low injection in the LED

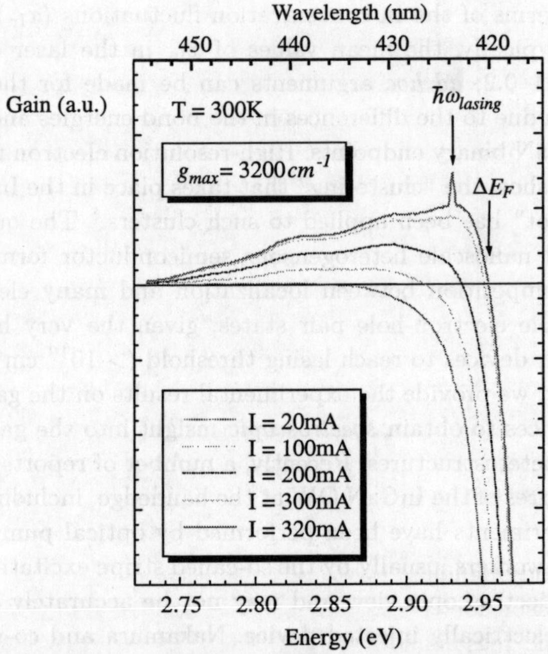

Fig. 1. Gain/absorption spectrum of a InGaN MQW diode laser at different injection levels at room temperature. The spectral position of laser emission and the quasi-Fermi level separation at threshold are indicated.

regime (I = 20 mA) to the lasing threshold and somewhat beyond. The spectral position of lasing slightly above threshold is indicated by an arrow as is the location of the quasi-Fermi level difference ΔE_F at approximately 2.995 eV at this injection level. The experiment and analysis also yield a self-consistent determination for the position of ΔE_F as a function of the injection current, graphed in Fig. 2. The vertical axis in Fig. 1 was calibrated from the threshold modal gain, measured for devices of different cavity reflectivities. Assuming a modal overlap factor of $\Gamma = 0.025$, based on the analysis of the passive waveguide performance, we obtain the peak gain at threshold of approximately 3200 cm^{-1} for the QW material.

The main result of significance in Fig. 1 is the pronounced extension of the gain spectra associated with the $n = 1$ QW transition into the low energy region. At threshold, finite gain is found as much as 200 meV below its peak position, indicating a degree of broadening which is most uncharacteristic of common semiconductor lasers. Note however, that the system reaches the transparency condition relatively easily, at levels of injection which are not very different from that of the conventional LED regime. With increasing current, gain builds up over the large spectral range, indicative of the participation of a corresponding range of electronic states. The position of the peak gain blue shifts somewhat at higher injection levels, but considerably less than anticipated from a one electron state-filling picture, possi-

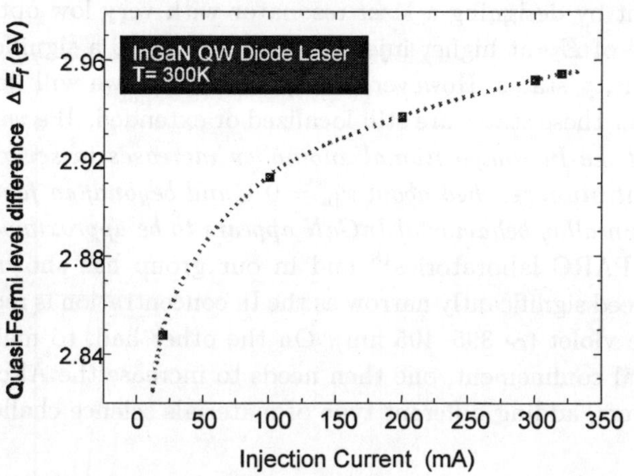

Fig. 2. Quasi-Fermi level as a function of injection in an InGaN MQW laser at room temperature.

bly due to many-body bandgap renormalization effects. Qualitatively, we may now understand one reason for the high *e*-*h* pair density required for laser operation, apart from extrinsic reasons such as unwanted optical losses. That is, while the spectrally integrated gain is, in fact, quite large, its peak value (determining the lasing threshold) is much diluted at the expense of the remarkably large broadening.

These observations provide an extension to recent arguments[5,9,15,16] that radiative recombination processes at the lowest interband transition in the InGaN QW are profoundly influenced by localized *e*-*h* pair states at room temperature, within an energy range which is nearly an order of magnitude larger than estimated for a simple random alloy. That is, the description of the system in terms of weak disorder, as usually applied to ternary and quaternary compounds in the III-V and II-VI semiconductors, is probably inapplicable. Whether this aspect of InGaN is intrinsic or subject to specific MOCVD growth conditions has yet to be established; suffice it to say that all available optical data on InGaN QWs and thin films to date display the striking extension of the bandedge states so that, for example, excitonic features in absorption at the $n = 1$ QW states have not been unambiguously identified. By contrast, gain spectroscopy performed on widegap ZnCdSe QW diode lasers by the same method described below shows very clearly the characteristic influence of the strong excitonic enhancement of the peak gain and an overall optical response at the $n = 1$ HH exciton with the pronounced Coulomb correlations in evidence. Such effects are clearly masked by the disorder contributions in the InGaN QW, making it difficult to isolate predicted many-body interactions[17] in the dense *e*-*h* system within the active region of the blue diode laser.

From Figs. 1 and 2, one sees how filling of the localized states is a necessary prerequisite prior to the buildup of a sufficient population inversion for threshold gain in the present devices. On the other hand, since transparency is reached at a rather low injection level ($n = p \sim 10^{18}$ cm^{-3}) it may be possible to reduce the threshold current by designing a laser resonator with very low optical losses. The near "clamping" of E_F at higher injection may be due to a significant increase in the effective density states. However, the question that we will defer to the next section is whether these states are still localized or extended. *We wish to emphasize that the issue of the In compositional anomalies increases in severity very rapidly as the In concentration reached about* $x_{\mathrm{In}} = 0.1$ *and beyond; in fact for* $x_{\mathrm{In}} \ll 0.1$, *the nearly random alloy behavior of InGaN appears to be approximated.* Very recent work at Xerox PARC laboratories[18] and in our group has shown how the gain spectra does indeed significantly narrow as the In concentration is reduced, for lasers operating in the violet ($\sim$ 395–405 nm). On the other had, to maintain adequate electronic/optical confinement, one then needs to increase the Al-concentration in the cladding layers, adding different type of materials science challenge.

3. On the Electronic Microstructure of InGaN Quantum Wells

In the previous sections we have frequently referred to the plethora of unusual features of the InGaN QW active medium which electronically powers the blue LEDs and diode lasers. Here we enlarge on the subject by reviewing by examples some of the still incomplete information that has been acquired about the relevant spatial texture and its impact on electron-hole transport in this material. It should be strongly underlined that the story is very much still at early evolving stages, even if high efficiency (10%) bright blue and green LEDs are being produced and sold by the millions by leading manufacturers such as Nichia, Hewlett-Packard, and Cree.

In case of the LEDs, there is general agreement that the pronounced compositional anomalies encountered in InGaN (in the sense of non-Poissonian cation distribution) are beneficial in enhancing the overall light emission efficiency. This is so because of the strong localization that accompanies the compositional fluctuations on the spatial scale of the electron and hole Bohr radii. Evidence for this has been acquired from the presence of soft absorption edges, large spectral redshifts ("stokes shifts") between absorption and luminescence, and the long (> nsec) room temperature lifetimes that show strong dispersion as well. However, the presence of the large average strain and local fluctuations in the strain, coupled with the known exceptionally large piezoelectric coefficients in GaN/AlGaN heterostructures,[19] make it difficult to assign specific contributions by these effects to the nature of near bandedge states in InGaN in the laser devices with any kind of quantitative certainty. Furthermore, the defect landscape from extended defects is subject to significant variations depending on the growth details, complicating an already camouflaged picture further. The case of the piezoelectric fields in particularly topical at this writing, in that the estimated piezofields (up to MV/cm) not only influence the energies of the conduction and valence band edges sizably ($\sim$ 100 meV) but can have a direct impact on the dilution of the electron-hole overlap and interaction. The former is of relevance to the one-electron matrix element while the latter influence the many-body electronic phenomena discussed in Sec. 5. Hangleiter and co-workers have studied the optical transitions in undoped InGaN QWs by time resolved photoluminescence to argue that the observed spectral shifts are consistent with the built-in piezoelectric fields.[20] In an InGaN MQW diode laser, the high electron hole pair density will, of course, significantly screen the piezoelectric contributions but so far there is no detailed experimental studies to address this question in the actual devices.

There have been a number of experimental reports, employing microscope and microprobe techniques that have provided real images of the In-aggregation in InGaN epitaxial thin films and quantum wells. At one end of the spatial spectrum, Kisielowski and co-workers have employed atomic resolution transmission electron microscopy scale to show evidence of clustering on the scale of a lattice constant in connection with studies of strain in the InGaN/GaN QWs.[21] At the other end of the apparent size distribution, both cathodoluminescence (CL)[22] and near field optical microscopy (NSOM)[23,24] have been used to acquire luminescence based images on a spatial scale on the order of 100 nm. Such a strikingly wide distribution of the cation aggregates is very strong contrast with the random alloy behavior, where spatial compositional (and crystal potential energy) fluctuations are confined to the atomic scale. We note in passing that the CL and the NSOM spatially resolved techniques have been used to also show how the presence of threading dislocations and larger extended state defects such as nanopipes has little evident impact on the radiative efficiency of InGaN QWs.[25,26] In other words, the conventional assumptions of such defects as highly efficient nonradiative recombination centers seem to be inapplicable in the nitride semiconductors.

Among the reports on high resolution microscopy of InGaN QWs and epilayers, Chichibu *et al.*[22] studied spectral variations of luminescence from InGaN/GaN QWs have by cathodoluminescence (CL) at low temperatures, under unspecified excitation conditions, to show experimental evidences for carrier localization at $T = 10$ K. In that work, mean the In-cluster size was estimated to be less then 60 nm, though the CL images revealed In-rich (deficient) areas up to half a micron in diameter. The NSOM is an attractive alternative to CL allowing subwavelength resolution in photoluminescence (PL) measurements with energetically direct and accurately measurable carrier injection into the QW, together with simultaneous topographic imaging. Comparison of the spatially resolved PL with other simultaneous optical microprobes, e.g., reflectivity or transmission, can provide additional useful information. Here we show an example of the use of *collection*-mode NSOM, where issues of carrier diffusion can be minimized when studying the local spectral variations in the PL emission on a sub-100 nm scale under high *e-h* pair injection such as encountered in a diode laser.[24]

Typical near-field PL images taken on four undoped 30 Å thick $In_xGa_{1-x}N$ QWs of with 90 Å thick GaN barriers are shown in Fig. 3 for three different wavelengths of emission ($x \approx 0.2$). The injection of *e-h* pairs ($\sim 10^{19}$ cm^{-3}) was resonant into the QWs by choice of the excitation wavelength. Figure 3(a) shows the NSOM

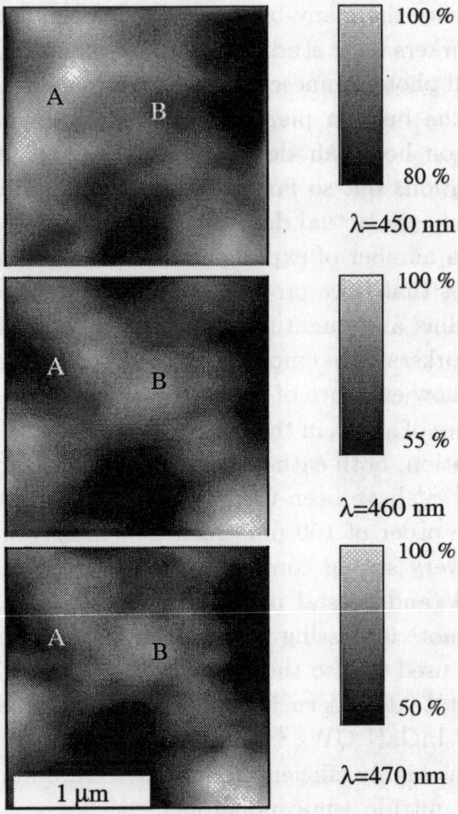

Fig. 3. Collection-mode NSOM PL from InGaN/GaN MQW for three different wavelengths. Traces (a), (b), and (c) correspond to PL NSOM images at $\lambda = 450$ nm $\lambda = 460$ nm, and at $\lambda = 470$ nm, respectively. Markers "A" and "B" highlight regions of different compositional content.

image recorded on the higher energy side of the far-field PL spectrum at $\lambda = 450$ nm ($\hbar\omega = 2.75$ eV), Fig. 3(b) was recorded at the center of the spectrum at $\lambda = 460$ nm ($\hbar\omega = 2.70$ eV), while Fig. 3(c) is the NSOM image taken at $\lambda = 450$ nm ($\hbar\omega = 2.64$ eV) on the lower energy side of the PL spectrum.[24] The images reveal darker and brighter regions of several hundred nm in. They strongly suggest that the light emission action in the room-temperature InGaN MQW lasers occurs from the extended states, once localized states are filled. Chichibu *et al.*[22] evoked electro-and photoluminescence in InGaN MQWs LEDs to to argue for recombination at certain potential minima while Nakamura and co-workers have proposed a "quantum dot" description for the non-random InGaN alloy, based on transmission electron microscopy (TEM) observations.[22] More recently, other band-tail producing mechanism driven by strain such as piezoelectric effects, have been discussed.[27] The gain spectroscopy on InGaN QW diode lasers discussed above shows how it is necessary to fill a large number of band tail states, over an energy range exceeding

100 meV, before the electron-hole pair quasi-Fermi level has reached a value at which the optical gain coefficient is sufficiently large for the onset of laser action ($g_{th} > 3000\ \text{cm}^{-1}$). The excitation and position dependence of the NSOM spectra in Figs. 3(a) and (b) gives another view of this state filling process. A significant part of the PL spectra in each particular point of the sample is broadened *homogeneously* (on the scale of the instrument resolution), and one can argue that all localized states and significant amount of extended states, perhaps "mixed" with higher energy localized states, contribute to the local radiative recombination under high injection levels. Also, sites with larger local band gap (e.g. In-deficient regions) can easily become interconnected through the carrier diffusion.

A different application of the NSOM imaging technique has been recently used to measure the electron-hole (ambipolar) diffusion in InGaN MQWs under optical injection.[28] The technique involves the setting up of an excitation interference grating with two blue laser beams, on a spatial scale of about 200 nm, and the direct imaging of the PL intensity variations with the NSOM fiber tip as the high spatial resolution light collector. A schematic of the experimental set up is shown in Fig. 4. In these types of experiments, one studies the grating contrast in the PL profiles, diminished due to carrier diffusion. Numerical fitting of such PL profiles with the solutions of the diffusion equation for electron-hole densities

$$\Delta n(x,t) = \Delta p(x,t) \propto \left(e^{-t/\tau} + \gamma_0 e^{-(t/\tau)(1+(KL_D)^2} \cos(2\pi x/\Lambda)\right)$$

yields the given diffusion length L_D. Here, τ is the (carrier density independent) recombination lifetime, γ_0 is the contrast parameter of the excitation grating, and Λ is the grating period. Vertikov *et al.*[28] studied a number of InGaN MQW samples grown at different laboratories to find that the diffusion lengths at room temperature could vary very widely under low *e-h* injection, presumably reflecting the individual microstructure of a particular MQW. On the other hand, in the high

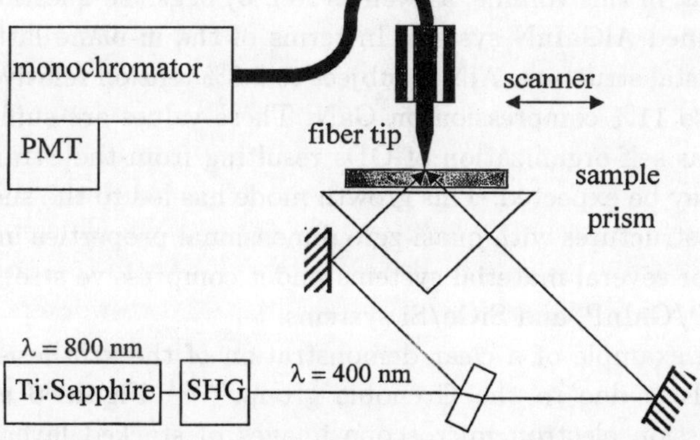

Fig. 4. Experimental schematic for studying *e-h* diffusion in InGaN QWs. The population grating induced by the two interfering blue excitation laser beams is imaged by the NSOM collector tip.

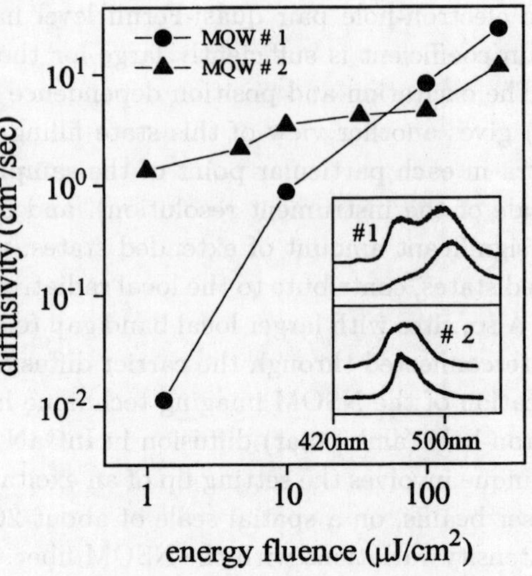

Fig. 5. Dependence of *e-h* diffusivity on injection level (excitation energy fluence).

injection regime typical of a blue diode laser ($> 10^{19}$ cm^{-3}) the values of diffusivity converge to those roughly expected for a "free" electron-hole gas, as shown in the data of Fig. 5 of two InGaN MQW samples with nominally similar thickness and In composition. Note the vastly dissimilar diffusivities for these contrasting samples at low injection, which are assumed to reflect the distinct individual "mesoscale" compositional fluctuation maps in the two samples.

4. Fabrication of Quantum Dots for Nitride Lasers

The striking recent success of using InAs quantum dots (QD) in reducing the threshold current density for infrared diode lasers (see articles by Bhattacharya and Zhukov *et al.* in this volume, as well as Ref. 6) begs the question for an analog within the strained AlGaInN system. In terms of the in-plane lattice constant of the wurtzite crystal structure, AlN is subject to 2.3% tension relative to GaN, while InN is subject to 11% compression on GaN. These values are sufficiently large so that spontaneous self-organization of QDs resulting from the Stranski-Krastanow growth mode may be expected. This growth mode has led to the successful demonstration of nanostructures with quasi-zero dimensional properties in semiconductor heteroepitaxy for several material systems under compressive stress, including the InAs/GaAs, InP/GaInP, and SiGe/Si systems.

A good case example of a clear demonstration of the synthesis of QDs in the nitride materials is due to the Grenoble group.[29–31] Figure 6 shows high resolution transmission electron microscope images of stacked layers of GaN dots, grown by using molecular beam epitaxy to deposit a series of ultrathin insertions (~ 3 monolayers) of GaN into AlN, the entire structure deposited on an AlN buffer

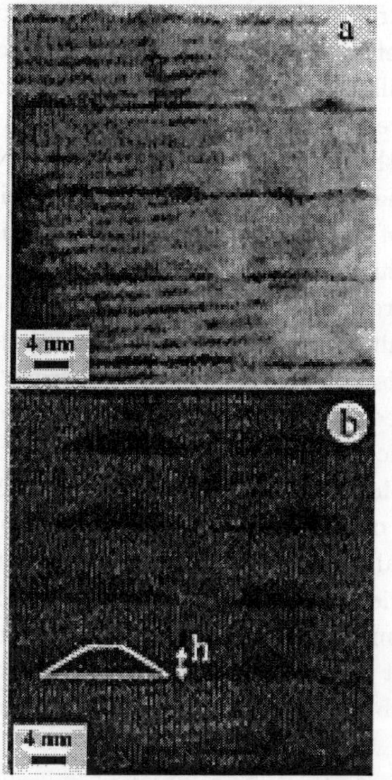

Fig. 6. Cross-sectional TEM view of GaN QDs, stacked in AlN matrix (Ref. 6).

layer on sapphire substrate.[31] Photoluminescence spectroscopy on such structures show anomalous spectral shifts which are attributed to the presence of very large (several MV/cm) piezoelectric fields in these structures.

Several other groups have also employed MBE or MOCVD growth techniques to create ordered GaN-based QD assemblies. Here we note work in which MOCVD growth was employed to form QDs in an $Al_{0.15}Ga_{0.85}N$/GaN heterostructure, deposited on AlN buffer layer on 6H-SiC (0001) substrate.[32] As the lattice mismatch between the constituents is small relative to the Stranski-Krastanow mode of growth, the QD formation was mediated in this work by a small amount of Si acting as antisurfactant. A key claim in this work was the observation of reduced exciton-longitudinal optical phonon interaction due to the reduced dimensionality of the electron/phonon system. (We note in passing that the Frohlich interaction in the nitrides is among the largest in any compound semiconductor.) Finally, the synthesis of GaN QDs in Si(111) substrate has been very recently led to the observation of room temperature photoluminescence from blue to orange, argued to originate from the interplay of QD size, strain, and piezofields.[33]

The insertion of multiple InGaN layers within a GaN matrix has been recently studied, with the aim of creating conditions suitable for lasing.[34] The MOCVD-based growth leads to strong modulation in the compositional spatial profile of InGaN in all three dimensions, although clear evidence for well defined quantum dot formation is still lacking. Nonetheless, under intense optical pumping, high gain conditions have been realized in these types of structures, leading to "surface lasing".[34] With the addition of one AlGaN/GaN distributed Bragg reflector, Krestnikov *et al.*[35] have obtained stimulated emission at theshold excitation intensities on the order of 400 kW/cm^2 and higher, which they assign to vertical cavity lasing due to the high gain in the InGaN system, even in the absence of a high-Q resonator. Such a resonator is generally viewed as a prerequisite for practical vertical cavity surface emitting lasers (see also Sec. 6 below).

While stimulated emission has been obtained from these types of nanoscale, compositionally modulated InGaN heterostructures, it is still unclear whether and how quantum dots participate in the formation of optical gain, given the non-random alloy characteristics of InGaN discussed in Secs. 2 and 3 above. At the same time, the present quality of the GaN/AlN QDs seems to be not quite adequate to support a required *e-h* pair density for optical gain due to the presence of defects. Nonetheless, work on nitride QDs, still at its infancy, is likely to produce important advances in the near term. From the viewpoint of optical gain, this is especially relevant and of significant potential impact, due to the enhancements promised by the electron-hole Coulomb interactions. We turn to the such excitonic effects next.

5. Gain Spectroscopy and Laser Emission in ZnSe-Based Quantum Wells: The Role of Excitonic Contributions

In terms of basic optical properties, the wide bandgap semiconductors are exceptional in the strength of the excitonic effects that dominate the linear optical response. Exciton binding energies exceeding $E_x > 40$ meV have been measured for ZnCdSe QWs and the nitride compounds are expected to possess considerably larger values. Noting that the exciton oscillator strength scales roughly as the square of E_x, the robustness of 2D excitons (against screening and thermal dissociation) implies a significant contribution to the optical gain even at room temperature. While the compositional anomalies in InGaN QWs have made it very difficult to isolate excitonic features in their optical spectra to date, the heteroepitaxial quality of II-VI structures has allowed a good deal of such spectroscopy to be performed. Thus we have chosen to include selected examples of this important subject in this chapter, using the recent experience with the II-VI materials as illustrative examples. In particular, both excitonic and biexcitonic spectrally distinct features have been clearly identified in ZnSe-based QWs up into a high density regime at cryogenic temperatures where stimulated emission and laser action commences. The entire subject is both broad and still developing, and will be further impacted by anticipated developments of lower dimensional quantum confining heterostructures such as quantum wires. In this section we first make some general comments about the many-body physics in the widegap semiconductors, following which specific illustrative examples are given of the multifaceted phenomena that has been discovered in the laboratory. We concentrate, in particular, on the room temperature ZnCdSe QW diode laser device which is demonstrably subject to pronounced *e-h* Coulomb effects.

In a high density electron-hole system (2D), the direct *e-h* Coulomb interaction is accompanied by exchange and correlation effects which profoundly increase in importance with the particle density. Physically, the two extreme limits are easily identified, for example, in a GaAs bulk crystal or quantum well: the low density bound state (exciton) regime and the high density electron-hole plasma. For the latter, the exchange and correlation effects are seen mainly through the bandgap renormalization effect which shifts the emission wavelength of such a semiconductor laser to longer wavelengths (by up to several tens of meV). In terms of the actual gain spectrum (lineshape or magnitude), however, the many-body effects induce only rather small corrections to calculations made in the one electron picture that usually include some phenomenological damping rates. Sophisticated theoretical approaches, converging to the semiconductor Bloch equations, have been applied to describe the gain spectra of various III-V lasers with impressive success[36] and these methods are presently also being tested with the widegap semiconductors.[17,37] While referring the reader elsewhere for expert discussion of the power of the semiconductor Bloch equations, we note here that the approach might be qualitatively described as having its starting point in the electron-hole plasma limit to which the *e-h* Coulomb interactions are added. In contrast, we take the viewpoint here where the starting point is taken at the opposite limit, i.e. a gas of noninteracting (or weakly interacting) gas of excitons subject to many-body interactions. This pathway is natural for an experimentalist when considering e.g. the ZnSe QW or future GaN QW laser in which the temperature is gradually increased from liquid helium to room temperature and where the role of bound *e-h* pair states at low temperatures is beyond dispute. Approaching the problem from this limit presents, however, significant challenges to the semiconductor Bloch equation approach, for example, due to the difficulty of including the screening by the bound states into the formalism. Yet evidence in the ZnCdSe QWs strongly suggest that the Coulomb correlations remain very potent, when compared e.g. with an GaAs laser.

When compared with conventional III-V lasers, the widegap semiconductors offer circumstances that are distinctly different simply because the exciton binding energies can be comparable or larger than the estimated "plasma" exchange/correlation contributions, as well as larger than the one electron Fermi energies, the optical phonon energy, and kT at room temperature. Hence a basic question, with applied significance, emerges about the imprint of the excitonic effects on optical gain, be it a ZnSe- or GaN-based quantum well laser. Thermodynamically, we may estimate the relative populations of the exciton and free *e-h* pair states as a function of the pair density in a fairly straightforward way by including phase space filling and exchange.[38] The purpose of the figure is partly pedagogical and the important aspects which are brought about by other excitonic complexes (biexcitons) as well as localization in weakly disordered media are not yet included. The figure also indicates the estimated pair density for a ZnCdSe QW laser at $T = 77$ K while comparing such a threshold density for a GaAs laser. Clearly at least qualitatively, the II-VI or a nitride laser can operate in a regime where the excitons (or even excitonic molecules) are the majority species. This type of thermodynamic argument lacks, however, any information about the relevant optical cross sections for the exciton and free pair radiative process; in fact the former is considerably enhanced by the Coulomb induced wavefunction overlap.

Chronologically, localized excitons were first evoked by the authors to define optical gain in the ZnCdSe QW,[13] with supporting and complementary evidence reported by other groups,[39,40] especially by Cingolani and co-workers.[41] Here we consider the experimental evidence in the case of a room temperature electrically injected diode laser in an attempt to gauge the physical role of the *e-h* pair correlations in a practical widegap semiconductor device.

Figure 7 compares gain/absorption spectra for a ZnCdSe[42] and GaAs SQW diode laser in the vicinity of the $n = 1$ QW HH exciton transition,[11] obtained in each case by the same experimental technique of correlating edge stimulated emission with top spontaneous emission.[10,12] This is the same experimental technique discussed in connection with the gain spectrosopy of InGaN MQW diode lasers above. Note how, as a function of increasing current injection, the HH and LH exciton features are rapidly bleached in the GaAs QW so that at laser threshold any semblance to excitonic resonances is absent. This is in striking contrast with the case for the II-VI laser, where a partially bleached HH exciton resonance and a nearly intact LH resonance are clearly present at laser threshold, with gain appearing some 60 meV below the HH exciton resonance. While in broad agreement with predictions of the semiconductor Bloch equations,[37] important differences remain concerning the details of the gain spectrum. In the authors' opinion the microsopic details of the system, in terms the appropriate many-body description remain theoretically incomplete for now. For some phenomenological ideas and interpretations, as well as experimental details leading to Fig. 7, the reader is referred to Ref. 42. The role of the Coulomb interactions remains practically relevant, for example, in the context of present efforts to fabricate wide bandgap semiconductor quantum wire lasers where the bound exciton states might be expected to be increasingly stable against dissociation and screening. In any event, the typical range of enhancement of the peak gain for room temperature ZnCdSe QW lasers is in the range of factor of three to four, clearly an important positive feature in designing low threshold devices. In considering the laser threshold condition, we remark that while the excitonic enhancements also imply an accelerated spontaneous (radiative) decay rate, the concentration of oscillator strength in the gain spectrum near the $n = 1$ HH exciton resonance specifically enhances the peak gain coefficient, whereas the spontaneous emission rate is proportional to the integral of the corresponding spontaneous emission spectrum. In particular with new exploratory II-VI and group III-nitride lasers, where nonradiative recombination may be appreciable, the excitonic enhancements to both spontaneous emission and optical gain are clearly important.

As yet another experimental illustration of the impact of the *e-h* Coulomb correlation at room temperature, we show recent results where ZnCdSe QW diode lasers have been studied in a magnetic field, applied perpendicular to the QW layer plane.[43] In this study $Zn_{1-x}Cd_xSe/Zn_{1-x}S_xSe/Zn_{1-x}Mg_xS_ySe_{1-y}$ SCH/SQW diode laser structures were used with a rather large Cd-fraction to ensure strong electron-hole confinement in the QW. This choice of quantum well design parameters was made in order to approach the quasi-two dimensional limit, i.e. to enhance the exciton binding energy, which for this concentration range of Cd is estimated to be at least 40 meV, based on earlier studies on the ZnCdSe/ZnSSe QW.[44]

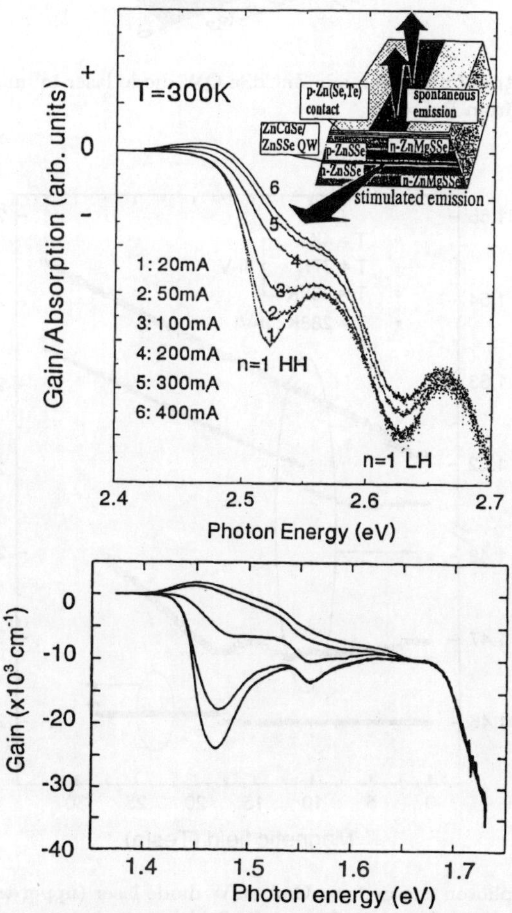

Fig. 7. Comparison of the gain-absorption spectra for SQW ZnCdSe and GaAs SCH diode lasers at room temperature, as a function of injection current (upper and lower traces, respectively), in the vicinity of the $n = 1$ HH and LH transition.

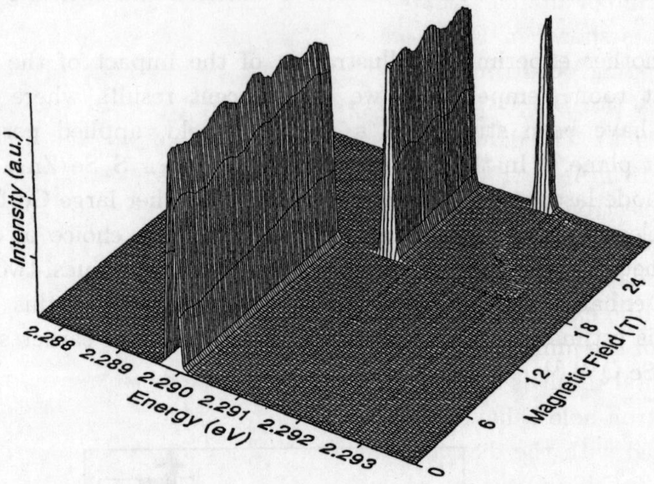

Fig. 8. Emission spectra of a single mode ZnCdSe QW diode laser in magnetic fields up to 29 T at room temperature (from Ref. 43)

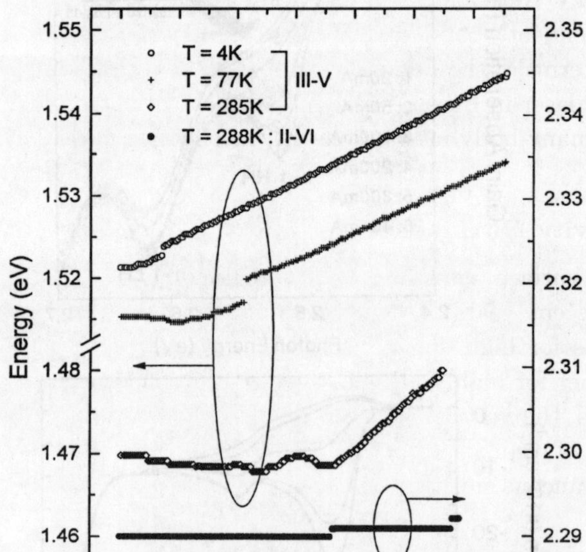

Fig. 9. The emission photon energy for a GaAs QW diode laser (upper traces) and ZnCdSe QW diode laser (lowest trace) as a function of magnetic field, oriented perpendicular to the layer plane.

The spectrum of an index guided diode laser devices operating in a single longitudinal mode is shown in Fig. 8 in magnetic fields up to 29 T,[43] where we see the mode hopping occurring as discrete jumps between the photon energies of $\hbar\omega = 2.2899$ eV (for $B = 0$–18 T), $\hbar\omega = 2.2901$ eV ($B = 18$–28 T), and $\hbar\omega = 2.2921$ eV ($B = 28$–29 T). The diode laser was operated well above threshold at injection level of $I = 18$ mA ($\sim 1.4I_{th}$, where $I_{th} = 13$ mA). From the measured room temperature electron-hole lifetime under comparable injection,[42] we calculate an e-h pair density of 1.9×10^{12} cm^{-2} in the ZnCdSe single QW. In a free carrier picture, a Landau shift, linear in the magnetic field and of the order of $\Delta E_L \sim 0.5$ meV/T would be expected (either due to the shift of the lowest conduction band Landau level or the jumps of the Fermi level between such occupied levels). Clearly, the experimental result is about one order of magnitude smaller and incompatible with free electron-hole behavior. On the other hand, the agreement is much better when compared with the diamagnetic shifts measured for the $n = 1$ HH exciton transition in ZnCdSe QWs, typically, $\Delta E_{dia} \sim 2$ μeV/T^2, with the reduced exciton mass of $\mu^* \sim 0.1$ m$_o$. Given the exciton binding energy of about 40 meV, this implies that at $B = 30$ Tesla, we are still in the regime where $E_x > \hbar\omega_c$, the electron cyclotron energy. We compare the behavior of the ZnCdSe QW diode laser with GaAs-based QW diode lasers in Fig. 9. In strong contrast, while exhibiting temperature dependent "Landau-level jumping" (oscillations) of the Fermi level at low B-fields (in terms of the Fermi-level), the spectral shifts in the emission of the GaAs QW diode laser in Fig. 9 evolve into a linear cyclotron-like shift, with some influence of the many-body interactions on the effective masses evident, however.

6. Vertical Cavity Nitride Devices

Vertical cavity, surface emitting semiconductor lasers (VCSEL) have gained importance in recent years as flexible alternatives for applications where beam quality, prospects for high density arrays, microelectronics integrability are at a premium. Whether for highly parallel optical interconnects, optical disk heads, or printers/scanners, the vertical cavity configuration offers many distinguishing advantages. In case of resonant cavity LEDs (RCLED), the quasi-beam like directionality in the spontaneous emission and enhancements to the radiative recombination can be useful.

The technological prospects for blue, green, and near ultraviolet VCSELs and RCLEDs are at very early research stages. At this writing, it is unclear what kind of a combination of epitaxial growth and device design/processing schemes might result in a technologically viable VCSEL, for instance. Still, there are ample fundamental physical reasons which suggest that microcavity emitters based on widegap semiconductors have special properties that offer opportunities both in terms of basic physics and device performance. These features originate from the strong light-matter coupling, fundamental to lower dimensional nitride and II-VI heterostructures, which is expected to endow a widegap semiconductor VCSEL with intertwining of the electronic (excitonic) and photonic degrees of freedom in a manner that exceeds the usual case in a laser (where the optical field is a perturbation).

As a basic measure of the strength of exiton-photon interaction in the linear optical regime, normal mode (Rabi) splittings of about 10 meV per QW have been recently measured in a ZnCdSe QW "green" microcavity, fabricated with dielectric DBR mirror stacks.[45] Sizable effects have also been seen in CdTe microcavities with as-MBE-grown DBRs.[46] Given the measured threshold carrier density in an optically pumped vertical cavity II-VI surface emitting laser[47] and the known optical characteristics of a ZnSe-based QWs at comparable densities, a II-VI microcavity can be expected to reach the lasing threshold before the composite oscillators is saturated by phase-space filling and screening of the excitonic resonance. In fact gain spectroscopy performed on ZnCdSe QW VCSEL structures[48] shows dramatically that the composite exciton-photon oscillator is not completely bleached at lasing threshold, emphasizing how in a widegap semiconductor microresonator laser the strong interaction between the QW excitons and the electromagnetic modes transcends the conventional perturbative regime.

The current challenges facing efforts to realize blue and violet RCLEDs and VCSELs in the AlGaInN material system have many parallels with those with the recent II-VI work. On one hand, the nitride *pn*-junction heterostructures have produced both excellent LEDs and strikingly robust diode lasers. On the other hand, the realization of a reasonably high quality vertical cavity is, if anything, even more challenging in the nitride compounds than in the equivalent II-VI widegap materials. There are two approaches for microcavity fabrication, based on *in situ*, as-grown distributed GaN/AlGaN multilayer Bragg reflectors (DBRs) and dielectric DBRs, respectively (and their hybrids). Optical experiments show some promise for each approach at this writing. We note that the low index of refraction contrast within the AlGaN alloy system makes it at first sight a major challenge to create low loss, high reflectivity DBR mirrors by direct MOCVD growth on a sapphire substrate. Nonetheless, Arakawa and co-workers[49,50] have incorporated a 43 period MOCVD grown $GaN/Al_{0.34}Ga_{0.66}N$ DBR mirror ($R \sim 0.98$) on sapphire into a hybrid vertical cavity structure, in which the second mirror comprises a 15 period ZrO_2/SiO_2 dielectric DBR ($R = 0.995$). The "active" medium in the resonator was a 2.5 λ thick InGaN MQW composed of 26 wells. Upon high excitation, pulsed pumping by a low repetition rate nitrogen laser, the authors as well as Krestnikov *et al.*[35] observe stimulated emission from the structure and attribute this to vertical cavity lasing (see comment below). The threshold for the onset of stimulated emission was very high, ~ 10 mJ/cm^2 in terms of the incident pulse energy density.

A different approach to the fabrication of relatively high quality vertical cavity has been demonstrated by Song *et al.*,[51,52] the objective being to create an all-dielectric DBR resonator. The idea involves the separation of a quantum well heterostructure from its sapphire substrate and its integration with high reflectivity, low loss dielectric mirrors. The process involves the initial flip-chip mounting of

Fig. 10. SEM cross sectional image of a InGaN MQW vertical cavity structure, equipped with dielectric DBRs.

the nitride substrate on an artificial host substrate (by various wafer bonding and related approaches) and the release of the sapphire substrate by the exposure of the structure to a single pulse of excimer laser radiation ($\lambda = 308$ nm).[53,54] The UV laser radiation is absorbed selectively near the GaN/sapphire interface and, once a critical pulse energy density is exceeded, leads to the physical separation of the sapphire over a macroscopic area due to forces that are presumed to originate from rapid thermal decomposition of the GaN. This allows the deposition of a second HfO_2/SiO_2 multilayer dielectric stacks to complete a vertical cavity structures, as shown in the cross sectional electron micrograph of Fig. 10. Song and co-workers have demonstrated a vertical cavity blue light emitting diode and observed stimulated emission in optically pumped structures that show a cavity Q-factor approaching 1000 in initial experiments. The spectrum of Fig. 11 (top) shows the emission spectrum at $T = 258$ K from one such structure, under quasi-continuous wave optical pumping.[52] The modes are well defined and exhibit a linewidth of approximately 0.5 nm under high resolution. Under optical pumping at relatively modest levels of quasi-cw excitation (average power ~ 30 mW), strong stimulated emission has been measured from these structures (Fig. 11, bottom), at frequencies corresponding to the modal positions, including a well defined threshold and the clear observation of linearly polarized far field emission.[52] In our experience, it has been very difficult to reach this type of "true vertical cavity lasing", the reason being that that such true VCSEL operation has to severely compete with in-plane lasing, the latter aided by the formation of "accidental" optical resonators defined by cracks induced by strained AlGaN layers.

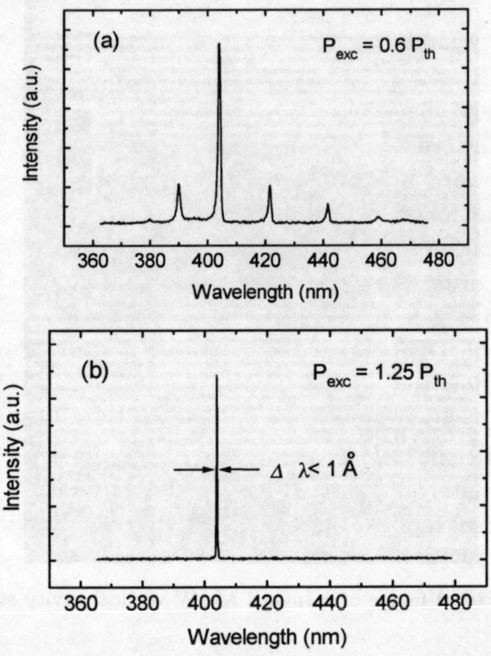

Fig. 11. Upper trace: Spontaneous emission spectra of the InGaN MQW VCSEL below threshold. Lower trace: Stimulated emission spectra above threshold under quasi-cw pumping conditions at $T = 258$ K.

Such competition between in-plane and true perpendicular vertical cavity lasing is a serious obstacle to further development of the nitride VCSELs. The nitride heterostructures contain extended defects such as the cracks that develop along principal crystalline axis in the case of AlGaN, once the composition of Al exceeds, say about 10%. In their recent work on surface mode lasing from stacked InGaN insertions within a GaN matrix, Krestnikov *et al.*[35] have noted that optical pumping conditions need to be sufficiently extreme so that the in-plane lasing saturates. If this is so, the implication is that the lasing threshold for a true VCSEL might be prohibitively high. Hence considerable further effort is required for improving the material homogeneity for a realistic VCSEL in the blue and near ultraviolet.

Finally, we mention rudimentary first efforts that have shown some promise but also highlighted the challenges for the realization of a electrical injection vertical cavity device in the nitride semiconductors. Song *et al.*[55] have fabricated demonstration devices that may form one approach for creating a resonant cavity LED. The method is again based on the separation of the sapphire substrate from the InGaN/GaN/AlGaN QW nitride heterostructure by excimer laser irradiation and the deposition of a high reflectivity dielectric DBR on the n-side of the device. The p-side uses the natural reflectivity of the contact metals (typically up to $R \sim 0.6$) to define the vertical cavity. The typical optical aperture defined by this construction was typically about 20 μm in diameter. The devices have been operated in a continuous mode up to a current density of 1 kA/cm^2, demonstrating both well defined vertical cavity modal structure as well as their robustness as rudimentary RCLEDs.

The current-voltage characteristics of such devices highlight a specific, difficult challenge for RCLEDs and, especially, blue VCSELs. The "turn-on" voltage at which significant forward bias curent (and LED emission) occurs is at least two volts higher than that for a device fabricated by conventional means. This issue is associated with the rather low resistivity of p-GaN and its alloys. The hole mobility in p-GaN is on the order of 1 cm^2/Vsec while the maximum reported free hole concentrations are $p < 1 \times 10^{18}$ cm^{-3}. Thus in case of vertical cavity structures that employ dielectric DBRs, the injection of holes requires some enhancement scheme which will allow the holes to reach the optically active region (below the DBR). In the case of a standard p-type GaN "current spreading layer", the lateral conductivity is insufficient to compete with vertical transport, until the device area approaches approximately 1 μm. Such a small optical aperture, in turn, implies prohibitive (transverse) modal losses for a VCSEL. Significant improvement in the operation of the RCLEDs has been obtained recently by an alternative approach of inserting transparent intracavity layers of indium-tin-oxide to the p-type contact side of the devices.[56]

Some progress has been made recently for reducing the lateral resistivity in GaN by employing concept that are related to modulation doping in conventional III-V semiconductors. Kozodoy *et al.*[57] have achieved high p-type conductivity in Mg-doped AlGaN/GaN superlattices. Effective hole concentrations exceeding 2.5×10^{18} cm^{-3} have been achieved and a lateral resistivity as low as 0.2 Ωcm is realized. In case of the nitride heterostructures, dielectric polarization and piezoelectric contributions significantly modulate the bandedge energies that can be exploited in these types of modulation doped schemes. While the value of the lateral resistivity reached by these authors is still about one order of magnitude less than required for a practical VCSEL, the "bandstructure engineered" approach suggests that further improvements maybe possible by optimizing the superlattice structure with specific nitride compounds.

Acknowledgments

This work was supported by the National Science Foundation. We acknowledge with gratitude our collaborators whose work has been featured in this article and in the reference list.

References

1. S. Nakamura, *Semic. Sci. Technol.* **14** (1999) R27.
2. J. Edmond, G. Bulman, H. S. Kong, M. Leonard, K. Doverspike, W. Weeks, J. Niccum, S. Sheppard, G. Negley, J. D. Brown, J. T. Swindell, T. Overocker, J. F. Schetzina, Y.-K. Song, M. Kuball, and A. Nurmikko, *Proceedings of the Second International Conference on Nitride Semiconductors ICSNS'97*, Tokushima, Japan, pp. 448–449.

3. M. Kneissel, D. P. Bour, C. G. van de Waale, L. T. Romano, J. E. Northrup, R. M. Wood, M. Teepe, and M. Johnson, *Appl. Phys. Lett.* **75** (1999) 581.
4. K. Katayama, H. Yao, F. Nakanishi, H. Doi, A. Saegusa, N. Okuda, T. Yamada, H. Matsubara, M. Irikura, T. Matsuoka, T. Takebe, S. Nishine, and T. Sirakawa, *Appl. Phys. Lett.* **73** (1998) 102.
5. Y. Narukawa, Y. Kawakami, Sz. Fujita, Sg. Fujita, and S. Nakamura, *Phys. Rev.* **B55** (1997) R1938.
6. D. Bimberg, M. Grundmann, and N. N. Ledentsov, *Quantum Dot Heterostructures* (Wiley Interscience, 1999).
7. G. Frankowsky, F. Steuber, V. Härle, F. Scholz, and A. Hangleiter, *Appl. Phys. Lett.* **68** (1996) 3746.
8. S. Nakamura and G. Fasol, *The Blue Laser Diode* (Springer, Berlin, 1997).
9. M. Kuball, E.-S. Jeon, Y.-K. Song, A. V. Nurmikko, P. Kozodoy, A. Abare, S. Keller, L. A. Coldren, U. K. Mishra, S. P. DenBaars, and D. A. Steigerwald, *Appl. Phys. Lett.* **70** (1997) 2580.
10. C. H. Henry, R. A. Logan, and F. R. Merritt, *J. Appl. Phys.* **51** (1980) 3042.
11. M. P. Kresler and C. Harder, *Appl. Phys. Lett.* **57** (1990) 123.
12. P. Blood, A. Kucharska, J. Jacobs, and K. J. Griffiths, *Appl. Phys.* **70** (1991) 1144.
13. J. Ding, T. Ishihara, M. Hagerott, H. Jeon, A. V. Nurmikko, N. Samarth, H. Luo, and J. Furdyna, *Phys. Rev. Lett.* **60** (1992) 1707.
14. Y.-K. Song, M. Kuball, A. V. Nurmikko, G. E. Bulman, K. Doverspike, S. T. Sheppard, T. W. Weeks, M. Leonard, H. S. Kong, H. Dieringer, and J. Edmond, *Appl. Phys. Lett.* **72** (1998) 1418.
15. S. Chichibu, T. Azuhata, T. Sota, and S. Nakamura, *Appl. Phys. Lett.* **69** (1996) 4188.
16. S. Chichibu, T. Azuhata, T. Sota, and S. Nakamura, *Appl. Phys. Lett.* **70** (1997) 2822.
17. W. W. Chow and S. W. Koch, *Appl. Phys. Lett.* **66** (1995) 3004.
18. M. Kneissl, C. G. van de Walle, D. P. Bour, L T. Romano, C. P. Master, J. E. Northrup, and N. M. Johnson, *Proc. Int. Conf. Luminescence*, Osaka, Japan (in press).
19. J. S. Im, H. Kollmer, J. Off, A. Sohmer, F. Scholz, and A. Hangleiter, *Phys. Rev.* **B57** (1998) R9435.
20. H. Kollmer, J. S. Im, S. Heppel, J. Off, F. Scholz, and A. Hangleiter, *Appl. Phys. Lett.* **74** (1999) 82.
21. C. Kisielowski, Z. Lilienthal-Weber, and S. Nakamura, *Jpn. J. Appl. Phys.* **36** (1997) 6932.
22. S. Chichibu, K. Wada, and S. Nakamura, *Appl. Phys. Lett.* **71** (1997) 2346.
23. P. A. Crowell, D. K. Yong, S. Keller, E. L. Hu, and D. D. Awschalom, *Appl. Phys. Lett.* **72** (1998) 927.
24. A. Vertikov, A. V. Nurmikko, K. Doverspike, G. Bulman, and J. Edmond, *Appl. Phys. Lett.* **73** (1998) 493.
25. J. Liu, N. R. Perkins, M. N. Horton, J. M. Redwing, M. A. Tischler, and T. F. Kuech, *Appl. Phys. Lett.* **69** (1996) 3519.
26. A. Vertikov, M. Kuball, A. V. Nurmikko, Y. Chen, and S.-Y. Wang, *Appl. Phys. Lett.* **72** (1998) 2645.
27. K. P. O'Donnell, T. Breitkopf, H. Kalt, W. van der Stricht, I. Moerman, P. Demeester, and P. G. Middleton, *Appl. Phys. Lett.* **70** (1997) 1843.
28. A. Vertikov, I. Ozden, and A. V. Nurmikko, *J. Appl. Phys.* **86** (1999) 4697.
29. B. Daudin, F. Widmann, G. Feuillet, Y. Samson, M. Arlery, and J. L. Rouviere, *Phys. Rev.* **56** (1997) R7069.
30. F. Widmann, B. Daudin, G. Feuillet, Y. Samson, and J. L. Rouviere, *J. Appl. Phys.* **83** (1998) 7618.

31. F. Widmann, J. Simon, B. Daudin, G. Feuillet, Y. Samson, J. L. Rouviere, N. T. Pelekanos, and G. Fishman, *Phys. Rev.* **B58** (1998) R15989.
32. P. Ramwall, S. Tanaka, S. Nomura, P. Riblet, and Y. Aoyagi, *Appl. Phys. Lett.* **75** (1999) 1935.
33. B. Damilano, N. Grandjean, F. Semond, J. Massies, and M. Leroux, *Appl. Phys. Lett.* **75** (1999) 962.
34. A. V. Sakharov, W. V. Lundin, I. L. Krestnikov, V. A. Semenov, A. S. Usikov, A. F. Tsatsul'nikov, Yu. G. Musikhin, M. V. Baidakova, Zh. I. Alferov, N. N. Ledentsov, A. Hoffmann, and D. Bimberg, *Appl. Phys. Lett.* **74** (1999) 3921.
35. I. L. Krestnikov, W. V. Lundin, A. V. Sakharov, V. A. Semenov, A. S. Usikov, A. F. Tsatsul'nikov, Zh. I. Alferov, N. N. Ledentsov, A. Hoffmann, and D. Bimberg, *Appl. Phys. Lett.* **75** (1999) 1192.
36. H. Haug and S. W. Koch, *Quantum Theory of the Optical and Electronic Properties of Semiconductors* (World Scientific, Singapore, 1993).
37. C. Ell and H. Haug, private communication.
38. R. Cingolani, L. Calcanile, G. Coli, R. Rinaldi, M. Lomoscolo, M. DiDio, A. Franciosi, L. Vanzetti, G. C. LaRocca, and D. Campi, *J. Opt. Soc. Am.* **B13** (1996) 1268.
39. K. Kawakami, I. Hauksson, J. Simpson, H. Stewart, I. Gailbraith, K. A. Prior, and B. C. Cavenett, *J. Cryst. Growth* **138** (1994) 759.
40. Zh. Alferov, S. Ivanov, P. Kop'ev, A. Lebedev, N. Ledentsov, M. Maximov, I. Sedova, T. Shubina, and A. Toropov, *Superlattices and Microstructures* **15** (1995) 65.
41. R. Cingolani, in *II-VI Blue/Green Light Emitters: Device Physics and Epitaxial Growth, Semiconductors and Semimetals*, Vol. 44, eds. R. L. Gunshor and A. V. Nurmikko (Academic Press, New York, 1996), pp. 163–226.
42. J. Ding, M. Hagerott, P. Kelkar, A. V. Nurmikko, D. C. Grillo, L. He, J. Han, and R. L. Gunshor, *Phys. Rev.* **B50** (1994) 5787.
43. Y.-K. Song, A. V. Nurmikko, C.-C. Chu, T. B. Ng, and R. L. Gunshor, *Appl. Phys. Lett.* **71** (1997) 2874.
44. N. T. Pelekanos, J. Ding, M. Hagerott, A. V. Nurmikko, H. Luo, N. Samarth, and J. Furdyna, *Phys. Rev.* **B45** (1992) 6037.
45. P. Kelkar, V. Kozlov, A. V. Nurmikko, C.-C. Chu, D. C. Grillo, J. Han, C. G. Hua, and R. L. Gunshor, *Phys. Rev.* **B52** (1995) R5491.
46. F. Quochi, G. R. Hayes, R. Andre, G. Bongiovanni, A. Mura, J. L. Staehli, and S. D. Le, *Phys. Rev. B* (1988).
47. H. Jeon, V. Kozlov, P. Kelkar, A. V. Nurmikko, C.-C. Chu, D. C. Grillo, J. Han, C. G. Hua, and R. L. Gunshor, *Appl. Phys. Lett.* **67** (1995) 1668.
48. P. Kelkar, V. Kozlov, A. V. Nurmikko, C.-C. Chu, D. C. Grillo, J. Han, C. G. Hua, and R. L. Gunshor, *Phys. Rev.* **B56** (1997) 7564.
49. T. Someya, K. Tachibana, J. Lee, T. Kamiya, and Y. Arakawa, *Jpn. J. Appl. Phys.* **37** (1998) L1424.
50. T. Someya, R. Werner, A. Forchel, M. Catalano, R. Cingolani, and Y. Arakawa, *Science* **285** (1999) 1905.
51. Y.-K. Song, H. Zhou, M. Diagne, I. Odzen, A. Vertikov, A. V. Nurmikko, C. Carter-Coman, R. S. Kern, F. A. Kish, and M. R. Krames, *Appl. Phys. Lett.* **74** (1999) 3441.
52. Y.-K. Song, H. Zhou, M. Diagne, A. V. Nurmikko, R. P. Schneider, C.-P. Kuo, C. Carter-Coman, R. S. Kern, F. A. Kish, and M. R. Krames, *Appl. Phys. Lett.* (in press).
53. M. K. Kelly, O. Ambacher, R. Dimitrov, R. Handschuh, and M. Stutzmann, *Phys. Status Solidi* **A159** (1997) R3.

54. W. S. Wong, T. Sands, and N. W. Cheung, *Appl. Phys. Lett.* **72** (1998) 599.
55. Y.-K. Song, H. Zhou, M. Diagne, A. V. Nurmikko, C. Carter-Coman, R. S. Kern, F. A. Kish, and M. R. Krames, *Appl. Phys. Lett.* **74** (1999) 3720.
56. Y.-K. Song, M. Diagne, H. Zhou, A. V. Nurmikko, and R. P. Schneider, Jr., *Electron. Lett.* (in press).
57. P. Kozodoy, M. Hansen, S. DenBaars, and U. Mishra, *Appl. Phys. Lett.* **74** (1999) 3681.

International Journal of High Speed Electronics and Systems, Vol. 9, No. 4 (1998) 1163–1188

ADVANCED CONCEPTS IN INTERSUBBAND UNIPOLAR LASERS

JEAN-PIERRE LEBURTON
Department of Electrical and Computer Engineering,
and
Beckman Institute, University of Illinois at Urbana-Champaign,
405 North Mathews Avenue, Urbana, IL 61801, USA

FRANCOIS H. JULIEN
Institut d'Electronique Fondamentale, Universite de Paris,
XI 91405 Orsay, France

YULI LYANDA-GELLER
Beckman Institute, University of Illinois at Urbana-Champaign,
405 North Mathews Avenue, Urbana, IL 61801, USA

Advanced concepts in intersubband optical emission for mid-infrared (MIR) and far-infrared (FIR) lasers in nanostructures are presented. Electronic tunability of low dimensional systems, and new simulation techniques are implemented for engineering electron scattering time with phonons and improving laser performances. We successively describe the operation of an optically pumped "quantum fountain laser" made of asymmetric coupled quantum well structures, a FIR quantum wire laser pumped by optic phonons and an electrically tunable, mode locked FIR quantum dot laser.

1. Introduction

The possibility of realizing solid state optical sources with tunable wavelengths in the mid- to far-infrared has stimulated research since the pioneering work of Kazarinov and Suris.[1] The large oscillator strengths combined with a moderate broadening associated with intersubband transitions are favorable for optical emission. The recent demonstration of the unipolar quantum cascade laser by Faist and collaborators has provided a regain of interest in this direction.[2,3]

Alternative intersubband quantum well (QW) laser schemes have been proposed with various designs,[4,5] and very recently, O. Gauthier-Lafaye and coworkers demonstrated mid-infrared (MIR) stimulated emission by optical pumping in asymmetric coupled QW structures.[6] In the meantime, intersubband optical transitions have also been investigated for MIR second harmonic generation,[7] far-infrared (FIR) and TeraHertz (THz) emission.[8,9] In this context quantum wires and quantum dots (QDs) have opened new horizons in the realization of MIR and FIR devices. It has already been emphasized that the full quantization of the electronic spectrum in QDs offers several advantages such as high differential gain, spectral purity and insensitivity to temperature for optical sources compared to 2D confined structures.[10] However, the most compelling reason to use QDs for FIR or MIR optical devices is certainly the tunability of the discrete 3D levels to forbid optic phonon emission when the separation between the levels is smaller or larger than the optic phonon energy.[11]

In this paper we present advanced concepts in intersubband emission which make use of the electronic tunability of low dimensional systems, and exploit novel properties associated with the reduction of the electron phase space for new principles of laser operations. We will successively describe the operation of a "quantum fountain laser" which is optically pumped, a proposal for a FIR quantum wire laser which is based on phonon pumping, and a FIR electrically tunable QD laser.

2. Quantum Fountain Laser (QFL)

The active region of the quantum cascade laser is a three-level system where population inversion is achieved between the third and the second levels if the lifetime of the former is longer than the latter. The ground state is the collector of electrons. The basic idea for optimizing the ratio between the level lifetimes is to design the QW structure such as the two lower states are separated by the optical phonon energy (resonance) E_{LO}, while the third state is placed at much higher energy. Hence, the resonance condition between the two lower levels insures the short lifetime of the second level. The main reason for this difference in the level lifetimes can be traced in the matrix element of optical phonon (POP) scattering rate which varies roughly as $1/q$ thereby favoring intersubband transitions with small phonon wave vector (Fig. 1).

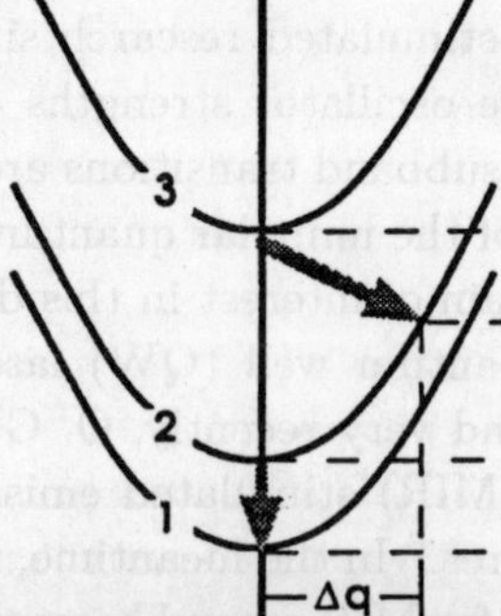

Fig. 1. Schematic representation of the subband dispersion of a three level system with $E_2 - E_1 = E_{LO}\hbar\omega_{\text{pop}}$ and showing the POP wavevector for intersubband transitions.

2.1. *QFL structure*

As an alternative and simpler design than the quantum cascade laser operating by electron injection, an optically pumped three-level system consists of an asymmetric highly coupled double quantum wells (ACQWs) (Fig. 2).[12] The structure is designed for electrons to be optically pumped from the ground state to the second excited state by circumventing the dipole selection rule between these states, while the optical emission takes place between the second and first excited states. The structure is n-doped in the barrier region to provide the necessary electron population in the ACQW ground state for intersubband optical pumping.

The structure[12–14] consists of two GaAs quantum wells with the well width of 78 Å and 49 Å, respectively, separated by a 17 Å $Ga_{0.78}Al_{0.22}As$ thin barrier. Each ACQW is separated from the next period by a 200 Å wide $Ga_{0.78}Al_{0.22}As$ barrier where the center was n-type doped with two-dimensional (2D) electron density of $1 \cdot 10^{11}$ cm^{-2}.

Figure 3 shows one period of the ACQW structure with the square of the first three electron wave functions at 77 K at equilibrium. The electron Fermi level E_f is 2.5 meV below the ground state E_{z_1}. The 1–3, 2–3 and 1–2 intersubband transition energies are 125 meV, 89 meV, and 36 meV, respectively, which is quite close to the experimental data.[12–14] At 4.2 K, all electrons stay in the ground state, and the electron Fermi level E_f increases to 3.5 meV above the ground state E_{z_1}. At the same time, a charge transfer between the two QWs takes place because of the thermal depopulation of the second subband which induces a dc built-in field and a red shift of the ground-to-first excited state intersubband transition with decreasing temperature.[15]

Because of the thermal population, the electron density on the first excited state has a very strong temperature dependence. In the meantime, because of the large electron population in the ground state, the Pauli exclusion principle may limit electron relaxation between the ground and first excited states.

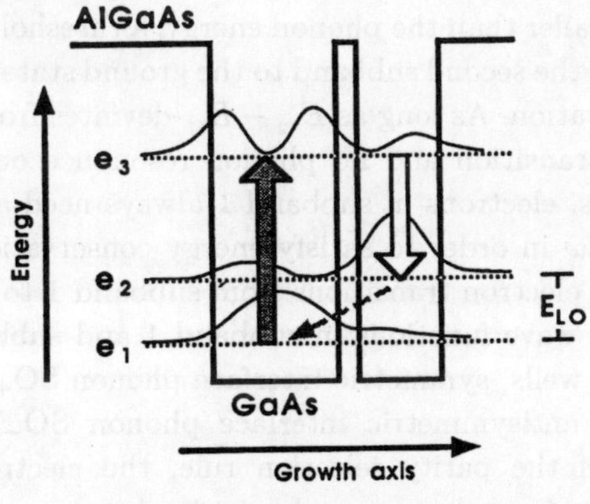

Fig. 2. Schematic plot of conduction band profile of the asymmetric coupled quantum well structure.

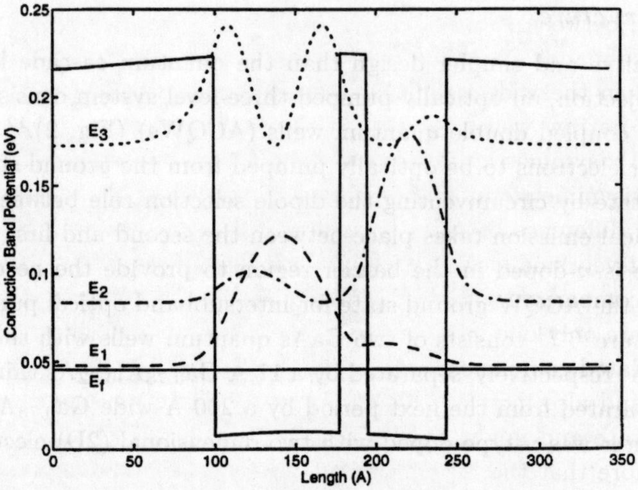

Fig. 3. Self-consistent conduction band potential, electron quasi-Fermi level and the square of the first three electron wavefunctions for the ACQW structure at 77 K.

2.2. *Electron-phonon interaction and intersubband scattering time*

A theoretical investigation of the electron dynamics of intersubband transitions in quantum well structures includes emission and absorption of confined (slab) and interface POP modes.[16,17] Figure 4 shows the electron-phonon scattering rates for LO confined modes and the two interface modes (SO_+ and SO_-) for emission processes among different subbands at 77 K. For electron transitions from level 2 to level 1, the parallel components of the kinetic energy in each subband are approximately equal, i.e. $E_{||_2} \approx E_{||_1}$ or $k_{||_2} \approx k_{||_1}$, because the subband energy separation is 36 meV which is close to the SO_+ phonon energy $\hbar\omega_+ = 38$ meV. Hence the electron-phonon scattering is close to resonance as shown in Fig. 4(a). However, this is only the case when the ASQW structure is optimized. For structures where $E_{z_2} - E_{z_1}$ is smaller than the phonon energy, a threshold energy $E^{th}_{||_2}$ for POP emission process from the second subband to the ground state is required in order to satisfy energy conservation. As long as $E_{z_2} - E_{z_1}$ deviates from the phonon energy, there is no vertical transition and no phonon resonance occurs. For the reverse POP emission process, electrons in subband 1 always need a large parallel energy $E_{||_1} \geq E_{z_2} - E_{z_1} + \hbar\omega$ in order to satisfy energy conservation. Hence there is no phonon resonance for electron transitions from subband 1 to subband 2.

Since the electron wave functions in subband 1 and subband 2 have the same parity (even) in both wells, symmetric interface phonon SO_+ modes have a larger scattering rate than antisymmetric interface phonon SO_- modes as shown in Fig. 4(a). Because of the parity selection rule, the electron-phonon scattering for antisymmetric interface phonon modes is dominant over symmetric interface phonon modes in the 1–3 intersubband transitions as shown in Fig. 4(b). Here the scattering rate between subbands $\{1,3\}$ is about twice as large as the scattering rate between subbands $\{2,3\}$, because the electron wave functions Ψ_1 and Ψ_3 are more localized in the wide well, while the electron wave function Ψ_2 is more localized in the narrow well as shown in Fig. 3. For electron transitions between subbands $\{2,3\}$, symmetric interface phonon modes are dominant in the narrow well, while antisymmetric interface phonon modes are dominant in the wide well. Therefore, the difference between symmetric and antisymmetric interface phonon modes are relatively small for electron transitions between subbands $\{2,3\}$ as shown in Fig. 4(c) in comparison with that for electron transitions between subbands $\{1,2\}$ and $\{1,3\}$ where only one interface phonon modes are dominant. Despite the fact that the stimulated gain increases with pumping intensity, the electron temperature for each subband also increases which leads to a large thermal population in the first excited state and might reduce or even destroy the population inversion. It appears therefore that the resonance condition between the two lower levels may be too stringent for operation at high temperature, since it cannot prevent thermal population of the second level which offsets the population inversion between the two higher levels. In that case, 'scattering time engineering' based on comprehensive simulation techniques which take into account the various phonon modes in the heterostructures and the self-consistency of the charge distribution with the quantized states in the QW structure can be used to enhance the ration between the carrier lifetimes and optimize the optical gain.[17,18]

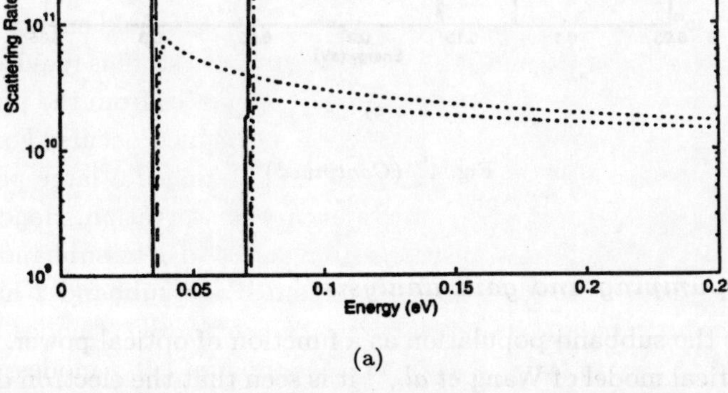

(a)

Fig. 4. Electron-phonon scattering rates of the total (solid line), the LO confined mode (dashed line), symmetric (dash-dotted line) and antisymmetric (dotted line) interface modes for emission process (a): between subbands 1 and 2, (b): between subbands 1 and 3, and (c): between subbands 2 and 3 at 77 K.

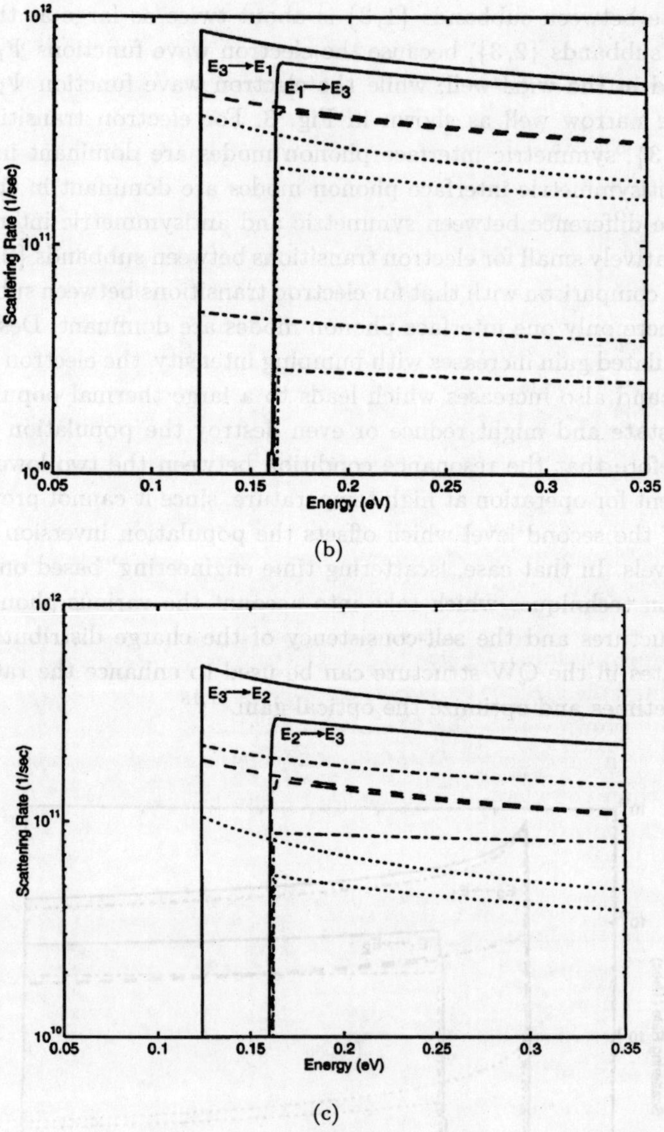

Fig. 4 (*Continued*)

2.3. *Optical pumping and gain analysis*

Figure 5 shows the subband population as a function of optical power. As obtained from the numerical model of Wang *et al.*,[18] it is seen that the electron density n_3 on subband 3 increases linearly with the optical power as observed experimentally[14] until about $1 \cdot 10^{10}$ cm^{-2}, and then slowly saturates for pumping intensities larger than 10^3 kW/cm^2; at the same time, the electron density n_2 on subband 2 remains

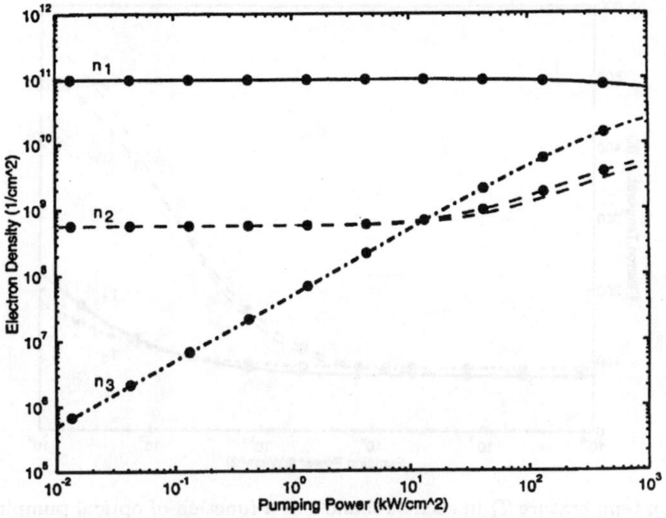

Fig. 5. Electron density n_i in each subband i as a function of optical pumping intensity under the steady state at 77 K. Lines with solid circles represent simulations with electron screening.

constant for low pumping intensity up to 10 kW/cm^2 and then gradually increases as photo-excited electrons cascade down from subband 3.

The population inversion between subbands $\{2, 3\}$ occurs with pumping intensity of 13 kW/cm^2 and above. Since the electron density n_3 on subband 3 increases much faster than the electron density n_2 on subband 2, the population inversion increases monotonically as the pumping intensity increases up to 10^3 kW/cm^2. For pumping intensity lower than 10 kW/cm^2, we see that the influence of screening on electron densities is negligible and mostly affects n_2. The weak screening effect is mostly due to the small electron density. The increase in the population of subband 2 with screening results from the fact the dielectric function is maximum at $q_{||} \approx 0$ which is also the condition for the 2–1 emission resonance that determines the relaxation rate of n_2 to subband 1.

In Fig. 6, we plot the electron temperature in each subband which increases very quickly at high pumping power, because the electron system in the ACQW structure is rapidly gaining energy from the optical field before phonon dissipation cools the system down. In particular, the electron temperature T_2 on subband 2 increases much faster than T_1 and T_3. This can be understood as follows: as shown in Fig. 4(a), the electron-phonon emission rate from subband 2 to subband 1 decreases quickly as one moves away from the phonon resonance, while other scattering processes remain practically constant. Hence at each time step, electrons with low parallel energy relax faster than electrons with high parallel energy from subband 2 to subband 1, thereby enhancing the high energy electron distribution in subband 2. Consequently, the electron temperature T_2 on subband 2 will increase more rapidly than in the other two subbands.

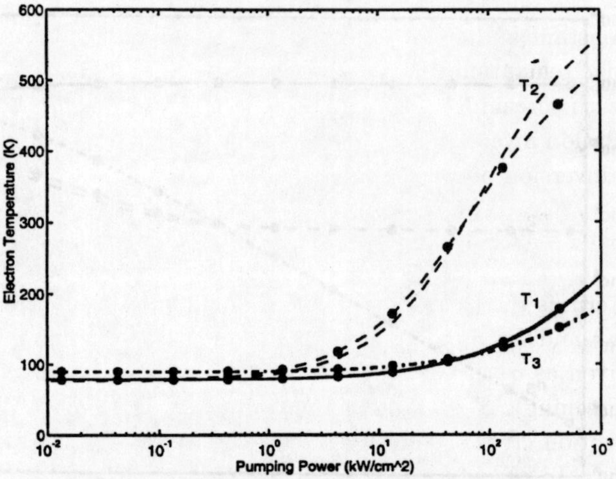

Fig. 6. Electron temperature T_i in each subband i as a function of optical pumping intensity under the steady state at 77 K. Lines with solid circles represent simulations with electron screening.

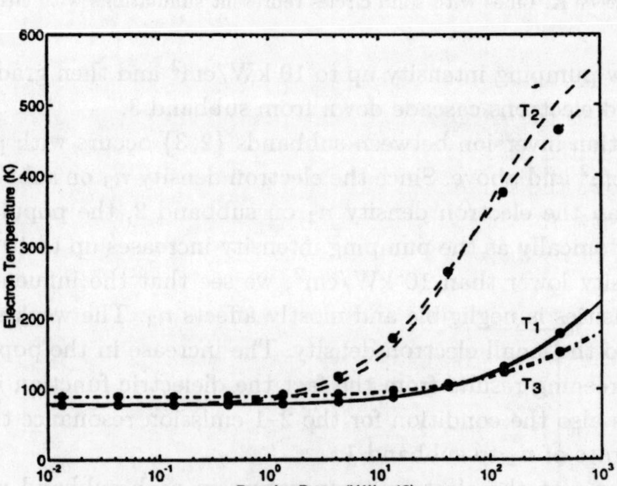

Fig. 7. Intersubband electron relaxation time $T_{ji}(j > i)$ as a function of optical pumping intensity under the steady state at 77 K. Lines with solid circles represent simulations with electron screening.

It is also seen that screening of the electron-phonon interaction has a nonmonotonic influence on T_2 as a function of the optical power. By reducing intersubband scattering, screening prevents high energy electrons in subband 2 to scatter down to subband 1, which increases T_2 at intermediate values of pumping intensity. At high pumping intensity, however, screening broadens the 2–1 resonance which makes the intersubband emission less energy dependent in comparison with the unscreened rates, thereby reducing the high energy population in n_2 and T_2 accordingly. In

Fig. 7, we show the effective relaxation time as a function of optical power. τ_{21} is practically constant and equal to about 0.27 ps (about 0.40 ps with screening), while the effective relaxation time τ_{32} from subband 3 to subband 2 is about 1.8 ps (about 1.9 ps with screening). One can use these time values to verify the optical power for population inversion $n_3 > n_2$. By using a simple rate equation formalism the population inversion between subbands 3 and 2 may be expressed as.[12]

$$\Delta n = n_3 - n_2 = (1 - \tau_{21}/\tau_{32})n_3 - \frac{m^* k T_2}{\pi \hbar^2} \ln\left[1 + \exp\left(\frac{E_{f_2} - E_{z_2}}{kT_2}\right)\right], \tag{1}$$

where the last term is the thermal population of subband 2 ($n_2 = 5.8 \cdot 10^8$ cm^{-2} at $T_2 = T_L = 77$ K) and E_{f_2} is the quasi-fermi level of level 2. The population of the subband 3 is given as $\sigma_{13} S \tau_3 n_1/\hbar\omega$, where $\tau_3 = (\tau_{31}^{-1} + \tau_{32}^{-1})^{-1} \approx 0.68$ ps (about 0.72 ps with screening), S is the pumping power, ω is the optical frequency and σ_{13} is the absorption cross section for the 1–3 transition which is estimated to be $1.5 \cdot 10^{-14}$ cm^2.[12] To reach $\Delta n = 0$ for population inversion between subbands 3 and 2, we found $S = 13$ kW/cm^2 (about 14 kW/cm^2 with screening) which is in agreement with our previous estimation.

The stimulated gain for the 3–2 emission can be expressed as[12]

$$G = \sigma_{32} \frac{\Delta n}{L}, \tag{2}$$

where $L = 144$ Å is the ACQW active-layer thickness,[20] the emission cross section σ_{32}[12–14] is calculated as $3.75 \cdot 10^{-14}$ cm^{-2}.

Figure 8 shows the stimulated gain at 4.2 K and 77 K as a function of optical pumping intensity. As the temperature decreases from 77 K to 4.2 K, the thermal population of n_2 decreases significantly and the population inversion can be

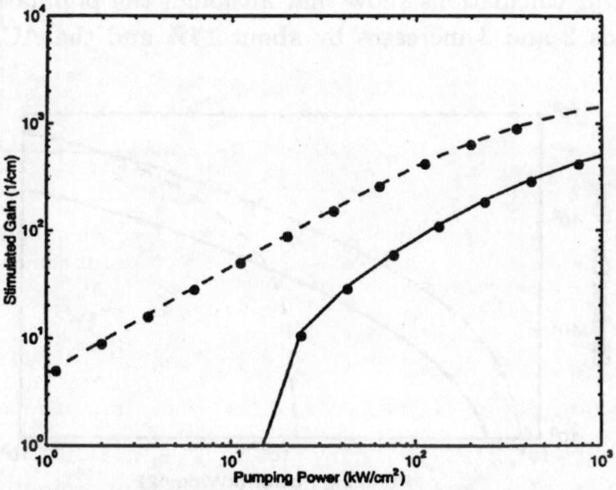

Fig. 8. Stimulated gain at 4.2 K (dashed lines) and 77 K (solid lines) for electron density $1 \cdot 10^{11}$ cm^{-2}. Solid dots represent simulations with electron screening. The first dot on the 77 K curve (solid line) falls below the horizontal axis.

achieved at very low pumping intensity. The high gain obtained at $T = 4.2$ K is directly related to the increase of the absorption cross section σ_{32} due to the reduction of the optical pumping linewidth Γ. It is seen that the theoretical gain exceeds 270 cm^{-1} for pumping power above 400 kW/cm^2. This G value appears sufficient for lasing in real structure where the optical confinement factor is only a few per cents at optimum. However, since the population inversion is linearly proportional to the electron density n_1 in subband 1 at given pumping power, the stimulated gain can be improved by increasing the n-type doping in the barrier region.

2.4. *Structure optimization*

As the electron density increases to $5 \cdot 10^{11}$ cm^{-2} in the original ACQW structure, the electron Fermi level E_f moves to 16 meV above the ground state energy E_{z_1} and the thermal population on the first excited state increases to $2 \cdot 10^{10}$ cm^{-2} at 77 K. Therefore, Pauli exclusion principle in the ground state and screening effects in the electron-POP interaction destroys the resonant relaxation process between the ground and first excited states.

In order to reduce the thermal population of subband 2 and achieve population inversion at a lower pumping power, the wide and narrow well widths of the ACQW structure can be reduced from 78 Å and 49 Å to 72 Å and 39 Å, respectively, and the Al concentration can be increased from 0.22 to 0.28. The 1–2 intersubband transition energy is increased to 47 meV, which leaves the electron-POP phonon relaxation process away from resonance.

In Fig. 9, it is shown that population inversion between subbands 2 and 3 is achieved at 12 kW/cm^2. However, the stimulated gain for the 3–2 emission has not been improved over the previous ACQW design at high pumping powers ($S >$ 750 kW/cm^2). Our calculations show that although the population inversion Δn between subbands 2 and 3 increases by about 14% and the ACQW thickness is

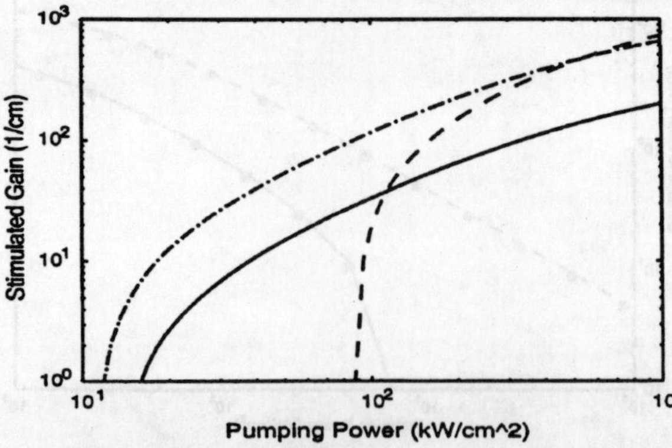

Fig. 9. Stimulated gain at 77 K for originally designed structure with electron density of $1 \cdot 10^{11}$ cm^{-2} (solid line) and $5 \cdot 10^{11}$ cm^{-2} (dashed line), and the improved ACQW structure with electron density of $5 \cdot 10^{11}$ cm^{-2} (dash-dotted line).

reduced by 5%, the dipole moment μ_{32} decreases in the new ACQW structure which in turn reduced the emission cross section σ_{32} by 22%. One simple solution to this problem is to reduce the 17 Å barrier to 14 Å to enhance the coupling between the wide and narrow wells. The energy separations between subbands $\{1,2\}$ and $\{2,3\}$ increase by 2 meV and 0.7 meV, respectively. Our calculations show the dipole moment μ_{32} increases from 22 eÅ to 24 eÅ and the emission cross section σ_{32} increases from $3.15 \cdot 10^{-14}$ cm^2 to $3.88 \cdot 10^{-14}$ cm^2. Hence a 23% increase in the stimulated gain is expected as the barrier is reduced to 14 Å. Any further reduction in the barrier thickness might cause the ground state wave function to penetrate significantly into the narrow well and increase the screening effect on the electron-POP relaxation process in the narrow well.

2.5. *Quantum fountain lasing action at 15.5 μm wavelength*

As seen in the last section, it is desirable to increase the carrier concentration in the wells for enhancing the stimulated gain and to design the structure with a larger energy spacing between the E_1 and E_2 subbands so that $E_2 - E_1 \approx E_f + E_{\mathrm{LO}}$ at equilibrium in order to maintain a low thermal population of the second subband, i.e. a low pump threshold for population inversion.

An active structure optimizing this figure of merit was designed. It consists of $\mathrm{Al_{0.35}Ga_{0.65}As}$ barriers and GaAs wells.[21] The thicknesses are 8.2 nm and 5.1 nm for the wide and narrow wells, respectively, 1.4 nm for the barrier separating the two wells. In order to improve the modal gain, the waveguide sample which was grown by molecular beam epitaxy contains 100 active structures separated by 20 nm thick $\mathrm{Al_{0.35}Ga_{0.65}As}$ barriers. The quantum wells were modulation doped resulting in a sheet carrier density of $3 \cdot 10^{11}$ cm^{-2}. In growth order, the waveguide structure consists of an n^+-doped GaAs substrate a 5 μm thick $\mathrm{Al_{0.9}Ga_{0.1}As}$ cladding layer, a 1 μm thick GaAs layer, the 3.49 μm thick multi-quantum well layer and a 2 μm thick GaAs cap layer. Four subbands are confined in the conduction band, the subband E_4 being of no practical use for the quantum fountain operation. The $E_1 - E_3$ and $E_2 - E_3$ intersubband energies are calculated at 119 meV and 80 meV, respectively. The associated dipole lengths z_{13} and z_{23} are equal to 1.5 and 2.5 nm, respectively. The waveguide is designed to be single-mode for TM propagating waves at both the pump and emission wavelengths. The overlap of the TM mode intensity within the 3.49 μm thick MQW layer is estimated to be $\Gamma_{\mathrm{MQW}} \approx 69\%$ at the emission wavelength.

Lasing action has been achieved by pumping with a CO_2 Transverse Electric Atmosphere laser. This laser can deliver 100 mJ infrared pulses, half of the energy being concentrated in a first 110 ns long pulse train. The samples were cleaved in 3×3 mm long bars and mounted in a variable temperature cryostat. Figure 10 shows the experimental configuration. The pump beam is focused on the 3 mm long side facet using an off-axis parabolic mirror which leads to $5 \cdot 0.4$ mm^2 spot size. The polarization of the pump beam is set perpendicular to the layer plane since the

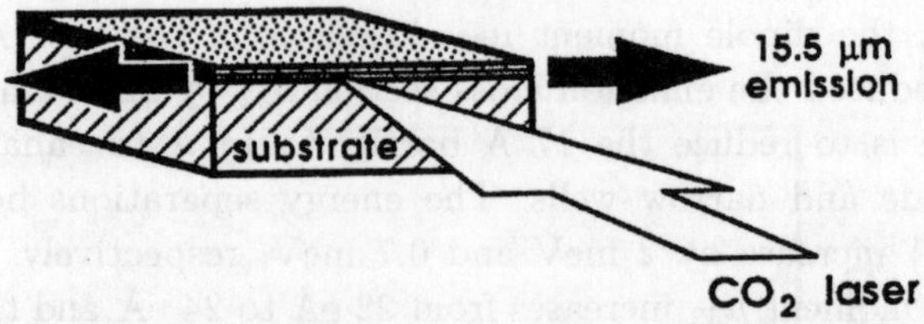

Fig. 10. Experimental arrangement for the Quantum Fountain intersubband laser. One side facet of the sample is optically pumped by the CO_2 laser. Laser oscillation at 15.5 μm wavelength occurs between the two cleaved facets perpendicular to the latter. In this simple configuration, the local gain rapidly decreases as the distance from the pumped facet increases because of the pump absorption and the emission is gain-guided.

$E_1 - E_3$ intersubband dipole moment is aligned along this direction. The intersubband emission is collected from one of the cleaved facets orthogonal to the pumped facet. The uncoated cleaved facets act as mirrors for the laser. The emission is detected with a broad-band Mercury-Cadmium-Telluride photodetector cooled at 77 K. In this simple configuration, i.e. without any mesa-etched ridge to provide lateral optical confinement, the laser emission is gain-guided and multimode transverse. In the weak intensity regime, the pump beam is totally attenuated over a 10 μm length. This length increases as the intensity of the pump beam gets closer to the saturation intensity, I_s, of the $E_1 - E_3$ intersubband absorption:

$$I_s = \frac{E_{13}}{2\sigma_{13}\tau_3} \approx 900 \text{ kW} \cdot \text{cm}^{-2}, \tag{3}$$

where E_{13} is the energy of $E_1 - E_3$ transition, $\tau_3 \approx 0.7$ ps the lifetime of the electrons in the E_3 subband and $\sigma_{13} \approx 1.53 \cdot 10^{-14}$ cm^{-2} the intersubband absorption cross-section.

The laser emission spectrum at 77 K shown in Fig. 11 was recorded with a dispersive infrared monochromator. As seen, the emission is peaked at 80 meV ($\approx$ 15.5 μm wavelength) with a multimode width of less than 1 meV. This strong narrowing of the emission is a first signature of lasing action. The laser emission vanishes when the pump is tuned out of resonance with the $E_1 - E_3$ intersubband transition. It also vanishes when the polarization of the pump beam is rotated in the layer plane, thus confirming the origin of the emission.

Note that the emission wavelength is the longest reported so far for a unipolar intersubband semiconductor laser. This is a direct consequence of the choice of an optical excitation of the quantum wells which allows to minimize the free-carrier absorption losses, in contrast to electrically pumped intersubband lasers.

The laser power versus the incident pump power is reported in Fig. 12 for different operating temperatures. The output power exhibits a threshold behavior which is characteristic of the lasing process. The threshold occurs around 10 kW at low temperature which corresponds to an incident pump intensity of 500 kW.cm^{-2}. The threshold power rapidly increases with temperature above 90 K. Indeed when the temperature is increased, the thermal population of subband 2 grows almost

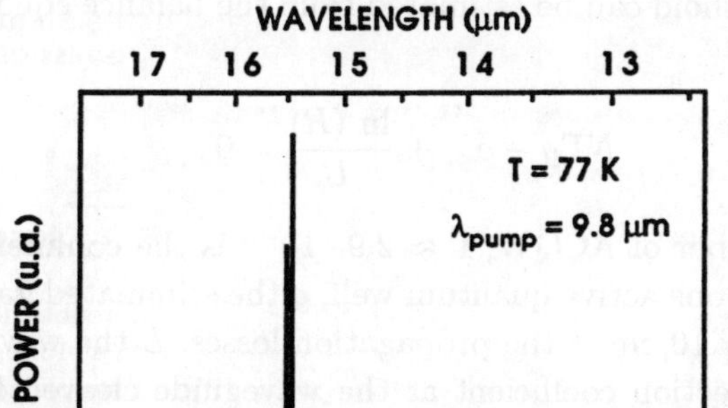

Fig. 11. Emission spectrum of the Quantum Fountain intersubband laser at 77 K.

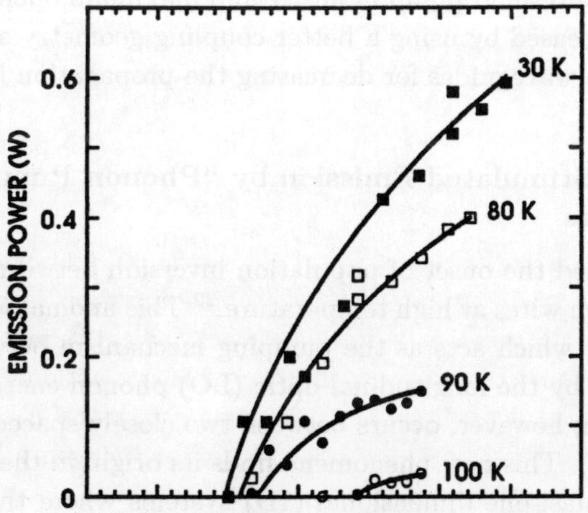

Fig. 12. Laser power at 15.5 μm wavelength versus pump power for different operating temperatures.

exponentially and higher pump intensities are required to maintain the population inversion. Lasing is observed up to 110 K for the investigated pump intensities. The output peak power reaches ≈ 0.6 W per facet at low temperature and ≈ 0.4 W at 77 K. We stress that lasing is observed in a sample with poor lateral optical confinement.

The gain at threshold can be estimated from the balance equation between gain and losses:

$$N\Gamma g - \alpha_{\rm pr} + \frac{\ln (R)}{L} = 0\,, \tag{4}$$

where N is the number of ACQW, $\Gamma \approx 2.9 \cdot 10^{-3}$ is the confinement factor of the optical mode within one active quantum well, g the stimulated gain for one coupled quantum well, $\alpha_{\rm pr} \approx 10\ {\rm cm}^{-1}$ the propagation losses, L the waveguide length and $R \approx 0.27$ is the reflection coefficient at the waveguide cleaved facets. The modal gain at threshold is estimated to be $N\Gamma g \approx 14.4\ {\rm cm}^{-1}$ for the 3 mm long waveguide. The gain g per quantum well at threshold is deduced to be $49.5\ {\rm cm}^{-1}$. The stimulated gain can also be separately evaluated using Eq. (2) accounting for a $2.5 \cdot 10^{10}$ electrons.cm^{-2} population inversion at the threshold intensity of 0.5 MW.cm^{-2}. The theoretical intersubband stimulated gain at threshold is calculated to be around $600\ {\rm cm}^{-1}$ which is about one order of magnitude larger than the $49.5\ {\rm cm}^{-1}$ value deduced from measurements. The discrepancy is mainly attributed to the uncertainty on the pump intensity which is effectively coupled in the mid-infrared waveguide and to an underestimation of the waveguide losses due to the poor lateral confinement in this first demonstration laser. The device performances in terms of threshold pump intensity and maximum operating temperature could be largely increased by using a better coupling geometry and by making use of single-mode ridge waveguides for decreasing the propagation losses.

3. Intersubband Stimulated Emission by "Phonon Pumping" in Quantum Wires

Briggs *et al.* predicted the onset of population inversion between subbands during transport in quantum wires at high temperature.[22] This anomalous effect is induced by POP absorption, which acts as the pumping mechanism between two resonant subbands separated by the longitudinal optic (LO) phonon energy $E = \hbar\omega_{\rm LO}$;[23,24] population inversion, however, occurs between two closely spaced off-resonant subbands ($\Delta E < \hbar\omega_{\rm LO}$). This new phenomena finds its origin in the absence of carrier thermalization in quasi-one dimensional (1D) systems where the carrier distribution deviates drastically from the maxwellian distribution, even in moderate electric fields.[25] In particular, the occupation of upper subbands is significantly enhanced by resonant POP scattering (RISOPS) for which the joint 1D density of states is maximum at the 1D subband edges ($k = 0$). The resulting distribution has a "jagged" profile characterized by sharp peaks at multiple of the POP energy $E = n\hbar\omega_{\rm LO}$ (Fig. 13). If an intermediate subband lies between the resonant subbands, its population can be significantly less than in an upper subband even at room temperature. Therefore, unlike conventional pumping mechanisms that require carrier injection at high energy,[26] population inversion arises as a natural consequence of carrier transport in quasi-one-dimensional (1D) systems.

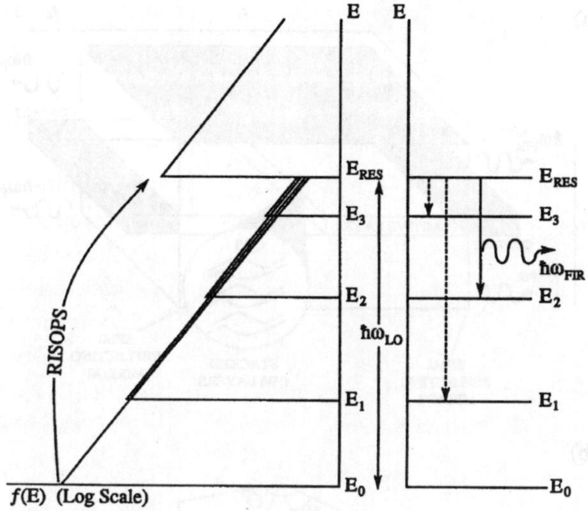

Fig. 13. Schematic profile of the distribution function $f(E)$ in quantum wires subject to a moderate electric field ($F \sim 100$ V/cm). Population inversion arises between the resonant levels $E_{\rm res}$ and the lower levels E_3, E_2, and E_1. Among the possible transitions (dashed lines), only the $E_{\rm res} \rightarrow E_2$ shows appreciable oscillator strength and population inversion to lead to stimulated emission.

Typically, population inversion by "phonon pumping" would arise between subbands separated by at most 20 meV, which is particularly interesting for FIR stimulated emission. An important condition for the observation of this new effect is the realization of highly confined 1D systems with well-controlled and reproducible wire structures. The fabrication of quantum wires by direct growth on patterned GaAs vicinal substrate offers a promising and convenient approach to obtain dense and regular arrays of 1D structures suited for FIR lasers.[27–30]

3.1. *Electronic model*

Among the different 1D confinement shapes obtained by regrowth on patterned substrate,[27–29] the sawtooth-like array of GaAs quantum wire structures obtained by growth on patterned (100) substrate proposed by Colas *et al.*,[28] appears particularly attractive because of its modeling simplicity and its potential for high density packing. A schematic cross section of the structure is shown in Fig. 14(a). The two-dimensionally confined wire regions are bounded by thinner quantum wells, and the confining layers of wide gap AlGaAs are modulation doped, which provides carriers in the 1D region. The array density is determined by the period of the perturbation on the substrate while the shape, the thickness and the relative dimension of the wires are controlled by the growth temperature and rate. Scanning electron microscopy (SEM) photographs, however, reveal rounded corner on the lower side of the quantum wires.[31] In the simplified version of a FIR laser, we assume electrical contacts are achieved on the front and back of the structure and

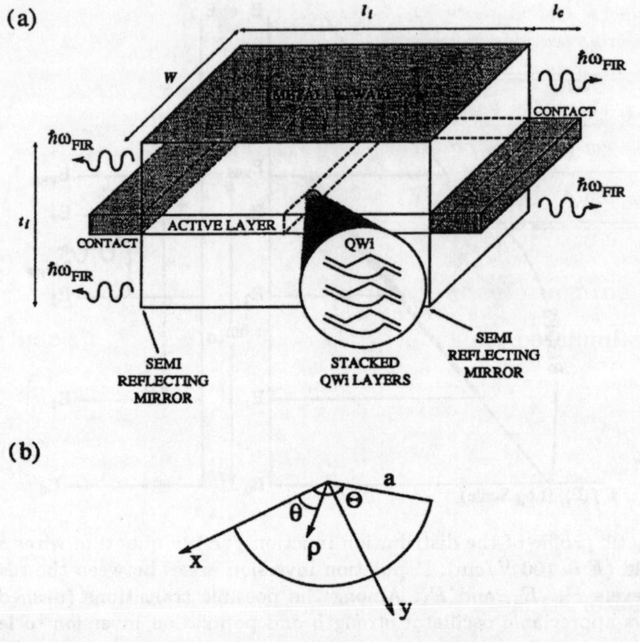

Fig. 14. (a) Schematic structure for FIR lasing device. Metallic plates for optical confinement are represented on the top, bottom, front, and back sides of the structure. Light emission occurs partially through the contact layers. (b) Schematic quantum wire cross section; the x and y axis are represented with the structure orientation.

optical confinement is realized by metallic walls on the lateral sides, top and bottom of the device.

In order to model the electronic properties of the system, we assume the wire cross section resembles a portion of a circle with radius a and angular aperture Θ (Fig. 14(b)). The eigenfunctions and energies are given by

$$\Psi(\rho,\theta) = \frac{2}{a}\sqrt{\frac{\rho}{\Theta}}\frac{J_{v_n}(x_{v_n m}\rho/a)}{|J_{v_n+1}(x_{v_n m})|}\sin(v_n\theta)\,, \tag{5a}$$

$$E_{nm} = \frac{\hbar^2 x_{v_n m}^2}{2ma^2}\,, \tag{5b}$$

where J_{v_n} are the Bessel functions of order $v_n = n\pi/\Theta$ and $x_{v_n m}$ is the mth zeros of J_{v_n}. We choose $\Theta = 8\pi/9$, which corresponds to the experimental results[31] and consider two wire configurations with $a = 237$ and $a = 283$ Å, which lead to POP resonances (see Table 1). In the former configuration, POP resonance occurs between the E_{11} and E_{12} levels, and population inversion (PI) arises between E_{12} and E_{31} separated by ~ 5 meV (the E_{21} state is too close to the ground state E_{11} to result in *PI*); in the latter case, the POP resonance occurs between E_{11} and E_{41} with *PI* between E_{41} and E_{12} (~ 9 meV) and E_{41} and E_{31} (~ 14 meV).

Table 1. Calculated energy levels E_{nm} (in meV) for the two quantum wire cross sections considered in this section.

	E_{11}	E_{21}	E_{31}	E_{12}	E_{41}	E_{22}
$a = 237$ Å	16.2	30.0	47.3	52.4	68.0	77.4
$a = 283$ Å	11.4	21.1	33.2	36.7	47.7	54.3

3.2. *Optical gain analysis*

The gain for stimulated emission between subbands with a population inversion ΔN reads

$$g(\lambda) = \frac{\Delta N}{A}\frac{e^2}{2\bar{n}\epsilon_0\, \mathrm{cm}_0} f_{if} \frac{\tau_2}{1 + 4\pi^2 c^2 \tau_2^2(\lambda^{-1} - \lambda_0^{-1})^2}\,, \tag{6}$$

where $\lambda_0 = hc/E_{if}$ is the wavelength at resonance for intersubband energy separation, E_{if}, c is the speed of light, h is the Planck constant, τ_2 is the intersubband scattering time, $A = 4\pi a^2/9$ is the wire cross section, and $\bar{n}$ is the index of refraction. The oscillator strength is given by

$$f_{if} = 2m_0 E_{if}/\hbar^2 |\langle f|\hat{e} \bullet \hat{p}|i\rangle|^2\,, \tag{7}$$

where m_0 is the free electron mass, $\hat{e}$ and $\hat{p}$ are the polarization and position vectors, respectively, and $|i\rangle$ and $|f\rangle$ are the initial and final states during the optical transition. For the $E_{12} \to E_{31}$ transition in the thinner ($a = 237$ Å) wire, the oscillator strength for x and y polarization are $f_{if}^x = 5.7 \cdot 10^{-3}$ and $f_{if}^y = 0.18$, respectively. If we assume $\Delta N = 10^5$ cm^{-1} and $\tau_2 = 2 \cdot 10^{-13}$ s, the gain at the peak value $g^x\lambda_0 = .22$ cm^{-1} and $g^y(\lambda_0) = 7.0$ cm^{-1}. These small values indicate that the nature of the states involved in the transition is not very favorable for stimulated emission. On the other hand, for the $E_{41} \to E_{31}$ transition in the thicker ($a = 283$ Å) wire, $f_{if}^x = 31.65$ and $f_{if}^y = 0.98$, with $g^x(\lambda_0) = 860$ cm^{-1} and $g^y(\lambda_0) = 27$ cm^{-1}, respectively, which is much more encouraging especially for x-polarization emission. From a general standpoint, optical transitions between electronic states where both n and m quantum numbers change are relatively weak, while transitions between states that conserve n or m are much stronger. This selection rule for intersubband transitions was already emphasized by Briggs *et al.*, in their original paper;[22] therefore we focus on the $E_{41} \to E_{31}$ transition in the wider wire.

In order to assess the potential of quantum wires for FIR stimulated emission, we now analyze the optical losses in the device configuration shown in Fig. 14 in a way similar to Borenstain and Katz.[32] Optical confinement is realized by the metallic walls on the sides of the device, and the active region consisting of the array of quantum wire structures of thickness t_a is comprised between two cladding layers of intrinsic GaAs layers of effective mode thickness $t_i = \lambda/2\bar{n} \sim 10 - 15$ μm. The total width of the structure is taken as $\omega = \lambda/2\bar{n}$. The most important losses for optical frequencies below the band gap are free carrier absorption and absorption by phonon interactions, which are calculated by using the general form of the complex dielectric constant

$$\epsilon(\omega) = (\bar{n} - ik) = \epsilon_\infty \left(1 + \frac{\omega_L^2 - \omega_T^2}{\omega_T^2 - \omega^2 + i\omega\gamma_{\text{ph}}} - \frac{\omega_p^2}{\omega(\omega - i\gamma_{\text{pl}})} \right) \tag{8}$$

with $\alpha = 2\omega k(\omega)/c$. Here, ω_L, ω_T, and ω_{pl} are the frequencies of the LO phonons, TO phonons, and the plasmons, respectively; γ_{ph} and γ_{pl} are the damping constants of the phonons and plasmons.

Given the wire orientation (Fig. 14), appreciable gain for x polarization will build up in the perpendicular direction compatible with the device structure, i.e., along the wire direction. At lasing threshold, gain equals the losses, which for FIR excitation, can be expressed by the following relation

$$\frac{A}{A_{\text{eff}}} g_{\text{th}}(\omega_0) = \frac{A}{A_{\text{eff}}} \alpha_{\text{qwi}} + \frac{t_i}{t_a}\alpha_i + \alpha_m + \frac{l_c}{l_i}\alpha_c + \frac{1}{L}\ln\left(\frac{1}{R}\right). \tag{9}$$

Here g_{th} is the gain at threshold, L is the cavity length and R the reflectivity; α_{qwi}, α_m, α_i, and α_c are the absorption coefficients for the quantum wires, the modulation doped, the intrinsic, and the contacting layers, respectively. A_{eff} is the quantum wire effective cross section in the active layer, i.e., $t_i \times l_i$/number of quantum wires. Since the radiations emitted by intersubband emission are polarized perpendicular to the direction of the wires, interaction with longitudinal 1D plasmons is forbidden so we assume $\alpha_{\text{qwi}} \sim 0$. Therefore, losses by plasmon absorption take place only in the modulation-doped AlGaAs and the GaAs contacting layers; the absorption in the intrinsic region is due to phonon only $\epsilon(\omega_{\text{pl}} = 0)$. By neglecting the last term in Eq. 9, we can express the population inversion at threshold ΔN_{th}

$$\Delta N_{\text{th}} = A_{\text{eff}} \left(\alpha_m + \frac{l_c}{l_i}\alpha_c + \frac{t_i}{t_a}\alpha_i \right) \Bigg/ \frac{e^2}{2\bar{n}\epsilon_0 \text{ c} m_0} f_{if}\tau_2 \,. \tag{10}$$

Here we used $A_{\text{eff}} = 10^{-10}$ cm^2, $\omega_L = 291.5$ cm^{-1}, $\omega_T = 269.2$ cm^{-1}, $\omega_{\text{pl}} = \sqrt{N_v e^2/m^*\epsilon_0\epsilon_s}/2\pi c$, where N_v is the carrier concentration per unit volume $\epsilon_s = 10.06$, $\gamma_{\text{ph}} = 2.3$ cm^{-1}, and $\gamma_{\text{pl}} = e/2\pi\mu m^* c = 28$ cm^{-1} with $\mu = 5 \times 10^3$ cm^2/V s. In Fig. 15, we show ΔN_{th} for the $E_{41} \to E_{31}$ transition as a function of the carrier concentration, N_d, in the modulation doped layer for x and y polarizations, and for different values of the length ratio l_c/l_i. Here we have assumed an aspect ratio $t_i/t_a = 50$, the contact layers have a thickness t_a and the doping concentration $n_c = 10^{18}$/cm^3. As expected from the values of the oscillator strengths, ΔN_{th} for y polarization is almost two orders of magnitude higher than for the x polarization, and increases rapidly with N_d. It is also seen that ΔN_{th} become independent of absorption in the contact layers for $l_c/l_i < 3 \cdot 10^{-3}$. However, the important result here is the relatively low population inversion at threshold, $\Delta N_{\text{th}} = 2 - 3 \cdot 10^5$/cm for $N_d \sim 10^{16}$/cm^3, which is quite reasonable. Monte Carlo simulation by Briggs *et al.*, have indicated the possibility of achieving 24% carrier occupation in the resonant upper subband against 4% only in the intermediate subband in certain configurations,[22] which would require an overall concentration of $1 - 1.5 \cdot 10^6$/cm to lead to stimulated emission. In order to analyze the gain dependence on the photon energy, we have plotted $\Delta N_{\rm th}$ as a function of the wavelength in Fig. 16. Here,

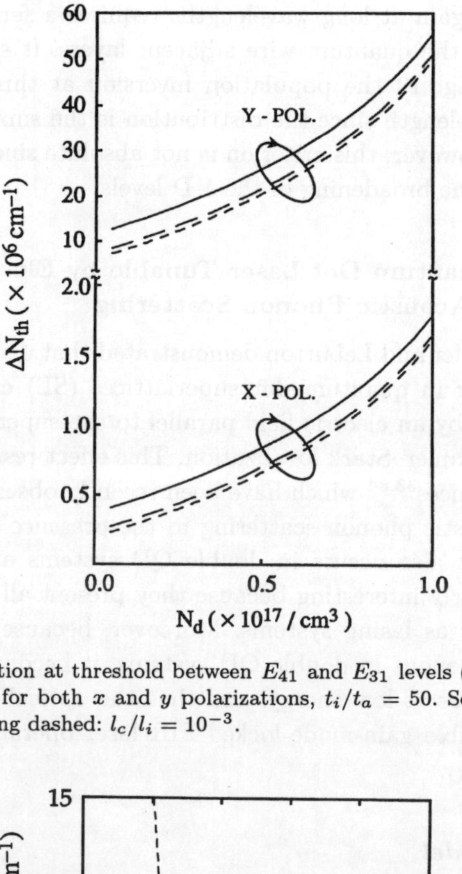

Fig. 15. Inverted population at threshold between E_{41} and E_{31} levels ($\Delta E = 14.5$ meV) in the thicker wire ($a = 283$ Å) for both x and y polarizations, $t_i/t_a = 50$. Solid: $l_c/l_i = 10^{-2}$, short dashed: $l_c/l_i = 3\ 10^{-3}$; long dashed: $l_c/l_i = 10^{-3}$.

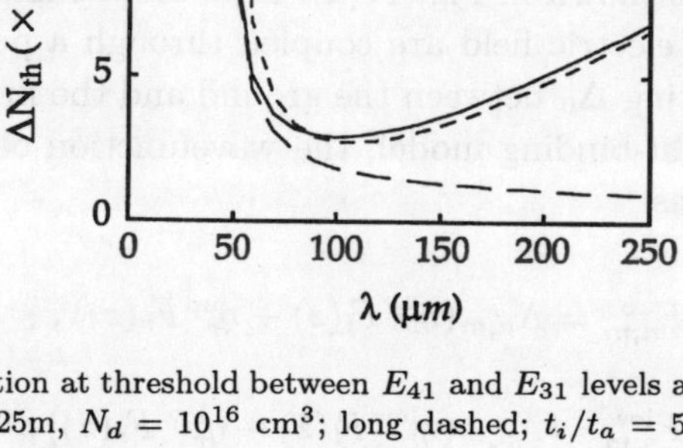

Fig. 16. Inverted population at threshold between E_{41} and E_{31} levels as a function of the energy separation. Solid: $t_a = 0.25$m, $N_d = 10^{16}$ cm^3; long dashed; $t_i/t_a = 50$, $N_d = 10^{16}$ cm^3; short dashed; $t_a = 0.25$ μm, $N_d = 0$.

$l_c/l_I = 3 \cdot 10^{-3}$. For $N_d = 10^{16}/\mathrm{cm}^3$, the two upper curves are roughly identical with a minimum around 100 μm. The large increase at short wavelengths is due to phonon absorption as λ approaches λ_T (Eq. (8)) and is less pronounced for constant t_a when t_i decreases with λ. The increase of ΔN_{th} at long wavelengths is due to plasmon absorption in the modulation doped layer as seen by comparison with the $N_d = 0$ (no plasmon) curve, which decreases monotonically with λ. Therefore, it appears that high gain at long wavelengths requires a serious reduction of free carrier absorption in the quantum wire adjacent layers. It should be pointed out that, according to Fig. 13 the population inversion at threshold is more easily achieved at long wavelength since the distribution is the smallest before threshold for POP emission. However, this criterion is not absolute since long wavelength PI could be limited by the broadening of the 1 D levels.

4. Far-Infrared Quantum Dot Laser Tunable by Electrical Modulation of Acoustic Phonon Scattering

Recently, Lyanda-Geller and Leburton demonstrated that dissipation due to acoustic phonon scattering in quantum dot superlattices (SL) can be tuned, and, in particular, quenched by an electric field parallel to the superlattice growth axis in the regime of the Wannier–Stark localization. This effect results in current oscillations with antiresonances[33,34] which have been recently observed experimentally.[35] Tunable rate of acoustic phonon scattering in the presence of electric field is not restricted to SLs but also occurs in double QD systems as shown below, these systems are particularly interesting because they present all advantages of double QW heterostructures as lasing systems. Moreover, because of the tunable dissipation of acoustic phonons in double QD systems, it becomes possible to control non-radiative processes at low energy, and thus the optical gain, thereby making a multi-spectral tunable gain-mode-locked FIR laser operating below the optical phonon frequency ω_{LO}.

4.1. *Electronic model*

A cascade structure made of elementary double dot cell and focus on electrons in the conduction band is shown in Fig. 17(a). Dots are assumed to have equal sizes, and in the absence of electric field are coupled through a potential barrier giving rise to an energy splitting Δ_0 between the ground and the first excited state of the double dot. In the tight-binding model, the wavefunction of the upper and lower states can be written as

$$\begin{aligned} F^{\mathrm{up}}_{n,m} &= \Lambda_{n,m}(\eta^{\mathrm{up}}_l F_l(z) + \eta^{\mathrm{up}}_r F_r(z))\,, \\ F^{\mathrm{low}}_{n,m} &= \Lambda_{n,m}(\eta^{\mathrm{low}}_l F_l(z) + \eta^{\mathrm{low}}_r F_r(z))\,, \end{aligned} \tag{11}$$

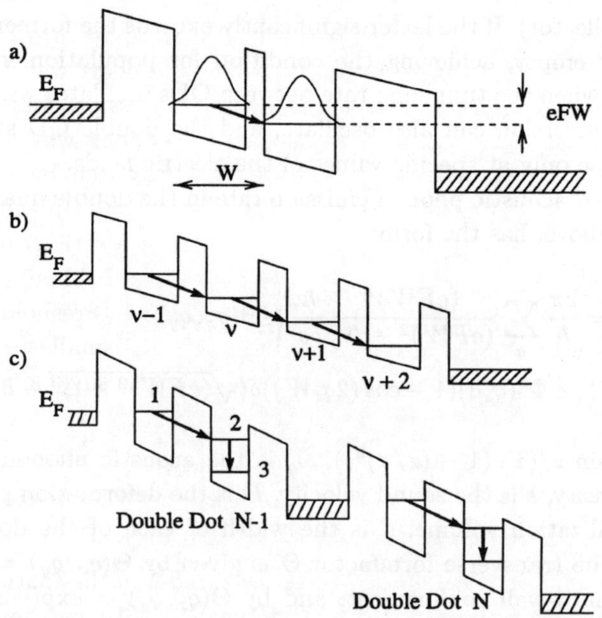

Fig. 17. (a) Double dot elementary cell. The arrow illustrates the interdot optical transitions; (b) Superlattice scheme; (c) Cascade scheme. Vertical arrow illustrates the intradot optical phonon emission.

where $\eta_{l,r}$ and $F_{l,r}(z)$ are the Wannier function amplitudes and longitudinal electron wavefunctions in the left and right dot, respectively, and $\Lambda_{n,m}$ is the wavefunction for transverse confinement.[36] For the sake of simplicity, we consider only the ground level in each of the dots. In the presence of a longitudinal electric field, the ground levels split with a splitting energy between the states which in a two level system[37] reads:

$$\delta E = 2\sqrt{(\Delta_0/2)^2 + (eFW/2)^2}\,, \tag{12}$$

where W is the distance between the QD centers in the longitudinal direction, and the Wannier amplitudes are given by $\eta_l^{\mathrm{low}} = \eta_r^{\mathrm{up}} = \sqrt{(\delta E + \Delta_0)/2\delta E}$ and $\eta_r^{\mathrm{up}} = \eta_l^{\mathrm{low}} = \sqrt{(eFW)^2/2(\delta E + \Delta_0)\delta E}$. At low electric fields, $(\Delta_0 \gg eFW)$ the down-state is localized in the right QD, and the up-state is localized in the left QD.

In general, the interdot transitions can be either nonradiative, resulting from absorption (emission) of an acoustic or optical phonon, or radiative, resulting from absorption (emission) of the energy difference δE of an electromagnetic field with the frequency $\omega = \delta E/\hbar$. The natural onset of population inversion, as for FIR emission in double quantum well structures,[5] occurs because of the dc current flowing through the external barriers: the left QD receives a supply of electrons from the two- or three-dimensional electrodes filled up to the Fermi energy E_F, and the electron population in the right QD depends on the difference between the interdot tunneling rate (central barrier) and the tunneling rate through the right barrier (collector). If the latter significantly exceeds the former, then the right well is virtually empty, achieving the condition for population inversion. As we shall see below, when the tunneling rate between QDs oscillates with electric fields, the population inversion can also oscillate, and the double QD structure can be engineered to lase only at specific values of the electric fields.

The calculated acoustic phonon emission rate in the double quantum dot structure described above, has the form:

$$M_{1,2} = \frac{2\pi}{\hbar} \sum_q \frac{(eFW)^2}{(eFW)^2 + \Delta_0^2} \frac{\hbar\omega_q}{\rho s^2 V} \Theta(q_x, q_y) \times \Phi^2(q_z d)(1 - \cos(2q_z W))\delta(\sqrt{(eFW)^2 + \Delta_0^2} - \hbar\omega_q)\,, \tag{13}$$

where $\Phi(z) = \sin z/(z \cdot (1 - (z/\pi)^2))$, ω_q is the acoustic phonon frequency, ρ is the material density, s is the sound velocity, D is the deformation potential energy, V is the normalization volume, d is the width of each of the dots in longitudinal direction. The transverse formfactor Θ is given by $\Theta(q_x, q_y) = \Phi^2(qa)\Phi^2(q_y a)$ in the case of hard-wall confinement, and by $\Theta(q_x, q_y) = \exp(-\alpha^2(q_x^2 + q_y^2))$ for parabolic confinement.[35] Figure 18 shows the calculated acoustic phonon emission rate for a double QD structure with transverse parabolic confinement characterized by $\omega_p = 7 \cdot 10^{12}$ Hz which corresponds to an effective QD size in the transverse direction $L = \sqrt{\hbar/m\omega_p} = 150$ Å, $W = 90$ Å, $\Delta_0 = 12.5$ meV.[36] Similarly to acoustic phonon emission rates in QDSL,[33–35] $M_{1,2}$ is an oscillating function of the electric field, and exhibits pronounced antiresonant minima at electric fields such as $(\sqrt{(eFW)^2 + \Delta_0^2} \sim \pi\hbar sj/W)$, which are due to destructive interferences of electron waves (amplitudes) in the double QD system (here j is an integer).

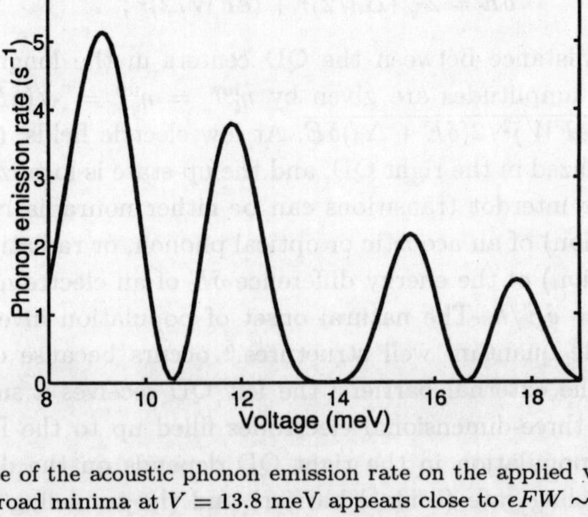

Fig. 18. Dependence of the acoustic phonon emission rate on the applied voltage in the double dot structure. The broad minima at $V = 13.8$ meV appears close to $eFW \sim \Delta_0$.

4.2. *Gain analysis*

The spontaneous emission rate in the double QD structure is given by $\frac{1}{\tau_{sp}} = \gamma_d f$, where the parameter γ_d characterizes the density of optical modes,[38] $d = 2$; 3 refers to the dimension of the optical cavity, and f is the oscillator strength. For interdot optical transitions, f is given by (see, e.g. Ref. 38):

$$f = m^* W^2 \Delta_0^2 / 4\hbar^2 \delta E\,, \tag{14}$$

where m^* is the effective electron mass. If the cavity dimensions are substantially larger than the wavelength of radiation, the density of optical modes has a three-dimensional (3D) character and γ_3 reads

$$\gamma_3 = \frac{e^2 \bar{n} (\delta E)^2}{3\pi\hbar^2 \epsilon_0 m_* c^3}\,, \tag{15}$$

where ϵ_0 is the dielectric constant of the vacuum, $\bar{n}$ is the refractive index. In the case of a cavity with the electromagnetic enclosures consisting of closely spaced conducting planar mirrors separated by a distance W_m along the growth direction, the optical mode density is two-dimensional (2D), and γ_2 reads

$$\gamma_2 = \frac{e^2 \bar{n} \delta E}{4\hbar \epsilon_0 W_m m_* c^2}\,. \tag{16}$$

The gain at frequency $\omega = \delta E/\hbar$ is given by

$$g(\omega) = \frac{\delta N}{t_m} \frac{\pi c^2 \tau}{\bar{n}^2 w^2 \tau_{sp}}\,, \tag{17}$$

where δN is the population inversion in units of inverse area, $\tau = \hbar/\Gamma$, Γ is the level broadening which determines the linewidth of the optical transition, $t_m = t_{ac} + t_c$, t_{ac} is the width of the active layer and t_c denotes the width of conducting and isolating layers, $t_m > 2\pi\hbar c / 2n\delta E$. The linewidth is a sum of three contributions, $\Gamma = \Gamma_R + \Gamma_L + \Gamma_i$ where $\Gamma_{R,L}$ are the rates of tunneling through external barriers into the buffer region or the contact, Γ_I is the total electron scattering rate. The critical issue in the QD structure is to construct a sufficiently wide collector barrier for Γ_R to be within the range of extremal acoustic phonon scattering rates shown on Fig. 18. Hence the parameters of the system can be adjusted to achieve the threshold gain for laser action only at the electric fields which correspond to the minima of the acoustic phonon scattering rate shown on Fig. 18. We note that small Γ_R lead to increase in τ and laser gain. At the same time, Γ_L should be maintained sufficiently large to provide a flux of electrons into the QD, by making the emitter barrier thinner than the collector barrier.

The maximal value of the population inversion in the antiresonances of acoustic phonon scattering rate is one electron per double dot. For $L = 150$ Å this value corresponds to $\delta N = 2.5 \cdot 10^{11}$ cm^{-2}. Calculation of the gain for this population inversion at $t_m = 10$ μm for 3D optical mode density yields $g \sim 18$ cm^{-1}

at $\hbar\omega = 20$ meV (~ 60 μm). For 2D optical mode density, the magnitude of the gain is $g \sim 85$ cm^{-1} for a distance between mirrors $W_m = 3$ μm. In the quantum cascade laser,[2] internal losses are as low as 9 cm^{-1}, but this device operates in the 4 to 5 μm range and benefits from low-loss waveguide confinement. A considerable source of loss in FIR laser is the free carrier absorption in the conducting layers. However, for quantum dot laser operating at low impurity and carrier concentration, ($\sim 2 \cdot 10^{16}$ cm^{-3}), the estimate of the free carrier absorption coefficient at wavelength 60 μm is about 20 cm^{-1}. This loss is acceptable in the case of 2D optical modes, although it is of the order of or higher than the lasing threshold for 3D optical modes. Therefore in the case of 3D optical modes a special design for the reduction of free carrier absorption is needed. For example, strips of doped GaAs inside an insulating matrix for current injection as suggested in Ref. 5, can help to overcome this difficulty. Thus, both quantum dot laser designs (3D and 2D optical modes) are feasible for achieving laser action at the discrete values of the applied voltage for which acoustic phonon scattering is minimum.

The threshold current should be adequate for nonradiative and spontaneous emission dissipation. If the linewidth of the laser is 0.5 meV, the threshold current density in the antiresonant regime (minima of phonon dissipation on Fig. 18) is $j_{tr} = 0.03$ A/cm^2 in the case of 3D optical mode density and $j_{tr} = 0.14$ A/cm^2 in the 2D case. The quantum efficiency of laser is determined by the ratio between radiative and nonradiative transition rates, $\eta = r_{nr}/(r_{nr} + r_{sp})$. For nonradiative dissipation at antiresonances, the quantum efficiency is $\eta \sim 0.2$ and $\eta \sim 0.5$, in the 3D case and 2D case, respectively.

Two alternative schemes for achieving the population inversion in QD structures are also possible. Aside from using the original scheme (Fig. 17(b)), a double dot scheme with current flow and emission of photons due to electron transition between excited levels is also possible (Fig. 17(c)).[33] The population inversion in this scheme arises due to fast resonant polar optical phonon transitions between excited and ground levels. The design on Fig. 17(c) illustrates the cascade principle:[2] repeated double QD cells are separated by buffer layers. In a laser system involving an array of QDs (Fig. 17c) using polar optical phonon resonance, the coherent light emission occurs for optical transitions between levels 1 and 2′ in the neighboring dots, which is followed by immediate resonant emission of a polar optical phonon. The process is then repeated in a pair of dots 2 and 3, 3 and 4 etc. We note that the optical transition $1 \to 3$ is still possible. However, no population inversion is maintained for this light emission, and it only lowers (almost negligibly) the quantum efficiency of $1 \to 2$ light emission. We also note that in the superlattice scheme (Fig. 17(b)), charge storage in SL cells can lead to the Coulomb blockade effects, high field domain formation and detuning from resonance for some part of the structure, but electric fields which lead to maximal population inversion in QD design prevent these effects. The main features of SL design are present in both cascade double dot schemes, discussed above, involving or not involving polar optical phonons.

Acknowledgments

This work was supported by the ARO grants No. DAAH04-95-1-0190 and DAAD19-99-1-0129. The authors are indebted to Ms. A. Jones for technical assistance in typing this manuscript.

References

1. R. F. Kazarinov and R. A. Suris, *Sov. Phys. Semicond.* **5** (1971) 797–800.
2. J. Faist, F. Capasso, D. Sivco, C. Sirtori, A. L. Hutchinson, S. N. G. Chu, and A. Y. Cho, *Science* **264** (1994) 553–556.
3. J. Faist, F. Capasso, C. Sirtori, D. Sivco, A. L. Hutchinson, and A. Y. Cho, *Appl. Phys. Lett.* **66** (1995) 538–540.
4. Q. Hu and S. Feng, *Appl. Phys. Lett.* **59** (1991) 2923–2925.
5. A. N. Korotkov, D. V. Averin, and K. K. Likharev, *Appl. Phys. Lett.* **65** (1994) 1865–1867.
6. O. Gauthier-Lafaye, S. Sauvage, P. Boucaud, F. H. Julien, R. Prazeres, F. Glotin, J. M. Ortega, V. Thierry-Mieg, R. Planel, J. P. Leburton, and V. Berger, *Appl. Phys. Lett.* **70** (1997) 3197–3199.
7. J. N. Heyman, K. Craig, B. Galdrikian, M. S. Sherwin, K, Campman, P. F. Hopkins, S. Fafard, and A. C. Gossard, *Phys. Rev. Lett.* **72** (1994) 2183–2186.
8. B. Xu, Q. Hu, and M. R Melloch, *Appl. Phys. Lett.* **71** (1997) 440–442.
9. P. F. Yuh and K. L. Wang, *Appl. Phys. Lett.* **51** (1997) 1404–1406.
10. Y. Arakawa and A. Yariv, *IEEE J. Quantum Electron.* **22** (1986) 1887–1899.
11. H. Benisty, C. M. Sotomayor-Torres, and W. Weisbuch, *Phys. Rev.* **B44** (1991) 10945–10948.
12. F. H. Julien, A. Sa'ar, J. Wang, and J. P. Leburton, *Electron. Lett.* **31** (1995) 838–839.
13. Z. Moussa, P. Boucaud, F. H. Julien, Y. Lavon, A. Sa'ar, V. Berger, J. Nagle, and N. Coron, *Electron. Lett.* **31** (1995) 912–913.
14. F. H. Julien, Z. Moussa, P. Boucaud, Y. Lavon, A. Sa'ar, J. Wang, J. P. Leburton, V. Berger, J. Nagle, and R. Planel, *Superlattices Microstruc.* **19** (1996) 69–79.
15. Y. Lavon, A. Sa'ar, J. Wang, J. P. Leburton, F. H. Julien, and R. Planel, *Appl. Phys. Lett.* **69** (1996) 197–199.
16. R. Lassnig, *Phys. Rev.* **B30** (1984) 7132–7137.
17. M. Dutta, M. A. Stroscio, and K. W. Kim, in *Quantum Based Electronic Devices and Systems*, eds. M. Dutta and M. A. Stroscio, World Scientific, 1998, pp. 281–312.
18. J. Wang, J. P. Leburton, Z. Moussa, F. H. Julien, and A. Sa'ar, *J. Appl. Phys.* **80** (1996) 1970–1978.
19. J. Wang, J. P. Leburton, F. H. Julien, and A. Sa'ar, *IEEE Photon. Technol. Lett.* **8** (1996) 1001–1003.
20. J. Wang, J. P. Leburton, F. H. Julien, and A. Sa'ar, *Superlattices Microstruc.* **20** (1996) 245–251. Notice that in this paper the theoretical gain has been calculated with respect to one ACQW period, i.e. by taking $L = 344$ Å, which includes the cladding barriers. The conversion factor between the data on Fig. 4 in Ref. 14 and the present result in Fig. 8 is therefore $\alpha = 344/144 = 2.39$.
21. O. Gauthier-Lafaye, P. Boucaud, F. H. Julien, S. Sauvage, S. Cabaret, J. M. Lourtioz, V. Thierry-Mieg, and R. Planel, *Appl. Phys. Lett.* **71** (1997) 3619–3621.
22. S. Briggs, D. Jovanovic, and J. P. Leburton, *Appl. Phys. Lett.* **52** (1989) 2012–2014. In this article, the driving fields is 100 V/cm.
23. S. Briggs and J. P. Leburton, *Superlattices Microstruc.* **5** (1989) 145–148.

24. D. Jovanovic, S. Briggs, and J. P. Leburton, *Phys. Rev.* **B42** (1990) 11108–11113.
25. J. P. Leburton, *Phys. Rev.* **B45** (1992) 11022–11030.
26. See, e.g., S. J. Allen, G. Brozak, E. Colas, F. DeRosa, P. England, J. Harbison, M. Helm, L. Florez, and M. Leadbeater, *Semicond. Sci. Technol.* **7** (1992) B1–B5.
27. P. M. Petroff, K. Ensslin, M. Miller, S. Chalmers, H. Weman, J. Merz, H. Kroemer, and A. C. Gossard, *Superlattices Microstruc.* **8** (1990) 35–39.
28. E. Colas, E. Kapon, S. Simhony, H. M. Cox, R. Bhat, K. Kash, and P. S. D. Lin, *Appl. Phys. Lett.* **55** (1989) 867–869.
29. E. Colas, S. Simhony, E. Kapon, R. Bhat, D. M. Hwang, and P. S. D. Lin, *Appl. Phys. Lett.* **57** (1990) 914–916.
30. S. S. Chang, S. Ando, and T. Fukui, *Surf. Sci.* **267** (1992) 214–217.
31. E. Colas (private communication).
32. S. I. Borentstain and J. Katz, *Appl. Phys. Lett.* **55** (1989) 654–656.
33. Y. Lyanda-Geller and J. P. Leburton, *Phys. Rev.* **B52** (1995) 2779–2783.
34. Y. Lyanda-Geller and J. P. Leburton, *Semicond. Sci. Technol.* **10** (1995) 1463–1469.
35. A. Nogaret, L. Eaves *et al.*, *Solid State Electron.* **42** (1998) 1489–1493; L. Eaves, private communication.
36. For hard-wall confinement in the direction transverse to the growth axis, the transverse wavefunction $\Lambda_{n,m} = (2/a) \sin (n\pi x/a) \sin (m\pi y/a)$, where a is the size of the dot in the transverse direction. In the case of parabolic transverse confinement, $\Lambda_{n,l} = C_{nl} \exp(-\alpha^2(x^2+y^2)/2)H_n(\alpha x)H_n(\alpha y)$, $C_{nl} = \alpha/(2^{n+l}n!l!)^{1/2}$, $\alpha = m\omega_p/\pi\hbar$, ω_p is the frequency characterizing the parabolic confinement of the dots.
37. L. D. Landau and E. M. Lifshits, *Quantum Mechanics*, Clarendon, Oxford, 1984.
38. J. Smet, C. Fonstad, and Q. Hu, *J. Appl. Phys.* **79** (1996) 9305–9320.

International Journal of High Speed Electronics and Systems, Vol. 9, No. 4 (1998) 1189–1209

OPTOELECTRONIC PROPERTIES OF STRAINED WURTZITE GaN QUANTUM-WELL LASERS

JIN WANG, J. B. JEON, K. W. KIM, and M. A. LITTLEJOHN

Department of Electrical and Computer Engineering,
North Carolina State University, Raleigh,
North Carolina 27695-7911, USA

Fundamental optical properties of strained wurtzite GaN quantum-well laser are calculated and evaluated near the threshold condition. The formalism is based on a self-consistent methodology that couples an envelope-function Hamiltonian for band structures with photon-carrier rate equations. Details of energy band structure, optical gain, and modulation response are studied comprehensively under the effects of strain-induced piezoelectric fields, bandgap renormalization, and the carrier capture processes. Comparisons between different approximations show that self-consistency is essential to accurately simulate pseudomorphically strained wurtzite GaN quantum-well lasers.

1. Introduction

There has been growing interest in wurtzite GaN quantum-well (QW) structures for blue-green lasers and other applications. Significant progress has been made to date in high-quality layer growth and *p*-type doping of this material system as illustrated by the recent report of continuous-wave operation of an InGaN laser diode.[1] At the same time, development of theoretical models for understanding the band structure, optical processes, and carrier dynamics has become of great importance for design and optimization of actual III–V nitride lasers. It is realized that wurtzite nitride materials have a few different characteristics from cubic materials in their electronic and optical behavior, even though they share the same tetrahedral atomic configurations. One prominent aspect is the recent piezoresistance measurements in *n*-type wurtzite GaN films which provide a clear evidence for large piezoelectric strain effect.[2,3] The strain-induced electric field for a GaN QW grown on an AlN substrate is estimated to be 4.7×10^3 kV/cm, which is more than one order of magnitude larger than previously reported piezoelectric fields in InGaAs/InAlAs superlattices grown along the [111] axis.[4,5] Another aspect to be considered is the large injected carrier densities needed in GaN QWs for population inversion because of the larger values for both electron and hole effective masses, compared to those for cubic materials. Hence, self-consistent effects and many-body phenomena may become more significant. The commonly used flat-band (FB) configurations would fail to provide accurate information on carrier dynamics for the lasing condition where large band-bending and strong carrier localization appear.

Many theoretical studies of III–V nitrides do not take into account these major physical effects along with an accurate treatment of valence band mixing beyond the effective mass approximation.[6–9] Based on an envelope-function (or $\mathbf{k}\cdot\mathbf{p}$) Hamiltonian and photon-carrier rate equations, this article presents a self-consistent simulation of the band structure, optical properties, modulation response, and carrier capture in pseudomorphically-strained GaN/AlGaN wurtzite QW structures with the influence of piezoelectric and many-body effects. To assess the significance of various physical effects, these results are compared with calculations employing simple approximations such as the flat-band conditions.

2. Theoretical Models

2.1. *Envelope-function Hamiltonian*

To include band mixing effects, the valence band states are described within an envelope-function approximation based on the Rashba–Sheka–Pikus (RSP) Hamiltonian for the top valence bands in the vicinity of the Γ-point.[10] The original 6×6 Hamiltonian is diagonalized with a set of basis states $|v_i,\pm\rangle$. This procedure yields the following upper (H_+) and lower (H_-) 3×3 Hamiltonian blocks:

$$H_\pm = -\begin{vmatrix} P+Q & \sqrt{3}\,R & \pm i\sqrt{\frac{3}{2}}\,S \\ \sqrt{3}\,R & P+Q+\frac{2}{3}\Delta_{\rm so} & \frac{\sqrt{2}}{3}\Delta_{\rm so}\pm i\sqrt{\frac{3}{2}}\,S \\ \mp i\sqrt{\frac{3}{2}}\,S & \frac{\sqrt{2}}{3}\Delta_{\rm so}\mp i\sqrt{\frac{3}{2}}\,S & P-2Q+\frac{1}{3}\Delta_{\rm so} \end{vmatrix} \begin{matrix} |v_1,\pm\rangle \\ |v_2,\pm\rangle \\ |v_3,\pm\rangle \end{matrix}\,, \tag{1}$$

where P, Q, R, and S are given by

$$\begin{aligned} P &= +\Delta_{\rm cr}/3 + \gamma_{1z}k_z^2 + \gamma_{1\perp}k_\perp^2 + \delta_{1z}\epsilon_{zz} + \delta_{1\perp}\epsilon_\perp + V_v(z)\,, \\ Q &= -\Delta_{\rm cr}/3 - 2\gamma_{3z}k_z^2 + \gamma_{3\perp}k_\perp^2 - 2\delta_{3z}\epsilon_{zz} + \delta_{3\perp}\epsilon_\perp\,, \\ R &= (\gamma_{2\perp} + 2\gamma_{3\perp})k_\perp^2/\sqrt{3}\,, \\ S &= 2(2\gamma_{2z} + \gamma_{3z})k_z k_\perp/\sqrt{3}\,. \end{aligned}$$

Here the z axis is taken along the c-axis [0001] direction and $\mathbf{k} = (\mathbf{k}_\perp, k_z)$ is the wave vector. We take values of the Luttinger-like parameters ($\hbar^2/2m_0$ is omitted for simplicity) $\gamma_{1z} = 2.47$, $\gamma_{1\perp} = 2.85$, $\gamma_{2z} \approx \gamma_{2\perp} = 1.29$ and $\gamma_{3z} \approx \gamma_{3\perp} = 0.95$ for GaN; $\gamma_{1z} = 1.54$, $\gamma_{1\perp} = 1.50$, $\gamma_{2z} \approx \gamma_{2\perp} = 0.78$ and $\gamma_{3z} \approx \gamma_{3\perp} = 0.63$ for AlN. The splitting energies for GaN and AlN are $\Delta_{\rm cr} = 12.5$ meV, $\Delta_{\rm so} = 17.9$ meV; and $\Delta_{\rm cr} = -214.9$ meV, $\Delta_{\rm so} = 18.9$ meV, respectively. For the spherical cubic approximations, the deformation potentials $\delta_{1z} = \delta_{1\perp} = -a_v$ and $\delta_{2z} = \delta_{2\perp} = \delta_{3z} = \delta_{3\perp} = -b/2$, and the valence band hydrostatic deformation potential a_v is estimated from its gap value $a_g = a_c - a_v \approx -8.16$ eV by using the empirical relation $a_v \approx -a_g/3$ and the shear deformation potential $b \approx -1.667$ eV for GaN.[10] For strained QW structures, the strain components in the QW are given by

$$\epsilon_{xx} = \epsilon_{yy} = \epsilon_\perp/2\,, \; e_{zz} = -(C_{13}/C_{33})\,\epsilon_\perp\,, \; \epsilon_{xy} = \epsilon_{yz} = \epsilon_{zx} = 0\,, \tag{2}$$

where C_{33} and C_{13} are the elastic stiffness constants. The barrier materials are assumed to be unstrained. The lattice constants for GaN and AlN are 3.189 Å and 3.112 Å, respectively. Under the effective mass approximation, the conduction band dispersion relation is given as

$$E_c = E_g + \hbar^2 k_z^2/2m_z^c + \hbar^2 k_\perp^2/2m_\perp^c + a_c(\epsilon_\perp + \epsilon_{zz})\,, \tag{3}$$

where m_z^c and $m_\perp^c$ are the effective masses along and perpendicular to the z axis, respectively, and E_g is the band-gap.[10] We take values $m_z^c = m_\perp^c = 0.20\ m_0$, $E_g = 3.44$ eV for GaN, and $m_z^c = m_\perp^c = 0.48\ m_0$, $E_g = 6.20$ eV for AlN. The parameters for AlGaN are obtained through interpolation.

2.2. *Piezoelectric effect*

The strain-induced electric polarization $\mathbf{P}$ is given by

$$\mathbf{P} = \mathbf{d} \cdot \mathbf{T}\,, \tag{4}$$

where the piezoelectric tensor $\mathbf{d}$ in wurtzite materials has the form

$$\begin{bmatrix} 0 & 0 & 0 & 0 & d_{15} & 0 \\ 0 & 0 & 0 & d_{15} & 0 & 0 \\ d_{31} & d_{31} & d_{33} & 0 & 0 & 0 \end{bmatrix} \tag{5}$$

and the stress tensor $\mathbf{T}$ in wurtzite materials can be expressed in a compact form in terms of the strain tensor ϵ and a stiffness tensor $\mathbf{C}$

$$\begin{pmatrix} T_{xx} \\ T_{yy} \\ T_{zz} \\ T_{yz} \\ T_{zx} \\ T_{xy} \end{pmatrix} = \begin{bmatrix} C_{11} & C_{12} & C_{13} & 0 & 0 & 0 \\ C_{12} & C_{11} & C_{13} & 0 & 0 & 0 \\ C_{13} & C_{13} & C_{33} & 0 & 0 & 0 \\ 0 & 0 & 0 & C_{44} & 0 & 0 \\ 0 & 0 & 0 & 0 & C_{44} & 0 \\ 0 & 0 & 0 & 0 & 0 & C_{44} \end{bmatrix} \begin{pmatrix} \epsilon_{xx} \\ \epsilon_{yy} \\ \epsilon_{zz} \\ \epsilon_{yz} \\ \epsilon_{zx} \\ \epsilon_{xy} \end{pmatrix}. \tag{6}$$

For a strain tensor ϵ given by Eq. (2), the stress tensor $\mathbf{T}$ is expressed by

$$T_{xx} = T_{yy} = \left(C_{11} + C_{12} - \frac{2C_{13}^2}{C_{33}} \right) \epsilon_{xx}\,, \quad T_{zz} = T_{yz} = T_{zx} = T_{xy} = 0\,, \tag{7}$$

and the strain-induced electric field $\mathbf{E}$ is found to be[3]

$$E_x = E_y = 0\,, \quad E_z = \frac{2d_{31}}{\epsilon_0 \epsilon_s}\left(C_{11} + C_{12} - \frac{2C_{13}^2}{C_{33}}\right)\epsilon_{xx}\,. \tag{8}$$

The piezoelectric constant for GaN is given by $d_{31} \approx -1.7\times10^{-12}$ m/V.[4] The elastic stiffness constants of GaN have been accurately determined to be $C_{11} = 390$ GPa, $C_{12} = 145$ GPa, $C_{13} = 106$ GPa, and $C_{33} = 398$ GPa by using Brillouin scattering methods.[11]

2.3. *Self-consistent band structure*

When carrier capture and escape rates in the QW region are nearly the same, the confined and continuum carriers can be considered in quasi-thermal equilibrium and be described by a single quasi-Fermi level. Then, with an assumption that the entire structure is charge neutral with constant electron and hole quasi-Fermi levels, the two-dimensional carrier densities along the z direction, $n(z)$ and $p(z)$, are evaluated from the following expressions

$$n(z) = \sum_i \frac{m_\perp^c k_B T}{\pi\hbar^2} \ln\left[1 + \exp\left(\frac{E_{fc} - E_i^c(0)}{k_B T}\right)\right] |\psi_i^c(z)|^2\,, \tag{9}$$

$$p(z) = \sum_{\sigma=\pm}\sum_j \int \frac{k_\perp dk_\perp}{2\pi} \frac{(|v_{\sigma j}^{(1)}(k_\perp, z)|^2 + |v_{\sigma j}^{(2)}(k_\perp, z)|^2 + |v_{\sigma j}^{(3)}(k_\perp, z)|^2)}{1 + \exp[E_{fv} - E_{\sigma,j}^v(k_\perp)]/k_B T}. \tag{10}$$

Here, ψ_i^c is the electron wave function corresponding to the ith energy E_i^c evaluated under the effective-mass approximation, $v_{\sigma j}^{(1)}$, $v_{\sigma j}^{(2)}$ and $v_{\sigma j}^{(3)}$ are components of the valence band envelope functions of the RSP Hamiltonian corresponding to the jth hole energy $E_{\sigma,j}^v$, and the index $\sigma = \pm$ denotes the upper (H_+) and lower (H_-) Hamiltonian blocks. As the electron-hole recombination rate increases, electron capture becomes considerably faster than escape to balance the loss through recombination and the electron quasi-Fermi level E_{fc} splits into a confined quasi-Fermi level E_f^{qw} and a continuum quasi-Fermi level E_f^{3d}. Also, the electron density $n(z)$ is the sum of confined and continuum electrons. The assumption of quasi-thermal equilibrium remains valid for holes in the valence band due to the lower barrier potential and more closely spaced energy levels that result in rapid transitions between continuum and confined holes.

The bandgap-narrowing effect is included in the local density approximation, and the potential energies $V_c(z)$ and $V_v(z)$ are given by

$$V_c(z) = -e\phi(z) + \delta V_c(z) + V_{xc}^c(z)\,, \qquad V_v(z) = +e\phi(z) + \delta V_v(z) + V_{xc}^v(z)\,, \tag{11}$$

where $\phi(z)$ is the solution of Poisson's equation for the strain-induced QW piezoelectric field; and $\delta V_c(z)$ and $\delta V_v(z)$ are the conduction and valence band-edge discontinuities, respectively.[12] For the local exchange-correction potential energy $V_{xc}(z)$, an expression parameterized by Gunnarson and Lundquist is used, i.e.[13]

$$V_{xc}^c(z) = -\left[1 + 0.0545 r_s \ln\left(1 + 11.4/r_s\right)\right](2/\pi\alpha r_s)Ry^* , \tag{12}$$

where

$$\alpha = (4/9\pi)^{1/3}, \quad r_s = [4\pi a_B^{*3} n(z)]^{-1/3} , \tag{13}$$
$$a_B^* = 4\pi\epsilon_0\epsilon_s\hbar^2/m_\perp^c e^2 , \; Ry^* = e^2/8\pi\epsilon_0\epsilon_s a_B^* .$$

The valence band local exchange-correction potential energy $V_{xc}^v(z)$ is found by simply replacing $n(z)$ in r_s by $p(z)$, and by approximating $m_\perp^c$ in a_B^* by the density of states effective mass of the lowest valence band energy state near $k_\perp = 0$.

2.4. *Optical gain*

The optical gain g and the spontaneous emission r^{sp} for QW structures are given by

$$g(\omega) = \frac{e^2}{n\epsilon_0 c m_0^2 \omega d} \sum_{\sigma=\pm} \sum_{i,j} \int k_\perp dk_\perp \langle M_{i\to j}^2 \rangle \times (f_i^c - f_j^v - 1)\,\delta(E_i^c - E_j^v - \hbar\omega) , \tag{14}$$

$$r^{sp}(\omega) = \frac{e^2}{n\epsilon_0 c m_0^2 \omega d} \sum_{\sigma=\pm} \sum_{i,j} \int k_\perp dk_\perp \langle M_{i\to j}^2 \rangle \times f_i^c f_j^v\,\delta(E_i^c - E_j^v - \hbar\omega) , \tag{15}$$

where n is the refractive index, c is the speed of light, d is the well width, f_i^c and f_j^v are Fermi functions for the conduction band ith confined energy state and valence band jth energy state, respectively, and $\langle M_{i\to j}^2 \rangle$ is the squared matrix element averaged over the angle $\varphi(= \arctan(k_y/k_x))$ as

$$\langle M_{c\to j}^2 \rangle = |P_\perp|^2 \left(\langle \psi_i^c | v_{\sigma j}^{(1)} \rangle^2 + \langle \psi_i^c | v_{\sigma j}^{(2)} \rangle^2 \right) \sin^2 \vartheta_{\mathbf{E}} + 2|P_z|^2 \langle \psi_i^c | v_{\sigma j}^{(3)} \rangle^2 \cos^2 \vartheta_{\mathbf{E}} . \tag{16}$$

Here $\vartheta_{\mathbf{E}}$ is the angle between the polarization vector $\mathbf{E}$ and the c axis. Constants $P_z \equiv \langle S|\hat{p}_z|Z\rangle$ and $P_\perp \equiv \langle S|\hat{p}_x|X\rangle$ are estimated by using Harrison's theory as $2|P_{z,\perp}|^2/m_0 \approx 24.6$ eV for cubic materials.[6] This theory provides a reasonable first estimation of P_z and $P_\perp$ for wurtzite materials. Kane's four-band model is not appropriate for GaN because of the strong mixing between the lowest conduction band Γ_{1c} and upper conduction bands Γ_{3c}, Γ_{6c}.[7] The overall spontaneous emission r^{sp} is an average of two TE-mode components and one TM-mode component. The radiative recombination rate U_{sp} is given as

$$U_{sp} = \int \left(\frac{n\omega}{\pi c} \right)^2 r^{sp}(\omega)\, d\omega . \tag{17}$$

Finally a broadening factor is included by replacing the δ-function with a modified-Lorentzian line-shape given by

$$\delta(E) \approx C \frac{\gamma}{E^2 + \gamma^2}, \tag{18}$$

where $\gamma \equiv \gamma(E) = \gamma_0 \, e^{-|E|/k_B T}$ and C is a normalization constant.[14] A previous study suggests that the overall shape for line broadening is important in the gain analysis. In particular, the shape of the low-energy tail affects significantly the gain magnitude. Although the Lorentzian line-shape is commonly used, it leads to negative gain values or photon energies below the effective bandgap under weak population inversion. Therefore, the Lorentzian line-shape is not a good choice for optical gain calculations. The low-energy tail in the modified-Lorentzian broadening function is approximated by an exponential decay, which tends to agree with experimental observations.

2.5. *Rate equation formalism*

For a given cavity length L, the longitudinal cavity mode wavelength must satisfy the resonance condition $\lambda_m = 2nL/m$, where λ_m is the wavelength of the mth Fabry–Perot mode. The simplified rate-equation (SRE) formalism is derived without net carrier capture, which means the confined and continuum carriers are considered to be in quasi-thermal equilibrium. Each Fabry–Perot mode grows according to the rate equations[15]

$$\frac{dS_m}{dt} = \Gamma\left(\frac{c}{n}\right) g_m S_m + \Gamma\left(\frac{c}{n}\right)\frac{r_m^{sp}}{V} - \frac{S_m}{\tau_{ph}}, \tag{19}$$

$$\frac{dN_{inj}}{dt} = \frac{J}{ed} - \frac{N_{inj}}{\eta_i \, \tau_{sp}} - \left(\frac{c}{n}\right)\sum_m g_m S_m, \tag{20}$$

where S_m is the photon density of the mth cavity mode, N_{inj} is the injected carrier density, Γ is the optical confinement factor which is evaluated by solving the Helmholtz equation, $V = Lwd/\Gamma$ is the effective volume of the optical mode, the optical mode gain g_m and the spontaneous emission r_m^{sp} for the mth cavity mode are calculated by Eqs. (14) and (15), $\tau_{sp} = U_{sp}/N_{inj}$ is the radiative recombination lifetime, η_i is the internal quantum efficiency, the photon lifetime τ_{ph} is defined as $\tau_{ph}^{-1} = (c/n)(\alpha_i + \alpha_m)$ with the internal loss α_i and the mirror loss $\alpha_m = \frac{1}{2L}\ln(\frac{1}{R_1 R_2})$, and J is the injected current density. The steady-state photon density S_m for each Fabry–Perot mode and the total output power P can be expressed by[16]

$$S_m = \frac{r_m^{sp}/V}{n/(c\Gamma\tau_{ph}) - g_m}, \qquad P = \sum_m \hbar\omega S_m V\left(\frac{c}{n}\right)\alpha_m. \tag{21}$$

For the single-mode small-signal analysis, the modulation response frequency $f_r = \omega_r/2\pi$ and the frequency response function $\left|\frac{s(\omega)}{j(\omega)}\right|$ are given by[16]

$$f_r = \frac{1}{2\pi}\sqrt{\left(\frac{c}{n}\right)\left(\frac{dg}{dN}\right)\frac{S_0}{\tau_{ph}}},$$

$$\left|\frac{s(\omega)}{j(\omega)}\right| \simeq \frac{(\Gamma\tau_{ph}/ed)\omega_r^2}{[(\omega^2-\omega_r^2)^2+\omega^2(1/\tau_{sp}+\tau_{ph}\omega_r^2)^2]^{1/2}}. \tag{22}$$

The previous rate-equation formalism is derived with the assumption that confined and continuum electrons are in quasi-thermal equilibrium and therefore $E_f^{3d} = E_f^{qw}$. This assumption should be justified by including the carrier capture and escape processes in a more generalized rate equation (GRE) formalism. The injected electrons in the QW laser can be classified as confined electrons localized inside the QW and continuum electrons localized in the barrier region. As the percentage of injected electrons leaks from the QW, a built-in electric field is formed and continuum electrons tend to relocalize around the QW region. The generalized photon-carrier rate-equations are[17]

$$\frac{dS_m}{dt} = \Gamma\left(\frac{c}{n}\right)g_m S_m + \Gamma\left(\frac{c}{n}\right)\frac{r_m^{sp}}{V} - \frac{S_m}{\tau_{ph}}, \tag{23}$$

$$\frac{dN_{qw}}{dt} = -\frac{N_{qw}}{\tau_{esc}} + \frac{N_{3d}}{\tau_{cap}} - \frac{N_{qw}}{\eta_i\,\tau_{sp}} - \left(\frac{c}{n}\right)\sum_m g_m S_m, \tag{24}$$

$$\frac{dN_{3d}}{dt} = -\frac{N_{3d}}{\tau_{cap}} + \frac{N_{qw}}{\tau_{esc}} + \frac{J}{ed}, \tag{25}$$

where N_{qw} is the confined carrier density inside the QW, N_{3d} is the continuum carrier density above the QW, τ_{cap} is the capture time from continuum states to confined states, and τ_{esc} is the escape time from confined states to continuum states. Here we point out that in the GRE formalism, the continuum carrier density N_{3d} should not be replaced by the carrier density N_b in the barrier region.[17] In fact, the carrier densities N_{3d} and N_b are related through thermionic emission across the heterojunction.[18] Another set of rate equations is required to describe the thermionic emission process and the drift-diffusion process in the barrier region. These thermionic and drift-diffusion processes are neglected in this article. The steady-state photon density S_m, the total output power P, and the injected current density J use the same expressions as in the SRE formalism. Also, the quasi-Fermi level E_f^{3d} for the continuum state electrons are determined by the requirement that the electron capture rate equals to the escape rate plus the radiative recombination and stimulated emission rate. At the lasing condition, as the injected current density increases, the stimulated emission rate increases. In order to balance the rate-equation, the quasi-Fermi level E_f^{3d} and the electrons in the barrier region increases. As a result, the confined electron density in the QW and the quasi-Fermi level E_f^{qw} decrease. To satisfy the Bernard–Duraffourg inversion condition, more injected carriers are needed. Thus a self-consistent adjustment on the band structure, the injected carrier density, the quasi-Fermi levels E_f^{qw} and E_f^{3d}, the optical gain g and spontaneous emission r^{sp}, the electron capture time τ_{cap} and escape time τ_{esc} is needed. For the single-mode small-signal analysis, the modulation response frequency $f_r = \omega_r/2\pi$ and the frequency response function $\left|\frac{s(\omega)}{j(\omega)}\right|$ are given by

$$f_r = \frac{1}{2\pi}\sqrt{\left(\frac{c}{n}\right)\left(\frac{dg}{dN_{inj}}\right)\frac{S_0}{\beta_1\ \tau_{ph}}}\,,$$

$$\left|\frac{s(\omega)}{j(\omega)}\right| \simeq \frac{(\Gamma\tau_{ph}/ed)\omega_r^2}{[(\omega^2-\omega_r^2)^2+\omega^2(\beta_2/\beta_1)^2]^{1/2}}\,, \tag{26}$$

where

$$\beta_1 = 1 + \frac{\tau_{3d}}{\alpha\,\tau_{qw}} + \frac{\tau_{3d}}{\tau_{sp}} + \tau_{3d}\left(\frac{c}{n}\right)\left(\frac{dg}{dN_{inj}}\right)S_0 \approx 1\,, \tag{27}$$

$$\beta_2 = \left(\frac{c}{n}\right)\left(\frac{dg}{dN_{inj}}\right)S_0\,(1+\tau_{3d}/\tau_{ph}) + 1/\tau_{sp}\,. \tag{28}$$

3. Comparison between Different Treatments

In this article, a pseudomorphically-strained GaN/$Al_{0.2}Ga_{0.8}N$ QW structure is studied at $T = 300$ K. The AlGaN barriers are assumed to be unstrained and have no piezoelectric field, while the GaN QW is under 0.48% compressive strain and the corresponding strain-induced electric field intensity is about 940 kV/cm.[4] An $Al_{0.4}Ga_{0.6}N$ cladding material is used for optical confinement. The refractive index n is taken to be 2.79 for GaN near the bandgap, and 2.1 for AlN.[19] The dielectric constant for a specific composition of AlGaN is calculated by using linear interpolation between these two values. The optimal optical confinement factor Γ of 3.02% is used for 60 nm $Al_{0.2}Ga_{0.8}N$ barriers. Based on recent experimental data, we assume the following structure parameters: 1000 μm cavity length (L), 5 μm thickness (w), high-reflection facet coating ($R_1 = R_2 =$50%), 54 cm^{-1} internal loss (α_i), 86% internal quantum efficiency (η_i), and negligible Auger recombination.[20] The optical gain for the threshold condition satisfies $\Gamma g_{\rm th} = \alpha_i + \alpha_m$ which gives $g_{\rm th} = 2020$ cm^{-1}. The results are presented in the following manner: the band structure, optical properties, and carrier capture and escape times under quasi-thermal equilibrium ($E_{fc} = E_f^{3d} = E_f^{qw}$) for models based on the FB condition, the self-consistent calculation with the local density approximation only (LDO, without piezoelectric fields), and the fully self-consistent solution (FSC). These discussions are followed by a detailed comparison between the SRE and GRE formalisms at the threshold condition to demonstrate the importance of complete self-consistency in laser simulations.

3.1. *Quantum-well band structure*

Figure 1 shows the (a) FB, (b) LDO and (c) FSC band structures for a 5 nm GaN QW with injected carrier density $N_{inj} = 2 \times 10^{19}$ cm^{-3}. The zero of energy is chosen as the unstrained bulk GaN heavy-hole bandedge at the Γ point. For simplicity, only three levels (E_1, E_2, HH_1) are plotted. The valence band dispersion relations for the three models are provided in Fig. 2. The top four energy states correspond to HH_1, LH_1, HH_2, and LH_2, respectively. The LDO band structure is very similar to the FB band structure. For both cases, the upper and lower Hamiltonians give the same dispersion due to symmetry of the structure along the z direction. However, for a large electric field inside the QW [Fig. 1(c)], the symmetry is broken for FSC and the upper and lower Hamiltonians give different dispersion relations. From basic quantum mechanics, it is known that for the perturbing electric field, the ground state energy shifts towards lower energy (a red shift) while the first excited state energy shifts toward higher energy (a blue shift). Therefore, the energy separations between ground state energies (HH_1, LH_1) and first excited state energies (HH_2, LH_2) increases as shown in Fig. 2(c).

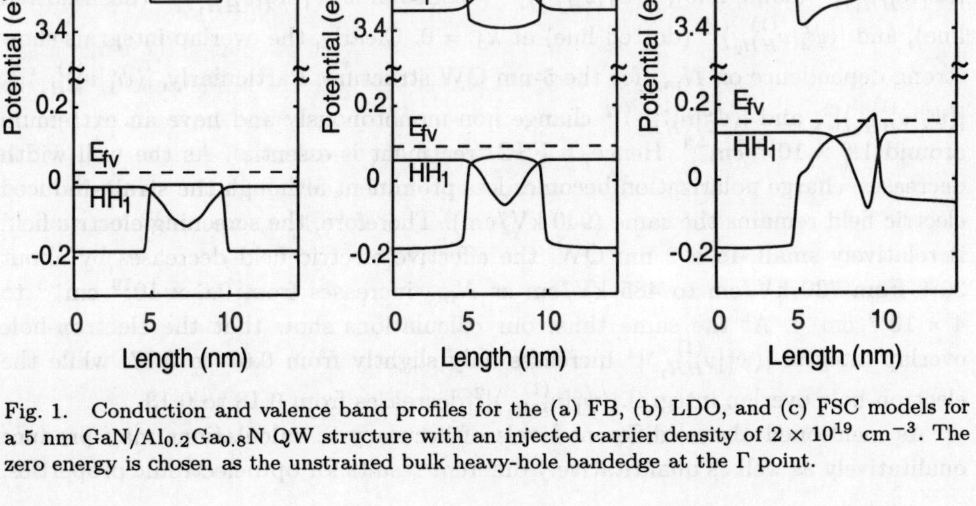

Fig. 1. Conduction and valence band profiles for the (a) FB, (b) LDO, and (c) FSC models for a 5 nm GaN/$Al_{0.2}Ga_{0.8}N$ QW structure with an injected carrier density of 2×10^{19} cm^{-3}. The zero energy is chosen as the unstrained bulk heavy-hole bandedge at the Γ point.

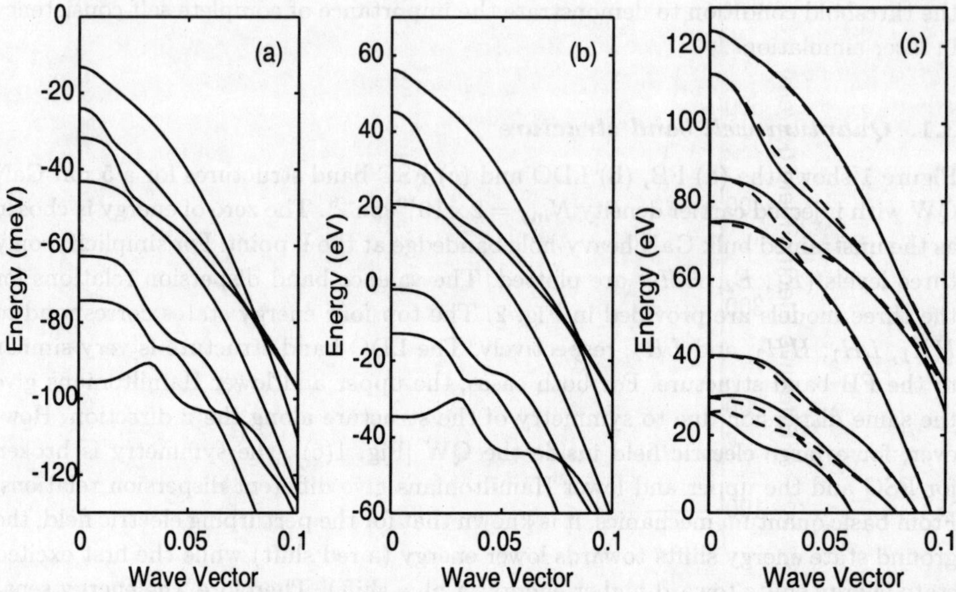

Fig. 2. Valence band dispersions for the (a) FB, (b) LDO and (c) FSC models for the 5 nm $GaN/Al_{0.2}Ga_{0.8}N$ QW structure with an injected carrier density of 2×10^{19} cm^{-3}.

Due to the large piezoelectric field, the electron and hole ground-state wave functions spatially separate and create a large screening electric field inside the QW that compensates the piezoelectric field. As N_{inj} increases from 0.6×10^{19} cm^{-3} to 4×10^{19} cm^{-3}, the effective electric field in the 5 nm QW is reduced from 600 kV/cm to 160 kV/cm as shown in Fig. 3(a). As a result, the electron-hole overlap integrals and the corresponding optical properties may vary significantly with N_{inj}. Figure 3(b) illustrates the FSC electron-hole overlap integrals $|\langle\psi_1^c|v_{HH_1}^{(1)}\rangle|^2$ (solid line), $|\langle\psi_1^c|v_{HH_2}^{(1)}\rangle|^2$ (dashed line), $|\langle\psi_2^c|v_{HH_1}^{(1)}\rangle|^2$ (dash-dotted line), and $|\langle\psi_2^c|v_{HH_2}^{(1)}\rangle|^2$ (dotted line) at $k_\perp = 0$. Clearly, the overlap integrals show strong dependence on N_{inj} for the 5-nm QW structure. Particularly, $|\langle\psi_1^c|v_{HH_2}^{(1)}\rangle|^2$, $|\langle\psi_2^c|v_{HH_1}^{(1)}\rangle|^2$, and $|\langle\psi_2^c|v_{HH_2}^{(1)}\rangle|^2$ change non-monotonously and have an extremum around 1.6×10^{19} cm^{-3}. Hence, a FSC treatment is essential. As the well width decreases, charge polarization becomes less prominent although the strain-induced electric field remains the same (940 kV/cm). Therefore, the screening electric field is relatively small. In a 2 nm QW, the effective electric field decreases by about 30% from 730 kV/cm to 485 kV/cm as N_{inj} increases from 0.6×10^{19} cm^{-3} to 4×10^{19} cm^{-3}. At the same time, our calculations show that the electron-hole overlap integral $|\langle\psi_1^c|v_{HH_1}^{(1)}\rangle|^2$ increases only slightly from 0.67 to 0.77, while the electron-hole overlap integral $|\langle\psi_1^c|v_{HH_2}^{(1)}\rangle|^2$ decreases from 0.18 to 0.13.

As mentioned above, different levels of accuracy in calculations may produce qualitatively as well as quantitatively different results for optoelectronic properties.

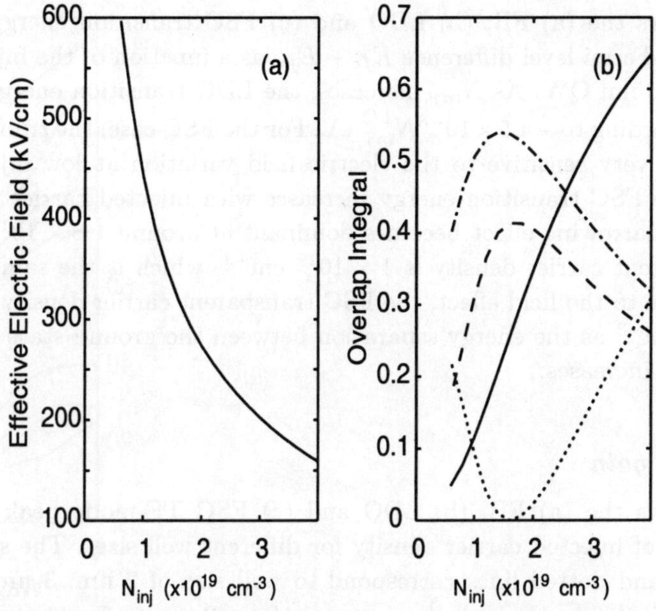

Fig. 3. (a) FSC effective electric field and (b) overlap integrals $|\langle\psi_1^c|v_{HH_1}^{(1)}\rangle|^2$ (the solid line), $|\langle\psi_1^c|v_{HH_2}^{(1)}\rangle|^2$ (the dashed line), $|\langle\psi_2^c|v_{HH_1}^{(1)}\rangle|^2$ (the dash-dotted line), and $|\langle\psi_2^c|v_{HH_2}^{(1)}\rangle|^2$ (the dotted line) as a function of the injected carrier density in the 5 nm GaN/Al$_{0.2}$Ga$_{0.8}$N QW structure.

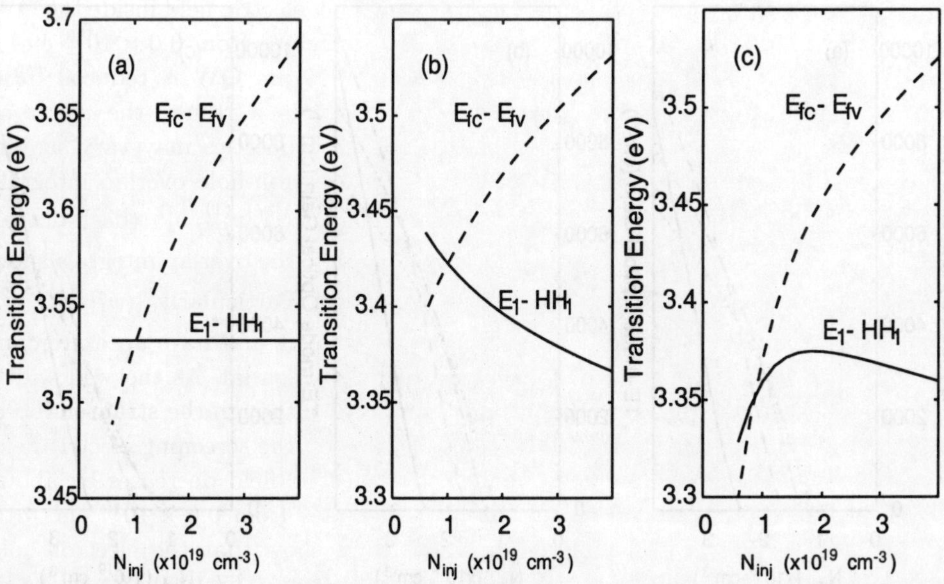

Fig. 4. Transition energy $E_1 - HH_1$ and the quasi-Fermi level difference $E_{fc} - E_{fv}$ as a function of the injected carrier density for the (a) FB, (b) LDO and (c) FSC models in the 5 nm GaN/Al$_{0.2}$Ga$_{0.8}$N QW structure.

Figure 4 shows the (a) FB, (b) LDO and (c) FSC transition energy $E_1 - HH_1$ and the quasi-Fermi level difference $E_{fc} - E_{fv}$ as a function of the injected carrier density for a 5-nm QW. As N_{inj} increases, the LDO transition energy $E_1 - HH_1$ decreases according to $-4.5 \times 10^{-8} N_{inj}^{1/3}$ eV. For the FSC case, the transition energy $E_1 - HH_1$ is very sensitive to the electric field variation at low injected carrier densities. The FSC transition energy increases with injected carrier density until the bandgap narrowing effect becomes dominant at around 1.8×10^{19} cm^{-3}. The LDO transparent carrier density is 1×10^{19} cm^{-3}, which is the same for the FB condition. Due to the field effect, the FSC transparent carrier density decreases to 0.87×10^{19} cm^{-3} as the energy separation between the ground state and the first excited states increases.

3.2. *Optical gain*

Figure 5 shows the (a) FB, (b) LDO and (c) FSC TE-mode peak optical gain as a function of injected carrier density for different well sizes. The solid, dashed, dash-dotted, and dotted lines correspond to well size of 2 nm, 3 nm, 4 nm, and 5 nm, respectively. As the well size increases from 2 nm to 5 nm, the transparent carrier density decreases from 2.4×10^{19} cm^{-3} to 1×10^{19} cm^{-3}. Also, the FSC transparent carrier density is slightly smaller than the density for the FB and LDO models as mentioned in the previous paragraph. The LDO peak optical gain is larger

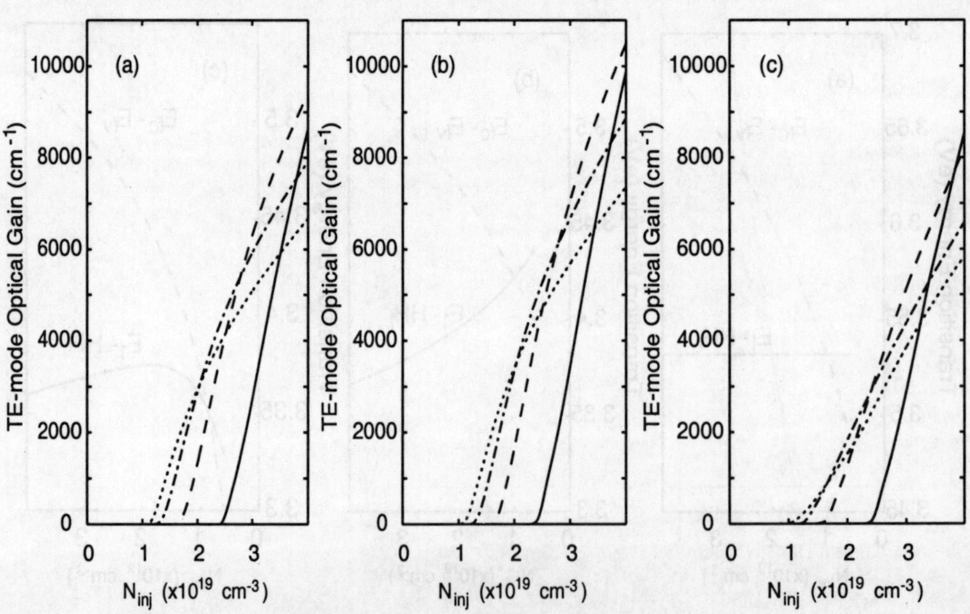

Fig. 5. (a) FB, (b) LDO and (c) FSC TE-mode peak optical gains as a function of the injected carrier density in the $GaN/Al_{0.2}Ga_{0.8}N$ QW structure with different well sizes. The solid, dashed, dash-dotted, and dotted lines correspond to well width of 2 nm, 3 nm, 4 nm, and 5 nm, respectively.

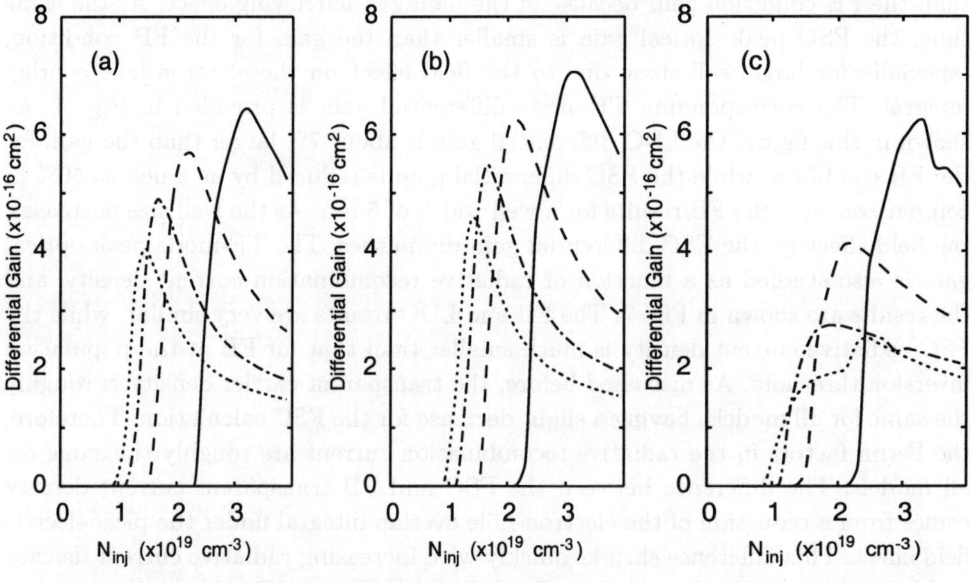

Fig. 6. (a) FB, (b) LDO and (c) FSC TE-mode differential gains as a function of the injected carrier density in the $GaN/Al_{0.2}Ga_{0.8}N$ QW structure with different well sizes. The solid, dashed, dash-dotted, and dotted lines correspond to well width of 2 nm, 3 nm, 4 nm, and 5 nm, respectively.

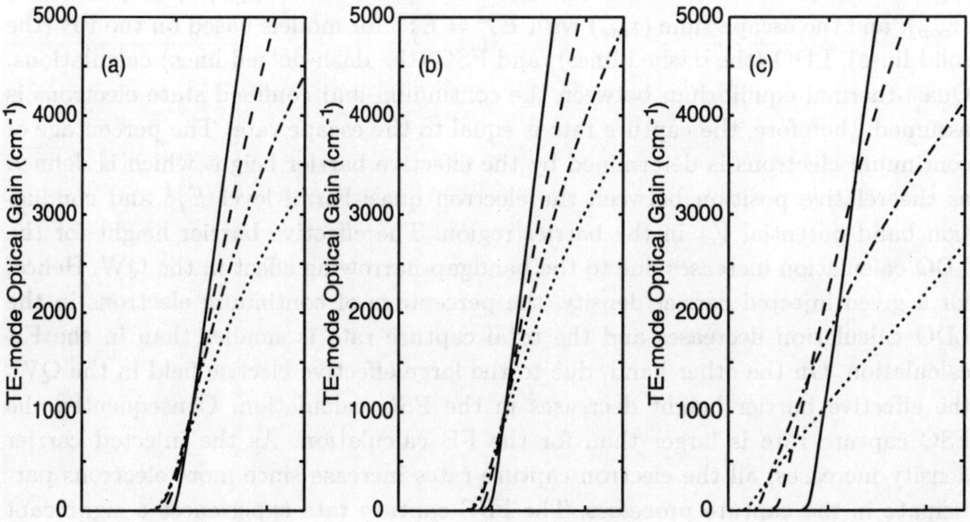

Fig. 7. (a) FB, (b) LDO and (c) FSC TE-mode peak optical gains as a function of the radiative recombination current density in the $GaN/Al_{0.2}Ga_{0.8}N$ QW structure with different well sizes. The solid, dashed, dash-dotted, and dotted lines correspond to well width of 2 nm, 3 nm, 4 nm, and 5 nm, respectively.

than the FB condition gain because of the bandgap narrowing effect. At the same time, the FSC peak optical gain is smaller than the gain for the FB condition, especially for large well sizes, due to the field effect on the electron-hole overlap integral. The corresponding TE-mode differential gain is provided in Fig. 6. As shown in this figure, the LDO differential gain is about 7% larger than the gain for the FB condition, while the FSC differential gain is reduced by as much as 50% in comparison with the FB results for a well width of 5 nm. As the well size decreases, the field effect on the FSC differential gain diminishes. The TE-mode peak optical gain is also studied as a function of radiative recombination current density, and the results are shown in Fig. 7. The FB and LDO results are very similar, while the FSC radiative current density is much smaller than that for FB at the population inversion threshold. As discussed before, the transparent carrier density is roughly the same for all models, having a slight decrease for the FSC calculation. Therefore, the Fermi factors in the radiative recombination current are roughly the same for all models. The difference between the FSC and FB transparent current density comes from a reduction of the electron-hole overlap integral under the piezoelectric field effect. This difference shrinks quickly with increasing radiative current density and becomes less pronounced than the difference shown in Fig. 5.

3.3. *Capture/emission processes and frequency response*

Figure 8 exhibits calculations of the electron capture rate (U_{cap}), the capture time (τ_{cap}), and the escape time (τ_{esc}) with $E_f^{3d} = E_f^{qw}$ for models based on the FB (the solid lines), LDO (the dashed lines), and FSC (the dash-dotted lines) calculations. Quasi-thermal equilibrium between the continuum and confined state electrons is assumed; therefore, the capture rate is equal to the escape rate. The percentage of continuum electrons is determined by the effective barrier height which is defined as the relative position between the electron quasi-Fermi level E_f^{3d} and conduction band potential V_{cb} in the barrier region. The effective barrier height for the LDO calculation increases due to the bandgap-narrowing effect in the QW. Hence, for a given injected carrier density, the percentage of continuum electrons in the LDO calculation decreases and the total capture rate is smaller than in the FB calculation. On the other hand, due to the large effective electric field in the QW, the effective barrier height decreases in the FSC calculation. Consequently, the FSC capture rate is larger than for the FB calculation. As the injected carrier density increases, all the electron capture rates increase since more electrons participate in the capture processes. The FSC capture rate experiences a significant increase for injected carrier densities between 1×10^{19} cm^{-3} and 1.8×10^{19} cm^{-3}. This is because a strongly localized continuum state is formed for large injected carrier densities due to increased band-bending. This phenomena leads to a very strong electron-phonon interaction. The hole capture rate is three orders of magnitude larger than the electron capture rate (Fig. 8) due to the larger density of states, lower effective barrier height and more confined energy states that serve as

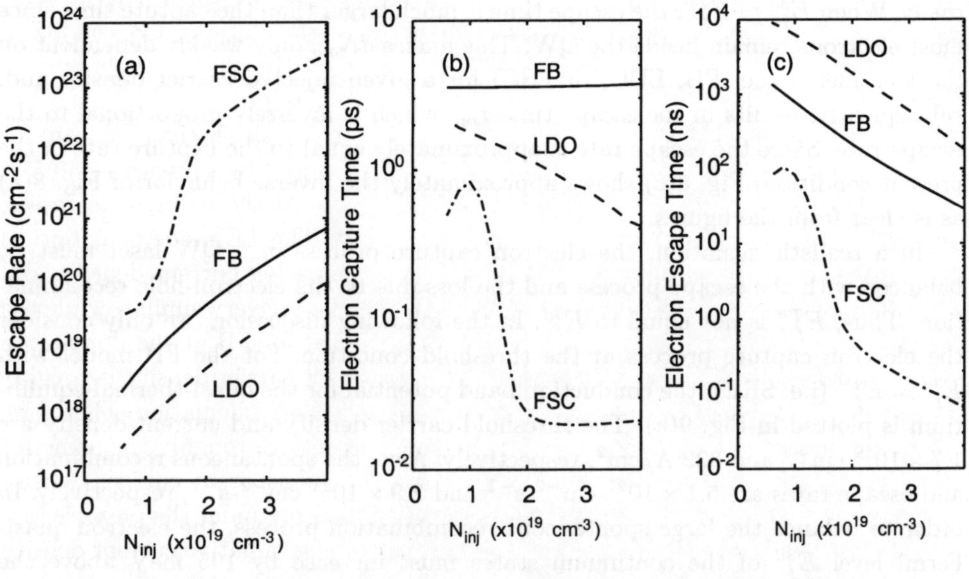

Fig. 8. (a) Electron capture rate, (b) capture time and (c) escape time in the 5 nm GaN/60 nm $Al_{0.2}Ga_{0.8}N$ QW structure with $E_f^{3d} = E_f^{qw}$ for FB (solid line), LDO (dashed line), and FSC (dash-dotted line).

capture channels for holes. Therefore we can very safely assume that the continuum and confined hole states in the valence band are in quasi-thermal equilibrium with each other.

The electron capture time τ_{cap} in Fig. 8(b) exhibits a different behavior compared to the capture rates. Since $1/\tau_{cap}$ is given as the capture rate U_{cap} divided by N_{3d}, the capture time is not determined by the number of available continuum carriers but by the intrinsic transition probability. Hence, the electron capture time for the FB condition varies only slightly as a function of injected carrier density. Due to the same reason, the average reduction in bandgap cannot explain the exponential decrease of capture time observed in the LDO model. Since the many-body effect is treated within the local density approximation, the bandgap narrowing effect varies according to the local electron density, which in turn is determined by the shape of the ground-state wave function. As a result, band-bending is formed for the barrier potential in the vicinity of the QW and varies according to the injected carrier density. The band-bending increases the localization of the continuum-state wave functions around the QW and therefore decreases the LDO capture time as shown in Fig. 8(b). A sharp decrease of the FSC capture time between carrier densities of 1×10^{19} cm^{-3} and 1.8×10^{19} cm^{-3} is again due to the fact that a strongly localized continuum state is formed above the barrier height. Our calculations show the FSC capture time is 20 fs when the carrier density is larger than 2×10^{19} cm^{-3}. This value is more than two orders of magnitude smaller than the FB capture time of about 3.6 ps. The characteristics for escape time shown in Fig. 8(c) can be explained easily. When $E_f^{3d} \approx E_f^{qw}$, the escape time is much larger than the capture time since most electrons remain inside the QW. This makes dN_{qw} only weakly dependent on the treatments (i.e. FB, LDO, or FSC) for a given injected carrier density and, subsequently, results in the escape time τ_{esc} which is inversely proportional to the escape rate. Since the escape rate is approximately equal to the capture rate in the present condition, Fig. 8(c) shows approximately the inverse behavior of Fig. 8(a) as is clear from the figures.

In a realistic situation, the electron capture process in a QW laser must be balanced with the escape process and the loss due to the electron-hole recombination. Thus, E_f^{qw} is not equal to E_f^{3d}. In the following discussion, we only consider the electron capture process at the threshold condition. For the FB model with $E_f^{3d} = E_f^{qw}$ (i.e. SRE), the conduction band potential for the quasi-thermal equilibrium is plotted in Fig. 9(a). The threshold carrier density and current density are 1.7×10^{19} cm^{-3} and 822 A/cm^2, respectively. Also, the spontaneous recombination and escape rates are 5.1×10^{21} cm^{-2}-s^{-1} and 2.9×10^{21} cm^{-2}-s^{-1}, respectively. In order to balance the large spontaneous recombination process, the electron quasi-Fermi level E_f^{3d} of the continuum states must increase by 195 meV above the quasi-Fermi level E_f^{qw}, which means that more than 30% of the electrons are in the continuum states and the structure is far from the threshold condition. After several self-consistent iterations under the GRE formalism, the FB threshold carrier density

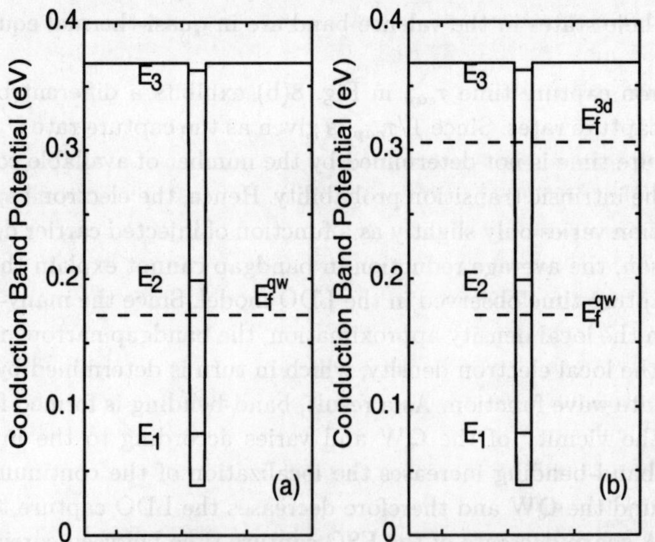

Fig. 9. FB conduction band potential profile of the 5 nm GaN/60 nm $Al_{0.2}Ga_{0.8}N$ QW structure under (a) the quasi-thermal equilibrium $E_f^{3d} = E_f^{qw}$ and (b) fixed capture and escape times calculated based on the FB condition.

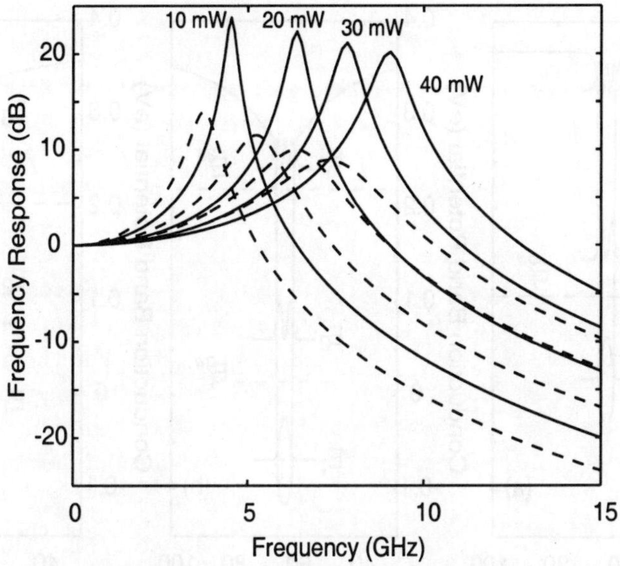

Fig. 10. FB modulation response of the 5 nm GaN/60 nm $Al_{0.2}Ga_{0.8}As$ QW structure for different optical output powers at 10, 20, 30, and 40 mW. The solid, and dashed lines correspond to the simplified and generalized rate-equation formalisms, respectively.

and current density are found to be 2.1×10^{19} cm^{-3} and 884 A/cm^2, respectively. About 33% of the electrons are in the continuum states and the electron quasi-Fermi level E_f^{3d} is 134 meV above E_f^{qw} as shown in Fig. 9(b). The spontaneous recombination, escape and capture rates are 5.5×10^{21} cm^{-2}-s^{-1}, 1.6×10^{19} cm^{-2}-s^{-1}, and 5.5×10^{21} cm^{-2}-s^{-1}, respectively. Our estimates of threshold current density represent theoretical limits and are compatible with recent experimental results.[1] The FB modulation response for the single GaN/AlGaN QW structure is shown in Fig. 10 for optical output powers of 10, 20, 30, and 40 mW. The solid and dashed lines correspond to the SRE and GRE formalisms, respectively. The GRE modulation response frequency decreases by 17% and the response function decreases by roughly 50% at the resonance in comparison with the SRE results. With a leakage of more than 30% of the injected electrons, a very large electric field will be generated by the continuum electrons in the barrier and the confined holes in the QW. Hence, a FB calculation cannot provide an accurate description of QW lasers.

Next, we discuss our results for the study of the carrier capture process in the LDO model for three different cases: (1) SRE formalism with $E_f^{3d} = E_f^{qw}$, (2) GRE formalism with fixed capture and escape times calculated from the band structure in the case 1, and (3) GRE formalism based on a self-consistent adjustment between the capture process and the band structure. Figure 11 shows the LDO conduction band potential for (a) case 1, (b) case 2, and (c) case 3. For case 1, the threshold carrier density and current density are 1.6×10^{19} cm^{-3} and 765 A/cm^2, respectively. The capture and escape times are evaluated as 1.2 ps and 3.1 ms, respectively. Using

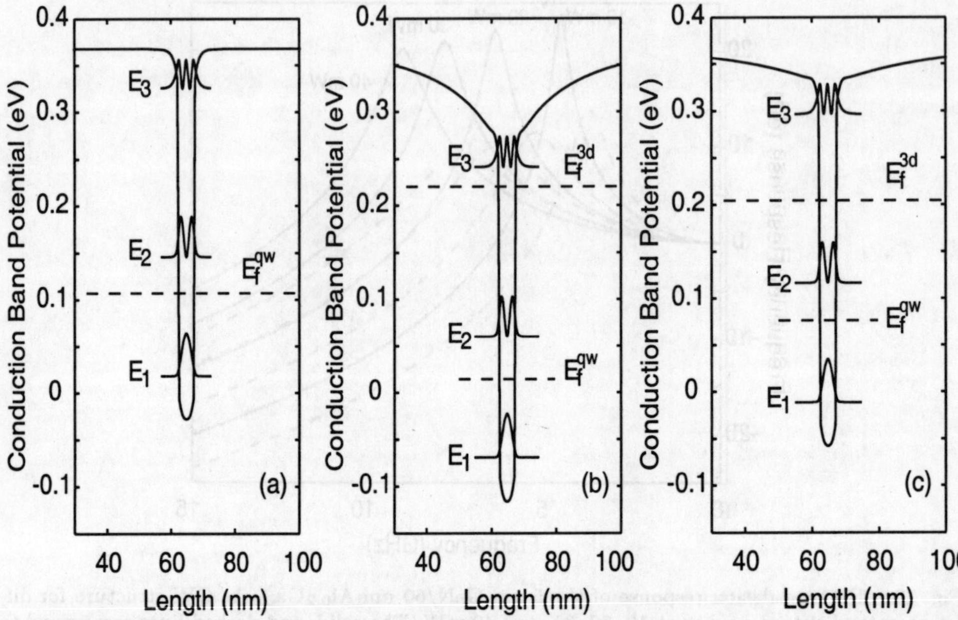

Fig. 11. LDO conduction band potential profile of the 5-nm GaN/60-nm $Al_{0.2}Ga_{0.8}N$ QW structure for (a) the quasi-thermal equilibrium $E_f^{3d} = E_f^{qw}$, (b) fixed capture and escape times calculated from the quasi-thermal equilibrium configuration, and (c) self-consistently adjusted capture and escape times based on the self-consistent band profile.

the capture and escape times from case 1, the conduction band structure is recalculated to balance the generalized rate equations. The resulting threshold carrier density and current density are 1.67×10^{19} cm^{-3} and 778 A/cm^2, respectively. The electron quasi-Fermi level E_f^{3d} moves 206 meV above E_f^{qw} and about 7% of the electrons leak out to the barrier region which alter the band structure significantly as shown in Fig. 11(b). Since the capture and escape times vary according to the shape of the band structure, a self-consistent adjustment of the capture process is performed and the final band structure is illustrated in Fig. 11(c). Here the threshold carrier density and current density are 1.61×10^{19} cm^{-3} and 766 A/cm^2, respectively. The electron quasi-Fermi level E_f^{3d} is only 128 meV above E_f^{qw} and less than 1.5% of the electrons leak out to the barrier region. The capture and escape times for case 3 are 75 fs and 0.17 ms, respectively, which is about 17% smaller than the capture and escape times used in case 2. By comparing band structures based on these models, it is clear that a self-consistent adjustment of the capture and escape process is essential for accurate modeling of QW lasers. The LDO modulation response for the single GaN/AlGaN QW structure is shown in Fig. 12 for different optical output powers at 10, 20, 30, and 40 mW. The solid, dashed and dotted lines correspond to cases 1, 2, and 3 respectively. Due to bandgap renormalization, the LDO lasing energy is more than 150 meV lower than the FB condition. Hence for a given output power, the LDO photon density S_0 is larger

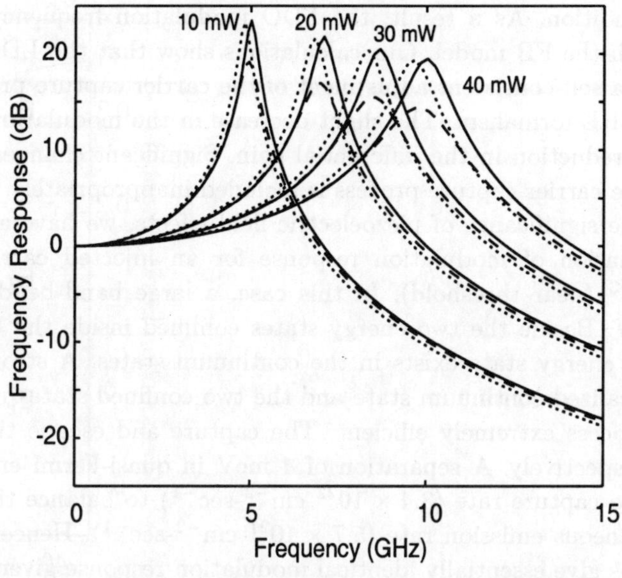

Fig. 12. LDO modulation response of the 5 nm GaN/60 nm $Al_{0.2}Ga_{0.8}As$ QW structure for different optical output powers at 10, 20, 30, and 40 mW. The solid, dashed and dotted lines correspond to the SRE formalism with $E_f^{3d} = E_f^{qw}$, GRE formalism with fixed capture and escape times calculated from the SRE band structure, and GRE formalism based on a self-consistent adjustment between the capture process and the band structure, respectively.

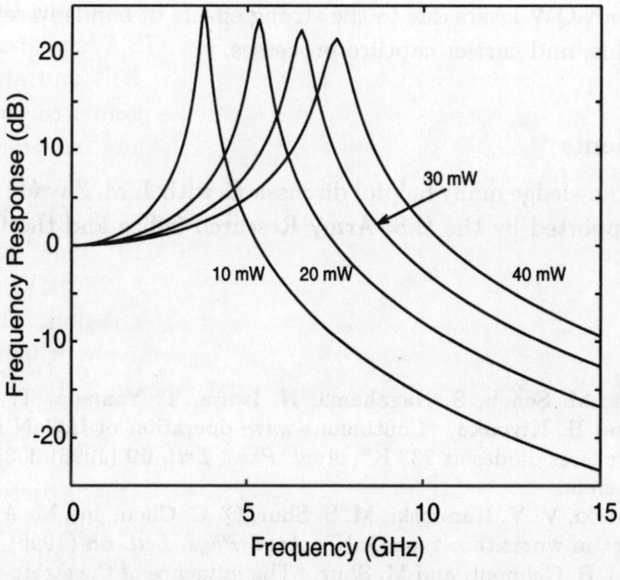

Fig. 13. FSC modulation response of the 5 nm GaN/60 nm $Al_{0.2}Ga_{0.8}As$ QW structure for different optical output powers at 10, 20, 30, and 40 mW.

than in FB condition. As a result, the LDO modulation frequency is higher in comparison with the FB model. Our calculations show that the LDO modulation responses with a self-consistent adjustment of the carrier capture process are very similar to the SRE formalism. The slight decrease in the modulation frequency is due to a small reduction in the differential gain. Significant changes are found in case 2 where the carrier capture process is included inappropriately.

To gauge the significance of piezoelectric field effects, we have also performed the FSC calculation of modulation response for an injected carrier density of 2.2×10^{19} cm^{-3} (near threshold). In this case, a large band-bending is formed around the QW. Beside the two energy states confined inside the QW, an additional localized energy state exists in the continuum states. A strong interaction between the localized continuum state and the two confined states makes the carrier capture process extremely efficient. The capture and escape times are 21 fs and 0.36 ns, respectively. A separation of 4 meV in quasi-Fermi energies will be sufficient for the capture rate (3.4×10^{22} cm^{-2}-sec^{-1}) to balance the escape rate plus the spontaneous emission rate (5.7×10^{21} cm^{-2}-sec^{-1}). Hence the SRE and GRE formalisms give essentially identical modulation response given in Fig. 13.

4. Summary

This article presents a detailed study of the band structure, optical properties, and modulation response of a pseudomorphically strained GaN QW structure based on an RSP envelope-function Hamiltonian and the photon-carrier rate equations. It is demonstrated that a fully self-consistent treatment is essential for accurate description of GaN QW lasers due to the strong effects of bandgap renormalization, piezoelectric fields, and carrier capture processes.

Acknowledgments

The authors acknowledge many helpful discussions with J. M. Zavada and M. Dutta. This work is supported by the U.S. Army Research Office and the Office of Naval Research.

References

1. S. Nakamura, M. Senoh, S. Nagahama, N. Iwasa, T. Yamada, T. Matsushita, Y. Sugimoto, and H. Kiyouku, "Continuous-wave operation of InGaN multi-quantum-well-structure laser diodes at 233 K", *Appl. Phys. Lett.* **69** (1996) 3034–3036; and the references therein.
2. A. D. Bykhovski, V. V. Kaminski, M. S. Shur, Q. C. Chen, and M. A. Khan, "Piezoresistive effect in wurtzite *n*-type GaN", *Appl. Phys. Lett.* **68** (1996) 818–819.
3. A. Bykhovski, B. Gelmont, and M. Shur, "The influence of the strain-induced electric field on the charge distribution in GaN-AlN-GaN structure", *J. Appl. Phys.* **74** (1993) 6734–6739.

4. G. Martin, A. Botchkarev, A. Rockett, and H. Morkoc, "Valence-band discontinuities of wurtzite GaN, AlN, and InN heterojunctions measured by x-ray photoemission spectroscopy", *Appl. Phys. Lett.* **68** (1996) 2541–2543.
5. D. L. Smith and C. Mailhiot, "Nonlinear optical response of [111] growth axis strained-layer superlattices", *J. Vac. Sci. Technol.* **5** (1987) 2060–2064.
6. A. T. Meney, E. P. O'Reilly, and A. R. Adams, "Optical gain in wide bandgap GaN quantum well lasers", *Semicond. Sci. Technol.* **11** (1996) 897–903.
7. S. L. Chuang, "Optical gain of strained wurtzite GaN quantum-well lasers", *IEEE J. Quantum Electron.* **32** (1996) 1791–1800.
8. M. Suzuki and T. Uenoyama, "Optical gain and crystal symmetry in III-V lasers", *Appl. Phys. Lett.* **69** (1996) 3378–3380.
9. P. Shah, V. Mitin, M. Grupen, G. H. Song, and K. Hess, "Numerical simulation of wide bandgap AlGaN/InGaN light-emitting diodes for output power characteristics and emission spectra", *J. Appl. Phys.* **79** (1996) 2755–2761.
10. Y. M. Sirenko, J. B. Jeon, B. C. Lee, K. W. Kim, M. A. Littlejohn, M. A. Stroscio, and G. J. Iafrate, "Hole scattering and optical transitions in wide bandgap nitrides: Wurtzite and zincblende structures", *Phys. Rev.* **B55** (1997) 4360–4375.
11. A. Polian, M. Grimsditch, and I. Grzegory, "Elastic constants of gallium nitride", *J. Appl. Phys.* **79** (1996) 3343–3344.
12. K. Inoue, H. Sakaki, J. Yoshino, and T. Hotta, "Self-consistent calculation of electronic states in AlGaAs/GaAs/AlGaAs selectively doped double-heterojunction systems under electric fields", *J. Appl. Phys.* **58** (1985) 4277–4281; G. Ji, T. Henderson, C. K. Peng, D. Huang, and H. Morkoc, "Analysis of energy states in modulation doped multiquantum well heterojunctions", *Solid-State Electron.* **33** (1990) 247–258.
13. O. Gunnarson and B. I. Lundquist, "Exchange and correlation in atoms, molecules, and solids by the spin-density-functional formalism", *Phys. Rev.* **B13** (1976) 4274–4298.
14. M. Asada, in *Quantum Well Lasers*, ed. P. S. Zory, Chap. 2, Academic Press, Boston, 1993.
15. T. P. Lee, C. A. Burrus, J. A. Copeland, A. G. Dentai, and D. Marcuse, "Short-cavity InGaAsP injection lasers: Dependence of mode spectra and single-longitudinal-mode power on cavity length", *IEEE J. Quantum Electron.* **18** (1982) 1101–1113.
16. S. L. Chuang, *Physics of Optoelectronic Devices*, Wiley, New York, 1995.
17. W. Rideout, W. F. Sharfin, E. S. Koteles, M. O. Vassell, and B. Elman, "Well-barrier hole burning in quantum well lasers", *IEEE Photon. Tech. Lett.* **3** (1991) 784–786.
18. M. Grupen, G. Kosinovsky, and K. Hess, "The effect of carrier capture on the modulation bandwidth of quantum well lasers", *Proc. IEDM* (1993) 23.6.1–23.6.4.
19. J. H. Edgar (ed.) *Properties of Group III Nitrides*, Electronic Materials Information Service Datareviews Series, IEE, London, 1994.
20. S. Nakamura, M. Senoh, S. Nagahama, N. Iwasa, T. Yamada, T. Matsushita, Y. Sugimoto, and H. Kiyoku, "Optical gain and carrier lifetime of InGaN multi-quantum well structure laser diodes", *Appl. Phys. Lett.* **69** (1996) 1568–1570.

International Journal of High Speed Electronics and Systems, Vol. 9, No. 4 (1998) 1211–1233

CARRIER CAPTURE IN SEMICONDUCTOR QUANTUM WELL LASERS: A QUANTUM TRANSPORT ANALYSIS

LEONARD F. REGISTER

Beckman Institute and Coordinated Science Laboratory,
University of Illinois at Urbana-Champaign,
Urbana, Illinois 61801, USA

A quantum transport-based analysis of the essential physics of carrier capture in semiconductor quantum wells is presented. First, the past progression of models of carrier capture by quantum wells is briefly reviewed. Then carrier capture is modeled using the Schrödinger Equation Monte Carlo (SEMC) quantum transport simulator. In addition to reproducing familiar effects, these simulations exhibit significant effects associated with partial phase-coherence of the carrier wave-function across the well which cannot be modeled via classical or perturbative Golden Rule calculations, and address fundamental transport limitations often overlooked in Golden Rule calculations. However, this analysis also points to simple changes that could significantly improve, although not perfect, the treatment of carrier capture via these latter more conventional approaches.

1. Introduction

Quantum wells (QWs) are now used routinely in the design of semiconductor lasers, providing enhanced spatial overlap of electron and hole distributions and allowing the carrier densities of states in energy within the well to be engineered through quantum confinement and strain effects on the band-structure of the well material. As a result, QWs allow lower threshold currents, higher efficiencies, faster modulation responses, and a wider range of operating frequencies than would otherwise be possible.[1] As discussed elsewhere in this book/issue, intersubband transitions in QWs even allow for lasing at far-infrared frequencies.

However, a bottleneck to fully realizing these advantages of quantum well (QW) lasers is the carrier capture process by the QW, itself. The efficiency of carrier capture affects both laser efficiency and modulation response, although not always in a trivial way.[2,3] Charge storage in the separate confinement region is required to replenish the electron-hole pairs that are converted into photons in the quantum well. However, both the charge stored in the separate confinement region and, often more importantly, that which escapes beyond the separate confinement region contribute to the laser's capacitance and its dark recombination current. Therefore, in a single QW laser, the more efficient carrier capture by the well and the less efficient transmission beyond the well, the better the laser. However, in a multiwell system — which both allows for increased net gain and increases the net capture efficiency of the QWs — the more uniform the carrier distribution among the wells, the better the laser performance also. Therefore, intermediate capture and/or transmission probabilities for each well may be optimum.

Despite the practical importance of carrier capture by QWs, this process remains one of the more poorly understood aspects of QW Laser operation, although not for a lack of attention. Initial attempts to model carrier capture into QWs relied on classical theory.[4] However, as to be expected given the "quantum well" label, it was subsequently demonstrated both theoretically via Fermi's Golden Rule-based calculations[5] and experimentally[6] that quantum interference effects in the free as well as bound carrier states could significantly affect capture by QWs. However, inelastic scattering is inherently accompanied by a loss of phase-coherence in the carrier wave-functions, and, thus, efficient capture by a significant loss of phase-coherence. Therefore, neither a classical analysis nor a perturbative quantum approach such as Fermi's Golden Rule may be appropriate in this application.[3] Indeed, as noted above, to achieve optimum performance in multi-well laser would *require* an intermediate degree of scattering as discussed above, and, thus, the failure of both classical and perturbative quantum treatments.

In this work, carrier capture by QWs is studied using the quantum transport simulator Schrödinger Equation Monte Carlo (SEMC).[7,8] The relevant feature of SEMC for this work is that it is designed to allow simulation of carrier transport — and in this case capture — with intermediate degrees of phase coherence allowing the gap between classical and quantum transport to be bridged; neither limit need be assumed *a priori*. Further, SEMC allows for a quantitatively accurate treatment of the long-range carrier–polar-optical-phonon interaction which is the dominate phonon scattering mechanism for carrier capture. The discussion of this work represents a more full discussion of work reported in a letter format elsewhere.[9] Some of this work also has been used to illustrate the SEMC method itself,[8] but in that latter work the simple ability to produce the results was emphasized rather than the physical significance of the results.

In the following sections, first a brief overview of classical, semiclassical, and Golden Rule treatments of carrier capture are presented. Then the SEMC method itself is briefly described; for a more complete description the interested reader is referred to Ref. 8. Then the essential physics of carrier capture by QWs, including fundamental yet often overlooked transport limitations and the effects associated with partial phase-coherence of the carrier wave-function across the well, is discussed and illustrated by SEMC simulation results for a model QW system.

2. Classical, Semiclassical and Golden Rule Analysis

The simplest treatment of carrier capture, and the one that if true would be optimal for single QW well lasers, would be capture simply by thermionic emission, as illustrated in Fig. 1(a). Every carrier from the separate confinement region *reservoir* incident on the QW *reservoir* would be captured by thermionic emission into the well; transport beyond the well could then only occur by subsequent thermionic emission out of the well according to the carrier distribution in energy within the well.[10] However, with lifetimes between scattering events greater than the nominal transient time across narrow wells, the presumption that every carrier is captured is difficult to justify, although thermionic emission theory can still represent a reasonable treatment of carrier capture into the separate confinement region of the laser.[2]

Accordingly, as early as 1978 it was shown experimentally that the efficiency of carrier capture appeared to diminished with decreasing well width.[4] To account for this behavior, the authors of Ref. 4 turned to Fermi's "age theory"[11] which, in this case, essentially acknowledges that the time required for thermalization may be significantly greater than the time required for carriers to traverse the well. However, the analysis remained semiclassical; as illustrated in Fig. 1(b), although it allowed for quasi-elastic transport across the well, it did not address the effects of phase-coherence that inherently accompany elastic transport. A semiclassical Monte Carlo analysis where the electrons are modeled as localized, would fall into the same basic category. One of the predictions of this model was that narrow wells would be too inefficient for laser operation. However, one of the authors of this work subsequently demonstrated lasing from "quantum wells" of a fraction of a lattice constant in width.[12]

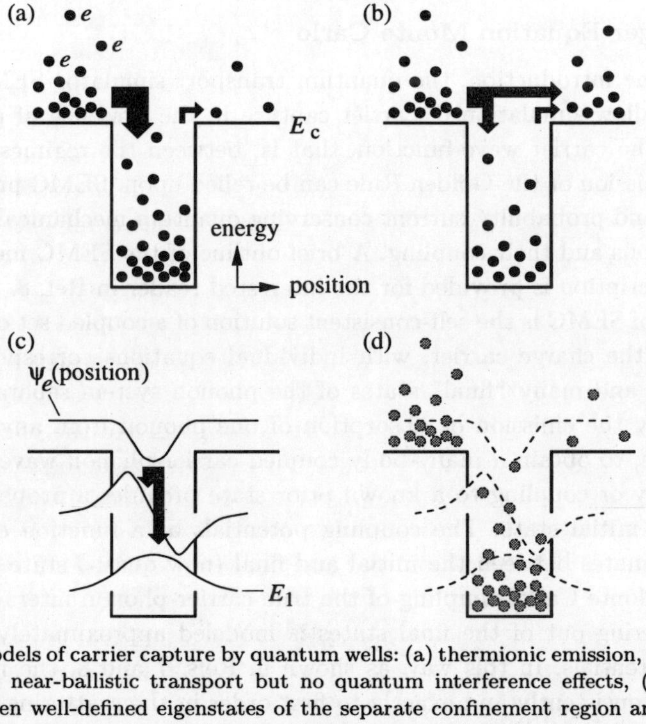

Fig. 1. Four models of carrier capture by quantum wells: (a) thermionic emission, (b) semiclassical with ballistic or near-ballistic transport but no quantum interference effects, (c) Golden Rule scattering between well-defined eigenstates of the separate confinement region and the QW, and (d) capture in the presence of partial carrier phase coherence, loosely depicted.

To address the effects of phase-coherence across *quantum* wells, Brum and Bastard turned to Golden Rule scattering between carrier eigenstates of the separate confinement region and of the QW, as illustrated in Fig. 1(c).[5] Results suggested that the capture rates could be subject to periodic oscillations with well width, roughly of an order of magnitude for electrons and between a factor of two and five for holes for the particular system they considered, due to resonances associated with the states of the separate confinement region above the well, and with the bound states of the QW. Experimental results later confirmed the existence of such oscillations for carrier capture by QWs, but the best fit to theory resulted from ignoring the theoretical oscillations for holes presuming that the phase-breaking inelastic scattering of holes was too strong to allow for resonant quantum effects.[6]

Despite their limitations, each of these past treatments has provided an improved understanding of carrier capture by QWs. Further, the limitation of these treatments of carrier capture to either phase-incoherent classical or pure phase-coherent quantum models of transport across the well has arisen in large part simply from the lack of any alternative. Only in recent years have quantum transport simulation tools been developed that allow simulation of transport with partial phase-coherence, as depicted very loosely in Fig. 1(d), and only still more recently with realistic modeling of scattering processes such as the long-range polar-optical phonon scattering process critical to carrier capture in QWs.[8,13,14]

3. Schrödinger Equation Monte Carlo

As noted in the introduction, the quantum transport simulator SEMC is used in this work to allow simulation of carrier capture in the presence of partial phase-coherence of the carrier wave-function, that is, between the regimes where either thermionic emission or the Golden Rule can be relied upon. SEMC provides a non-perturbative and probability current conserving quantum mechanical treatment of carriers, phonons and their coupling. A brief outline of the SEMC method follows; a detailed description is provided for the interested reader in Ref. 8.

The basis of SEMC is the self-consistent solution of a coupled set of Schrödinger equations for the charge carrier, with individual equations corresponding to the "initial" state and many "final" states of the phonon system separated from the initial state by the emission or absorption of one phonon from any of the many phonon modes, to obtain a many-body coupled carrier-phonon wave-function. An open boundary or coupling to a known prior state provides a probability current source to the initial state. The coupling potentials as a function of carrier and phonon coordinates between the initial and final (now quasi-) state equations are obtained by Monte Carlo sampling of the true carrier-phonon interactions.[15] Subsequent scattering out of the final states is modeled approximately via complex self-energy potentials. In this way, as shown in Refs. 7 and 8, the initial state is obtained self-consistently and exactly to first order in the scattering from it, both real and virtual. Second and higher order scattering contributions to the initial

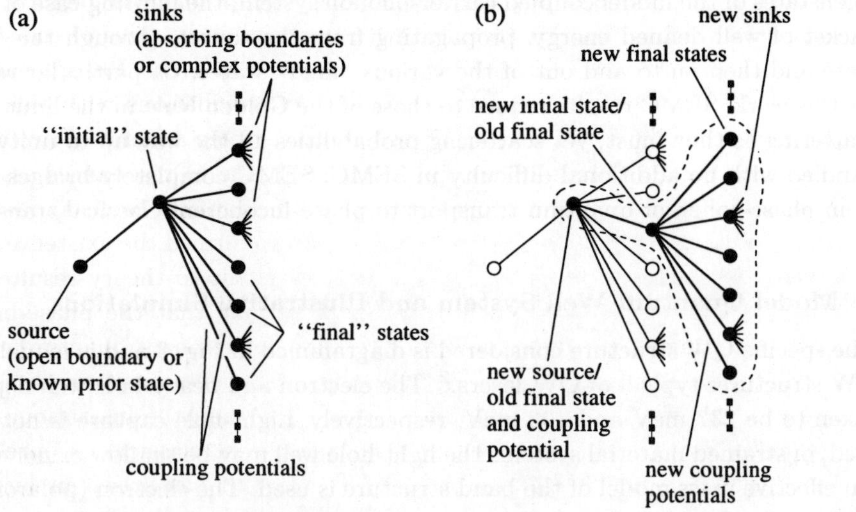

Fig. 2. (a) Phonon states considered in one SEMC calculation, showing "initial" and "final" states and coupling potentials (which are functions of both the carrier and phonon coordinates) along with the source and sinks terms. For every phonon state there is a corresponding Schrödinger equation for the charge carrier that must be solved self-consistently. (b) A second sequential SEMC calculation, where the old initial state becomes the source, a new initial phonon state is selected stochastically from the old final states and a new set of final states and associated coupling potentials is generated to replace the old complex potential functions, with each new final state having its own complex potential function.

state are treated within the accuracy of the assumed complex self-energy potentials in the final states, and are small in the case at hand.[16] Any observable of interest — carrier probability density, phonon emission or absorption probability, capture probability, etc. — is obtained from this many-body wave-function. This basic coupled system of equations is depicted in Fig. 2(a). As depicted in Fig. 2(b), this procedure also can be repeated sequentially to trace carriers through an unlimited number of scattering "events"; the old initial state becomes the source, a new initial phonon state is selected stochastically from the old final states and a new set of final states and associated coupling potentials is generated to replace the complex potential functions, with each new final state having its own complex potential function. The selected old final state/new initial state is then re-obtained exactly to first-order, and so forth.

In the sampling of the true carrier-phonon interactions and this ability to trace the carrier "trajectory" through many scattering events, SEMC represents a quantum analog of semiclassical Monte Carlo — whence comes the "Schrödinger Equation Monte Carlo" moniker — and output from semiclassical Monte Carlo could be used as input to SEMC and vice versa. However, in contrast to semiclassical Monte Carlo, in SEMC the carriers (and phonons) are treated as quantum mechanical particles and scattering is nonlocal in both position and time. Indeed, the calculations of this work are actually time-independent so that the solutions are energy eigenstates of the model coupled carrier-phonon system, the limiting case of a wave-packet of well defined energy, propagating from the source, through the "initial" state and then on to and out of the various "final" states. Of particular relevance for this work, SEMC results reduce to those of the Golden Rule in the limit of weak scattering as they must, yet scattering probabilities all the way up to unity can be handled with no additional difficulty in SEMC; SEMC completely bridges the gap from phase-coherent quantum transport to phase-incoherent classical transport.

4. Model Quantum Well System and Illustrative Simulations

The specific QW structure considered is diagrammed in Fig. 3 and is modeled after QW structures typical of QW lasers.[2] The electron and heavy-hole well depths are taken to be 237 meV and 103 meV, respectively. Light-hole capture is not considered; in strained material systems the light-hole well may be shallow or nonexistent. An effective mass model of the band structure is used. The electron (polaron) effective mass to free electron mass ratios in the barrier and well materials are 0.068 and 0.053, respectively, and for heavy-holes the effective mass ratios are 0.47 and 0.31, respectively. The latter effective mass is chosen to be representative of the density of states of the strained material near the top of the heavy-hole well rather than at the Γ-point. Bulk GaAs longitudinal-optical phonon modes are assumed, which should be a reasonable approximation given a common predominance of GaAs in both well and barrier alloy materials, and the existence of conservation rules for total carrier scattering from confined, interface, half-plane, and bulk modes.[17,18] The complex potential functions used to model subsequent scattering — real and virtual — in the final states are taken to be the imaginary and real components of

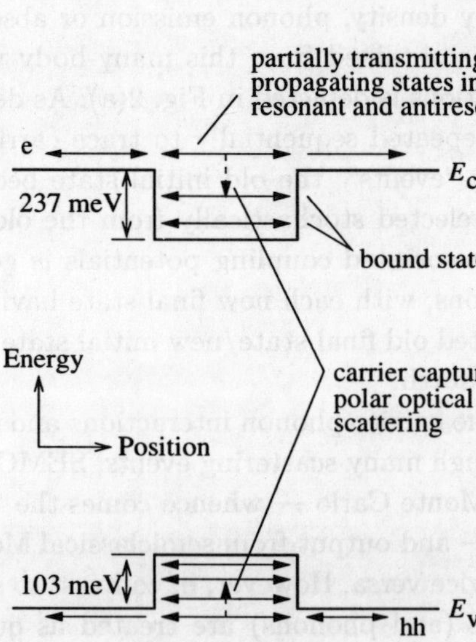

Fig. 3. Quantum well model.

the bulk-material self-energies as a function of material and carrier energy. Such a simple model is of course insufficient for precise quantitative analysis of any specific QW system. However, the single parabolic electron and hole bands of significantly differing effective masses and the single phonon energy help clarify the essential physics of interest in this work, and within the limits of this model the calculations of polar-optical-phonon scattering and resulting carrier capture by the well to follow are quantitatively correct. The requirements of more precise analysis of specific structures are discussed in Ref. 8.

An example of the SEMC calculated electron probability density across such a 15 nm well before and after capture is shown in Fig. 4. For this example a thermal distribution of incident electrons weighted by their interface normal components of velocity is stochastically sampled, and a separate sequential SEMC calculation is performed for each. The lattice and temperature, along with the carrier temperature, is taken to be 77 K for this example so that, for clarity, only optical phonon emission is significant. For the electron distribution, an ensemble provided by semiclassical Monte Carlo simulation could be used as easily. Figures 4(a)–(c) show, respectively, the self-consistently obtained electron probability density as a function of well normal position and well parallel momentum in the SEMC "initial" state averaged over the thermal distribution prior to phonon emission (after the first SEMC calculation for each electron with a boundary source), after emission of one optical phonon (after the second sequential SEMC calculation using the first "initial" state as a source), and after emission of four phonons (after the fifth sequential SEMC calculation). The resonance structure above the well and the subband structure within the well are clearly apparent in this figure. The loss of carrier energy to the phonon system is also clearly apparent; after emission of four phonons, the highest lying well subband is no longer reachable. Note that the probability of scattering from the incident electron states is small on average as evidenced by the scale change on the probability density axis between Figs. 4(a) and (b), while the probability of scattering from bound states within the simulation region subsequently is unity.

Figure 5 provides an illustrative example of transmission, reflection and capture probabilities calculated by SEMC, in this case, for a thermal distribution of incident heavy-holes at 300 K as a function of well width, after one iteration of SEMC per carrier and after four iterations of SEMC per carrier. For the purposes of calculation, any carrier trapped in a bound final state or any carrier in a propagating final state that is absorbed above the well by the complex self-energy potential representing subsequent scattering during the last iteration of SEMC is considered captured. Otherwise, however, the region of carrier-phonon coupling extends about 2.5 nm to either side of the well, which allows for overlap of the coupling potentials with at least most of the barrier penetration by the bound state wave-functions such as evident in Fig. 4. As a result, reflection and transmission are, specifically, from the vicinity of the well and include both coherent contributions and contributions after scattering. In this simulation, well widths of up to 20 nm are considered,

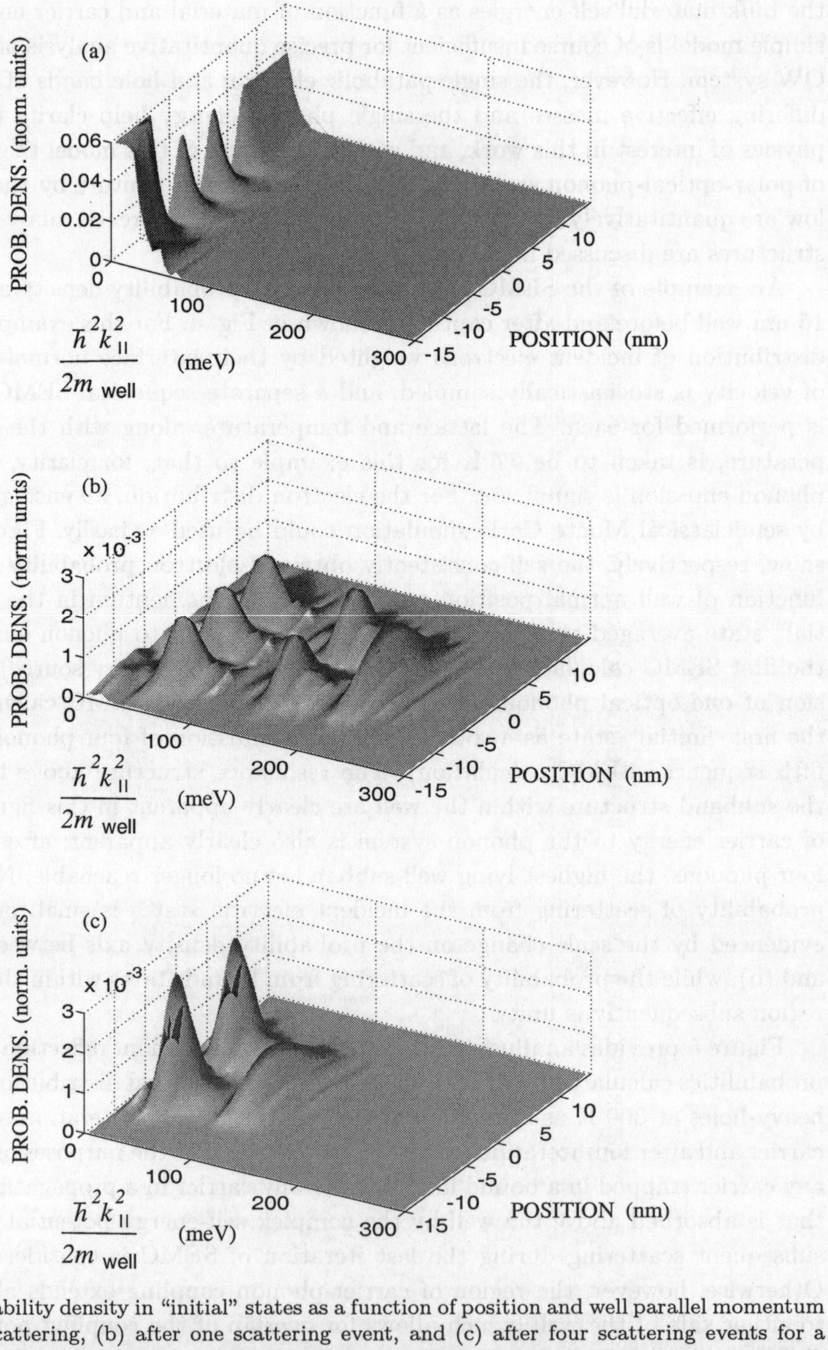

Fig. 4. Probability density in "initial" states as a function of position and well parallel momentum (a) prior to scattering, (b) after one scattering event, and (c) after four scattering events for a thermal distribution of electrons at 77 K incident from the left on a 15 nm wide quantum well. (Well-parallel energies are different within and outside of the well for fixed well-parallel momentum because of different effective masses.) (Reprinted with permission from Ref. 8. © World Scientific Publishing Company.)

and up to 50 nm later. This consideration of well widths larger than normally encountered in QW lasers serves two purposes: It allows the limiting effects of physical trends established in more common well widths to be observed. It also

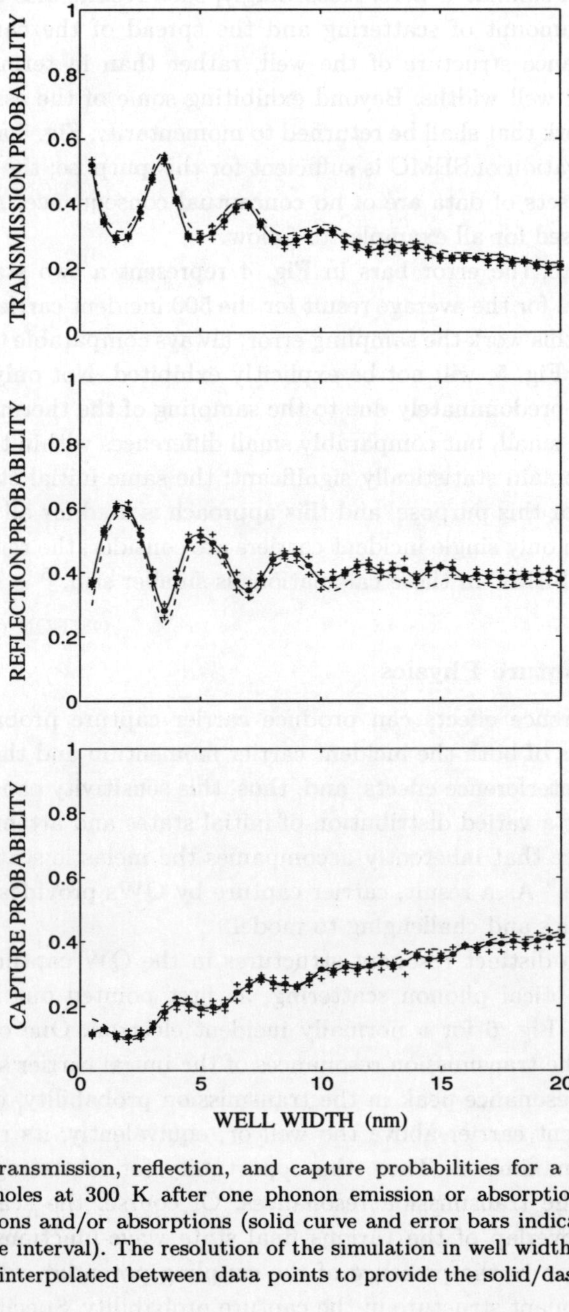

Fig. 5. Average transmission, reflection, and capture probabilities for a thermal distribution of incident heavy-holes at 300 K after one phonon emission or absorption (dashed curve) or four phonon emissions and/or absorptions (solid curve and error bars indicating a two standard deviation confidence interval). The resolution of the simulation in well width is 0.5 nm. Parabolic splines are used to interpolated between data points to provide the solid/dashed curves as an aid to the eye.

allows for the possibility that the behavior exhibited for this model system for relatively large well widths could be manifested in more common well widths due to stronger scattering — resulting from carrier-carrier[19] or other additional scattering mechanisms and/or greater densities of wells states — and/or more energetic carrier distributions than simulated here. Accordingly, such results also should be viewed in terms of the amount of scattering and the spread of the carrier distribution relative to resonance structure of the well, rather than in terms of electrons or holes and specific well widths. Beyond exhibiting some of the essential physics of interest in this work that shall be returned to momentarily, Fig. 5 also demonstrates that only one iteration of SEMC is sufficient for this purpose; the small differences between the two sets of data are of no conceptual consequence. Accordingly, only one iteration is used for all examples to follow.

As a final note, the error bars in Fig. 4 represent a two standard deviation confidence interval for the average result for the 500 incident carriers simulated. For the remainder of this work the sampling error, always comparable to or smaller than that exhibited in Fig. 5, will not be explicitly exhibited. Not only is the sampling error in Fig. 5 — predominately due to the sampling of the thermal distribution of "initial" states — small, but comparably small differences within the data sets (and between them) remain statistically significant; the same initial states are used for each data point for this purpose, and this approach is used for all similar examples that follow. When only single incident carriers are consider, the remaining sampling error inherent to SEMC in these calculations is smaller still.[15]

5. Essential Capture Physics

Quantum interference effects can produce carrier capture probabilities that are sensitive functions of both the incident carrier momentum and the well geometry.[5] These quantum interference effects, and, thus, this sensitivity can also be obscured by averaging over a varied distribution of initial states and attenuated by the loss of phase-coherence that inherently accompanies the inelastic scattering that leads to carrier capture.[9] As a result, carrier capture by QWs provides a system that is both rich in physics and challenging to model.

There are two distinct resonant structures in the QW capture probability for carriers due to optical phonon scattering, as first pointed out in Ref. 5 and as exhibited here in Fig. 6 for a normally incident electron. One of these is clearly associated with the transmission resonances of the initial carrier state and is easily explained. At a resonance peak in the transmission probability, the probability of finding the incident carrier above the well or, equivalently, its reside time above the well also is maximized. Thus, the opportunity for scattering within the well is enhanced at the transmission resonances. Of course, the scattering rate also depends on the overlap of the various final state wave-functions with the initial state wave-function in the presence of the coupling potential, which is the source of the second resonant structure in the capture probability. Specifically, the second

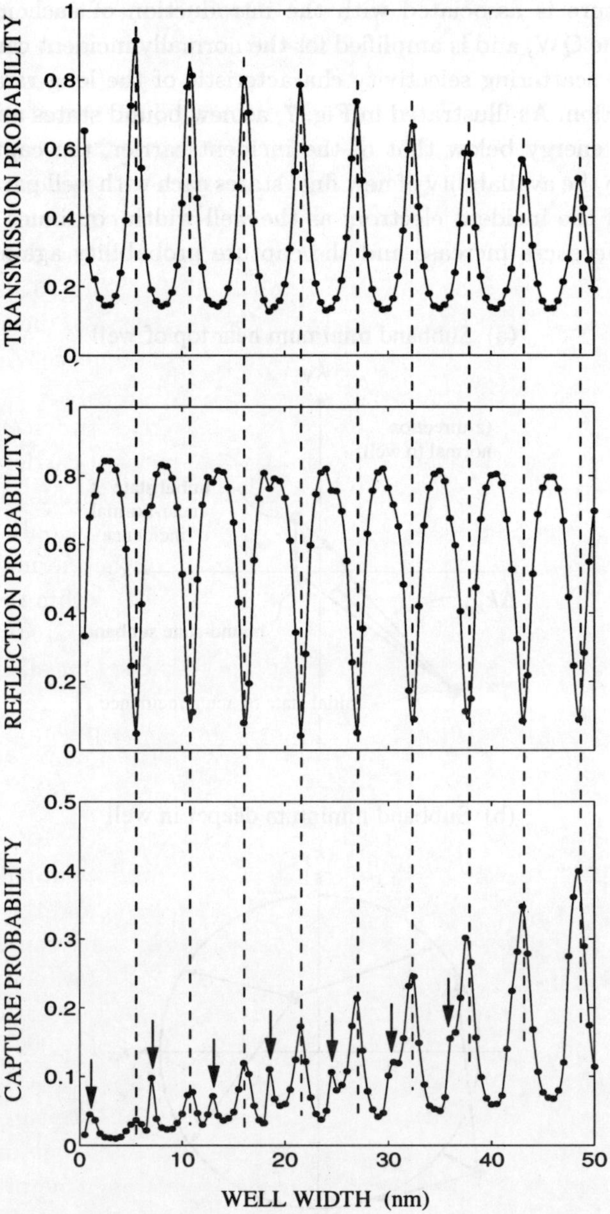

Fig. 6. (a) Transmission, (b) reflection, and (c) capture probabilities at a temperature of $T = 300$ K for a normally incident electron of $\frac{1}{2}k_BT \simeq 13$ meV energy. Two resonant structures are apparent. One, as indicated by the vertical dotted lines, is associated with the transmission resonances for the electron and dominates in wide wells; the other, indicated by the arrows, is associated with the introduction of new subbands into the quantum well and, in this case, dominates in narrow wells such as used in QW lasers. Note that with the 0.5 nm resolution, the peaks of narrow resonances may not be well resolved even with the parabolic interpolation between data points.

resonant structure is associated with the introduction of each new bound-state subband into the QW, and is amplified for the normally incident carrier of Fig. 6 by the small-angle scattering selectivity characteristic of the long-range polar-optical phonon interaction. As illustrated in Fig. 7, as new bound states appear in the well one phonon in energy below that of the incident carrier, the capture probability increases due to the availability of new final states each with well parallel momentum close to that of the incident electron; as the well width continues to increase, the momentum differences increase and the capture probability again decreases until

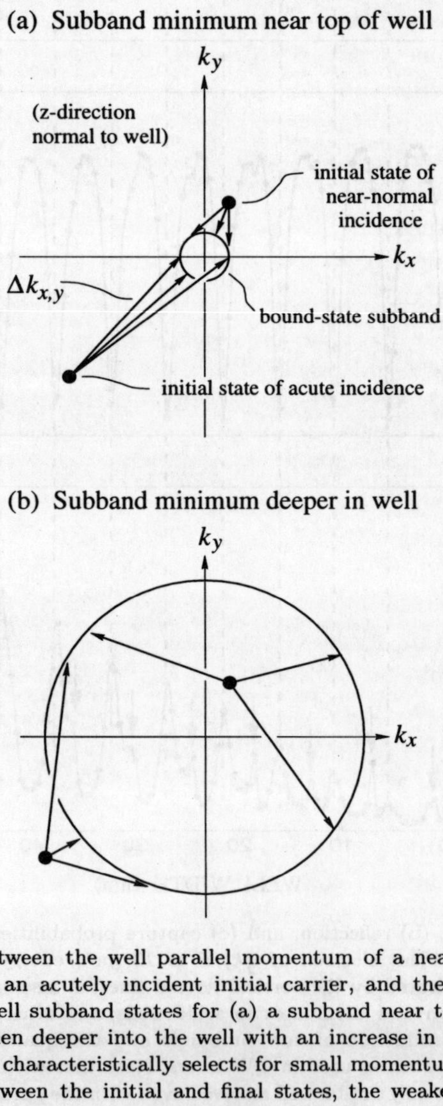

Fig. 7. Relationships between the well parallel momentum of a near-normally incident initial carrier state and that of an acutely incident initial carrier, and the well parallel momenta of final energy conserving well subband states for (a) a subband near the top of the well and (b) after the subband has fallen deeper into the well with an increase in well width. Because polar-optical phonon scattering characteristically selects for small momentum changes, the greater the momentum difference between the initial and final states, the weaker the coupling. Thus, the capture rate for the carrier of near normal incidence will be much more sensitive to the subband position in the well than that of the acutely incident carrier.

another subband appears in the well. In the example of Fig. 6, the capture resonance structure associated with the final states in narrow QWs such as used in lasers overcomes and even suppresses that associated with the transmission resonances in the initial state, although the latter does eventually overcome the former as it continues to grow with increasing well width.

However, when the incident carrier has a significant amount of well-parallel momentum, the resonant structure associated with the transmission resonances becomes more important even in narrow wells; the small angle selectivity actually serves to attenuate the resonant structure associated with the introduction of each new bound-state subband into the well. Again as illustrated in Fig. 7, when new bound states first become available in the well, the momentum differences between the acutely incident carrier and these bound states are large and, thus, the coupling weak. Therefore, there is no abrupt step in the capture probability, as even momentum randomizing scattering would cause. As the magnitudes of the incident and bound state well parallel momentums approach and then pass each other, the magnitudes of the momentum differences between the incident and most bound states remain large. The result is a lower and broader and therefore less noticeable resonance structure, as per the example of Fig. 8. If a thermal distribution of incident carriers at 300 K is considered — a conservative model for lasers given the injection of high-energy electrons into the separate confinement region — as in Fig. 9, there is no obvious resonance structure associated with the bound final state subbands remaining, both because of the significant transverse momenta of many of the incident carriers and because much of the remaining resonant structure is obscured by averaging over the distribution of incident carriers. Note that due to the more energetic and varied initial carrier distribution considered here, for both of these reasons, the results of Fig. 9 contrast somewhat with the results of Ref. 5 where the bound-state resonant structure remains more apparent. However, the small — but still potentially important for device optimization — shift of the peaks nominally associated with transmission resonances toward larger well widths in the narrower wells in Fig. 9 still attests to the effects of introducing new subbands into the well; if one replaces the coupling from the initial states to the bound final states of the QW subbands in SEMC with simple uniform complex absorbing potentials in the initial states above the well (results not shown), this shift is not observed.

As the electrons with their small effective mass proved ideal for demonstrating the effects of quantum interference in both the initial and final states on carrier capture in this model system, the holes with their large mass are ideal for demonstrating the attenuation and obscuring of these quantum interference effects. These latter effects are not insignificant for the electrons, but they are far more self-evident for the holes. For the holes, both the reside time above the well and the density of final states within the well are relatively larger and, thus, the phase breaking inelastic scattering is stronger. Further, capture, transmission and reflection probabilities oscillate more rapidly with the incident carrier's energy, so the effects of averaging over an energetic distribution of incident carriers are more pronounced.

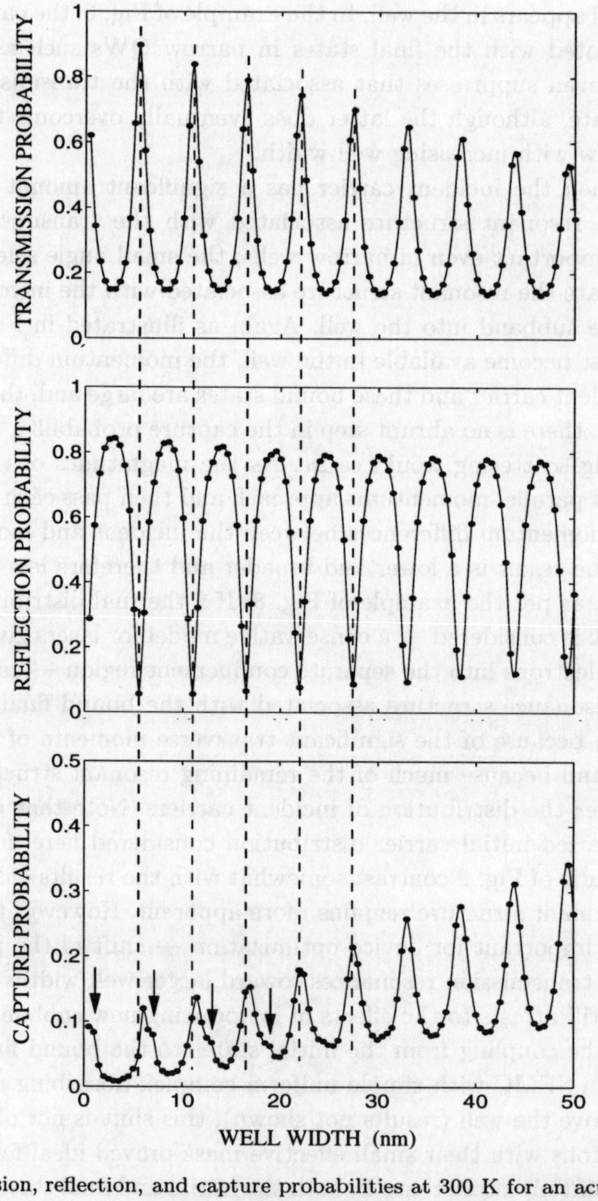

Fig. 8. Transmission, reflection, and capture probabilities at 300 K for an acutely incident electron with a well normal component of energy of $\frac{1}{2}k_BT$ and a well parallel component of energy of k_BT. Only a weak resonant structure associated with the introduction of new subband states into the well is evident. However, a slight shift of the peaks nominally associated with transmission resonances toward larger well widths, i.e. toward those associated with new subbands, also can be observed for the narrower wells.

Figure 10 shows the initial and final state probability densities for a hole acutely incident on a 21.5 nm well. This particular well width is considered because the reflection probability is only 0.014% (0.00014) in the absence of inelastic

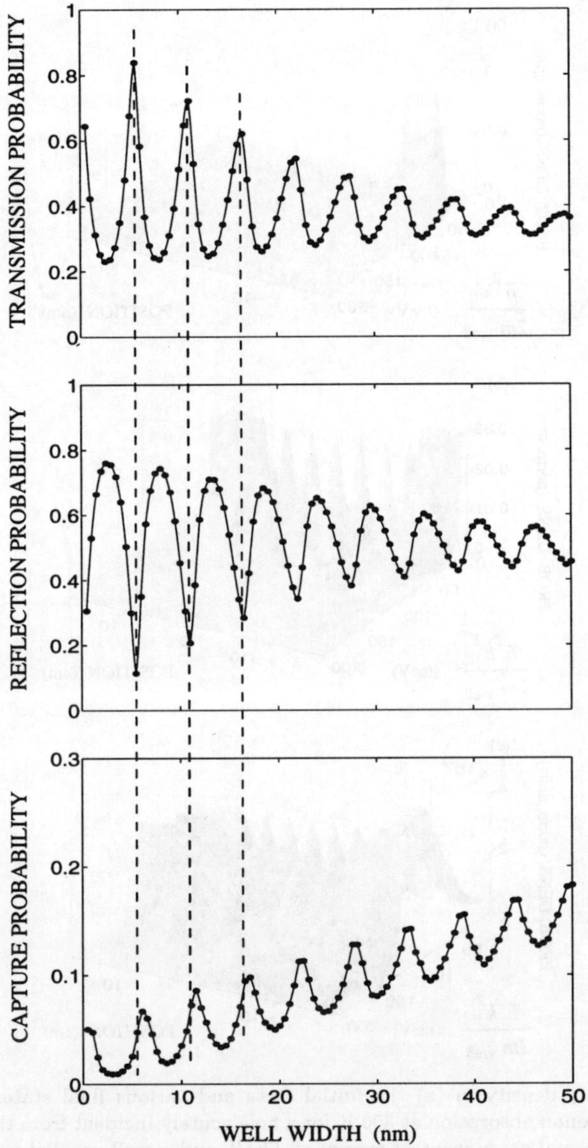

Fig. 9. Average transmission, reflection, and capture probabilities for a thermal distribution of electrons at 300 K. No apparent resonant structure associated with the introduction of new subband states into the wells is evident, but the tale-tale shift of the peaks nominally associated with transmission resonances toward larger well widths in the narrower wells remains.

scattering due to quantum interference effects across the well, which is a fortunate close approach to zero given the 0.5 nm resolution used in these simulations. The attenuation of the initial state wave-function across the well and, even more clearly, the formation of a standing-wave pattern in front of well indicating coherent

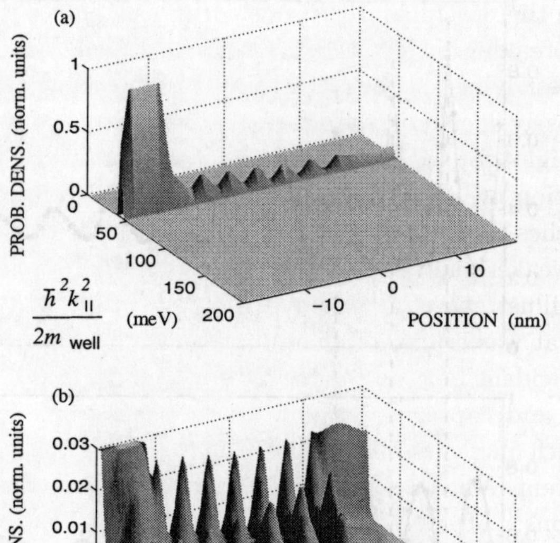

Fig. 10. Probability density in (a) the initial state and various final states after (b) phonon emission and (c) phonon absorption at 300 K for a hole acutely incident from the left on a 21.5 nm well, with a well normal component of energy of $\frac{1}{2}k_BT$ and a well parallel component of energy of k_BT. Some of the peaks in the density probability have been cutoff to allow the remainder to be seen more clearly. Not all of the scattering leads to capture, particularly in (c); propagating final states are identified by their continuation through the simulation region boundaries. Note the attenuation of the initial state wave-function across the well and the formation of a standing-wave in front of the well indicating coherent reflection; in the absence of scattering, the reflection probability from the well is essentially zero and, accordingly, the probability density in front of the well is constant. Also note that due to subsequent scattering (via the approximation of bulk self-energies in this case in contrast to Fig. 4) the lower lying subbands in the quantum well are not well resolved.

reflection testify to the destruction of quantum interference effects across the well due to the phase-breaking inelastic scattering. Also note that due to subsequent scattering, the lowest lying subbands in the QW which have the smallest energy separation and, loosely speaking, the longest "round-trip" times across the QW — clearly visible in Fig. 4 for electrons in a narrower well even after averaging over a thermal distribution of incident states — are not well resolved. However, when the carrier approaches the bottom of these subbands after subsequent energy loss, scattering will be weaker and the subbands should be more well defined. Figure 10 provides a precise illustration of the physics of capture in the presence of partial phase-coherence that was crudely depicted previously in Fig. 1(d).

For the same incident hole as considered for Fig. 10, Fig. 11 shows the transmission, reflection and capture probabilities as a function of well width, where the attenuation with increasing well width of the quantum interference induced resonances is apparent. Also shown in Fig. 11 for comparison are the results of perturbative calculations[20] after re-normalization ($C' = C/(T+R+C)$, etc.) so that the sum of the capture, reflection and transmission probabilities are again unity; in contrast to SEMC, the perturbative calculations do not conserve probability current. The large differences between the two sets of results, including the much larger oscillations in the latter, demonstrate that the attenuation of the resonances in the former is not due to the simple bleeding-off of the transmission and/or reflection probability into the capture probability. As for Fig. 10, note the removal of the zeros in the reflection probability in the former, where increased inelastic scattering produces increased capture and increased reflection simultaneously; for large well widths, the reflection probability converges toward 32%.

If a thermal distribution of incident holes at 300 K is averaged over, the quantum interference induced oscillations in the transmission, reflection and capture probabilities that remain after the phase-breaking scattering are largely obscured by around 10 nm, as seen in Fig. 12, and still more energetic initial carrier distributions are likely as noted previously. However, one non-classical effect remains prominent far beyond 10 nm, the reflection of over a third of the incident carriers from the well. Further, at 20 nm the transmission of approximately a third of the remaining incident holes beyond the well testifies to the still significant ballistic and, thus, phase-coherent transport across the well.

The simulation results above, as has been pointed out previously, were provided by the SEMC quantum *transport* simulator. One simple consequence is that probability current is conserved and, in particular, the rate of carrier capture (the capture probability times the rate of incident carriers) cannot exceed the rate at which carriers are incident on the well. This bit of the essential physics of carrier capture would not be discussed here because of its seeming self-evidence if it were not one that can be overlooked readily in Golden Rule calculations of carrier capture. In that context, this seemingly simple feature of this analysis is as important as any quantum interference effect, or attenuation thereof, exhibited above. Figure 13 shows the perturbative results for hole capture, along with those

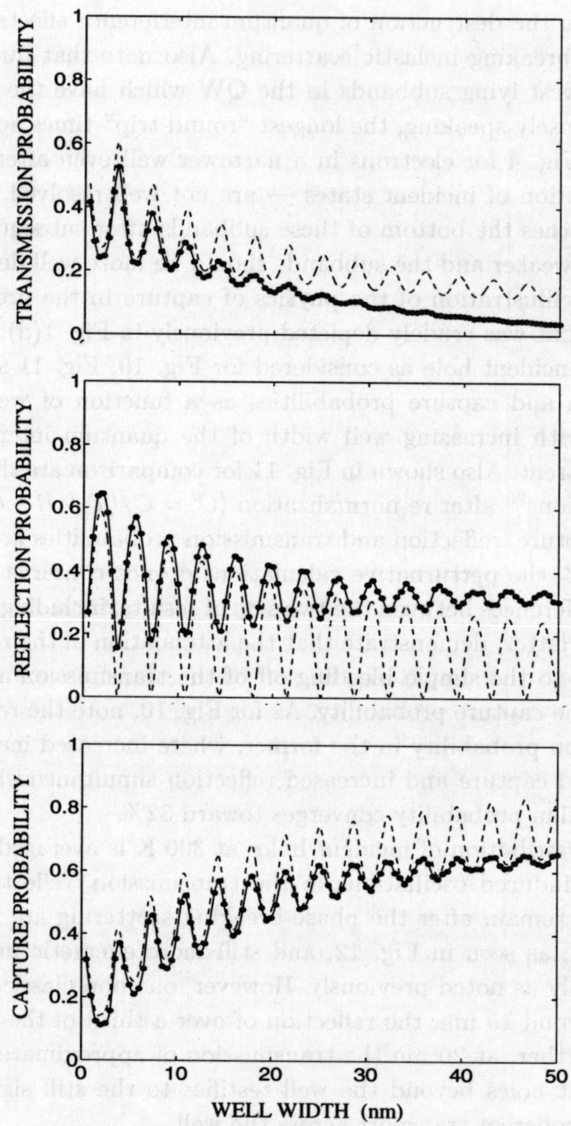

Fig. 11. Transmission, reflection, and capture probabilities at 300 K for an acutely incident hole with a well normal component of energy of $\frac{1}{2}k_BT$ and a well parallel component of energy of k_BT (data points and solid curve). Also shown for comparison, are the results of re-normalized perturbative calculations. By comparison, note that the obvious attenuation of the quantum interference effects in the former results cannot be interpreted as simply a bleeding-off of the transmission and/or reflection probability into the capture probability.

obtained by SEMC for comparison, for the same incident monoenergetic hole considered in Fig. 11 above. The perturbative capture rate can readily exceed the rate of incident carriers, and it can be strongly exaggerated before then. Although the need to at least re-normalized these results to conserved probability current, as was

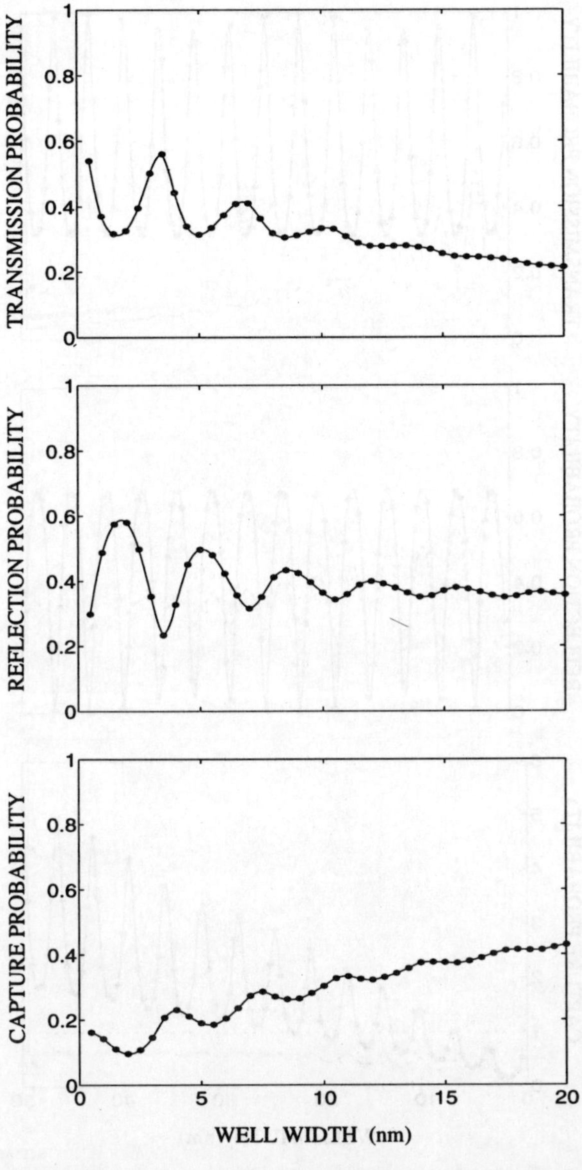

Fig. 12. Transmission, reflection, and capture probabilities at 300 K for a thermal distribution of incident holes. The quantum interference induced oscillations in the transmission, reflection and capture probabilities still evident in Fig. 11 are largely obscured by 10 nm.

done for Fig. 11, is obvious here, this is not usually the case in Golden Rule calculations. Typically, although there is no inherent restriction to this approach, the Golden Rule is employed to calculate the scattering rates between standing-wave eigenstates of the separate confinement region and of the QW as in original work of Ref. 5. Beyond the problem that the standing-waves eigenstates states of the

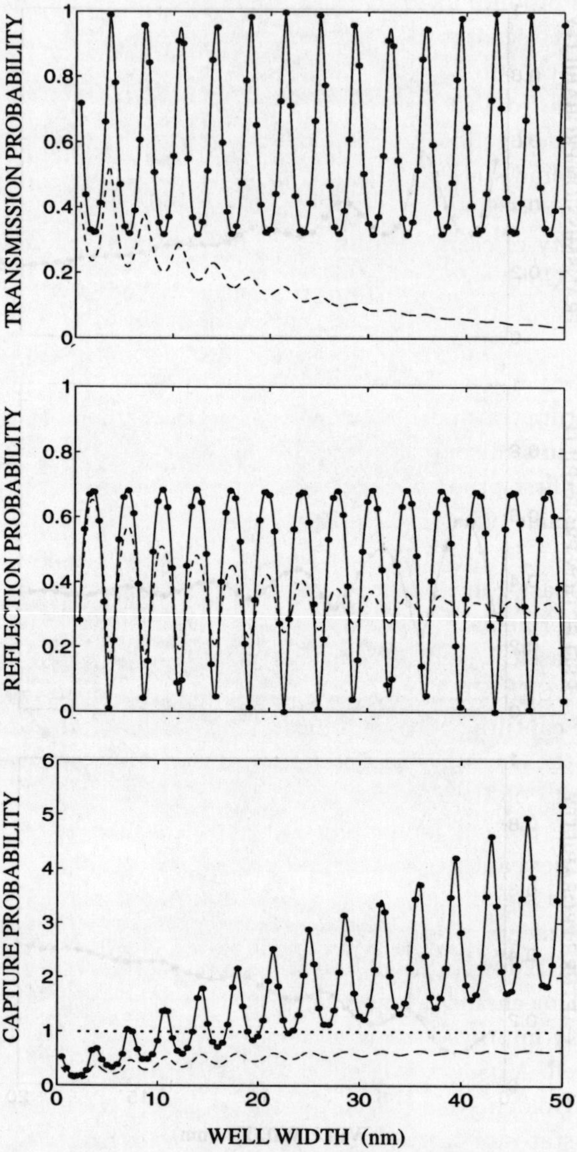

Fig. 13. Perturbative transmission, reflection, and capture probabilities at 300 K for an acutely incident hole with a well normal component of energy of $\frac{1}{2}k_BT$ and a well parallel component of energy of k_BT (data points and solid curve). Also shown for comparison, are the SEMC results of Fig. 11. Note that the calculated capture probability can readily exceed unity.

relatively broad separate confinement region are not likely to be well resolved in the presence of scattering, as for the lowest hole subbands in Fig. 10, there is no obvious rate of carrier incidence on the well, let alone transmission or reflection rates, to which to compare the capture rate. Further, not only may such Golden

Rule calculations produce a capture rate that exceeds that justifiable by the number of carriers in the separate confinement region in the vicinity of the well, but the calculated carrier capture does not inherently deplete the carriers in the separate confinement region from the vicinity of the well; the entire extended initial eigenstate of the separated confinement region simply may be assumed to be uniformly depleted as the bound state is uniformly filled, which ignores the issue of transport to the well. In short, without even considering the effects of dephasing, to the extent that the probability of carrier capture by a QW is strong, it also can be strongly exaggerated by a Golden Rule analysis.

6. Conclusion

The SEMC quantum transport simulator was used to simulate the essential physics of carrier capture in semiconductor QW lasers. The results of these simulations provide an improved first-principles understanding of this critical process. In addition to exhibiting familiar effects, these simulations addressed fundamental transport limitations that can be overlooked in Golden Rule calculations and exhibited effects of partial phase-coherence of the carrier wave-function on capture that are significant for intermediate capture efficiencies and that cannot be modeled via Golden Rule or classical/semiclassical treatments.

Further, while SEMC provided an efficient tool for this study of the essential physics of carrier capture,[21] the identification and characterization of this essential physics provides the necessary first step toward the development of still much more efficient treatments of these effects appropriate for full-scale device modeling.[22] By identifying the present limitations of classical/semiclassical and Golden Rule treatments of carrier capture, this quantum transport analysis also points to simple changes that could significantly improve, although not perfect, these latter more conventional treatments of capture and, thus, extend their range of applicability. The results of Fig. 12 for large well widths (strong scattering) suggest that allowing for the quantum mechanical reflection from the individual QW heterointerfaces would significantly improve otherwise semiclassical models of transport across and capture by the well. A comparison of Figs. 11 and 13 illustrates the potentially vast improvement in Golden Rule calculations possible by calculating scattering from the propagating states of carriers incident on the well, rather than from standing-wave states of the separate confinement region as more typical. This change both addresses the issue of transport to and across the well and makes the required re-normalization of the capture rate, along with transmission and reflection rates, self-evident.

Acknowledgments

The author thanks Prof. Karl Hess, Dr. Mathew Grupen and Mr. Fabiano Oyafuso for many discussion on the subject of carrier capture by QWs and its relevance to laser operation. This material is based upon work supported by the United States Army Research Office under Contract No. DAAG55-98-1-0306.

References

1. G. P. Agrawal (ed.), *Semiconductor Lasers: Past, Present and Future*, AIP Press, Woodbury, NY, 1995.
2. M. Grupen and K. Hess, "Simulation of carrier transport and nonlinearities in quantum-well laser diodes", *IEEE J. Quantum Electron.* **34** (1998) 120–140; errata, **34** (1998) 384.
3. K. Hess, private communications.
4. H. Shichijo, R. M. Kolbas, and N. Holonyak, Jr., "Carrier collection in a semiconductor quantum well", *Solid State Comm.* **27** (1978) 1029–1032.
5. J. A. Brum and G. Bastard, "Resonant carrier capture by semiconductor quantum wells", *Phys. Rev.* **B33** (1986) 1420–1423.
6. P. W. M. Blom, C. Smit, J. E. M. Haverkort, and J. H. Wolter, "Carrier capture into a semiconductor quantum well", *Phys. Rev.* **B47** (1993) 2072–2081.
7. L. F. Register and K. Hess, "Numerical simulation of electron transport in mesoscopic structures with weak dissipation", *Phys. Rev.* **B49** (1994) 1900–1907.
8. L. F. Register, "Schrödinger equation Monte Carlo: Bridging the gap from quantum to classical transport" in *Selected Topics in Electronics and Systems, Vol. 14: Quantum-Based Electronic Devices and Systems*, ed. P. K. Tien, World Scientific, Singapore, 1998, pp. 251–279, and in the *Int. J. High Speed Electron. Syst.* **9** (1998) 251–279.
9. L. F. Register and K. Hess, "Simulation of carrier capture in semiconductor quantum wells: Bridging the gap from quantum to classical transport", *Appl. Phys. Lett.* **71** (1997) 1222–1224.
10. K. Hess, *Advanced Theory of Semiconductor Devices*, Prentice Hall, Englewood Cliffs, New Jersey, 1988, pp. 143, 144.
11. E. Fermi, *Nuclear Physics*, University of Chicago Press, Chicago, 1950, pp. 187–191.
12. S. D. Benjamin, T. Zhang, Y. L. Hwang, M. S. Mytych, and R. S. Kolbas, "Vertical cavity surface emitting laser with submonolayer thick InAs active layer", *Appl. Phys. Lett.* **60** (1992) 1800–1802.
13. P. Roblin and W.-R. Liou, "Three-dimensional scattering-assisted tunneling in resonant-tunneling diodes", *Phys. Rev.* **B47** (1993) 2146–2161.
14. R. Lake, G. Klimeck, R. Bowen, and D. Jovanovic, "Single and multiband modeling of quantum electron transport through layered semiconductor devices", *J. Appl. Phys.* **81** (1997) 7845–7869.
15. More precisely, each carrier-phonon coupling potential is generated by Monte Carlo sampling to have on average the same spatial correlation function as the complete set of relevant true carrier-phonon interactions, which is far more efficient than but otherwise equivalent to sampling the carrier-phonon coupling potentials of the true phonon modes one at a time. Further, for eigenstate calculations such as performed in this work, the spatial correlation function of the set of true carrier-phonon coordinates can be used directly in the calculation of the "initial" state, eliminating sampling error there. This latter option is relatively new to SEMC although noted in Ref. 8. As a result, many of the results displayed in this work are subject to less sampling noise than seen in either Ref. 9 or Ref. 8. However, to find the final states in the presence of the "initial" state still requires the stochastic generation of the coupling potentials.
16. See, for example, R. P. Feynman *Statistical Mechanics: A Set of Lectures*, Addison-Wesley, Redwood City, CA, 1972, p. 226.

17. N. Mori and T. Ando, "Electron–optical-phonon interaction in single and double heterostructures", *Phys. Rev.* **B40** (1989) 6175.
18. L. F. Register, "Microscopic basis for a sum rule for polar-optical-phonon scattering of carriers in heterostructures", *Phys. Rev.* **B45** (1992) 8756–8759.
19. K. Kálna, M. Moško, and F. M. Peters, "Electron capture in quantum wells via electron-electron and optic phonon scattering", *Appl. Phys. Lett.* **68** (1996) 117–119.
20. For this purpose, SEMC is used with precisely the same coupling potentials and final-state self-energy shift and broadening for control, but without coupling back from the final states to the initial state. In this limit, SEMC results reduce to Golden Rule results except for the final state broadening and the small sampling error.
21. The results of Fig. 9, the most time-consuming to obtain, required 50,000 separate SEMC simulations of carrier capture — numbers reminiscent of semiclassical Monte Carlo analysis — yet can be reproduced in a few days on current personal computers.
22. G. Baraff, "Model for the effect of finite phase-coherence length on resonant transmission and capture by quantum wells", *Phys. Rev.* **B58** (1998) 13799–13810.

International Journal of High Speed Electronics and Systems, Vol. 9, No. 4 (1998) 1235–1264

QUANTUM STATE ENGINEERING BASED ON ELECTROMAGNETIC ANALOGIES AND NUMERICAL METHODS FOR SEMICONDUCTOR INTERSUBBAND LASERS

EMMANUEL ANEMOGIANNIS, ELIAS N. GLYTSIS, and
THOMAS K. GAYLORD
School of Electrical and Computer Engineering,
and Microelectronics Research Center,
Georgia Institute of Technology, Atlanta, Georgia 30332, USA

The design of semiconductor intersubband lasers using quantum state engineering based on electromagnetic analogies is a relatively new area. There has been little systematic effort in this direction as yet. The use of these analogies allows laser researchers to apply tools and methodologies found in the mature field of electromagnetic optics. The purpose of this paper is to describe the procedures by which intersubband quantum-well semiconductor lasers (or detectors) may be designed using these methodologies.

1. Introduction and Motivation

In the last two decades there has been an intense interest in optical transitions between intersubband energy levels in semiconductor quantum-well devices. The reason for this surge of interest is that intersubband transitions in quantum wells produce large oscillator strengths[1] while the energy levels and the operating wavelength are largely independent of the material used. Quantum-state engineering is the discipline dealing with the design of quantum-mechanical devices according to which spatial variation of the composition and/or the doping of the semiconductor can produce devices with the desired quasibound states and intersubband-transition specifications. A success story of quantum-state engineering has been the design of quantum well infrared photodetectors (QWIP's).[2] QWIP's consist of single or multiple quantum-wells and use bound-to-bound state transitions or the more efficient bound-to-quasibound state transitions. QWIP's have reached a high degree of maturity realizing two-dimensional imaging focal plane arrays of high complexity.[3] On the other hand the development of intersubband lasers lags behind their intersubband detector counterparts. The first intersubband laser was described in Ref. 4 and the first room-temperature intersubband laser was reported in Ref. 5, but there are still many technological problems to be solved before intersubband lasers are widely available for general use. In addition to the necessity of producing needed energy difference between the two intersubband energy levels (eigenstates) involved in lasing, it is imperative to initiate and maintain population inversion between them. Generally, the lower the energy level the longer the time electrons reside in it. Population inversion can be accomplished by tailoring the lifetime of electrons residing in these two quasibound states in such a way that the electron lifetime of the upper state is much longer than the electron lifetime of the lower state. In addition, population inversion is maintained only if the electron lifetime of the lower state is very short (much shorter than the optical phonon emission time) in order for the electrons to be depleted in time. Advanced intersubband laser designs with vertical electron transitions utilize a multi-quantum well (MQW) region which acts as an electron injector to the laser active region and as a Bragg reflector to enhance the carrier confinement, i.e. lifetime of the upper energy state. The drawback is that these lasers need a relatively large number of layers (400 layers for the quantum cascade laser structure of Ref. 6 and approximately 30 layers for the single non-cascaded laser of Ref. 7) the great percentage of which are used for the Bragg reflector implementation. In the present work we show that the necessary confinement in the upper eigenstate can be attained by a proper quantum-state engineering design of the active laser region, permitting further simplification of the electron-injector structure. Moreover, the proposed methodology can be used in the design of newly demonstrated optically-pumped intersubband lasers.[8]

2. Electromagnetic Analogies

Ballistic electrons are quantum-mechanical deBroglie waves and they have been observed to exhibit the wave effects of reflection, refraction, diffraction and interference. Even though there are fundamental differences between the electrons and the photons (see Table 1), there are cases where electrons and photons can be described by similar physical laws.[9,10] Semiconductor materials such as GaAs, InSb, InP, GaSb, InAs, CdTe, and $Al_xGa_{1-x}As$ posses a non-degenerate conduction energy-band. For this band where there is no coupling to other bands, the motion of the electron can be described by an effective-mass single-band time-dependent Schrödinger equation. In addition, if the electron energy is not a function of time or position, then the original equation simplifies to a time-independent Schrödinger

Table 1. Comparison of the characteristics of electrons and photons.

Property	Electron	Photon
Mass	m_0	0
Charge	e	0
Spin	1/2	1
Pauli Exclusion Principle	Yes	No
Energy vs. Wavevector (dispersion)	Parabolic	Linear
Wave-function representation	Scalar	Vector

equation which is similar to the scalar Helmholtz equation. The scalar Helmholtz equation describes the propagation of either TE or TM electromagnetic waves in a dielectric medium. The formal analogies are between the time-independent electron-wave amplitude function $\psi(\mathbf{r})$ and the transverse electric field $E(\mathbf{r})$ of a TE electromagnetic wave and the transverse magnetic field $H(\mathbf{r})$ of a TM electromagnetic mode. These have been developed in Ref. 10.

Generally phase effects like path differences and wave interference for both electromagnetic and electron waves can be described by the phase propagation constant. The equivalent phase propagation constant of an electron wave having an electron wavelength λ_i in the ith semiconductor layer is

$$\kappa_i = \frac{2\pi}{\lambda_i} = \frac{\sqrt{2m_i^*(E - V_i)}}{\hbar}, \tag{1}$$

where E is the electron energy and $E - V_i = (KE)_i$ is the total kinetic energy of the electron in the ith layer having effective mass m_i^*. From Eq. (1), the electron wavelength *inside* the ith layer is given as

$$\lambda_i = \frac{h}{\sqrt{2m_i^*(KE)_i}}. \tag{2}$$

An electron phase refractive index n_i for the ith layer can be defined by

$$\kappa_i = \frac{2\pi n_i}{\lambda_0}, \tag{3}$$

where λ_0 is the electron's wavelength in the 0th (reference) layer. Equation (2) allows the designer to calculate the physical length of the ith semiconductor layer, d_i, as a function of the electron kinetic energy $(KE)_i$ in the same layer. Usually, an electron-wave filter consists of a combination of layers having widths $d_i = \lambda_i/4 \Leftrightarrow \kappa_i d_i = \pi/2$, and/or $d_i = \lambda_i/2 \Leftrightarrow \kappa_i d_i = \pi$.

2.1. *Bragg reflector filter design*

The design algorithm of an electron-wave filter for a specific total energy E is comprised by the following steps: (a) Specify the two layer regions m_0^*, V_0, m_N^*, V_N with $E > V_0$ and determine n_0, n_N. (b) Specify the total number of layers in the electron-wave filter. The majority of optical filters are synthesized by a combination of a high **H** and a low **L** refractive index layers having lengths $\lambda_H/4$ and $\lambda_L/4$ respectively. Therefore, two layers can be chosen with m_i^*, V_i, n_i for $i = 1, 2$ such that Eqs. (1)–(3) are consistent. (c) Follow standard filter design algorithms or standard filter configurations found in Refs. 11 and 12 to synthesize the electron-filter. The design of an energy bandpass electron-wave filter under bias requires a more complicated procedure because the electric field alters the potential of individual layers and, in general, a self-consistent solution must be found as described in Ref. 13.

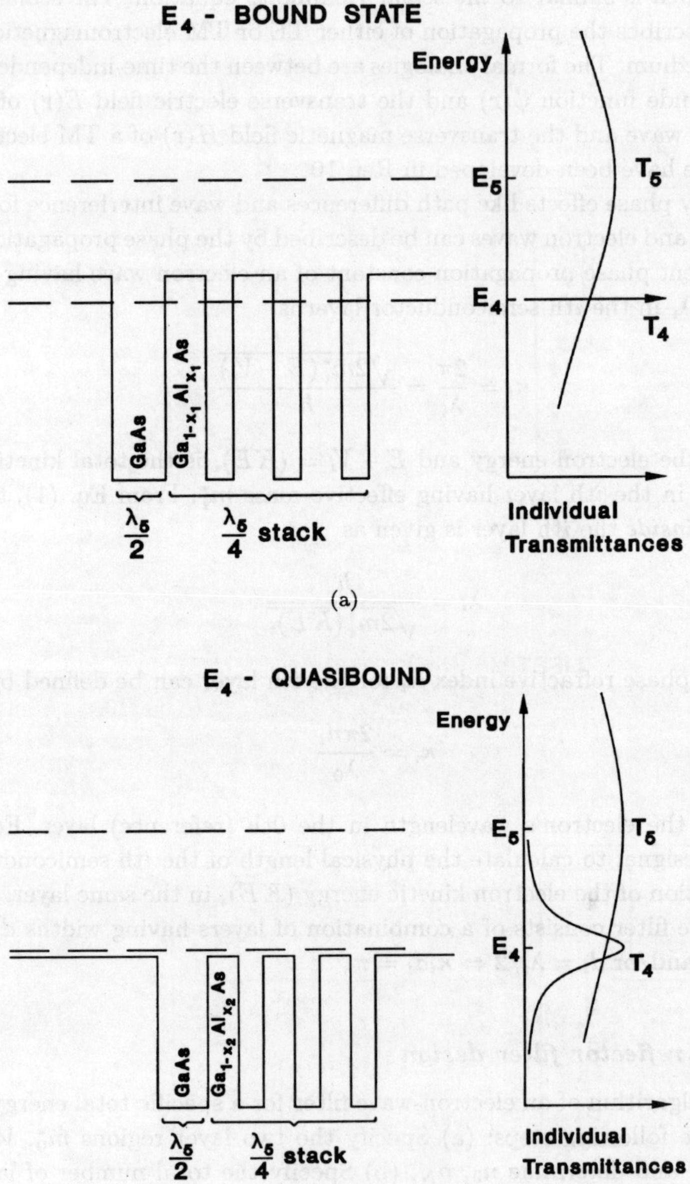

Fig. 1. Design steps of an unbiased electron-wave filter and the corresponding state lifetimes (in fsec): (a) $\tau_5 = 7$, $\tau_4 = \infty$, (b) $\tau_5 = 5.5$, $\tau_4 = 63$, (c) $\tau_5 = 22$, $\tau_4 = 82$, (d) $\tau_5 = 89$, $\tau_4 = 84$, (e) $\tau_5 = 169$, $\tau_4 = 83$, and (f) $\tau_5 = 207$, $\tau_4 = 83$.

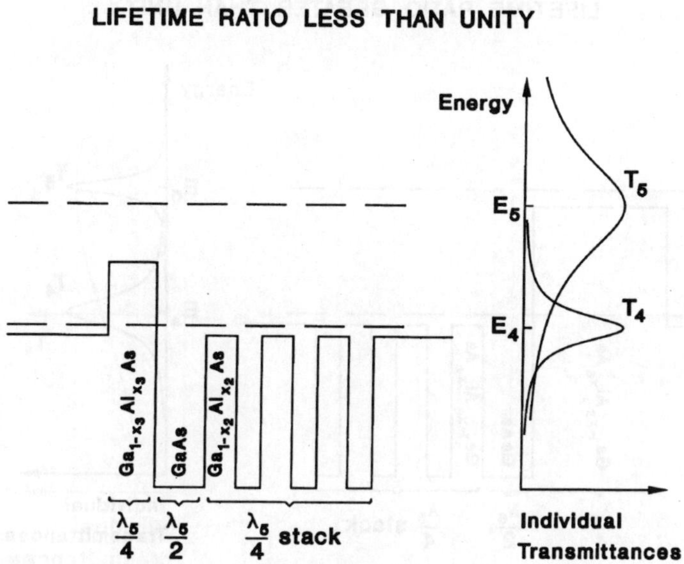

(c)

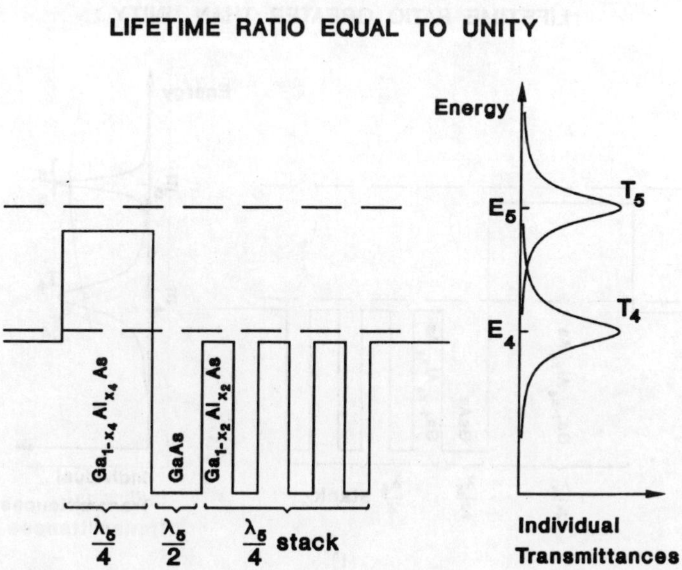

(d)

Fig. 1. (*Continued*)

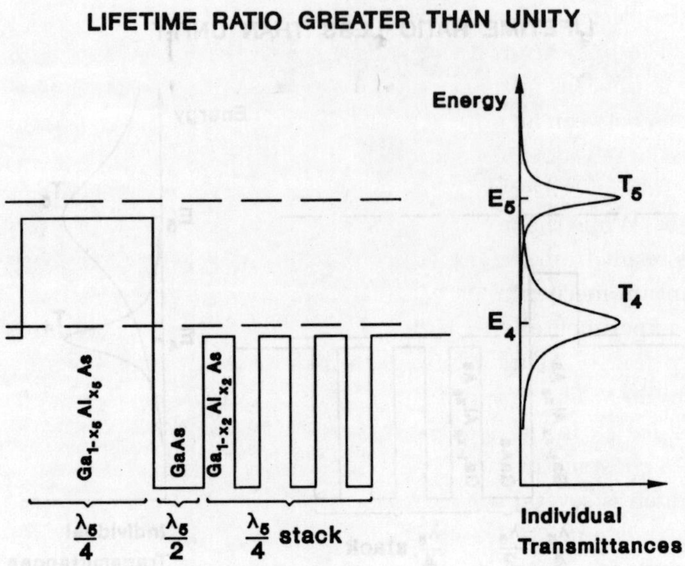

(e)

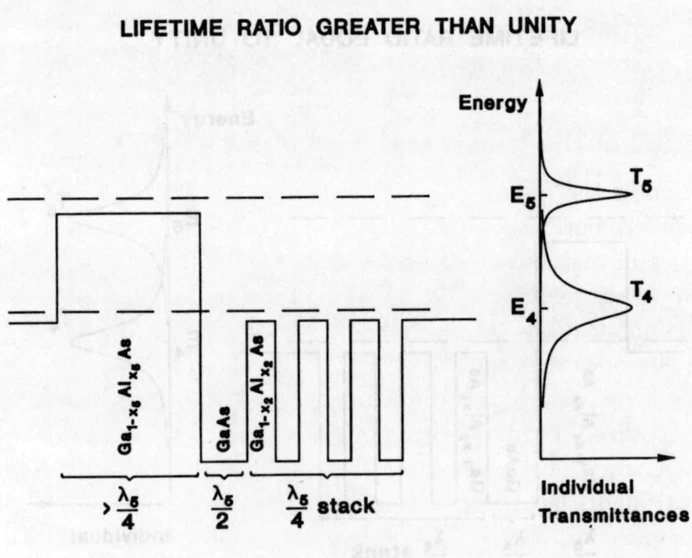

(f)

Fig. 1. (*Continued*)

Equations (1)–(3) are sufficient for the design of any kind of electron-wave energy-filter from their optical equivalents. A primary difference will be that optical filters have a periodic wavelength-spectrum response, while the electron-wave filters do not. This is due to the fact that the electron total-energy is primarily a parabolic function of the electron wavevector, i.e. electrons inherently experience dispersion effects, while the photon total energy is a primarily linear function of the photon wavevector, i.e. photons primarily do not experience dispersion effects.[10] Therefore, the equivalence between an optical and a electron-wave filter is exact for one energy only, E. This largely does not pose any difficulties since the wave-electron filters used for intersubband-laser design requires increased confinement at the upper single lasing-state. While the use of $\lambda_i/4$ layers to increase the lifetime of an eigenstate is an obvious result from the application of equivalent electromagnetic optics principles to quantum mechanics, it has also been proven theoretically in Refs. 14 and 15, reported experimentally in Refs. 16 and 17, and applied to an intersubband-laser device in Ref. 6. As an example case, we present the design steps of a simple electron-wave filter. The purpose of the filter is to increase the confinement of the quasibound state E_5 in the continuum (see Fig. 1). The quarter-wavelength layers contribute to a constructive interference *only* near the energy level E_5, while the left thick-layer which is several wavelengths wide acts as an infinite barrier. The function of this layer can be understood from its optical equivalent. Optically, this thick layer corresponds to a very low refractive index medium which has high reflectivity and confines light within the structure. Additional quarter-wavelength layers in the structure will increase the E_5 lifetime. The intuitive design of this electron-wave filter and its characteristics show the promise of quantum-state engineering for high performance devices.

3. Formulation of the Numerical Methods

The application of the quantum state engineering techniques presented in Sec. 2 is straightforward but has some limitations: (a) The electron-wave filters are comprised of semiconductor layers having lengths $d_i = \lambda_i/4$ and/or $d_i = \lambda_i/2$. Electron-wave filters may have improved characteristics for arbitrary layer thicknesses. Recent publications (e.g. Ref. 18) showed an improved optical-filter performance when the optical layers have thicknesses different than $\lambda_i/4$ and/or $\lambda_i/2$. (b) There might be cases where the pairs $n_H \Leftrightarrow V_H$ and/or $n_L \Leftrightarrow V_L$ for a given E can not be realized, i.e. there is not a semiconductor material to realize the required kinetic energy for a given n. (c) The design of a biased electron-wave filter needs a special algorithm which includes self-consistency loops.[13] To avoid all of these shortcomings new design algorithms must be used which optimize the MQW structure in a direct way, i.e. to find the thickness and the potential energies of the semiconductor layers such that the resulting MQW structure has predefined eigenstate characteristics. This design problem is an optimization problem, where the optimum state of the MQW structure is defined, and a globally convergent optimization algorithm, such as simulated annealing or the genetic algorithm, determines the layer thicknesses and the compositions of the MQW structure such that its characteristics are close to optimum. While the above mentioned optimization process is global in nature, i.e. it converges to the global minimum rather than

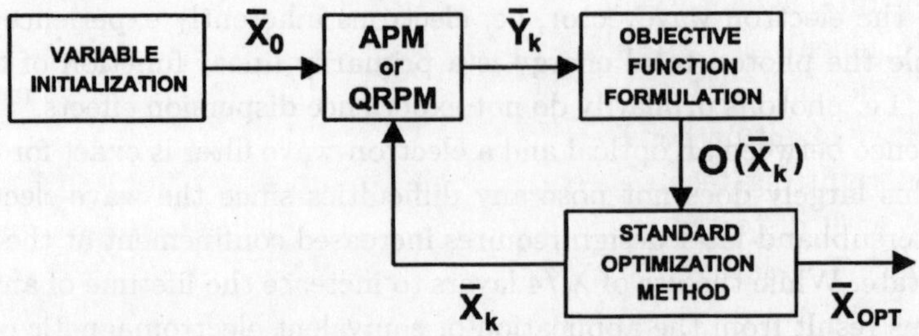

Fig. 2. Generic optimization scheme. $\overline{\mathbf{X}}$ is a vector whose entries are the MQW structure parameters to be optimized and $\overline{\mathbf{Y}}$ is the complex eigenstates vector. $\mathbf{O}(\overline{\mathbf{X}})$ is the difference between the system state due to $\overline{\mathbf{X}}$ and the optimum (desired) system state.

to a local function minimum, the optimization process requires a method which determines the MQW eigenstates at each iteration step (see Fig. 2). In this Section we present two numerical methods which are capable of determining the eigenstates of a MQW structure.

A complete definition of an electron-wave eigenstate includes its energy level and its lifetime. Fabry–Perot-like reflections of the electrons at quantum heterostructure boundaries produce Lorentzian-shape transmission probability peaks at the eigenenergy levels E. The width of each Lorentzian resonance is related to the eigenstate lifetime τ, i.e. the time that electrons remain confined inside the heterostructure at the given eigenenergy. The electron lifetime is $\tau = \hbar/\Gamma$ [19], where Γ is the full-width-at-half-maximum (FWHM) of the Lorentzian resonance, and it dictates the fastest possible time response of the device. An alternative approach to (E, τ) determination is the numerical solution of Schrödinger's equation in the complex plane. The real part of the complex solution corresponds to E and the imaginary part corresponds to $\Gamma/2$.[19] The numerical methods presented here are the argument principle method (APM) and the quantum reflection pole method (QRPM), both of which efficiently and accurately extract quasibound states and their lifetimes. Both methods use the transfer matrix approach to discretize the conduction-band profile of the heterostructure. The use of the transfer matrix formulation allows the user of the numerical methods to detect quasibound states in heterostructures with arbitrary geometry. The APM is a purely numerical method which extracts zeros of an analytic function in the complex plane. Therefore, by using the transfer matrix approach a dispersion equation is formed whose zeros correspond to the complex eigenenergies. APM is guaranteed to find all complex eigenenergies within a user defined complex contour without any initial information about the locations of the eigenenergies. Besides its quantum mechanical applications the accuracy and effectiveness of APM have been extensively tested on multilayer optical waveguides.[20,21] QRPM also utilizes the transfer matrix method to formulate the reflection coefficient of the quantum heterostructure. QRPM monitors the phase of the reflection coefficient denominator as a function of the potential energy a differentiation of which produces Lorentzian peaks corresponding to the quasibound state eigenenergies. QRPM's distinct features are that it provides the number of eigenenergies within an energy range and that the denominator of the reflection coefficient can be given by an analytical formula, even for cases of arbitrary geometry quantum heterostructures. For the case of overlapping Lorentzians a simple curve-fitting technique can be used to measure the correct FWHM of Lorentzians with distorted shape or which are completely screened by the spectrum of adjacent wide Lorentzians.

3.1. *Complex eigenenergies*

For conduction electrons in a single band with no zero-field splitting, the envelope wavefunction $\Psi(\mathbf{r},t)$, is described by the effective-mass Schrödinger's equation[22–25]

$$-\frac{\hbar^2}{2}\boldsymbol{\nabla}\cdot\left(\frac{\boldsymbol{\nabla}\Psi(\mathbf{r},t)}{m^*(\mathbf{r})}\right)+V(\mathbf{r})\Psi(\mathbf{r},t)=j\hbar\frac{\partial}{\partial t}\Psi(\mathbf{r},t)\,, \tag{4}$$

where $\mathbf{r}$ is the space position vector, $V(\mathbf{r})$ is the potential energy at $\mathbf{r}$, $m^*(\mathbf{r})$ is the spatially varying effective electron mass, $\hbar$ is the Planck's constant divided by 2π, and $j=\sqrt{-1}$. The steady-state solutions of Eq. (4) have the form[24]

$$\Psi(\mathbf{r},t)=\psi(\mathbf{r})\exp\left(-j\frac{E}{\hbar}t\right)\,, \tag{5}$$

where E is defined as the eigenenergy of the system, and $\psi(\mathbf{r})$ is the spatially-varying envelope wavefunction.

The possible solutions of Eq. (4), i.e. various eigenenergies E, correspond to energy levels or states which can be classified with respect to the behavior of the carriers residing in them. As shown in Fig. 3, a quantum structure can support three different kinds of states. Type 1 quasibound states are defined as the states for which the residing carriers have classically free propagation through and/or over the left and the right heterostructure barriers. States above all the heterostructure band edges are classified as type 1 quasibound state class, and in the literature these states are also referred as "virtual" or "classically free quasibound" or "continuum" or "extended" states. Type 2 quasibound states are defined as the states in which carriers have classically free propagation to the right (left) and have absolutely bounded propagation to the left (right) by an infinitely thick half space, as opposed to partially bound propagation due to a finite thickness barrier that allows tunneling. In the literature, type 2 are also called simply "quasibound" states. Bound states are the states in which the carriers are absolutely bounded within infinitely thick left and right barriers.

When an electron leaks out from a quantum heterostructure its energy eigenvalues are no longer discrete. However, it is possible for the electron to need a relatively long time to leak out from the quantum structure. An example would be an electron confined in a quantum well surrounded by high and wide energy barriers. The particle remains in the quasibound state for a time which is related

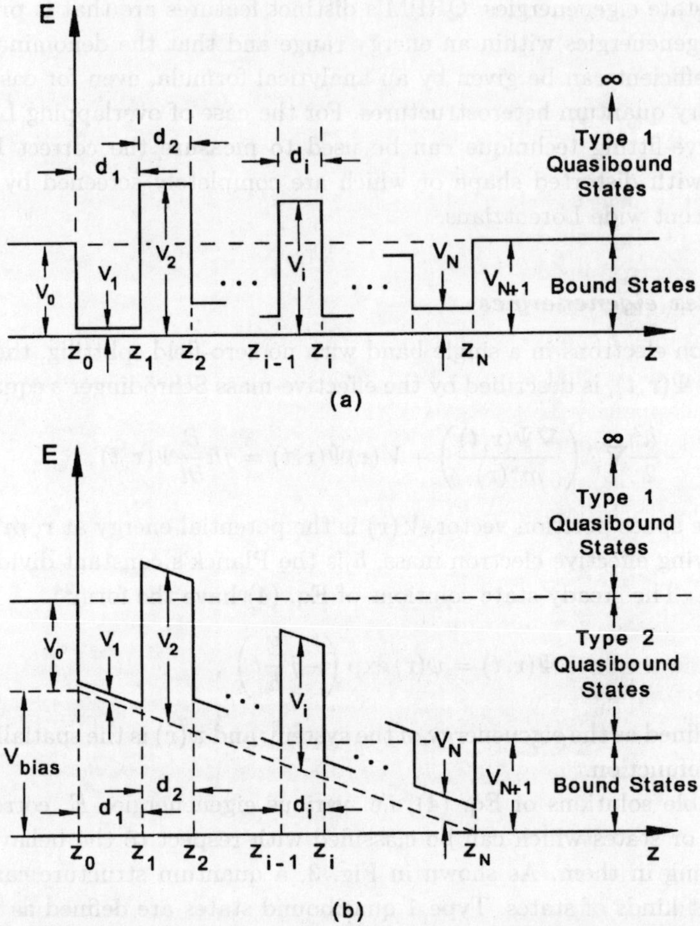

Fig. 3. Conduction-band edge of quantum structure consisting of N layers, showing energy ranges for bound states and for type 1 and type 2 quasibound states. (a) Unbiased structure with $V_0 = V_{N+1}$ and (b) The same structure under bias V_{bias}.

to the lifetime of the state.[22] The energy spectrum of these states becomes quasi-discrete since it consists of a series of broadened energy levels the width of which are related to the lifetime of the quasibound state.[22] Type 1 quasibound states can also be characterized as quasi-discrete with a corresponding broadened energy level with width inversely related to the lifetime of the state. In these cases the quantum reflectivity for the electron is provided through the differing effective mass and/or potential energy layers.[26,27] For quasibound energy levels the envelope wavefunction approximation (Eq. (4)) must be used with boundary conditions that require an electron (or a particle in general) to leak out to infinity.[22] However, a boundary condition of this type is complex[22] and the energy eigenvalues also become complex having the form $\tilde{E} = E_r - jE_i$ where $E_i > 0$. The physical significance of the complex energy eigenvalues, $\tilde{E}$, can be seen from the form of the steady-state solution (Eq. (5))

$$\Psi(\mathbf{r},t) = \psi(\mathbf{r}) \exp\left(-j\frac{E_r}{\hbar}t\right) \exp\left(-\frac{E_i}{\hbar}t\right) . \tag{6}$$

From Eq. (6) it is observed that the electron state-occupation probability decreases as a function of time as $\exp(-2E_i t/\hbar)$ for any $\mathbf{r}$. Consequently, the probability of finding the electron inside the structure decreases exponentially as a function of time and the imaginary part of the complex energy can be related to the lifetime of the state, τ, as $\tau = \hbar/2E_i$. The lifetime, τ, of an eigenstate is defined as the time that carriers reside at that state (energy level) in the heterostructure, before they leak out.[19] The quantity $2E_i$ is equal to the full width at half maximum (FWHM), Γ, of the broadened energy state E_r.[19]

Substitution of Eq. (6) into Eq. (4) produces the single-band effective-mass time-independent Schrödinger equation with complex eigenenergies,

$$\frac{-\hbar^2}{2}\boldsymbol{\nabla} \cdot \left(\frac{\boldsymbol{\nabla}\psi(\mathbf{r})}{m^*(\mathbf{r})}\right) = [\tilde{E} - V(\mathbf{r})]\psi(\mathbf{r}) . \tag{7}$$

Therefore, under the envelope wavefunction approximation, the solution of the time-independent Schrödinger equation in the complex plane leads to the determination of the complex eigenenergy, $\tilde{E}$. Its real part, $\Re e[\tilde{E}] = E_r$, corresponds to the heterostructure broadened energy level, while its imaginary part, $\Im m[\tilde{E}] = E_i$, corresponds to the width of the level and to the eigenstate lifetime as $\tau = \hbar/2E_i = \hbar/\Gamma$. Bound energy states have real eigenenergies due to the exponentially decaying wavefunctions outside the structure (boundary conditions at $\pm\infty$). Thus, the bound states have infinite lifetimes τ as expected.

The lifetimes τ that are calculated from the imaginary part of the complex eigenenergies can be related to the wave-packet transit-time, τ_{tr}, from the time-dependent solutions of Schrödinger's equation, as $\tau_{\text{tr}} = 2\tau = \hbar/E_i = 2\hbar/\Gamma$.[28,29] The factor of 2 can be quantitatively attributed to a time τ needed for the wavefunction to build-up inside the structure (assuming an externally incident wave packet), and to a time τ needed for the wavefunction to leak out from the structure.

An interesting observation can be made relative to Eq. (7) and the complex energies that describe the quasibound states. These states are leaky since they correspond to probability current flow away from the quantum heterostructure. The wavefunctions that correspond to these complex eigenenergies are similar to the leaky modes in optical waveguides.[30,31] These wavefunctions, however, do not satisfy the radiation conditions at infinity (finite wavefunction) and consequently, do not belong to the proper eigenvalue spectrum.[31] Thus, the operator described in Eq. (7) is not Hermitian under the boundary conditions of the leaky wavefunctions. The solution of Eq. (7) using the transfer matrix formulation and the derivation of the complex transcendental equation are described in the next subsection.

3.2. *Transfer matrix formulation of a heterostructure*

All the methods described in this work utilize the transfer-matrix approach which discretizes the conduction-band potential-profile of the heterostructure and imposes boundary conditions satisfying the appropriate effective mass continuity between adjacent layers. Let us consider the one-dimensional problem of electron carriers in a region characterized by an arbitrary potential energy profile $V(z)$ under a bias, V_{bias}, as is shown in Fig. 3. $V(z)$ can always be divided into $i = 0, \ldots, N+1$ layers with a piecewise constant potential energy V_i. The discretized time-independent single-band effective-mass parabolic-Schrödinger's equation (Eq. (7)), for the ith region with constant potential energy V_i and constant effective mass m_i^* can be written as

$$\frac{d^2}{dz^2}\,\psi_i(z) + \kappa_i^2\psi_i(z) = 0 \quad \text{for } z_{i-1} \le z \le z_i \tag{8}$$

with

$$\kappa_i = \pm\sqrt{\frac{2m_i^*}{\hbar^2}(E - V_i)}\,, \tag{9}$$

where $\psi_i(z)$ represents the envelope wavefunction in the ith layer, E is the eigenstate energy level, and κ_i defines the wavenumber in the same layer along the z-direction. The solution of Eq. (8) can be written as a superposition of a right and a left traveling waves

$$\psi_i(z) = A_i \exp[j\kappa_i(z - z_{i-1})] + B_i \exp[-j\kappa_i(z - z_{i-1})] \quad \text{for } z_{i-1} \le z \le z_i\,. \tag{10}$$

According to Eq. (9), both signs of Eq. (10) are valid. The boundary conditions for $\psi(z)$ at the interface between layers i and $i+1$ at position $z = z_i$ where $i = 1, 2, \ldots N$, (see Fig. 3) are

$$\begin{aligned} \psi_i(z_i) &= \psi_{i+1}(z_i)\,, \\ \frac{1}{m_i^*}\frac{d}{dz}\,\psi_i(z_i) &= \frac{1}{m_{i+1}^*}\frac{d}{dz}\,\psi_{i+1}(z_i)\,. \end{aligned} \tag{11}$$

After using Eqs. (10) and (11) the envelope wavefunction expansion coefficients, A, B, between two consecutive layers are related as

$$\begin{aligned} \begin{pmatrix} A_i \\ B_i \end{pmatrix} &= \begin{pmatrix} \frac{1}{2}\left[1 + \frac{\kappa_{i+1}m_i^*}{m_{i+1}^*\kappa_i}\right]\exp[-j\kappa_i d_i] & \frac{1}{2}\left[1 - \frac{\kappa_{i+1}m_i^*}{m_{i+1}^*\kappa_i}\right]\exp[-j\kappa_i d_i] \\ \frac{1}{2}\left[1 - \frac{\kappa_{i+1}m_i^*}{m_{i+1}^*\kappa_i}\right]\exp[+j\kappa_i d_i] & \frac{1}{2}\left[1 + \frac{\kappa_{i+1}m_i^*}{m_{i+1}^*\kappa_i}\right]\exp[+j\kappa_i d_i] \end{pmatrix} \\ &\quad \times \begin{pmatrix} A_{i+1} \\ B_{i+1} \end{pmatrix} = \mathcal{Q}_{i,i+1}\begin{pmatrix} A_{i+1} \\ B_{i+1} \end{pmatrix}, \end{aligned} \tag{12}$$

with d_i representing the thickness of the ith layer. Recursive application of Eq. (12) gives the wavefunction expansion coefficients at the left ($i = 0$) and at the right ($i = N + 1$) heterostructure boundaries as

$$\begin{aligned}\begin{pmatrix} A_0 \\ B_0 \end{pmatrix} &= \mathcal{Q}_{0,1}\mathcal{Q}_{1,2}\mathcal{Q}_{2,3}\cdots\mathcal{Q}_{N,N+1}\begin{pmatrix} A_{N+1} \\ B_{N+1} \end{pmatrix} \\ &= \begin{pmatrix} q_{11}(E) & q_{12}(E) \\ q_{21}(E) & q_{22}(E) \end{pmatrix}\begin{pmatrix} A_{N+1} \\ B_{N+1} \end{pmatrix},\end{aligned} \tag{13}$$

with $d_0 = 0$. All the numerical methods discussed in this work utilize Eq. (13) but with different boundary conditions. APM assumes one wavefunction component at the incident region and one wavefunction component at the output region. QRPM assumes a forward and a backward wavefunction component in the incidence region and one transmitted wavefunction component in the output region.

3.3. *APM formulation*

The APM is a numerical method which guarantees the extraction of all complex roots of a nonlinear equation without prior knowledge of their number or their location in the complex plane. Therefore, in order to apply the APM algorithm we must formulate an equation having as roots the heterostructure's complex eigenenergies. This complex equation can be derived from Eq. (13) by assuming only one kind of eigenfunctions present at the two heterostructure boundary regions that propagate away from it, i.e.

$$\begin{aligned}\psi_0(z) &= B_0 \exp[j\kappa_0(z - z_0)] \qquad \text{for } z < z_0\,, \\ \psi_{N+1}(z) &= A_{N+1} \exp[-j\kappa_{N+1}(z - z_N)] \qquad \text{for } z > z_N\,.\end{aligned} \tag{14}$$

In other words, APM uses Eq. (11) with $A_0 = 0$ and $B_{N+1} = 0$ and the resulting complex equation is $q_{11}(\tilde{E}) = 0$. Generally, $\Re e[\kappa_i]$ can be positive or negative, due to the sign of the square root. This sign duality does not pose a problem since the expression of $\psi(z)$ inside the heterostructure region includes both the $\pm\kappa_i$ terms. Care must be taken for the correct sign selection at the heterostructure boundaries where only one term of $\pm\kappa_0$ and $\pm\kappa_{N+1}$, exists. Since for bound states the forward propagating wave decays exponentially in the $+z$-direction and the backward propagating wave decays exponentially in the $-z$-direction, the correct sign can be selected for κ_i for $i = 0, N + 1$, which dictates the decaying amplitude characteristics of the two traveling waves in both heterostructure boundary regions. In the case of quasibound states a similar procedure is required. Specifically, Eqs. (12) are still valid, but the wavefunction solutions in regions 0 and $N + 1$ (outside the heterostructure) must be chosen to correspond to leaky waves.[22,30] The sign of κ_0 (the square root in Eq. (9)) has to be selected in such a way that it corresponds to a leaky wave that travels towards the $-z$-direction for the left boundary region, and/or the sign of κ_{N+1}, to correspond to a leaky wave propagating towards $+z$-direction for the right boundary region. These leaky wavefunctions in the boundary regions increase exponentially as $z \to -\infty$ and $z \to +\infty$ for the left and right boundary regions respectively.[22,30] Consequently, the complex zeros of $q_{11}(\tilde{E}) = 0$ correspond to the quasibound eigenstates of the heterostructure given that the aforementioned boundary conditions for leaky waves have been used.

The APM,[32] relies on the generalized argument principle theorem from complex analysis. This theorem relates the number of zeros of an analytic function to a contour integral, and it has the form[33]

$$\sigma_P = \frac{1}{j2\pi} \oint_C z^P \frac{f'(z)}{f(z)}\, dz = \sum_{i=1}^{n} \zeta_i^P , \tag{15}$$

where $f(z)$ is an analytic function inside and on the simple closed contour C and on which no zeros are located. In addition, ζ_i^P are the $i = 1, \ldots, n$ zeros of $f(z)$ inside C raised to the Pth power. The quantity σ_P is their summation and $f'(z)$ denotes the derivative of $f(z)$ with respect to z. For the case that $f'(z)$ is not analytically given, it can be calculated numerically via the Cauchy's theorem,[33] according to which the first derivative of an analytic function $f(z)$ at the point $z = z_0$, $f'(z_0)$ is given by

$$f'(z_0) = \frac{1}{j2\pi} \oint_D \frac{f(z)}{(z - z_0)^2}\, dz , \tag{16}$$

where D is any simple path which encloses the point z_0 and $f(z)$ is analytic inside and on D. Efficient numerical methods for the calculation of the integrals (15) and (16) are given in Ref. 20.

The application of APM to the dispersion equation $f(z) = q_{11}(\tilde{E}) = 0$, is straightforward. The energy range for which the heterostructure energy levels are of interest, is enclosed in rectangular contours which exclude the two potential energy levels, $V_0 + V_{\text{bias}}$, V_{N+1}, which correspond to the heterostructure boundary regions (see Fig. 4). The vertical contour segments can approach the two aforementioned energy levels arbitrarily closely. The path integral of Eq. (15) evaluated for $P = 0$ gives the number of zeros, $\sigma_0 = n$, of $q_{11}(\tilde{E}) = 0$ enclosed by C. Also for $P = 1, \ldots, n$, the same integral produces the summations $\sigma_1, \ldots, \sigma_n$ which can be used to evaluate the coefficients of a polynomial $S(\tilde{E})$ of degree n, which has the same roots, $\tilde{E}_1, \tilde{E}_2, \ldots, \tilde{E}_n$, as the function $q_{11}(\tilde{E}) = 0$ inside C. The approximation polynomial, $S(\tilde{E})$, can be written as

$$S(\tilde{E}) = \sum_{k=0}^{n} c_k\, \tilde{E}^k , \tag{17}$$

with $c_n = 1$. The coefficients c_k are given via the Newton recursive formulas,[34,35]

$$(n-k)\, c_k + \sigma_1 c_{k+1} + \sigma_2 c_{k+2} + \cdots + \sigma_{n-k} c_n = 0 \quad \text{for } k = n-1, \ldots, 0 . \tag{18}$$

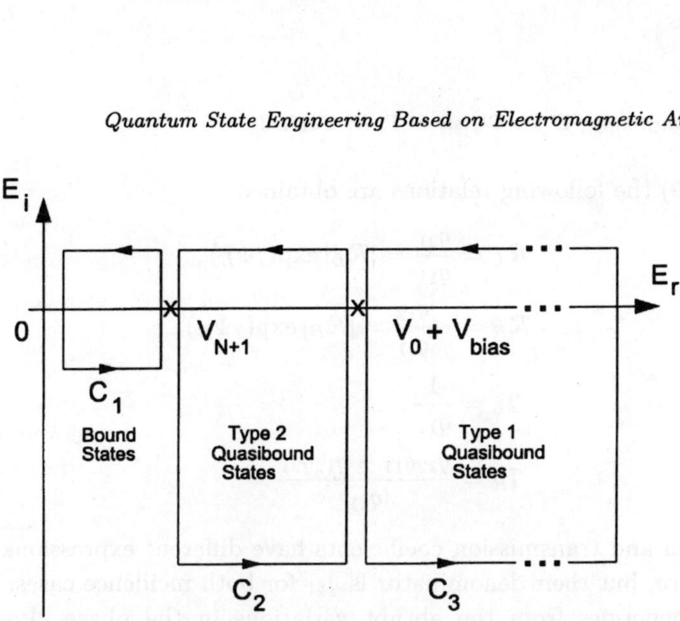

Fig. 4. Complex energy plane representation of the APM operation. The X's correspond to the two singularity points at $E = V_{N+1}$ and $E = V_0 + V_{\text{bias}}$. C_1, C_2, and C_3 are the integration contours which enclose regions to be searched for complex energy eigenvalues for bound, type 2 quasibound, and type 1 quasibound (virtual) states.

The polynomial, $S(\tilde{E})$, can be solved by well known techniques,[35] which in general are more efficient and accurate for low-degree polynomials. Therefore the maximum number of zeros enclosed by C must be low, i.e. $n \leq 6$, as proposed in Refs. 32, 34, and 35. Following the above procedure, the initial difficult problem of finding the complex zeros of an arbitrary function, $q_{11}(\tilde{E}) = 0$, was transformed to the problem of finding the complex zeros of the polynomial, $S(\tilde{E})$, for which a variety of reliable numerical methods exist. By suitably dividing the region where the zeros, $\tilde{E}_j$, are located and using Eq. (15) with $P = 0$, the degree of $S(\tilde{E})$ can be set such that the polynomial root solver can be optimally effective. In practice, the zeros of the polynomial $S(\tilde{E})$, $\tilde{E}'_j$, do not coincide with the zeros, $\tilde{E}_j$, of the initial function $q_{11}(\tilde{E}) = 0$, due to errors introduced by the numerical evaluation of integrals of Eqs. (15) and (16). For this reason, a further refinement must be done on the roots of $q_{11}(\tilde{E}) = 0$, applying a zero-finding numerical method[35,36] with initial conditions $\tilde{E}'_j$.

3.4. *QRPM formulation*

The reflection coefficient $\mathcal{R}$ and the transmission coefficient $\mathcal{T}$ can be obtained from Eq. (13) for both the left and right incident wavefunction

$$\begin{pmatrix} 1 \\ \mathcal{R}_L \end{pmatrix} = \begin{pmatrix} q_{11} & q_{12} \\ q_{21} & q_{22} \end{pmatrix} \begin{pmatrix} \mathcal{T}_L \\ 0 \end{pmatrix}, \qquad \begin{pmatrix} 0 \\ \mathcal{T}_R \end{pmatrix} = \begin{pmatrix} q_{11} & q_{12} \\ q_{21} & q_{22} \end{pmatrix} \begin{pmatrix} \mathcal{R}_R \\ 1 \end{pmatrix}. \tag{19}$$

From Eq. (19) the following relations are obtained

$$\begin{aligned} \mathcal{R}_L &= \frac{q_{21}}{q_{11}} = |\mathcal{R}_L| \exp(j\Phi_L)\,, \\ \mathcal{R}_R &= -\frac{q_{12}}{q_{11}} = |\mathcal{R}_R| \exp(j\Phi_R)\,, \\ \mathcal{T}_L &= \frac{1}{q_{11}}\,, \\ \mathcal{T}_R &= \frac{q_{22}q_{11} - q_{12}q_{21}}{q_{11}}\,. \end{aligned} \tag{20}$$

The reflection and transmission coefficients have different expressions for left and right incidence, but their denominator is q_{11} for both incidence cases. The QRPM detects eigenenergies from the abrupt variations in the phase, Φ_{QRPM}, of the transfer-matrix element $q_{11} = |q_{11}| \exp(j\Phi_{\text{QRPM}})$. The zeros of $q_{11}(E) = 0$, specify the eigenstates of the heterostructure under the appropriate boundary conditions for the outgoing wavefunctions. This can easily be understood if the incident wave is zero, i.e. by replacing 1 with 0 in the LHS of Eq. (19). According to Eq. (10) there are always two counter propagating plane-waves wavefunctions in each layer except in the left and the right heterostructure boundary barriers. For the case of right incidence there are the incident and the reflected counter propagating waves at the right barrier and one transmitted wavefunction at the left barrier. Similarly, for left incidence there is only one transmitted wavefunction at the right barrier. For type 1 quasibound states the transmitted wave is a propagating wave and for type 2 quasibound states the transmitted wave is an evanescent wave.

Generally, the peaks of the $\mathcal{R}$ and $\mathcal{T}$ magnitudes and the peaks of their phase derivatives with respect to energy have Lorentzian shapes (see App. A). But there are some cases where this general rule does not apply. Recently, Zhao *et al.*,[37] have shown that the phase and magnitude of the reflection coefficient $\mathcal{R}$ may have a peculiar shape. In fact, the right (left) $\mathcal{R}$ coefficient phase can have a sharp monotonic shape where as the left (right) $\mathcal{R}$ can have a zig-zag variation. In this subsection a simple model of $\mathcal{R}$ (and $\mathcal{T}$) is given which provides satisfactory explanations to the above mentioned $\mathcal{R}$ characteristics. The eigenenergies of a quantum heterostructure correspond to the poles of its reflection coefficient $\mathcal{R}$. In addition it is well known that the behavior of $\mathcal{R}$ as a function of energy does not have any abrupt changes except at the eigenenergy levels. Therefore it is natural to conclude that the left and the right $\mathcal{R}$ (and $\mathcal{T}$) can be accurately modeled as a rational polynomial having the form

$$\mathcal{R}(E) = A(E)\frac{\prod_{j=1}^{M}(E - Z_j)}{\prod_{i=1}^{N}(E - P_i)} \simeq \frac{\prod_{j=1}^{M}(E - Z_j)}{\prod_{i=1}^{N}(E - P_i)} \quad \text{for } E_{\min} < E < E_{\max}\,, \tag{21}$$

where P_i and Z_j represent complex poles and complex zeros whose real part is within the energy range of interest $[E_{\min},\ E_{\max}]$, and $A(E)$ is a slowly varying complex function of energy which can be neglected for phase calculations. Equation (21) is well known from signals and systems theory (see e.g. Ref. 38). This equation was used successfully in Ref. 39 to generate the magnitude of $\mathcal{T}$ for symmetric heterostructures. The accuracy of Eq. (21) on highly asymmetric heterostructures has been proven in Ref. 40. Since the global matrix elements which correspond to the numerator of $\mathcal{R}$ are complex conjugates,[41] i.e. $q_{21} = q_{12}^*$, the $\mathcal{R}$ zeros for a left and for a right incident wavefunction form conjugate pairs. Moreover, the poles of $\mathcal{R}$ correspond to the heterostructure complex eigenenergies and they must have the form $P = \tilde{E} = E_r - jE_i$ with $E_i \geq 0$.[42] The model equation (21) suggests that the use of $\mathcal{R}$ to detect eigenstates is impaired by its numerator zeros Z. If these zeros are positioned between the real energy axis and the poles then they can screen the effects of poles causing inaccurate results when $\mathcal{R}_R$ or $\mathcal{R}_L$ are determined. Therefore, the use of the denominator of $\mathcal{R}$ for eigenenergy detection will always give better results compared to the use of the total $\mathcal{R}$. In addition, Eq. (21) permits the use of methods from electrical circuit analysis to understand the magnitude and phase variations of $\mathcal{R}$. It is well known that a pole or a zero of a function contributes a total phase shift of π to that function. The relative position of $\Im m[P]$ and $\Im m[Z]$ dictates the abruptness of the phase and the magnitude variation, with the most rapid variation being for the case of $\Im m[P] = \Im m[Z] = 0$. These results are well known from the Bode plot representation in circuit analysis[43] and explain the observed differences between $\mathcal{R}$ phase variations for left and right incident wavefunctions.[37] If $\Re e[Z] \simeq \Re e[P] = E_r$ and $\Im m[P] \times \Im m[Z] < 0$ then both π phase shifts will be added and the total phase will change by 2π and the derivative of $\mathcal{R}$ phase with respect to energy or the $\mathcal{R}$ magnitude will have a Lorentzian shape centered at E_r, while for the case of $\Im m[P] \times \Im m[Z] > 0$ the zero will add a phase of π while the pole will subtract a phase of π, resulting a zig-zag variation about zero total phase variation.

The accurate modeling of the reflection (or the transmission) coefficient as a rational polynomial function has many advantages. Since the QRPM utilizes the denominator of Eq. (21), each heterostructure eigenenergy will contribute phase π to the total phase variation of Φ_{QRPM}. Therefore, the total number of eigenfunctions, N, for an energy range $[E_{\min},\ E_{\max}]$ is the result of the total phase variation divided by π, i.e., $N = |\Phi_{\text{QRPM}}(E_{\min}) - \Phi_{\text{QRPM}}(E_{\max})|/\pi$. In addition, Eq. (21) provides analytic expressions for $|q_{11}|$, Φ_{QRPM}, and for $d\Phi_{\text{QRPM}}/dE$. The analytic expression for $d\Phi_{\text{QRPM}}/dE$ (see App. B) allows us to apply curve fitting (optimization) techniques to resolve Φ_{QRPM} spectra consisting of overlapping Lorentzians. Since the eigenenergy levels correspond to the individual Lorentzian peaks, the lifetime of all eigenenergies can be numerically found by fitting the analytic equation of $d\Phi_{\text{QRPM}}/dE$ derived from the denominator of Eq. (21) to the values of $d\Phi_{\text{QRPM}}/dE$ calculated from the transfer-matrix formulation (element q_{11}). The curve fitting problem will have as unknown parameters the FWHM ($= 2E_i$) of the Lorentzians. There will be cases where the number of states (deduced from $\Phi_{\rm QRPM}$) is larger than the number of observed Lorentzians. When this happens the E_r's of the "hidden" short-lifetime quasibound states can also be included in the set of the unknown parameters of the optimization problem. Finally, it is important to note that for a type 2 quasibound state, the reflection probability is unity ($R = |\mathcal{R}|^2 = 1$) meaning that $\mathcal{R}$ cannot be zero. Therefore for this energy range, Eq. (21) does not have zeros and $\mathcal{R}$ is an all-pole function.

For the numerical implementation of QRPM, the energy range of interest is divided into energy intervals ΔE of approximately 1 meV, and $\Phi_{\rm QRPM}$ is calculated at both ends of ΔE. In order to be able to detect abrupt phase jumps which are related to quasibound states with long lifetime, the total phase variation $\Delta\Phi_{\rm QRPM}$ within the energy interval ΔE must be less than π. For the case that $\Delta\Phi_{\rm QRPM} \geq \pi$ the interval ΔE is divided into two new subintervals and $\Phi_{\rm QRPM}$ is calculated again at their edges. The subdivision of energy intervals continues until the total phase variation is $\Delta\Phi_{\rm QRPM} < \pi$. For accurate FWHM measurements, the $\Phi_{\rm QRPM}$ must be calculated in a way that $\Delta\Phi_{\rm QRPM} \approx 0.1\pi$ between consecutive energy values. After $\Phi_{\rm QRPM}$ is calculated at discrete energy values the second derivative of $\Phi_{\rm QRPM}$ with respect to E is calculated numerically. Potentials which make the second derivative equal to zero correspond to the eigenenergy real parts (Lorentzian peaks), while the Lorentzian FWHMs are calculated via the use of the first-order derivative of $\Phi_{\rm QRPM}$ with respect of E.

3.5. *Application to an example case*

In this example the performance of our methods was tested on a symmetric heterostructure having two closely spaced adjacent eigenenergies among others. Even though this heterostructure does not pose any difficulties for the APM, it is a challenging case for QRPM which relies on the Lorentzian representation of states. Differences between the values of APM and QRPM are expected due to the fact that each numerical method solves the same eigenvalue problem subject to different boundary conditions. APM assumes monotonic eigenstate magnitude behavior in the heterostructure boundary regions, while QRPM assumes monotonic wavefunction magnitude behavior in the output barrier and oscillatory wavefunction magnitude behavior in the input barrier. Table 2 provides the physical characteristics of the heterostructure as well as the results for each of the numerical methods, where (according to APM) the first two quasibound states have real energy difference $\Delta E_r \approx 0.37$ meV. The QRPM initial results are shown in Fig. 5. The total $\Phi_{\rm QRPM}$ variation was more than 5π (but less than 6π) because of the phase contribution of higher order states above the 400 meV. In fact for the energy interval $100 < E < 366$ meV the phase variation is $\Delta\Phi_{\rm QRPM} = 5.1\pi$, revealing 5 eigenstates for this energy range. The derivative of $\Phi_{\rm QRPM}$ with respect to E is also given in Fig. 5 and it has only 4 peaks. The approximate position of the hidden Lorentzian peak is close to 130 meV, since at that energy level the Φ_{QRPM} has a 2π step. The exact position of the screened Lorentzian can be found by the application of curve fitting. Since the second and third peaks (states $E_3^{(1)}$ and $E_4^{(1)}$) correspond to single $\Delta\Phi_{QRPM} = \pi$ variations these peaks and their corresponding FWHM's measured from Fig. 5 are used as constants in the curve fitting program along with the energy level of the last Lorentzian peak (state $E_5^{(1)}$). We do not use the FWHM of the $E_5^{(1)}$ state as a constant since from the Φ_{QRPM} plot we know that there is significant overlap between $E_5^{(1)}$ and $E_6^{(1)}$. For the curve-fitting process, the Nelder–Mead algorithm with simple bounds (see Ref. 44) was used. The variable bounds for this specific example were set as $125 < \Re e[E_1^{(1)}] < 135$ meV, $-30 < \Im m[E_1^{(1)}] < 0$ meV, $125 < \Re e[E_2^{(1)}] < 135$ meV, $-30 < \Im m[E_2^{(1)}] < 0$ meV, and $-30 < \Im m[E_5^{(1)}] < 0$ meV, while the initial (guess) values for all variables were set at the midpoint of their bounds. The equations of App. B were also utilized for the analytic calculation of the derivative $d\Phi_{QRPM}/dE$. The results of the optimization are given in Table 2 and show very good agreement with the APM results. Figure 6 shows the comparison between the original Φ_{QRPM} spectrum and the fitted Φ_{QRPM}.

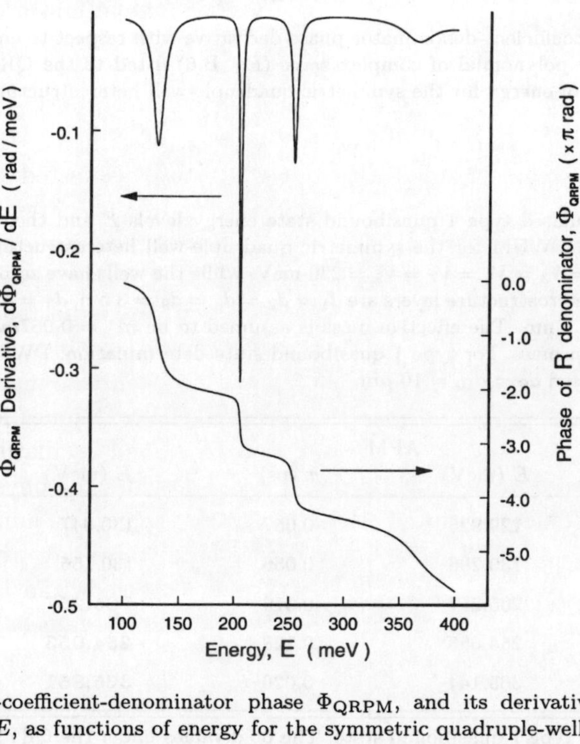

Fig. 5. Reflection-coefficient-denominator phase Φ_{QRPM}, and its derivative with respect to energy, $d\Phi_{QRPM}/dE$, as functions of energy for the symmetric quadruple-well heterostructure.

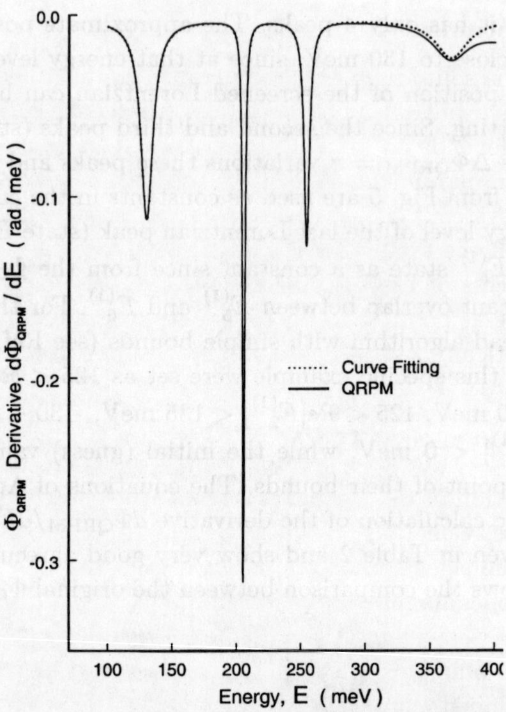

Fig. 6. Reflection-coefficient-denominator phase derivative with respect to energy, $d\Phi_{\text{QRPM}}/dE$, and the fifth-degree polynomial of complex zeros (Eq. B.6) fitted to the QRPM's $d\Phi_{\text{QRPM}}/dE$ curve as a function of energy for the symmetric quadruple-well heterostructure.

Table 2. Calculated type 1 quasibound state energy levels E and their corresponding lifetimes $\tau = \hbar/\text{FWHM}$ for the symmetric quadruple-well heterostructure. The barrier heights are $V_1 = V_3 = V_5 = V_7 = V_9 = 230$ meV, while the wells have zero potential. The widths of the heterostructure layers are $d_1 = d_2 = d_8 = d_9 = 3$ nm, $d_3 = d_5 = d_7 = 5$ nm, and $d_4 = d_6 = 1$ nm. The effective mass is assumed to be $m^* = 0.067m_0$, where m_0 is the free electron mass. For type 1 quasibound state determination, PWM and modified DOS method used $d_0 = d_{10} = 10$ μm.

Energy	APM		QRPM	
level	E (meV)	τ (ps)	E (meV)	τ (ps)
$E_1^{(1)}$	129.925	0.057	130.047	0.058
$E_2^{(1)}$	130.296	0.056	130.156	0.055
$E_3^{(1)}$	205.257	0.319	**205.257**	**0.316**
$E_4^{(1)}$	254.053	0.128	**254.053**	**0.125**
$E_5^{(1)}$	365.144	0.020	**365.861**	0.023

$E^{(1)}$ represents type 1 quasibound state. The bold values under the QRPM column were used as constants in the optimization process.

4. Design and Optimization of MQW Structures

A prerequisite for the solution of any optimization problem is the appropriate definition of an objective (merit) function, $\mathbf{O}(\overline{\mathbf{X}})$, representing the design goals. The objective function depends on system parameters, $\overline{\mathbf{X}} = [x_1, \ldots, x_P]^T$, where P is the number of system parameters. The same function is a measure of the disagreement between the system's ideal (desired) output state, $\mathcal{F}_d(\overline{\mathbf{X}})$, and the output state produced when the system's parameters have arbitrary values, $\mathcal{F}(\overline{\mathbf{X}})$. Optimization is the process of finding the parameter values, $\overline{\mathbf{X}}_{\mathrm{OPT}}$, which produce a system state as close as possible to the ideal output state and where the objective function is asymptotically zero. The mathematical representation of the optimization procedure is

$$\mathbf{O}(\overline{\mathbf{X}}_{\mathrm{OPT}}) = \min\{\mathbf{O}(\overline{\mathbf{X}})\} = \min\{\mathcal{F}(\overline{\mathbf{X}}) - \mathcal{F}_d(\overline{\mathbf{X}})\}\,. \tag{22}$$

An intelligent definition of the merit function is crucial for the convergence of the optimization process. Generally, the MQW structure is designed to operate in a "fuzzy" sense, i.e. the eigenstates have real and imaginary parts within a range of acceptable values depending on the device specifications. Other eigenstate properties like total length, scattering times, dipole matrix elements, or a self-consistent loop for conduction-band profile compensation due to the application of electric field can also be included. The schematic diagram of the optimization procedure is given in Fig. 2. This is a generic optimization scheme since any numerical method that solves a MQW structure and produces the complex eigenstates accurately and fast can be substituted for APM/QRPM. The elements of $\overline{\mathbf{X}}_k$ may include any combination of layer thicknesses, d_i, or layer potential energies V_i, and k is the optimization iteration step. Possible constraints on the value range of the elements of $\overline{\mathbf{X}}_k$ may be integrated into the optimization algorithm to be used. $\overline{\mathbf{X}}_k$ is the input data to the APM/QRPM algorithm which calculates the eigenenergies vector $\overline{\mathbf{Y}}_k = [\tilde{E}_1, \ldots, \tilde{E}_q]_k^T$ where q is the total number of the supported eigenstates. The APM and the QRPM algorithms described in Sec. 3, can be used for the design/optimization of any MQW structure under the single-band with no zero-field splitting effective-mass Schrödinger's equation. A non-parabolic effective-mass model (see App. C) can also be used. In this Section we design the active region of an $\mathrm{Al}_x\mathrm{Ga}_{1-x}\mathrm{As}$ intersubband laser using a combination of the simulated annealing algorithm and the APM.[21] The design of the emitter/active-region/collector MQW system is beyond the scope of the present work.

Besides the explicit eigenstate features like lifetime and energy difference the maximization of the spatial overlap between the two lasing eigenstates is critical for an efficient intersubband laser. According to the Fermi's golden rule,[45] the intersubband-transition gain between two orthogonal eigenstates Ψ_H and Ψ_L is proportional to the intersubband dipole matrix element

$$z_{LH} = \int_0^L \Psi_L z \Psi_H^* dz\,, \tag{23}$$

were L is total length of the MQW structure excluding the left and the right barrier thicknesses. The representation of quasibound states as leaky wave functions (see Sec. 3.1) implies that the quasibound wave functions of any given heterostructure are not orthogonal, but in Ref. 46 it has been shown that the overlap integral (23) gives results close to the experimental values of z_{LH} even for the case of non-orthogonal eigenfunctions. Therefore, in the proposed optimization scheme we maximize z_{LH} performing the integration of Eq. (23) analytically. In this work a limited but important set of MQW features are optimized including: (a) The energy difference between the upper eigenstate E_H and the lower eigenstate E_L, $\Delta E = E_H - E_L$. (b) The lifetimes ratio of the upper eigenstate, τ_H and the lower eigenstate, τ_L. (c) The definition of an energy region within which only the two lasing eigenstates are allowed, effectively reducing the possibility of unwanted transitions. (d) The dipole matrix element z_{LH}. The MQW structure optimization parameters were the Al composition x under the constraint that $0 \leq x \leq 0.45$ and the thicknesses of the individual layers. The number of MQW structure layers N, the applied bias $V_{\text{bias}} = 225$ meV, as well as the left and the right barrier aluminum compositions $x = 0.2$, i.e. $V_0 = V_{N+1} = 154.628$ meV, were set as constants. Therefore, the $P\ (=2N)$ system parameters were $\overline{\mathbf{X}} = [x_1, \ldots, x_N, V_1, \ldots, V_N]^T$. The desired state of the MQW structure $\mathcal{F}_d(\overline{\mathbf{X}})$ was defined as $\Delta E = 100 \pm 5$ meV, $\tau_H/\tau_L \geq 4$, $z_{LH} \geq 3.5$ nm, and only two eigenenergies for the energy range $V_0 + V_{\text{bias}} \pm 0.8\Delta E = 380 \pm 85$ meV were to be allowed. The objective function $\mathbf{O}(\overline{\mathbf{X}})$ for the specific example cases was $\mathbf{O}(\overline{\mathbf{X}}) = A + B + C + D + F + G$ where the terms A, B, C, D, F and G were defined as

$$
\begin{aligned}
A &= \begin{cases} 10^{-3}|\Delta E - 100| & \text{if } |\Delta E - 100| \leq 5 \text{ meV} \\ 10|\Delta E - 100| & \text{otherwise} \end{cases} \\
B &= \begin{cases} 10^5 \tau_L/\tau_H & \text{if } \tau_H/\tau_L \leq 1 \\ 0 & \text{if } \tau_H/\tau_L \geq 4 \\ 10^3(4 - \tau_H/\tau_L) & \text{otherwise} \end{cases} \\
C &= \begin{cases} 0 & \text{if } z_{LH} \geq 3.5 \text{ nm} \\ 10^3(3.5 - z_{LH}) & \text{otherwise} \end{cases} \\
D &= \begin{cases} 0 & \text{if } \tau_H \geq 250 \text{ fsec} \\ 10^2(250 - \tau_H) & \text{otherwise.} \end{cases} \\
F &= 10(\{\text{number of type 1 quasibound states}\} - 1) \\
G &= 10(\{\text{number of type 2 quasibound states}\} - 1)
\end{aligned}
\tag{24}
$$

In addition, when there were no states within the predetermined energy range 380 ± 85 meV the values $\Delta E = 0$, $z_{LH} = 0$, $\tau_H = 10^{-6}$, and $\tau_L = 1$ were used in Eq. (24). The weights used in Eq. (24) were found empirically after some exploratory runs of the optimization program.

For an arbitrary input vector $\overline{\mathbf{X}}_0$ and $N = 8$ layers the optimization process of Fig. 2 gave: $E_H = 405.15$ meV, $\tau_H = 305$ fsec, $E_L = 303.03$ meV, $\tau_L = 52$ fsec, $z_{LH} = 3.81$ nm. The resulting MQW structure along with the plots of the two eigenfunctions are given in Fig. 7. Also for an arbitrary input vector $\overline{\mathbf{X}}_0$ and $N =$ 9 layers we had the following results $E_H = 400.32$ meV, $\tau_H = 262$ fsec, $E_L =$ 296.08 meV, $\tau_L = 33$ fsec, $z_{LH} = 3.78$ nm (see Fig. 8). In addition, to show that the thickness of the ith layer is not an integer multiple of the corresponding $\lambda_i/2$ or $\lambda_i/4$ lengths, in the captions of Figs. 7 and 8 we report the quotients of the divisions ϕ_i over π and $\pi/2$. The quantity ϕ_i is the total phase shift that an electron-wave undergoes propagating inside a layer, and it is calculated as $\phi_i = \int \kappa_i(z)zdz$ for $0 \leq z < d_i$. The quotients are integer numbers when the total phase shifts ϕ_i are multiples of $\lambda_i/2$ and/or multiple of $\lambda_i/4$. For cases where $V_i > E$ the

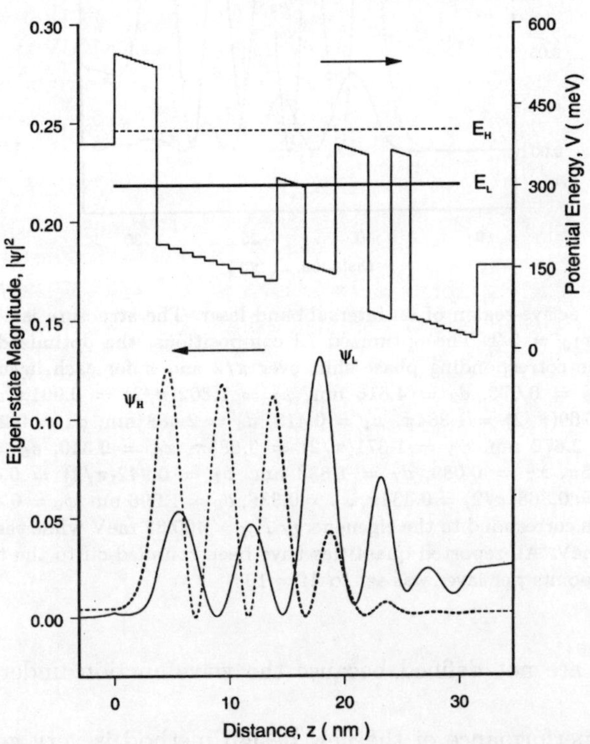

Fig. 7. Optimized active-region of an intersubband laser. The structure is comprised of $N = 8$ layers with $x_0 = x_9 = 0.2$. The optimized Al compositions, the optimized layer thicknesses, and the ratio of the corresponding phase shift over $\pi/2$ and π for each layer were $x_1 = 0.418$, $d_1 =$ 3.636 nm, $x_2 = 0.001$, $d_2 = 10.482$ nm, $\phi_2 = 4.518(\pi/2) = 2.259\pi$, $x_3 = 0.255$, $d_3 = 2.493$ nm, $\phi_3 = 0.743(\pi/2) = 0.371\pi$, $x_4 = 0.074$, $d_4 = 2.628$ nm, $\phi_4 = 1.210(\pi/2) = 0.605\pi$, $x_5 = 0.384$, $d_5 = 2.865$ nm, $\phi_5 = 0.553(\pi/2) = 0.277\pi$, $x_6 = 0.013$, $d_6 = 2.019$ nm, $\phi_6 = 1.063(\pi/2) =$ 0.531π, $x_7 = 0.417$, $d_7 = 1.652$ nm, $\phi_7 = 0.353(\pi/2) = 0.176\pi$, $x_8 = 0.036$, $d_8 = 5.796$ nm, $\phi_8 = 3.203(\pi/2) = 1.601\pi$, The quotient results correspond to the eigenenergy $E_H = 405.15$ meV while the other eigenenergy was $E_L = 303.03$ meV. All reported quantities have been rounded-off to the third decimal digit. The discretization points per layer was $M = 10$.

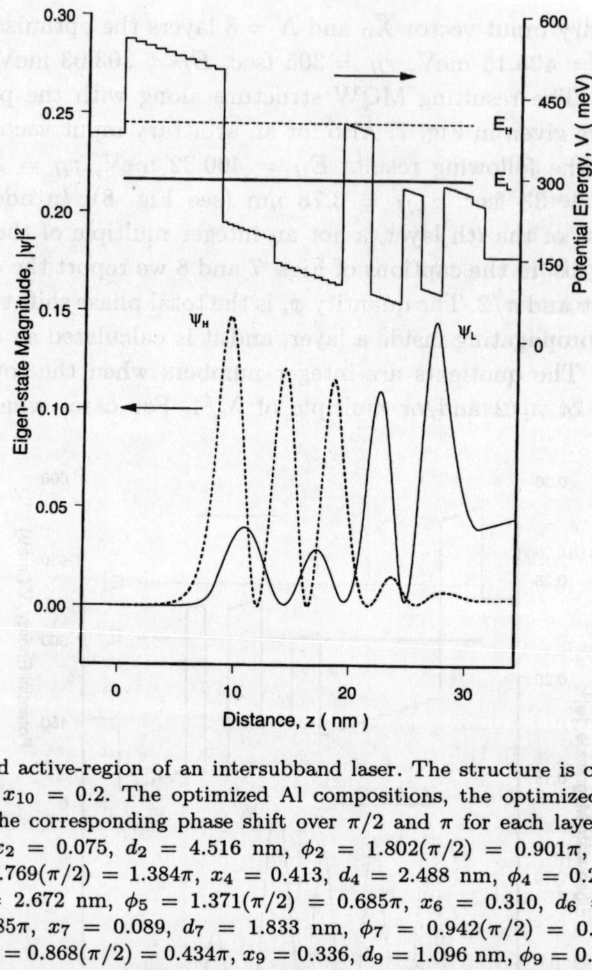

Fig. 8. Optimized active-region of an intersubband laser. The structure is comprised of $N = 9$ layers with $x_0 = x_{10} = 0.2$. The optimized Al compositions, the optimized layer thicknesses, and the ratio of the corresponding phase shift over $\pi/2$ and π for each layer were $x_1 = 0.432$, $d_1 = 8.527$ nm, $x_2 = 0.075$, $d_2 = 4.516$ nm, $\phi_2 = 1.802(\pi/2) = 0.901\pi$, $x_3 = 0.024$, $d_3 = 5.906$ nm, $\phi_3 = 2.769(\pi/2) = 1.384\pi$, $x_4 = 0.413$, $d_4 = 2.488$ nm, $\phi_4 = 0.209(\pi/2) = 0.105\pi$, $x_5 = 0.033$, $d_5 = 2.672$ nm, $\phi_5 = 1.371(\pi/2) = 0.685\pi$, $x_6 = 0.310$, $d_6 = 1.646$ nm, $\phi_6 = 0.570(\pi/2) = 0.285\pi$, $x_7 = 0.089$, $d_7 = 1.833$ nm, $\phi_7 = 0.942(\pi/2) = 0.471\pi$, $x_8 = 0.350$, $d_8 = 2.481$ nm, $\phi_8 = 0.868(\pi/2) = 0.434\pi$, $x_9 = 0.336$, $d_9 = 1.096$ nm, $\phi_9 = 0.421(\pi/2) = 0.210\pi$. The quotient results correspond to the eigenenergy $E_H = 400.32$ meV while the other eigenenergy was $E_L = 296.08$ meV. All reported quantities have been rounded-off to the third decimal digit. The discretization points per layer was set to $M = 10$.

phase quotients are not defined because the wavefunction undergoes magnitude variations only.

Clearly, the performance of the new design method is very good because the designed MQW structures have the following characteristics: (a) τ_L several orders of magnitude shorter that the optical phonon scattering time $\simeq 1$ ps, (b) lasing eigenstates vertically positioned, (c) the upper state lifetime much larger than the lower state lifetime, $\tau_H > \tau_L$, and (d) large z_{LH} value which increases the laser power efficiency. An issue than needs to be addressed is the stability of the resulting solution. The computer produces thicknesses and Al compositions which are accurate to the 16th decimal digit and this kind of fabrication accuracy is unattainable. This is a frequent issue encountered in every solution produced via an optimization process and is called solution stability. Solution stability can be achieved such that small variations for example $\pm 5\%$ from the produced $\overline{\mathbf{X}}_{\mathrm{OPT}} = [x_1, \ldots, x_P]^T$ values will not affect the MQW structure properties. According to Fig. 2, a method to force stable solutions is to introduce random perturbations at each iteration to the vector $\overline{\mathbf{X}}_k$. Each time the optimization method produces a new $\overline{\mathbf{X}}_k$, each of its elements $x_1, \cdots, x_P$ is randomly perturbed ℓ times producing ℓ different $\overline{\mathbf{X}}_k$ vectors. The APM or QRPM will be executed ℓ times producing ℓ different $\overline{\mathbf{Y}}_k$. There will be ℓ different objective functions the average of which will be the vector $\mathbf{O}(\overline{\mathbf{X}})$ which will be the input vector to the optimization algorithm. In this way every $\mathbf{O}(\overline{\mathbf{X}})$ scalar value will carry stability information. The higher the number ℓ the more stable the solution will be. This solution-stability enhancement is straightforward but was not implemented in the current work.

5. Conclusions

Powerful tools that are available to the quantum-state engineer for the design of electron-wave filters are the analogies that exist between electromagnetic optics and quantum-mechanical electron-wave optics. Unfortunately, due to their dispersive nature the electron-wave filters are most similar to their optical counterparts only near a single energy. Therefore, a systematic procedure for designing generalized electron-filters having a desired energy-spectrum response for more than one energy level has been needed. In this work we presented such a method of designing generalized electron-wave energy filters under the single-band with no zero-field splitting effective-mass approximation. First, we described two numerical methods the APM and the QRPM that are highly accurate in extracting complex eigenenergies of MQW structures of arbitrary geometry. Second, we demonstrated a method which by using the APM or the QRPM automatically set the thicknesses and the compositions of a biased MQW structure layers such that the resulting MQW structure had a predefined set of properties. The designed MQW structure was not a simple energy filter, but the active region of an intersubband laser having lasing states with predefined energy difference, predefined minimum lifetime ratio, and predefined minimum dipole matrix element. Third, we proposed a methodology using the design algorithm presented that can produce sets of thicknesses and compositions that are insensitive to fabrication tolerances.

Appendix A. Reflection and Transmission Coefficients near a Quasibound State

According to Eq. (21), the complex reflection coefficient $\mathcal{R}_L$ or $\mathcal{R}_r$ in the vicinity of an eigenenergy (quasibound state) $\tilde{E} = E_r + jE_i$ (with $E_i \leq 0$), can be expressed as

$$\mathcal{R}(E) = |\mathcal{R}| \exp(j\Phi) = \frac{A(E)}{E - \tilde{E}}, \tag{A.1}$$

where $A(E) \simeq A = A_r + jA_i$, is a slowly varying function. The magnitude of $\mathcal{R}$ is given as

$$|\mathcal{R}|^2 = \mathcal{R}(E)\mathcal{R}(E)^* \approx \frac{A_r^2 + A_i^2}{(E - E_r)^2 + E_i^2}\,, \tag{A.2}$$

while the phase and the phase derivative of $\mathcal{R}$ are expressed as

$$\begin{aligned} \Phi &= \tan^{-1}\left(\frac{\Im m[\mathcal{R}(E)]}{\Re e[\mathcal{R}(E)]}\right) \approx -\tan^{-1}\left(\frac{A_r E_i + A_i(E_r - E)}{A_i E_i + A_r(E - E_r)}\right), \\ \frac{d}{dE}\Phi &\approx \frac{E_i}{(E - E_r)^2 + E_i^2}\,. \end{aligned} \tag{A.3}$$

Equations (A.2) and (A.3) show that both $\mathcal{R}$ magnitude squared and phase derivative have Lorentzian shapes with peaks at $E = E_r$ and FWHM $= 2E_i$. In addition, the phase derivative is independent of the unknown quantities A_r and A_i.

Appendix B. Approximate Analytical Expressions of Φ_{QRPM} and $d\Phi_{\mathrm{QRPM}}/dE$

According to Eq. (19) the denominator of the reflection coefficient can be approximated as a Nth-degree polynomial

$$\begin{aligned} q_{11} = |q_{11}|\exp(j\Phi_{\mathrm{QRPM}}) &\simeq (E - \tilde{E}_N)(E - \tilde{E}_{N-1})\cdots(E - \tilde{E}_1) \\ &= E^N + r_{N-1}E^{N-1} + \cdots + r_0\,, \end{aligned} \tag{B.1}$$

where $\tilde{E}_j$ is the jth complex eigenvalue and N is the number of the eigenvalues for the particular energy range. In order to find the analytic expression of the quantity $d\Phi_{\mathrm{QRPM}}/dE$ we must collect all the coefficients of Eq. (B.1) with the same power of E. This can be done by using Vieta's formulas which provide relationships between the roots of a polynomial, E_j, and its complex power coefficients, r_k. For the example case of the symmetric quadruple-well heterostructure, Vieta's relations for a fifth-degree polynomial are

$$\begin{aligned} r_4 &= -(\tilde{E}_1 + \tilde{E}_2 + \tilde{E}_3 + \tilde{E}_4 + \tilde{E}_5)\,, \\ r_3 &= \tilde{E}_1\tilde{E}_2 + \tilde{E}_1\tilde{E}_3 + \tilde{E}_1\tilde{E}_4 + \tilde{E}_1\tilde{E}_5 + \tilde{E}_2\tilde{E}_3 + \tilde{E}_2\tilde{E}_4 + \tilde{E}_2\tilde{E}_5 + \tilde{E}_3\tilde{E}_4 \\ &\quad + \tilde{E}_3\tilde{E}_5 + \tilde{E}_4\tilde{E}_5\,, \\ r_2 &= -(\tilde{E}_1\tilde{E}_2\tilde{E}_3 + \tilde{E}_1\tilde{E}_2\tilde{E}_4 + \tilde{E}_1\tilde{E}_2\tilde{E}_5 + \tilde{E}_1\tilde{E}_3\tilde{E}_4 + \tilde{E}_1\tilde{E}_3\tilde{E}_5 + \tilde{E}_1\tilde{E}_4\tilde{E}_5 \\ &\quad + \tilde{E}_2\tilde{E}_3\tilde{E}_4 + \tilde{E}_2\tilde{E}_3\tilde{E}_5 + \tilde{E}_2\tilde{E}_4\tilde{E}_5 + \tilde{E}_3\tilde{E}_4\tilde{E}_5)\,, \\ r_1 &= \tilde{E}_1\tilde{E}_2\tilde{E}_3\tilde{E}_4 + \tilde{E}_1\tilde{E}_2\tilde{E}_3\tilde{E}_5 + \tilde{E}_1\tilde{E}_2\tilde{E}_4\tilde{E}_5 + \tilde{E}_1\tilde{E}_3\tilde{E}_4\tilde{E}_5 + \tilde{E}_2\tilde{E}_3\tilde{E}_4\tilde{E}_5\,, \\ r_0 &= -\tilde{E}_1\tilde{E}_2\tilde{E}_3\tilde{E}_4\tilde{E}_5\,. \end{aligned} \tag{B.2}$$

From Eqs. (B.1) and (B.2), the real and the imaginary part of q_{11} can be approximated as

$$\begin{aligned} \Re e[q_{11}] &\simeq E^5 + \Re e[r_4E^4 + r_3E^3 + r_2E^2 + r_1E + r_0]\,, \\ \Im m[q_{11}] &\simeq \Im m[r_4E^4 + r_3E^3 + r_2E^2 + r_1E + r_0]\,, \end{aligned} \tag{B.3}$$

as well as their derivatives with respect to E

$$\begin{aligned} \frac{d}{dE}\Re e[q_{11}] &\simeq 5E^4 + \Re e[4r_4E^3 + 3r_3E^2 + 2r_2E + r_1]\,, \\ \frac{d}{dE}\Im m[q_{11}] &\simeq \Im m[4r_4E^3 + 3r_3E^2 + 2r_2E + r_1]\,. \end{aligned} \tag{B.4}$$

Therefore, using Eqs. (B.3) and (B.4), both the phase of the reflection denominator $q_{11}(E)$

$$\Phi_{\text{QRPM}} = \tan^{-1}\left(\frac{\Im m[q_{11}]}{\Re e[q_{11}]}\right), \tag{B.5}$$

and its derivative with respect to E

$$\begin{aligned} \frac{d}{dE}\Phi_{\text{QRPM}} &\approx \frac{\Re e[q_{11}]^2}{\Re e[q_{11}]^2 + \Im m[q_{11}]^2} \\ &\quad \times \frac{\Re e[q_{11}]\dfrac{d}{dE}\Im m[q_{11}] - \Im m[q_{11}]\dfrac{d}{dE}\Re e[q_{11}]}{\Re e[q_{11}]^2}\,, \end{aligned} \tag{B.6}$$

can be expressed analytically.

Appendix C. Non-Parabolic Conduction-Band Effective-Mass of $Al_xGa_{1-x}As$

The corrected non-parabolic effective-mass of $Al_xGa_{1-x}As$ can be given as[47]

$$m^* = m_0\left[\frac{2X(x) + (KE)}{(KE) + E_G(x)} + \frac{X(x)}{(KE) + E_G(x) + SO(x)} + 1\right]^{-1}, \tag{C.1}$$

where x is the Al composition, (KE) is the electron kinetic energy in meV, and

$$\begin{aligned} X(x) &= 6061.54x + 7544.44(1-x)\ \text{meV}\,, \\ E_G(x) &= 3018x + 1424(1-x)\ \text{meV}\,, \\ SO(x) &= 290x + 340(1-x)\ \text{meV}\,. \end{aligned} \tag{C.2}$$

References

1. L. C. West and S. J. Eglash, "First observation of an extremely large-dipole infrared transition within the conduction band of a GaAs quantum well", *Appl. Phys. Lett.* **46** (1985) 1156–1158.
2. B. F. Levine, "Quantum-well infrared photodetectors", *J. Appl. Phys.* **74** (1993) R1–R81.
3. S. D. Gunapala, J. K. Liu, J. S. Park, M. Sundaram, C. A. Shott, T. Hoelter, T. Lin, S. T. Massie, P. D. Maker, R. E. Muller, and G. Sarusi, "9 μm cutoff 256×256 GaAs/Al_xGa_{1-x}As quantum well infrared photodetector hand-held camera", *IEEE Trans. Electron. Devices* **44** (1997) 51–57.
4. J. Faist, F. Capasso, D. L. Sivco, C. Sirtori, A. L. Hutchinson, and A. Y. Cho, "Quantum cascade laser", *Science* **264** (1994) 553–556.
5. J. Faist, F. Capasso, C. Sirtori, D. L. Sivco, J. N. Baillargeon, A. L. Hutchinson, S. G. Chu, and A. Y. Cho, "High power mid-infrared ($\lambda \sim 5$ μm) quantum cascade lasers operating above room temperature", *Appl. Phys. Lett.* **68** (1996) 3680–3682.
6. J. Faist, F. Capasso, C. Sirtori, D. L. Sivco, A. L. Hutchinson, and A. Y. Cho, "Vertical transition quantum cascade laser with Bragg confined excited state", *Appl. Phys. Lett.* **66** (1995) 538–540.
7. C. Gmachl, F. Capasso, A. Tredicucci, D. L. Sivco, A. L. Hutchinson, S. G. Chu, and A. Y. Cho, "Non-cascaded intersubband injection lasers at $\lambda \approx 7.7$ μm", *Appl. Phys. Lett.* **73** (1998) 3830–3832.
8. O. Gauthier-Lafaye, P. Boucaud, F. H. Julien, S. Sauvage, S. Cabaret, and J. M. Lourtioz, "Long-wavelength (≈ 15.5 μm) unipolar semiconductor laser in GaAs quantum wells", *Appl. Phys. Lett.* **71** (1997) 3619–3621.
9. G. N. Henderson, T. K. Gaylord, and E. N. Glytsis, "Ballistic electron transport in semiconductor heterostructures and its analogies in electromagnetic propagation in general dielectrics", *Proc. IEEE* **79** (1991) 1643–1659.
10. T. K. Gaylord, G. N. Henderson, and E. N. Glytsis, "Application of electromagnetics formalism to quantum-mechanical electron wave propagation in semiconductors", *J. Opt. Soc. Am.* **B10** (1993) 333–339.
11. H. A. Macleod, *Thin-Film Optical Filters*, 2nd Ed., McGraw-Hill, New York, 1986.
12. A. Thelen, *Design of Optical Interference Coatings*, McGraw-Hill, New York, 1989.
13. E. N. Glytsis, T. K. Gaylord, and K. F. Brennan, "Theory and design of semiconductor electron-wave interference filter/emitters", *J. Appl. Phys.* **66** (1989) 6158–6167.
14. D. R. Herrick, "Construction of bound states in the continuum for epitaxial heterostructure superlattices", *Physica* **B85** (1997) 44–50.
15. F. H. Stillinger, "Potentials supporting positive-energy eigenstates and their application to semiconductor heterostructures", *Physica* **B85** (1997) 270–276.
16. C. Sirtori, F. Capasso, J. Faist, D. L. Sivco, S. G. Chu, and A. Y. Cho, "Quantum wells with localized states at energies above the barrier height: A Fabry–Perot etalon filter", *Appl. Phys. Lett.* **61** (1992) 898–900.
17. C. Sirtori, J. Faist, F. Capasso, D. L. Sivco, S. G. Chu, and A. Y. Cho, "Narrowing of the intersubband absorption spectrum by localization of continuum resonances in a strong electric field", *Appl. Phys. Lett.* **62** (1993) 1931–1933.
18. P. G. Verly, A. V. Tikhonravov, and M. K. Trubetskov, "Efficient refinement algorithm for the synthesis of inhomogeneous optical coatings", *Appl. Opt.* **36** (1997) 1487–1495.
19. J. Price, "Resonant tunneling properties of heterostructures", *Superlattices Microstruct.* **2** (1986) 593–596.
20. E. Anemogiannis and E. N. Glytsis, "Multilayer waveguides: Efficient numerical analysis of general structures", *J. Lightwave Technol.* **10** (1992) 1344–1351.

21. E. Anemogiannis, E. N. Glytsis, and T. K. Gaylord, "Optimization of multilayer integrated optics waveguides", *J. Lightwave Technol.* **12** (1994) 512–518.
22. L. D. Landau and E. M. Lifshitz, *Quantum Mechanics, Non-Relativistic Theory*, Pergamon Press, London, 1958.
23. M. Altarelli, "Band Structure, Impurities and Excitons in Superlattices," in *Heterostructures and Semiconductor Superlattices*, eds. G. Allan, G. Bastard, N. Boccara, M. Lannoo, and M. Voos, Springer-Verlag, New York, 1986, pp. 12–37.
24. G. Bastard, "Superlattice band structure in the envelope-function approximation", *Phys. Rev.* **B24** (1981) 5993–5997.
25. G. Bastard and J. Brum, "Electronic states in semiconductor heterostructures", *IEEE J. Quantum Electron.* **22** (1986) 1625–1644.
26. T. K. Gaylord, E. N. Glytsis, and K. F. Brennan, "Semiconductor superlattice interference filter design", *J. Appl. Phys.* **65** (1989) 2535–2540.
27. G. N. Henderson, T. K. Gaylord, and E. N. Glytsis, "Ballistic electron transport in semiconductor heterostructures and its analogies in electromagnetic propagation in general dielectrics", *Proc. IEEE* **79** (1991) 1643–1659.
28. S. Collins, D. Lowe, and J. R. Barker, "A dynamic analysis of resonant tunneling", *J. Phys. C: Solid State Phys.* **20** (1987) 6233–6243.
29. J. Price, "Theory of resonant tunneling in heterostructures", *Phys. Rev.* **B38** (1988) 1994–1998.
30. T. Tamir and F. Y. Kou, "Varieties of leaky waves and their excitation along multilayer structures", *IEEE J. Quantum Electron.* **22** (1986) 544–551.
31. R. E. Collin, *Field Theory of Guided Waves*, 2nd Ed., IEEE Press, New York, 1991.
32. L. M. Delves and J. N. Lyness, "A numerical method for locating the zeros of an analytic function", *Math. Comp.* **21** (1967) 543–560.
33. R. V. Churchill and J. W. Brown, *Complex Variables and Applications*, McGraw-Hill, New York, 1984.
34. L. C. Botten, M. S. Graig, and R. C. McPhedran, "Complex zeros of analytic functions", *Comput. Phys. Commun.* **29** (1983) 245–259.
35. A. S. Householder, *The Numerical Treatment of a Single Nonlinear Equation*, McGraw-Hill, New York, 1970.
36. FORTRAN subroutine DZANLY from the International Mathematics and Statistics Library (*IMSL MATH/LIBRARY, Version 1.1*) Houston, Texas, 1989.
37. X. D. Zhao, H. Yamamoto, K. Taniguchi, and Z. M. Chen, "Resonant tunneling in periodic multiple-barrier structures with compound-barrier unit", *Superlattices Microstruct.* **19** (1996) 279–289.
38. A. Papoulis, *Signal Analysis*, McGraw-Hill, New York, 1977.
39. Z. Shao and W. Porod, "Resonance formalism for the transmission probability of symmetrical multiberrier resonant-tunneling structures", *Phys. Rev.* **B51** (1995) 1931–1934.
40. E. Anemogiannis, E. N. Glytsis, and T. K. Gaylord, "Quantum reflection pole method for determination of quasibound states in semiconductor heterostructures", *Superlattices Microstruct.* **22** (1997) 481–496.
41. L. Coldren and S. Corzine, *Diode Lasers and Photonic Integrated Circuits*, John Wiley & Sons, New York, 1995.
42. E. Anemogiannis, E. N. Glytsis, and T. K. Gaylord, "Bound and quasibound state calculations for biased/unbiased semiconductor quantum heterostructures", *IEEE J. Quantum Electron.* **29** (1993) 2731–2740.
43. A. Papoulis, *Circuits and Systems: A Modern Approach*, Holt, Rinehart and Wiston, 1980.

44. FORTRAN subroutine DBCPOL from the International Mathematics and Statistics Library, "IMSL MATH/LIBRARY", Houston, 1989, pp. 872–875.
45. S. L. Chuang, "Physics of Optoelectronic Devices", Wiley Series in Pure and Applied Optics, 1995, Equation (9.6.6), p. 375.
46. C. Sirtori, F. Capasso, J. Faist, and S. Scandolo, "Nonparabolicity and a sum rule associated with bound-to-bound and bound-to-continuum intersubband transitions in quantum wells", *Phys. Rev.* **B50** (1994) 8663–8674.
47. L. C. West and G. N. Henderson, "Non-parabolic effective-mass calculation for $Al_xGa_{1-x}As$", personal communication, 1990.

International Journal of High Speed Electronics and Systems, Vol. 9, No. 4 (1998) 1265–1277

ADVANCED SEMICONDUCTOR LASERS: PHONON ENGINEERING AND PHONON INTERACTIONS

M. DUTTA and M. A. STROSCIO
US Army Research Office, P.O. Box 12211, Research Triangle Park, NC 27709-2211, USA

This account presents theoretical results on the tailoring of carrier scattering by polar-optical phonons in novel quantum-well lasers such as the tunneling injection laser and the quantum cascade laser. These scattering rates are tailored by engineering semiconductor heterostructures on a scale sufficiently small that both the phonon amplitudes and the phonon envelopes are modified by dimensional confinement. As background information on the properties of confined phonons, this account provides a brief survey of phonon confinement effects in dimensionally-confined structures.

1. Introduction

The engineering of phonons in nanostructures has emerged as a technique for tailoring carrier interactions in electronic and optical devices with quantum-well, quantum-wire, and quantum-dot active regions. Recent attempts to tailor carrier interactions in such structures have focused on engineering the properties of both acoustic and optical phonons. In this account, emphasis is placed on the use of dimensional confinement to tailor the spatial characteristics and frequencies of optical phonons in quantum-well-based semiconductor lasers. Special emphasis is placed on advancing the current understanding of how phonon confinement plays a role in device structures such as the quantum cascade laser and the tunneling injection laser. To put these recent developments in context, this account will survey briefly the general progress in phonon engineering in nanostructures; following this general introductory discussion, this article will focus primarily on the role of optical-phonon effects in semiconductor lasers.

2. Recent Developments in Engineering of Acoustic and Optical Phonons in Nanostructures

In recent years, acoustic phonons have been quantized for a variety of nanoscale and mesoscopic structures in numerous publications;[1–10] these developments make it possible to assess the role of electron–acoustic-phonon scattering in limiting the performance of mesoscopic and nanoscale electronic devices. There is also an expanding literature on the tailoring of optical phonon modes to engineer scattering rates[11–34] in a variety of semiconductor devices and structures; these structures include quantum wells, superlattices, quantum wires with cylindrical and rectangular cross sections, and quantum dots with spherical, cylindrical and rectangular boundaries. Reference 11 provides a review of about 238 papers dealing with theories and experiments on the properties of these confined phonon modes; Ref. 11 provides references to the early seminal works on phonon confinement as well as the more recent works dealing with the application of these modes in electronic and optoelectronic devices.

Quantized acoustic phonons have been studies for the two cardinal boundary conditions of classical acoustics: free boundaries where the phonon displacements are unrestricted and allowed to balance all normal traction forces to zero; and clamped boundaries where phonon displacements are required to vanish at the boundaries. Scattering rates have been calculated for free-standing quantum-well structures[5] fabricated of non-polar semiconductor materials. For the case of quantum wires, scattering rates have been calculated for the case of infinitely long quantum wires and, as appropriate for this case, the acoustic phonons have been quantized in only the lateral dimensions.

The design of quantum structures to modify scattering rates through the tailoring of optical phonon modes has received attention by a growing number research groups in recent years.[11–34] In this account, a number of these efforts are surveyed for the case of engineering scattering times in quantum-well lasers.

3. Essential Features of Tailoring of Optical Phonon Modes in Quantum Structures

Models for determining the properties of confined optical modes in zero-, one-, and two-dimensional structures have been developed during the last two decades. Leburton has discussed[12] efforts to engineer scattering rate in quantum structures based on the tailoring of phonon modes in such structures.

One of the principal differences between bulk phonon modes and those of dimensionally-confined phonons is the large difference between the optical-phonon frequency of the bulk optical-phonon mode for a quantum-well material and the actual energies of the high-frequency interface optical-phonon modes in the quantum well. As a specific example, a AlAs-GaAs-AlAs quantum well or a GaAs-AlAs superlattice has a symmetric interface optical-phonon with an energy of about 50 meV in the GaAs. Thus, this symmetric interface phonon has an energy that is significantly larger than the 36-meV of the bulk GaAs optical phonon. For quantum wells with dimensions of about 40 Angstroms or less, it is by now appreciated widely that interface optical-phonon modes can be the modes making the dominant contribution to polar-optical-phonon scattering; such a situation is illustrated in Ref. 15 for the case of short-period AlAs-GaAs superlattice. As a result of energy conservation it is clear that such interface phonon modes play a special role in the case of interface–phonon-assisted transitions of carriers between quantum states occupied by carriers.

Leburton *et al.*[19–23] have examined optically pumped intersubband scattering in coupled quantum-well lasers and they have demonstrated that interface–phonon-assisted transitions are important in such structures; clearly, the findings of these workers represent an important manifestation of phonon confinement effects. These results indicate an important feature of phonon interactions in optical devices: the proper treatment of optical-phonon confinement in optical systems such as semiconductor laser should depend critically on the detailed energy spectrum of the phonon spectrum. Indeed, semiconductor lasers have precisely defined bound states and phonon-assisted transitions between such states occur only if the phonon energy is well matched to the carrier transition energy. This is illustrated further by recent calculations[24,31] which are relevant to a number of novel current-driven semiconductor intersubband lasers.[32–34] In a particular embodiment of such a laser structure, the energy levels E_1 and E_2 are separated by an energy equal to the optical-phonon energy; this situation is depicted[35] in Fig. 1.

Clearly, phonon-assisted transitions between levels E_2 and E_3 depend strongly on the phonon energy involved in the transition; as a specific case, for $Al_xGa_{1-x}As$-based structures, there are interface modes in the GaAs regions with energies close to the AlAs optical-phonon energy of 50 meV. Fermi golden rule calculations comparing the phonon-assisted transition rates for a related structure are presented in Ref. 24 for two cases: AlAs-like interface-phonon–assisted transitions where the phonon energy is close to 50 meV, and GaAs confined-phonon–assisted transitions where the phonon energy is close to 36 meV. These Fermi golden rule calculations reveal that the interface-phonon–assisted transition are dominant for small quantum well dimensions of approximately 40 Angstroms. Such dimensions are typical of these used in the novel lasers of Refs. 32–34 as well as those of the far-infrared devices of Refs. 36–39. Novel lasers in this group include both the unipolar quantum cascade laser and the tunnelling injection laser. It is shown in Ref. 25 that the gain of such a semiconductor laser system depends critically on using the confined, interface and half-space phonons which result from phonon confinement. In recent numerical calculations, J.-P. Sun *et al.*[29,31] have confirmed the results of Ref. 24 without a number of approximations made in Ref. 24 so that an analytical calculation could be performed. This important consequence of optical-phonon confinement provides an exciting and useful application of optical-phonon confinement in semiconductor

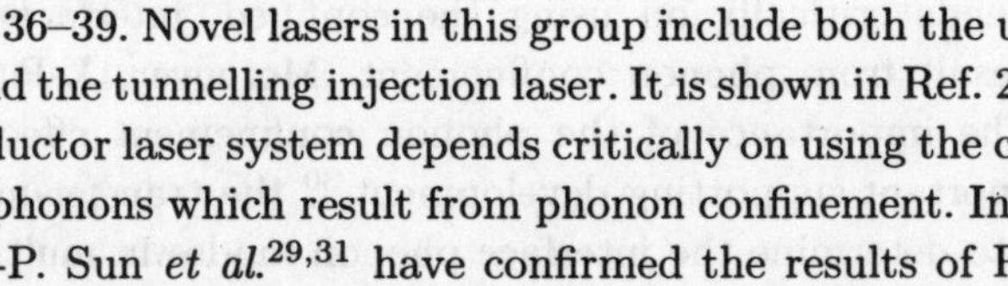

Fig. 1. Step quantum-well laser with E_1 and E_2 separated by the LO-phonon energy.

nanostructures. To be sure, the recent attention on this field is caused, in part, by the successful room-temperature infrared operation of intersubband lasers by the team of Ref. 34.

The recent advances in gallium nitride-based semiconductor lasers have stimulated interest in the properties of confined optical phonons in polar wurtzite material systems. The previous discussion in this article on phonon effects has been focussed on structures with cubic and zinc blende crystal structures. In recent years it has become clear that device applications requiring wide bandgap semiconductor can be realized by using III–V nitride-based compound semiconductor structures with wurtzite crystal structure. Recently, it has been shown[40] that the Frohlich interaction in such structures is similar to the in cubic structures but there is an angular dependence — with respect to the wurtzite c-axis — in the scattering matrix elements; it was further shown that scattering rates based on carrier–TO-phonon interactions can be modified substantially relative to the case of cubic structures. The confined and interface phonon modes for wurtzite based heterostructures were recently presented in Ref. 41. With these modes, it is possible to investigate all of the previously-discussed confined optical-phonon effects in wurtzite materials. This opens a wide area of study that has not been explored previously.

4. Optical Phonon Confinement in Heterostructure Semiconductor Lasers

Recently, there have been several detailed theoretical calculations of the gain and phonon-assisted carrier transition rates in semiconductor heterostructures. In particular, Refs. 25 and 26 demonstrated that the gain of such a semiconductor laser system depends critically on using the confined, interface and half-space phonons which result from phonon confinement. Moreover, J.-P. Sun *et al.*[29,31] have confirmed the importance of the phonon confinement effects predicted in Ref. 24. In an important supporting development,[30] the transfer-matrix technique has been applied to determine the interface phonon modes in multi-layer semiconductor heterostructures; the results of this transfer matrix model demonstrate that phonon-confinement effects lead to major modifications in phonon amplitudes when quantum-well or barrier widths have dimensions of approximately 50 Angstrom or less. Kisin *et al.*[25,26] have demonstrated that modeling the spectrum of energies for confined, interface, and half-space phonons is essential to correctly determining intersubband optical gain in quantum-well lasers. In this Section, these are related results will be illustrated in greater detail.

As in Refs. 24, 29 and 31, the simplified single-step quantum-well structure of Fig. 2 is used to assess the importance of interface–phonon-assisted transitions in quantum-well lasers; the width of the inner well is taken to be 15 Angstroms and that of the outer well is 420 Angstroms. The potentials of the inner and outer wells are taken to be 0.2 and 0.3 eV, respectively. The longitudinal-optical-phonon-assisted transition rates between E_2 and E_1 for this structure are determined for

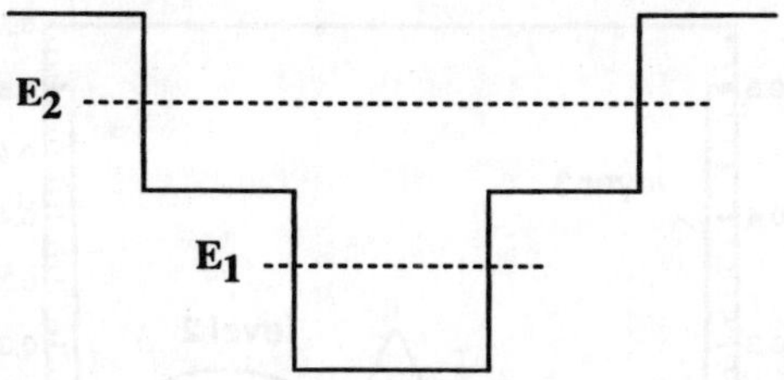

Fig. 2. Single-step quantum-well laser structures.

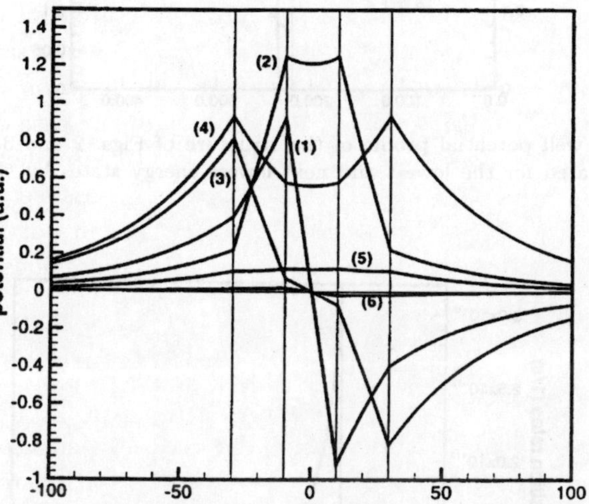

Fig. 3. Electrostatic potential for some typical interface phonon modes in the structure of Fig. 2 with material compositions as indicated in the text. Because the structure is symmetric, the phonon modes are either symmetric or antisymmetric. As in Ref. 31, the mode frequencies are: (1) 35.86 meV, (2) 34.91 meV, (3) 34.26 meV, (4) 34.01 meV, (5) 33.17 meV, (6) 33.07 meV, respectively. The heterointerfaces are represented by vertical lines. The phonon wavevector is taken so that $qa = 0.5$.

two cases: a rate of 3.0×10^{13} sec^{-1} with a half width of 6 meV is obtained for the case when an antisymmetric interface optical phonon is emitted in the electron transition from E_2 to E_1; and a rate of 1.2×10^{12} sec^{-1} with a half width of 6 meV is obtained when a bulk GaAs LO phonon is emitted in the electron transition from E_2 to E_1. As discussed in Refs. 29 and 31, the Fermi golden rule is used to calculate these rates. Figure 3 depicts the electrostatic potentials for some typical interface phonon modes for the structure of Fig. 2 with the heterolayer widths as indicated in Fig. 3 and with the material layers as follows: a central well of GaAs is surrounded by $Al_{0.25}Ga_{0.75}As$ barriers which are in turn surrounded by $Al_{0.4}Ga_{0.6}As$.

For the heterostructure of Figs. 2 and 3, the quantum-well potential profile and the electronic wavefunctions associated with the two lowest-energy states are shown[31] in Fig. 4.

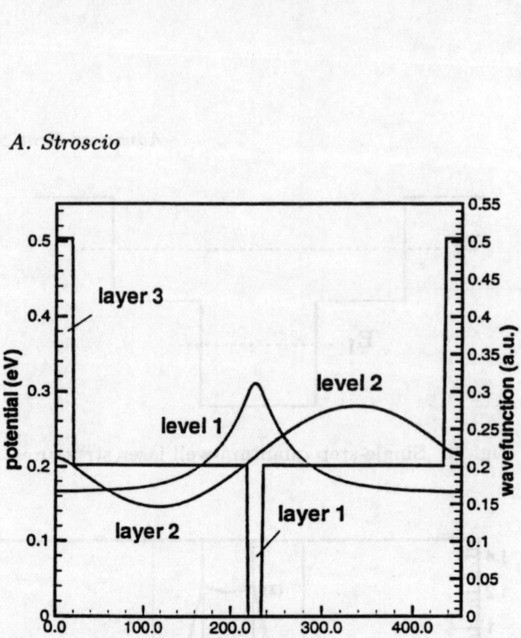

Fig. 4. Quantum well potential profile of the structure of Figs. 2 and 3; the electronic wavefunctions are indicated for the lowest and next lowest energy states by "level 1" and "level 2" respectively.

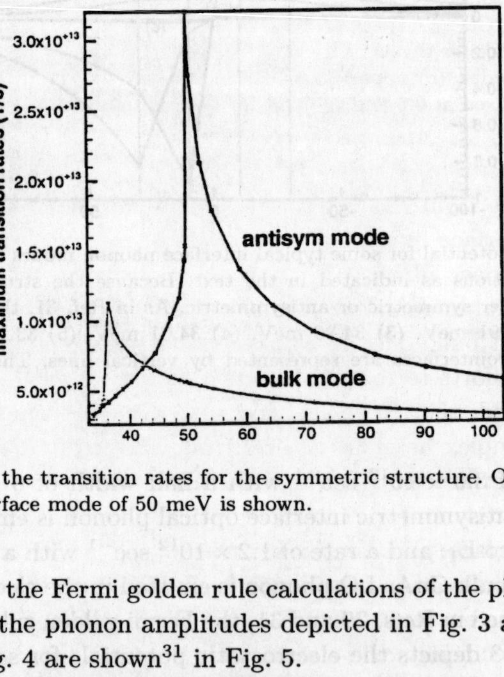

Fig. 5. Comparison of the transition rates for the symmetric structure. Only the contribution of one antisymmetric interface mode of 50 meV is shown.

These results for the Fermi golden rule calculations of the phonon-assisted electron transitions for the phonon amplitudes depicted in Fig. 3 and for the electron wavefunctions of Fig. 4 are shown[31] in Fig. 5.

The results depicted[31] in Fig. 5 demonstrate that longitudinal-optical-phonon-assisted transition rates can be enhanced significantly when E_2–E_1 is equal to the energy of the interface optical phonon rather than the energy of a bulk optical phonon. This result is expected since it is known that interface phonons may play a dominant role when the quantum well has a thickness of about 40 Angstroms or less.[15,24,26,30,31] For such narrow heterolayers, the exponential decay of the interface

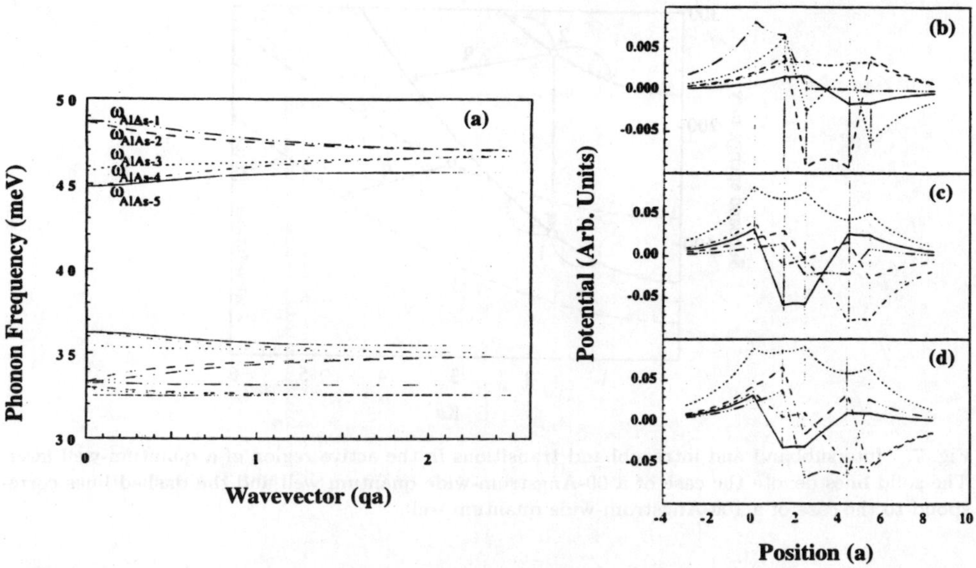

Fig. 6. (a) Phonon dispersion curves for interface modes in a five-interface $Al_yGa_{1-y}As$-GaAs-$Al_xGa_{1-x}As$-GaAs-$Al_xGa_{1-x}As$-GaAs heterostructure with $y = 0.25$ and $x = 0.6$. Electrostatic potentials versus distance for the same five-interface heterostructure are depicted for three cases: (b) lower GaAs-like modes, (c) upper GaAs-like modes, and (d) AlAs-like modes. The value of the wavevector, qa, is fixed at 0.5 and the unit for the position is a.

modes with distance from the heterointerface is not the dominant effect and the scattering from interface modes may be larger than that from the other phonon modes. As has just been illustrated, the simplified single-step quantum-well structure of Fig. 2 is useful in gaining insights into the relative importance of interface–phonon-assisted and bulk–phonon-assisted transitions. Detailed calculations may require additional efforts to model scattering processes where phonon spectra are very complex. Indeed, the interface–phonon mode spectrum is significantly more complicated for complex heterostructures; this is illustrated[30] in Fig. 6 which depicts the fifteen interface modes of a five-interface heterostructure. References 28 and 31 have considered more complicated phonon spectra to gain additional insights into the role of interface–phonon-assisted transitions in quantum-well lasers.

The importance of confined phonon effects, including the confinement-induced change in the phonon energy spectrum, is also evident in the optical gain of a semiconductor laser. Specifically, the optical gain of such a quantum-well semiconductor laser[26] depends on the integral of the phonon-broadened line shape function. Reference 26 considers the case of an $Al_{0.3}Ga_{0.7}As$/Ga As/$Al_{0.3}Ga_{0.7}As$ double-heterostructure quantum-well laser having a well width of 60 Angstroms. Reference 26 employs a four-band Kane model and the confined phonon modes of the dielectric continuum model[11] to predict gain curves that depend on the energy of the laser transition and the phonon mode spectrum. Figure 7 illustrates[26] the

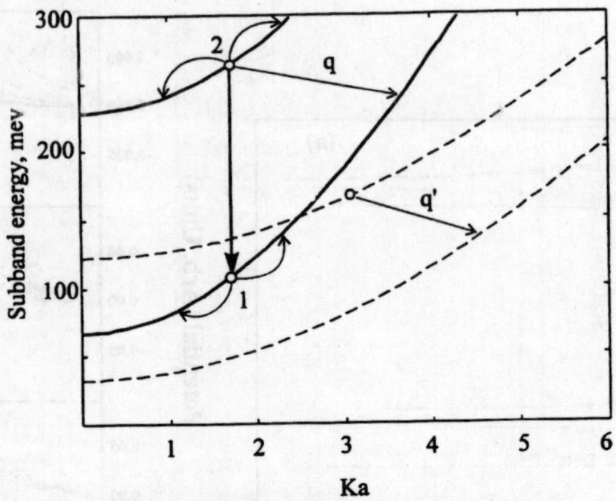

Fig. 7. Intersubband and intrasubband transitions in the active region of a quantum-well laser. The solid lines denote the case of a 60-Angstrom-wide quantum well and the dashed lines correspond to the case of a 100-Angstrom-wide quantum well.

subband energies and typical transitions for a gallium arsenide quantum well surrounded by $Al_{0.3}Ga_{0.7}As$ barriers; the quantum-well parameters were taken to be: $E_g = 1.4$ eV, $m_e = 0.067$, conduction band offset equals 0.3 eV, and valence band offset equals 0.15 eV. In Fig. 7 the solid lines represent a quantum-well width of 60 Angstroms and the dashed lines represent a 100-Angstrom-wide quantum well. From Fig. 7 it is clear that it is essential to include the nonparabolicity of the electron energy spectrum when calculating phonon-assisted transition.

As shown in Ref. 26, the intrasubband scattering rates for the quantum-well system of Fig. 7 depend strongly on the energies of the various confined optical phonons[11] in the quantum well. This result is expected on the basis of the enormous spread in the optical phonon energies in the GaAs-AlAs system: symmetric interface optical phonons have energies of approximately 50 meV and 33 meV, antisymmetric interface optical phonons have energies of approximately 45 meV and 36 meV, confined optical phonons in the GaAs layer have energies ranging from 33 meV to 36 meV, and the half-space modes in a given heterolayer have energies close to those of the bulk modes of that layer. Accordingly, the phonon emission threshold varies from 33 meV to 50 meV depending on the particular confined optical phonon mode participating in the emission process. This situation is similar to that for carrier-phonon interactions governing carrier transport in dimensionally-confined polar semiconductors.[11] In Fig. 8 the various thresholds for the modes may be identified readily; in Fig. 8, is^+ denotes the high-frequency symmetric interface optical phonon mode, is^- denotes the low-frequency symmetric interface optical phonon mode, c represents the confined modes, and b represents the barrier or half-space modes. As discussed in Ref. 26, the optical gain for a quantum well laser may be represented as an integral containing the line shape function which depends on the energies of the various scattering mechanisms contributing to the line shape.

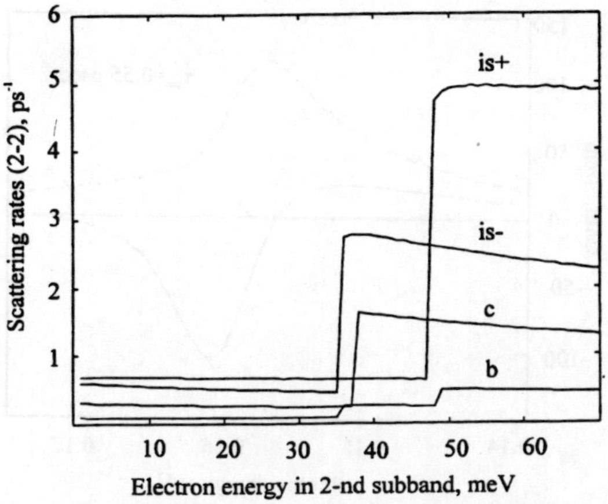

Fig. 8. Room temperature scattering rates for optical-phonon-assisted intersubband transitions in the second subband as a function of energy for the various phonons in a dimensionally-confined gallium arsenide quantum-well with a width of 60 Angstroms. As discussed in the text, is^{+} identifies the high-frequency symmetric interface phonon mode, is^{-} represents the low-frequency symmetric interface phonon mode, c denote the confined mode, and the barrier, or half-space, mode is denoted by b; after Ref. 26.

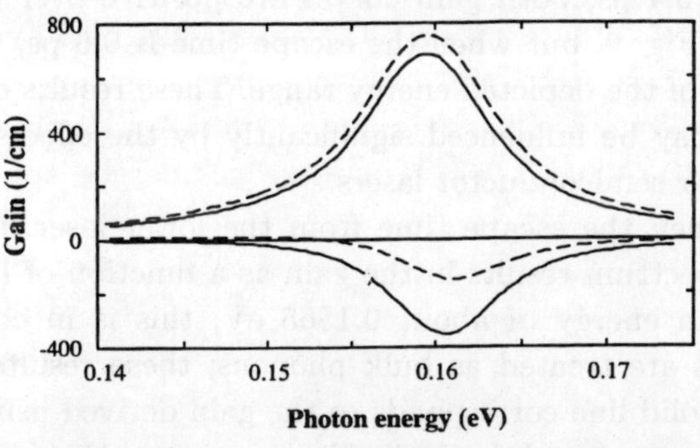

Fig. 9. Gain versus photon energy for quantum-well laser for two phonon models: dashed lines – – – bulk GaAs LO phonons; solid lines —— full spectrum of confined, interface, and half-space phonon modes. The two curves exhibiting positive gain were calculated with an escape time of 0.4 ps and the two curves exhibiting maximum negative gain near 0.16 eV were calculated for an escape time of 0.6 ps.

Since one of the major scattering mechanisms for polar semiconductor quantum-well lasers is carrier-phonon scattering, it is clear that replacing the bulk optical phonon with a rich spectrum of confined, interface, and half-space optical phonons is likely to have an influence on the optical gain of a polar-semiconductor quantum well laser. These results are shown[26] in Fig. 9 where the dashed lines represent the gain when only the bulk GaAs longitudinal-optical phonon is taken into account and the solid line represents the case where the full spectrum of confined, interface, and half-space phonon modes is taken into account. To include the effects of depopulation of the lower laser level, the escape time of the electron from the lower laser level

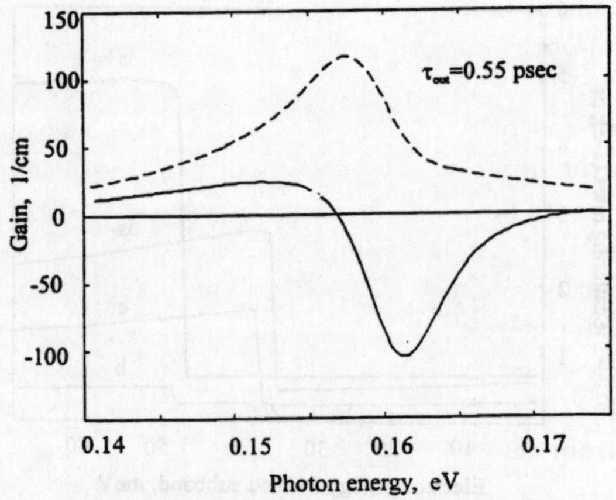

Fig. 10. Gain for the same quantum well-laser as that of Fig. 9 but with the escape time from the lower laser level taken as 0.55 ps. The solid curve denotes the case where the gain calculation considers the full confined phonon spectrum and the dashed line corresponds to the use of a single bulk GaAs longitudinal-optical phonon in the gain calculation.

is taken to have representative values of 0.4 ps and 0.6 ps. For the case where the escape time is 0.4 ps, both gain curves are positive over the entire range of energy depicted in Fig. 9, but when the escape time is 0.6 ps, the gain curves are negative over most of the depicted energy range. These results demonstrate clearly that optical gain may be influenced significantly by the effects of optical-phonon confinement in polar semiconductor lasers.

For the case when the escape time from the lower laser level is 0.55 ps, the complex phonon spectrum results in the gain as a function of laser photon energy changing sign at an energy of about 0.1565 eV; this is in contrast to the case where the phonons are treated as bulk phonons; these results[26] are depicted in Fig. 10 where the solid line corresponds to the gain derived using the full complex confined phonon spectrum and the dashed line represents the case where the phonon spectrum is approximated by the bulk GaAs longitudinal-optical phonon.

This account has emphasized the role of phonons on determining the characteristics of quantum-well lasers; however, it is widely recognized that for sheet carrier densities in excess of 10^{12} cm^{-2} that electron-plasmon interactions can play an important role in intersubband carrier transitions. Recently, it was shown[27] that for the case of inverted subband populations that such plasmon effects may play a significant role in carrier scattering at lower sheet densities in the range of a few times 10^{11} cm^{-2}; these results have not yet been used to include plasmon effects in the gain calculations for quantum-well lasers.

5. Conclusion

Theoretical findings indicate that the characteristics of polar-semiconductor quantum-well lasers with heterolayers as thin or thinner than 50 Angstroms are influenced by confinement-induced modifications in the frequencies and envelopes of confined longitudinal-optical phonons. These results are especially significant since recently-proposed novel quantum-well including the tunneling injection laser and the quantum cascade laser are usually designed so that they contain heterolayers with thicknesses on the order of or less than 50 Angstroms. Moreover, calculations of the gains of such lasers must incorporate the complex energy spectra of confined optical phonons. The role of interface optical phonons is found to be of special significance in the quantum cascade laser where the heterostructures are engineered so that phonon-assisted carrier transitions are used to depopulate the lower laser level; indeed, in such cases the theory shows that it is essential to considered the fact that in many heterostructures the interface phonon frequencies may be substantially different from the bulk phonon frequency.

Acknowledgments

The authors gratefully acknowledge the extremely beneficial collaborations with Prof. G. Haddad, Dr. J. P. Sun, Prof. G. J. Iafrate, Prof. M. A. Littlejohn, Prof. K. W. Kim, Dr. SeGi Yu, Prof. M. Shur, Prof. V. V. Mitin, Prof. S. Das Sarma, Prof. A. Petrou, Prof. S. Luryi, Prof. G. Belenky, and Dr. M. Kisin. This effort was supported by the U.S. Army Research Office.

References

1. M. A. Stroscio, K. W. Kim, S. Yu, and A. Ballato, "Quantized acoustic phonon modes in quantum wires and quantum dots", *J. Appl. Phys.* **76** (1994) 4670–4673.
2. S. Yu, K. W. Kim, M. A. Stroscio, and G. J. Iafrate, "Electron–acoustic-phonon scattering rates in cylindrical quantum wires", *Phys. Rev.* **B51** (1994) 4695–4698.
3. M. A. Stroscio and K. W. Kim, "Piezoelectric scattering of carriers in confined acoustic modes in cylindrical quantum wires", *Phys. Rev.* **B48** (1993) 1936–1939.
4. N. Bannov, V. Mitin, and M. Stroscio, "Confined acoustic phonons in semiconductor slabs and their interactions with electrons", *Physica Status Solidi* **B183** (1994) 131–138.
5. N. Bannov, V. Aristov, V. Mitin, and M. A. Stroscio, "Electron relaxation times due to deformation-potential interaction of electrons with confined acoustic phonons in a free-standing quantum well", *Phys. Rev.* **B51** (1995) 9930–9938.
6. Yuri M. Sirenko, M. A. Stroscio, and K. W. Kim, "Elastic vibrations of microtubules in a fluid", *Phys. Rev.* **E53** (1996) 1003–1011.
7. M. A. Stroscio, Yu. M. Sirenko, S. Yu, and K. W. Kim, "Acoustic phonon quantization in buried waveguides and resonators", *J. Phys: Condensed Matter* **8** (1996) 2143–2151.
8. G. J. Iafrate and M. A. Stroscio, "Application of quantum-based devices: Trends and challenges", *IEEE Trans. Electron Devices* **43** (1996) 1621–1625.
9. M. A. Stroscio, G. J. Iafrate, K. W. Kim, S. Yu, V. Mitin, and N. Bannov, "Scattering of carriers from confined acoustic modes in nanostructures", *Proc. 1993 Int. Device Res. Symp.*, University of Virginia Engineering Academic Outreach Publication, Charlottesville, 1993, pp. 873–875.
10. Y. Lyanda-Geller and J.-P. Leburton, "Antiresonant hopping conductance and negative magnetoresistance in quantum-box superlattices", *Phys. Rev.* **B52** (1995) 2779–2783.

11. M. Dutta, M. A. Stroscio, and K. W. Kim, "Recent developments on electron-phonon interactions in structures for electronic and optoelectronic devices", in *Quantum-Based Electronic Devices and Systems, Selected Topics in Electronics and Systems,* **14** (World Scientific, Singapore, New Jersey, London, Hong Kong, 1998).
12. J.-P. Leburton, "Dissipation and scattering time engineering in quantum devices", in *Proc. Int. Conf. Quantum Devices*, eds. K. Ismail, S. Bandyopadhyay, and J.-P. Leburton (World Scientific, Singapore, 1997), pp. 242–252; this review paper contains an excellent set of additional references.
13. M. A. Stroscio, K. W. Kim, G. J. Iafrate, M. Dutta, and H. L. Grubin, "Reduction and control of inelastic longitudinal-optical phonon scattering in nanoscale and mesoscopic device structures", *Proc. 1991 Int. Device. Res. Symp.*, University of Virginia Engineering Academic Outreach Publication, Charlottesville, 1991, pp. 87–89.
14. M. A. Stroscio, K. W. Kim, G. J. Iafrate, M. Dutta, and H. Grubin, "Dramatic reduction of the longitudinal-optical phonon emission rate in polar-semiconductor quantum wires", *Phil. Mag. Lett.* **65** (1992) 173–176.
15. M. A. Stroscio, G. J. Iafrate, K. W. Kim, M. A. Littlejohn, H. Goronkin, and G. Maracas, "Transitions from LO-phonon to SO-phonon scattering in short-period AlAs-GaAs superlattices", *Appl. Phys. Lett.* **59** (1991) 1093–1096.
16. A. R. Bhatt, K. W. Kim, M. A. Stroscio, G. J. Iafrate, M. Dutta, H. L. Grubin, R. Haque, and X. T. Zhu, "Reduction of LO-phonon interface modes using metal-semiconductor heterostructures", *J. Appl. Phys.* **73** (1993) 2338–2245.
17. M. Dutta, H. L. Grubin, G. J. Iafrate, K. W. Kim, and M. A. Stroscio, "Metal-encapsulated quantum wire for enhanced charge transport", Patent Number 5,264,711, November 23, 1993.
18. M. Dutta, G. J. Iafrate, K. W. Kim, and M. A. Stroscio, "Phonon modulated switch", Patent Number 5,374,831, December 20, 1994.
19. J. L. Educato, J.-P. Leburton, P. Boucaud, P. Vagos, and F. H. Julien, "Influence of interface phonons on intersubband scattering in asymmetric coupled quantum wells", *Phys. Rev.* **47** (1993) 12949–12952.
20. F. H. Julien, A. Sa'ar, J. Wang, and J.-P. Leburton, "Optically pumped intersubband emission in quantum wells", *Electron. Lett.* **31** (1995) 838–839.
21. J. Wang, J.-P. Leburton, Z. Moussa, F. H. Julien, and A. Sa'ar, "Simulation of optically pumped mid-infrared intersubband semiconductor laser structures", *J. Appl. Phys.* **80** (1996) 1970–1978.
22. J. Wang, J.-P. Leburton, F. H. Julien, and A. Sa'ar, "Design and performance optimization of optically-pumped mid-infrared intersubband semiconductor lasers", *IEEE Photon. Tech. Lett.* **8** (1996) 1001–1003.
23. O. Gauthier-Lafaye, S. Sauvage, P. Boucaud, F. H. Julien, R. Prazeres, F. Glotin, J.-M. Ortega, V. Thierry-Mieg, R. Planel, J.-P. Leburton, and V. Berger, "Intersubband stimulated emission in GaAs/AlGaAs quantum wells: Pump-probe experiments using a two-color free electron laser", *Appl. Phys. Lett.* **70** (1997) 3197–3199.
24. M. A. Stroscio, "Interface–phonon-assisted transitions in quantum well lasers", *J. Appl. Phys.* **80** (1996) 6864–6867.
25. M. V. Kisin, V. B. Gorfinkel, M. A. Stroscio, G. Belenky, and S. Luryi, "Influence of complex phonon spectrum of heterostructure on gain lineshape in quantum cascade laser", *Abstract CThS4 at CLEO '97*, Baltimore, Maryland, May 22, 1997.
26. M. V. Kisin, V. B. Gorfinkel, M. A. Stroscio, G. Belenky, and Serge Luryi, "Influence of complex phonon spectra on intersubband optical gain", *J. Appl. Phys.* **82** (1997) 2031–2038.

27. M. V. Kisin, M. A. Stroscio, G. Belenky, and S. Luryi, "Electron-plasmon relaxation in quantum wells with inverse subband occupation", *Appl. Phys. Lett.* **73** (1998) 2075–2077.
28. M. V. Kisin, M. A. Stroscio, G. Belenky, V. B. Gorfinkel, and S. Luryi, "Effects of interface phonon scattering in three-interface heterostructures", *J. Appl. Phys.* **83** (1998) 4816–4822.
29. J.-P. Sun, H. B. Teng, G. I. Haddad, M. A. Stroscio, and G. J. Iafrate, "Intersubband relaxation in step quantum well structures", International Conference on Computational Electronics, University of Notre Dame, 28–30 May 1997, private communication.
30. S. Yu, K. W. Kim, M. A. Stroscio, G. J. Iafrate, J. P. Sun, and G. I. Haddad, "Transfer matrix method for interface optical-phonon modes in multiple-interface heterostructure systems", *J. Appl. Phys.* **82** (1997) 3363–3367.
31. H. B. Teng, J. P. Sun, G. I. Haddad, Michael A. Stroscio, S. Yu, and K. W. Kim, "Phonon assisted intersubband transitions in step quantum well structures", *J. Appl. Phys.* **84** (1998) 2155–2164.
32. H. C. Sun, L. Davis, S. Sethi, J. Singh, and P. Bhattacharya, "Properties of a tunneling injection quantum-well laser: Recipe for a 'cold' device with a large modulation bandwidth", *IEEE Photon. Tech. Lett.* **5** (1993) 870–872; X. Zhang, A. L. Gutierrez-Aitken, D. Klotzkin, P. Bhattacharya, C. Caneau, and R. Bhat, "0.98-micrometer multiquantum well tunneling injection lasers with ultra-high modulation bandwidths", *Electron. Lett.* **32** (1996) 1715–1716; also see the references in this paper.
33. C. Y. Sung, T. B. Norris, A. Afzali-Kushaa, and G. I. Haddad, "Femtosecond intersubband relaxation and population inversion in stepped quantum well", *Appl. Phys. Lett.* **68** (1996) 435–437; also see the references in this paper.
34. J. Faist, F. Capasso, C. Sirtori, D. L. Sivco, J. N. Baillargeon, A. L. Hutchinson, S.-N. G. Chu, and A. Y. Cho, "High power mid-infrared quantum cascade lasers operating above room temperature", *Appl. Phys. Lett.* **68** (1996) 3680–3682; also see the references in this paper.
35. G. Belenky, M. Dutta, V. B. Gorfinkel, G. I. Haddad, G. J. Iafrate, K. W. Kim, M. Kisin, S. Luryi, M. A. Stroscio, J. P. Sun, H. B. Teng, and S. Yu, "Tailoring of optical modes in nanoscale semiconductor structures: Role of interface-optical phonons in quantum-well lasers", *Physica* **B263–264** (1999) 462–465.
36. Q. Hu and S. Feng, "Feasibility of far-infrared lasers using multiple semiconductor quantum wells", *Appl. Phys. Lett.* **59** (1991) 2923–2925.
37. J. H. Smet, C. G. Fonstad, and Q. Hu, "Intrawell and interwell intersubband transitions in multiple quantum wells for far-infrared sources", *J. Appl. Phys.* **79** (1996) 9305–9320.
38. B. Xu, Q. Hu, and M. R. Melloch, "Electrically pumped tunable terahertz emitter based on intersubband transition", *Appl. Phys. Lett.* **71** (1997) 440–442.
39. I. Lyubomirsky and Q. Hu, "Energy level schemes for far-infrared quantum well lasers", *Appl. Phys. Lett.* **73** (1998) 300–302; see also M. Dutta and M. A. Stroscio, "Comment on energy level schemes for far-infrared quantum well lasers", *Appl. Phys. Lett.* **74** (1999) 2555.
40. B. C. Lee, K. W. Kim, M. Dutta, and M. A. Stroscio, "Electron-optical-phonon scattering in wurtzite crystals", *Phys. Rev.* **B56** (1997) 997–1000.
41. B. C. Lee, K. W. Kim, M. A. Stroscio, and M. Dutta, "Optical phonon confinement and scattering in wurtzite heterostructures", *Phys. Rev.* **B58** (1998) 4860–4865.